DEVELOPMENTS IN FOOD SCIENCE

MORE EFFICIENT UTILIZATION OF FISH AND FISHERIES PRODUCTS

Proceedings of the International Symposium on the occasion of the 70th anniversary of the Japanese Society of Fisheries Science, held in Kyoto, Japan, 7-10 October 2001

Edited by
M. SAKAGUCHI

Division of Applied Biosciences, Graduate School of Agriculture, Kyoto University, Kyoto, Japan

2004

ELSEVIER

Amsterdam - Boston - Heidelberg - London - New York - Oxford
Paris - San Diego - San Francisco - Singapore - Sydney - Tokyo

ELSEVIER B.V.	ELSEVIER Inc.	**ELSEVIER Ltd**	ELSEVIER Ltd
Sara Burgerhartstraat 25	525 B Street, Suite 1900	**The Boulevard, Langford Lane**	84 Theobalds Road
P.O. Box 211, 1000 AE	San Diego, CA 92101-4495	**Kidlington, Oxford OX5 1GB**	London WC1X 8RR
Amsterdam, The Netherlands	USA	**UK**	UK

© 2004 Elsevier Ltd. All rights reserved.

This work is protected under copyright by Elsevier Ltd, and the following terms and conditions apply to its use:

Photocopying
Single photocopies of single chapters may be made for personal use as allowed by national copyright laws. Permission of the Publisher and payment of a fee is required for all other photocopying, including multiple or systematic copying, copying for advertising or promotional purposes, resale, and all forms of document delivery. Special rates are available for educational institutions that wish to make photocopies for non-profit educational classroom use.

Permissions may be sought directly from Elsevier's Rights Department in Oxford, UK: phone (+44) 1865 843830, fax (+44) 1865 853333, e-mail: permissions@elsevier.com. Requests may also be completed on-line via the Elsevier homepage (http://www.elsevier.com/locate/permissions).

In the USA, users may clear permissions and make payments through the Copyright Clearance Center, Inc., 222 Rosewood Drive, Danvers, MA 01923, USA; phone: (+1) (978) 7508400, fax: (+1) (978) 7504744, and in the UK through the Copyright Licensing Agency Rapid Clearance Service (CLARCS), 90 Tottenham Court Road, London W1P 0LP, UK; phone: (+44) 20 7631 5555; fax: (+44) 20 7631 5500. Other countries may have a local reprographic rights agency for payments.

Derivative Works
Tables of contents may be reproduced for internal circulation, but permission of the Publisher is required for external resale or distribution of such material. Permission of the Publisher is required for all other derivative works, including compilations and translations.

Electronic Storage or Usage
Permission of the Publisher is required to store or use electronically any material contained in this work, including any chapter or part of a chapter.

Except as outlined above, no part of this work may be reproduced, stored in a retrieval system or transmitted in any form or by any means, electronic, mechanical, photocopying, recording or otherwise, without prior written permission of the Publisher.
Address permissions requests to: Elsevier's Rights Department, at the fax and e-mail addresses noted above.

Notice
No responsibility is assumed by the Publisher for any injury and/or damage to persons or property as a matter of products liability, negligence or otherwise, or from any use or operation of any methods, products, instructions or ideas contained in the material herein. Because of rapid advances in the medical sciences, in particular, independent verification of diagnoses and drug dosages should be made.

First edition 2004

Library of Congress Cataloging in Publication Data
A catalog record is available from the Library of Congress.

British Library Cataloguing in Publication Data
A catalogue record is available from the British Library.

ISBN: 0 08 044450 4
ISSN: 0167-4501

∞ The paper used in this publication meets the requirements of ANSI/NISO Z39.48-1992 (Permanence of Paper).
Printed in The Netherlands.

DEVELOPMENTS IN FOOD SCIENCE

Volume 1	J.G. Heathcote and J.R. Hibbert
	Aflatoxins: Chemical and Biological Aspects
Volume 2	H. Chiba, M. Fujimaki, K. Iwai, H. Mitsuda and Y. Morita (Editors)
	Proceedings of the Fifth International Congress of Food Science and Technology
Volume 3	I.D. Morton and A.J. MacLeod (Editors)
	Food Flavours
	Part A. Introduction
	Part B. The Flavour of Beverages
	Part C. The Flavour of Fruits
Volume 4	Y. Ueno (Editor)
	Trichothecenes: Chemical, Biological and Toxicological Aspects
Volume 5	J. Holas and J. Kratochvil (Editors)
	Progress in Cereal Chemistry and Technology. Proceedings of the VIIth World Cereal and Bread Congress, Prague, 28 june-2 July 1982
Volume 6	I. Kiss
	Testing Methods in Food Microbiology
Volume 7	H. Kurata and Y. Ueno (Editors)
	Toxigenic Fungi: Their Toxins and Health Hazard. Proceedings of the Mycotoxin Symposium, Tokyo, 30 August-3 September 1983
Volume 8	V. Betina (Editor)
	Mycotoxins: Production, Isolation, Separation and Purification
Volume 9	J. Holló (Editor)
	Food Industries and the Environment. Proceedings of the International Symposium, Budapest, Hungary, 9-11 September 1982
Volume 10	J. Adda (Editor)
	Progress in Flavour Research 1984. Proceedings of the 4th Weurman Flavour Research Symposium, Dourdan, France, 9-11 May 1984
Volume 11	J. Holló (Editor)
	Fat Science 1983. Proceedings of the 16th International Society for Fat Research Congress, Budapest, Hungary, 4-7 October 1983
Volume 12	G. Charalambous (Editor)
	The Shelf Life of Foods and Beverages. Proceedings of the 4th International Flavor Conference, Rhodes, Greece, 23-26 July 1985
Volume 13	M. Fujimaki, M. Namiki and H. Kato (Editors)
	Amino-Carbonyl Reactions in Food and Biological Systems. Proceedings of the 3rd International Symposium on the Maillard Reaction, Susuno, Shizuoka, Japan,1-5 July 1985
Volume 14	J. Skoda and H. Škodová
	Molecular Genetics. An Outline for Food Chemists and Biotechnologists.
Volume 15	D.E. Kramer and J. Liston (Editors)
	Seafood Quality Determination. Proceedings of the International Symposium, Anchorage, Alaska, U.S.A., 10-14 November 1986
Volume 16	R.C. Baker. P. Wong Hahn and K.R. Robbins
	Fundamentals of New Food Product Development
Volume 17	G. Charalambous (Editor)
	Frontiers of Flavor. Proceedings of the 5th International Flavor Conference, Porto Karras,Chalkidiki, Greece, 1-3 July 1987
Volume 18	B.M. Lawrence, B.D. Mookherjee and B.J. Willis (Editors)
	Flavors and Fragrances: A World Perspective. Proceedings of the 10th International Congress of Essential Oils, Fragrances and Flavors, Washington, DC, U.S.A., 16-20 November 1986
Volume 19	G. Charalambous and G. Doxastakis (Editors)
	Food Emulsifiers: Chemistry, Technology, Functional Properties and Applictations
Volume 20	B.W. Berry and K.F. Leddy
	Meat Freezing. A Source Book

Volume 21	J. Davídek, J. Velis̆ek and J. Pokorn´y (Editors) Chemical Changes during Food Processing
Volume 22	V. Kyzlink Principles of Food Preservation
Volume 23	H. Niewiadomski Rapeseed. Chemistry and Technology
Volume 24	G. Charalambous (Editor) Flavors and Off-flavors '89. Proceedings of the 6th International Flavor Conference, Rehymnon, Crete, Greece, 5-7 July 1989
Volume 25	R. Rouseff (Editor) Bitterness in Foods and Beverages
Volume 26	J. Chelkowski (Editor) Cereal Grain. Mycotoxins, Fungi and Quality in Drying and Storage
Volume 27	M. Verzele and D. De Keukeleire Chemistry and Analysis of Hop and Beer Bitter Acids
Volume 28	G. Charalambous (Editor) Off-Flavors in Foods and Beverages
Volume 29	G. Charalambous (Editor) Food Science and Human Nutrition
Volume 30	H.H. Huss, M. Jakobsen and J. Liston (Editors) Quality Assurance in the Fish Industry. Proceedings of an International Conference, Copenhagen, Denmark, 26-30 August 1991
Volume 31	R.A. Samson, A.D. Hocking, J.I.Pitt and A.D. King (Editors) Modern Methods in Food Mycology
Volume 32	G. Charalambous (Editor) Food Flavors, Ingredients and Composition. Proceedings of the 7th International Flavo Conference, Pythagorion, Samos, Greece, 24-26 June 1992
Volume 33	G. Charalambous (Editor) Shelf Life Studies of Foods and Beverages. Chemical, Biological, Physical and Nutritional Aspects
Volume 34	G. Charalambous (Editor) Spices, Herbs and Edible Fungi
Volume 35	H. Maarse and D.G. van der Heij (Editors) Trends in Flavour Research. Proceedings of the 7th Weurman Flavour Research Symposium, Noordwijkerhout, The Netherlands, 15-18 June 1993
Volume 36	J.J. Bimbenet, E. Dumoulin and G. Trystram (Editors) Automatic Control of Food and Biological Processes. Proceedings of the ACoFoP III Symposium, Paris, France, 25-26 October 1994
Volume 37A+B	G. Charalambous (Editor) Food Flavors: Generation, Analysis and Process Influence Proceedings of the 8th International Flavor Conference, Cos, Greece, 6-8 July 1994
Volume 38	J.B. Luten, T. Børresen and J. Oehlenschläger (Editors) Seafood from Producer to Consumer, Integrated Approach to Quality Proceedings of the International Seafood Conference on the occasion of the 25th anniversary of the WEFTA, held in Noordwijkerhout, The Netherlands, 13-16 November 1995
Volume 39	D. Wetzel and G. Charalambous † (Editors) Instrumental Methods in Food and Beverage Analysis
Volume 40	E.T. Contis, C.-T. Ho, C.J. Mussinan, T.H. Parliment, F. Shahidi and A.M. Spanier (Editors) Food Flavors: Formation, Analysis and Packaging Influences Proceedings of the 9th International Flavor Conference The George Charalambous Memorial Symposium
Volume 41	G. Doxastakis and V. Kiosseoglou (Editors) Novel Macromolecules in Food Systems
Volume 42	M. Sakaguchi (Editor) More Efficient Utilization of Fish and Fisheries Products

PREFACE

This volume consists of the papers presented at the international symposium "More Efficient Utilization of Fish and Fisheries Products (MEUFFP)" in Kyoto, Japan, October 7-10, 2001. This symposium was one of the satellite symposiums of the Commemorative International Symposium held in Yokohama to celebrate the 70th Anniversary of the Japanese Society of Fisheries Science. It arose from the wise decision to exclude from the Post Harvest: Science and Technology session of the Yokohama's symposium many aspects of MEUFFP so that we could consider them in more detail here.

In the beginning of the 21st century, we should open up our minds in this symposium and take a close look at possibilities for MEUFFP, because the exploitation of new food resources for human consumption is bound to approach a dead end. In addition, currently increasing environmental deterioration is predicted to cast a gloomy shadow on our globe.

This international symposium allowed to gather many researchers and industrial representatives to discuss such a broad spectrum of information as zero emission, resources availability, sustainable utilization of the resources, bioactive and functional components in aquatic organisms, utilization of wastes, seafood quality, surimi technologies, processing and safety. The latter 3 items connect to production of a driving force for achieving MEUFFP. Almost all of these items are, of course, closely related to seafood science and technology; several papers are involved in aquaculture areas and also medical and environmental fields, all of which would be highly evaluated to meet the current situation.

There would be numerous personal and scientific links with fellow readers together with much scientific and technical information not only in the academic side but in the industries. Research, development and even trade association, for example, are important for sharing the information with the authors of this volume.

I gratefully acknowledge the secretarial work by Dr. Takashi Hirata (Kyoto University, Japan) in organizing as well as scientific aspects. Thanks are also greatly due to Prof. Koretaro Takahashi (Hokkaido University, Japan) and Dr. H. Allan Bremner (Allan Bremner & Associates). Both of them contributed outstandingly as vice-organizers before and during the symposium.

Also I take this opportunity to thank following chairpersons in the symposium sessions: Y. Akahane (Fukui Prefectural University, Japan), T. Børresen (Technical University of Denmark, Denmark), C. M. Lee (University of Rhode Island, USA), M. T. Morrissey (Oregon State University, USA), F. S. Noguchi (Maruha Corporation, Japan), H. W. Rehbein (Federal Research Center for Fisheries, Germany), N. K. Sorensen (Norwegian Institute of Fisheries and Aquaculture Ltd., Norway), K. Takahashi (Hokkaido University, Japan), M. Tanaka (Tokyo University of Fisheries, Japan), M. Tejada (Ciudad University, Spain), J. Weerasinghe (Food Science Australia, Australia),

C. H. Xue (Ocean University of Qingdau, China) and P. G. Zhou (Shanghai Fisheries University, China).

My thanks are due to the review of proceedings papers by T. Børresen, H. Allan Bremner, M. T. Morrissey, N. K. Sorensen, K. Takahashi, C. H. Xue, P. G. Zhou and also to the reviewing and editorial works by local members including K. Sato (Kyoto Prefectural University) and T. Kawai (Shiono Koryo Kaisha, Ltd.).

I would like to express my sincere gratitude to the invited and general speakers, presenters of the poster session, and all the participants. The sponsorship and donation by the public organizations and supporting companies are greatly appreciated. In addition, some other donors contributed the symposium expenses and provided seafood for the reception. Finally, special thanks are also due to Ms. Yoshiko Tonari who took on many assistant secretarial works for the symposium holding.

Morihiko Sakaguchi
Graduate School of Agriculture
Kyoto University
Japan

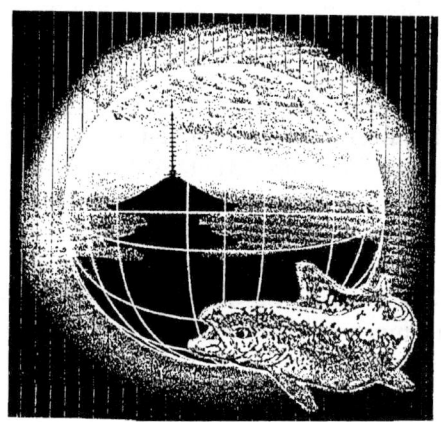

ORGANIZING AND SCIENTIFIC COMMITTEE

Prof. Yoshiaki Akahane
Department of Maine Bioscience
Faculty of Biotechnology
Fukui Prefectural University
Obama, Fukui 917-0003, Japan

Prof. Toger Børresen
Danish Institute for Fisheries Research
Department of Seafood Research
Technical University, Bldg. 221
DK-2800 Lyngby, Denmark

Dr. H. Allan Bremner
Allan Bremner and Associates
21 Carrock Court, Mount Coolum
QLD 4573, Australia

Dr. Yutaka Fukuda
JIRCAS
1-1Ohwashi, Tsukuba, Ibaraki 305-8686
Japan

Prof. Nobuhiro Fusetani
Graduate School of Agriculture and
Life Sciences
The University of Tokyo
1-1-1 Yayoi, Bunkyo, Tokyo 113-8657
Japan

Prof. Isao Hayashi
Division of Applied Biosciences
Graduate School of Agriculture
Kyoto University
Kitashirakawa, Sakyo, Kyoto 606-8502
Japan

Dr. Takashi Hirata
Division of Applied Biosciences
Graduate School of Agriculture
Kyoto University
Kitashirakawa, Sakyo, Kyoto 606-8502
Japan

Prof. Yoshiaki Ito
Faculty of Agriculture
Kochi University
Nankoku, Kochi 783-8502, Japan

Dr. Shoji Kimura
Central Research Institute, Maruha
Corporation
16-2 Wadai, Tsukuba, Ibaraki 300-4295
Japan

Prof. Yasuo Makinodan
Faculty of Agriculture
Kinki University
3327-204 Naka-machi, Nara, Nara 631-8504, Japan

Prof. Teisuke Miura
Graduate School of Fisheries Sciences
Hokkaido University
3-1-1 Minato, Hakodate, Hokkaido 041-8611, Japan

Prof. Katsuyoshi Mori
Graduate School of Agricultural Science
Tohoko University
1-1 Tsutsumidori Amamiya-machi, Aoba
Sendai 981-8555, Japan

Prof. Michael T. Morrissey
Seafood Laboratory
Department of Food Science and Technology
Oregon State University
2001 Marine Drive, Rm 253
Astoria, OR 97103, USA

Dr. Fujio Nishioka
National Research Institute of Fisheries Science
2-12-4 Fukuura, Kanazawa, Yokohama 236-8648, Japan

Prof. Jae-Hyeun Pyeung
Div. Marine Bioscience/Inst. Marine Industry
Gyeongsang National University
Tongyoung 650-160, Korea

Prof. Morihiko Sakaguchi
Division of Applied Biosciences
Graduate School of Agriculture
Kyoto University
Kitashirakawa, Sakyo, Kyoto 606-8502
Japan

Dr. Mikio Satake
Central Research Laboratory, Nippon Suisan Kaisha, Ltd., 559-6 Kitano-machi, Hachioji, Tokyo 192-0906, Japan

Dr. Nils K. Sorensen
Norwegian Institute of Fisheries & Aquaculture Ltd., Fiskeriforskning
Breivika, 9291 Tromso, Norway

Prof. Takeshi Suzuki
Department of Food Science and Technology
Tokyo University of Fisheries
4-5-7 Konan, Minato, Tokyo 108-8477
Japan

Prof. Koretaro Takahashi
Graduate School of Fisheries Sciences
Hokkaido University
3-1-1 Minato, Hakodate, Hokkaido 041-8611, Japan

Prof. Changhu Xue
Department of Food Science and Technology, Faculty of Fisheries
Ocean University of Qingdao
5 Yushan, Qingdao, 266003, China

Prof. Peigen Zhou
College of Food Science and Technology
Shanghai Fisheries University
Shanghai, 200090, China

Contents

PREFACE vii

ORGANIZING AND SCIENTIFIC COMMITTEE ix

OPENING REMARKS

More efficient utilization of fish and fisheries products toward zero emission
 Torger Børresen 3

SUSTAINABLE AND MORE EFFICIENT UTILIZATION OF RESOURCES

Future demand for, and supply of, fish and shellfish as food
 U. Wijkstrom 9

Bycatch and discard fish in the Japanese fisheries
 Yoh Watanabe 25

Intrinsic and extrinsic factors affecting efficient utilization of marine resources
 Michael T. Morrissey and Gilbert Sylvia 37

Utilization of Antarctic krill for food and feed
 Bunji Yoshitomi 45

Total utilization of squids: its zero emissions approach
 Teisuke Miura 55

Utilization of the resources of lantern fish as fisheries products
 Satoshi F. Noguchi 63

Pacu (*Piaractus mesopotamicus*) a new fish species in Israeli aquaculture: possibility of utilization
 A. Gelman, V. Drabkin, O. Sachs, K. Chechic, I. Gabay and L. Glatman 75

BIOACTIVE AND FUNCTIONAL COMPONENTS

Medical applications of fisheries byproducts
 Koretaro Takahashi 87

Separation and physiological functions of anserine from fish extract
 Kazuaki Kikuch, Yoshiharu Matahira and Kazuo Sakai 97

Bone and lipid metabolism in the bone modeling of rats administered with the bonito docosahexaenoic acid, the viscera vitamin D and the cuttlefish calcium
 Keiko Yoshioka, Azusa Seki, Ai Yamada, Mamoru Fujita, Yasuhiko Sakaguchi, Koichi Nagafusa and Shun Wada 107

Value-adding to Australian marine oils
   Peter D. Nichols, Ben D. Mooney and Nicholas G. Elliott 115

Utilization of marine invertebrates as resource for bioactive metabolites: isolation
of new mycalolides and calyculins
   Shigeki Matsunaga and Nobuhiro Fusetani 131

Antioxidative activities of low molecular fucoidans from kelp *Laminaria japonica*
   Changhu Xue, Lei Chen, Zhaojie Li, Yuepiao Cai, Hong Lin and Yu Fang 139

Properties and utilization of shark collagen
   Yoshihiro Nomura 147

Moderation of chemoinduced cancer by water extract of dried shark fin: anticancer
effect of shark cartilage
   Kenji Sato, Naho Murata, Masahiro Tsutsumi, Masami ShimizuSuganuma,
   Kazuhiro Shichinohe, Tsukasa Kitahashi, Kazunari Nishimura, Yasushi Nakamura
   and Kozo Ohtsuki 159

Polyphenolic compounds from seaweeds: distribution and their antioxidative effect
   Joko Santoso, Yumiko Yoshie and Takeshi Suzuki 169

Possible utilization of the pearl oyster phospholipids and glycogen
as a cosmetic material
   Satoshi Kanoh, Kaoru Maeyama, Risa Tanaka, Tsukasa Takahashi,
   Mayumi Aoyama, Mie Watanabe, Koichi Iida, Seishi Ueda, Maki Mae,
   Keiji Takagi, Kenji Shimomura and Eiji Niwa 179

UTILIZATION OF WASTES

Extractive components of fish sauces from waste of the frigate mackerel surimi
processing and a comparison with those of several Asian fish sauces
   Yasuhiro Funatsu, Ken-ichi Kawasaki and ShiroKonagaya 193

Development of "Katsuobushi" -like dried fish using underutilized over-matured
chum salmon
   Tsutomu Abe, Nobuo Tachibana and Masaaki Tanaka 203

Volatile compounds in the hepatic and muscular tissues of common carp, Japanese
flounder, Spanish mackerel and skipjack
   XingAn Song, Takashi Hirata, Tetsuo Kawai, Norifumi Niizeki and
   Morihiko Sakaguchi 209

Purification and some properties of cuttlefish ink polyphenol oxidase
   Peigen Zhou, Xiaoyu Qi and Xiaoxian Zheng 223

Biological activities of the components from scallop shells
  Y. C. Liu, K. Uchiyama, N. Natsui and Y. Hasegawa 233

Usefulness of waste algae as a feed additive for fish culture
  Heisuke Nakagawa 243

Disposal and recycling of fisheries plastic wastes: fishing net and expanded polystylene
  Haruyuki Kanehiro 253

Shell nurseries: artificial reefs using waste shells
  Keiichi Katayama, Minoru Tahara, Ken Tsumura and Hiroshi Kakimoto 263

SEAFOOD QUALITY AND THE RELATED ASPECTS

Quality issues and more efficient utilization of the fish catch
  H. Allan Bremner 275

Occurrence of the sea urchin with pulcherrimine, a novel bitter amino acid
  Yuko Murata, Masahito Yokoyama, Tatsuya Unuma, Noriko U. Sata,
  Ryuji Kuwahara, Masaki Kaneniwa and Ichiro Oohara 289

How do handling and killing methods affect ethical and sensory quality of farmed Atlantic salmon ?
  Nils K Sorensen, Torbjorn Tobiassen and Mats Carlehoeg 301

Autoxidation activation energies of docosahexaenoic acid ethyl ester and docosahexaenoic triglyceride
  Hidefumi Yoshii, Takeshi Furuta and Pekka Linko 309

Hemocyanin-related reactions induce blackening of freeze-thawed prawns during storage
  Kohsuke Adachi, Takashi Hirata, Katsunori Nagai, Atsushi Fujio and
  Morihiko Sakaguchi 317

SURIMI TECHNOLOGIES

Surimi gel preparation and texture analysis for better quality control
  Jae W. Park 333

Gelation of threadfin bream surimi as affected by thermal denaturation, transglutaminase and proteinase(s) activities
  J. Yongsawatdigul and J. Park 343

Oxidation during washing of fish meat induces a decrease in gel forming ability
  Dusadee Tunhun, Yoshiaki Itoh, Katsuji Morioka and Satoshi Kubota 357

Gelforming characteristics of various fish caught in Tosa bay
Akira Nomura, Atsushi Obatake and Yoshiaki Itoh 375

Acceptability and its improvement of kamaboko gel derived from silver carp surimi
*Xichang Wang, Takashi Hirata, Yutaka Fukuda, Masato Kinoshita
and Morihiko Sakaguchi* 387

SEAFOOD PROCESSING AND SAFETY

Innovative techniques for traditional dried fish products
*Tiong Ngei Kok, Soon Eong Yeap, Woan Peng Lee, How Kwang Lee and
Guat Theng Ang* 397

Functional fish protein isolates prepared using low ionic strength, acid solubilization/precipitation
S.D. Kelleher, Y. Feng, H. Kristinsson, H.O. Hultin and D.J. McClements 407

Control of sugar content in fish fillets by soaking and cold preservation
Tooru Ooizumi, Katsuhiko Tsuruhashi, Yohei Miono and Yoshiaki Akahane 415

Properties of proteases responsible for degradation of muscle proteins
during anchovy sauce fermentation
Yeung Joon Choi, MinSoo Heu, HeungRak Kim and JaeHyeung Pyeun 425

Microbial risk assessment of Persian caviar during processing and cold storage
Vadood Razavilar and Sohrab Rezvani 441

Some bacterial pathogens in the intestine of cultivated silver carp and common carp
Afshin Akhondzadeh Basti, Taghi Zahrae Salehi and Saied Bokaie 447

PARTICIPANTS 453

SPONSORS AND DONORS 457

AUTHOR INDEX 459

KEYWORD INDEX 461

OPENING REMARKS

MORE EFFICIENT UTILIZATION OF FISH AND FISHERIES PRODUCTS TOWARD ZERO EMISSION

Torger Børresen
Danish Institute for Fisheries Research, Department of Seafood Research, Soeltofts Plads, Building 221, Technical University, DK-2800 Kgs Lyngby, Denmark.

It is a great honor and a pleasure for me to make this introductory address to you. First of all I have to congratulate the organizers for choosing such a very important topic for this symposium. It is a very good time to address this theme, putting emphasis on more efficient utilization of fish and fisheries products going toward a situation with zero emission from the processing industry.

It is a fact that the known fisheries resources in the open oceans and in the coastal zones are fully exploited, or even exploited to a higher extent than certain stocks can tolerate. It is therefore important that further management of these resources is done very cautiously, and that sustainable development of all marine species is secured in the future.

Even so, it is also important that everything being brought on deck on commercial fishing vessels is being utilized optimally, and that nothing remains as waste fractions. This requires that the catch handling be done properly from the moment the fish leaves the water. Any quality lost at this critical stage can never be regained. According to estimates made by the FAO, as much as 25-30 million tons of the worlds total fisheries catch are lost due to improper handling of the catch or during processing. Proper catch handling applies to what we refer to as the main products, but it is just as important, or even more so, if we are to obtain the best economy from the fractions not utilized today. Concerning e.g. the components in the gut fraction this is crucial, as some of them contain digestive material or enzymes that may create vigorous reactions if not chilled immediately. It may also be very important to separate the individual components as soon as possible, as one component may influence the quality of the other. These conditions must be considered taken care of when we are to discuss full utilization of the fish being caught. The same applies to the next steps in the processing chain, where also fractions may occur that today are considered waste fractions. All operations must be performed such that quality is not lost. Thus, quality assurance is critical in all steps.

If we go back to the consideration of limited marine resources from traditional fisheries in the future, one could be concerned about how seafood products can be satisfying market requirements. Seafood is highly appreciated not only for its health benefits, which will become even more important in the future, but also for its culinary values. Seafood is increasingly being valued for its delicatessen features, and for that reason prices are steadily raised on many markets, particularly within the restaurant sector. Many fractions today considered waste fractions, will in the future be highly relevant for this well paying market.

According to statistics of potential new developments within the aquatic sector for supplying market needs, FAO has pointed to the possibility of releasing the resources within aquaculture production. The potential is large, and it is believed that it can be developed at a rate that will satisfy the market needs as this is being increased in the coming years. The total production from traditional fisheries is believed to be limited to a maximum of 100 million tons or less on a world basis. The market needs, believed to reach more than 20 million tons more than this during the next decade, can only be met if aquaculture is being developed properly. In this connection it should be remembered that aquaculture is providing not only fish species, but also all kind of other aquatic resources needed.

However, aquaculture production will also yield fractions that may be considered waste fractions, just as it is seen in the traditional fisheries. In some cases it may become even more important to develop new strategies for the utilization of all fractions here, as this industry may otherwise be 'buried' in material that today goes into waste fractions, and this may in turn create a hurdle for further development. Again, it is very important to emphasize the necessary proper handling of the fish throughout all stages, starting as soon as the fish or, other specimen, is taken out of the water. Actually this may be easier in aquaculture than in traditional fisheries, as it may be easier to control and manage each step throughout the processing chain. However, all factors, including the energy input needed should be considered.

When it comes to the actual new products that may be developed from the new fractions, I am convinced that we will see a lot of interesting proposals for this at the present symposium. The possibilities have never been better than today for new developments. It is therefore this symposium is so timely. I am not the only one having this opinion. It is rewarding to read the major headline under the 'Fisheries by-products' column of this years' May issue of the magazine Seafood International. This simply states: "There's gold in marine waste". The article is recording the main results presented at a conference in Norway this spring. I participated in the planning of the conference and was of course present when it was launched. The presentations were overwhelmingly positive to the development of a new industry from raw materials formerly considered to be waste. I am convinced that the same will be the case at the present symposium.

When speaking of the so-called bioprocessing industry, I am particularly thinking of the vast opportunities that exist within molecular biology and biotechnology. Until now it has been the utilization of down stream processing, development of DNA technology and the mapping of genes that has received most attention. This will probably continue to be very important, particularly concerning the fast screening techniques made available through e.g. the micro array analytical techniques. However, even more important will be the development of post-genomic techniques. These address the expression of the gene products, of which the proteins are the most important. Due to the interesting development within detailed protein analysis, particularly two-dimensional gel electrophoresis techniques combined with mass spectroscopy analysis, a much more dynamic picture can be obtained. Finally, the importance of bioinformatics also needs to be mentioned. This discipline is utilizing the power of computers in modeling biological structures and molecular reactions. The importance of this cannot be overestimated.

These are just examples of potential new techniques of great importance. When combined with already developed knowledge within e.g. nutrition research, new avenues such as 'nutrigenomics' is opening up. Several similar developments can be imagined in the future, so it is necessary for seafood researchers to keep an open eye on the new research results obtained within the biosciences.

I am mentioning these factors, as I firmly believe they will have a great impact also on the improved utilization of present day waste fractions. This does not only mean that valuable material can be extracted for biotechnological application, as has been the more traditional way of thinking until now. I rather think in the direction that biotechnological techniques may open up new ways of processing. Enzyme reactors are only a traditional example in this direction.

The ultimate goal must be to establish processing techniques that make it economically feasible to produce high priced end products to an extent similar to what has already been seen in many areas within agriculture. For the seafood area, this is the challenge that will make possible a sustainable and full utilization of all the biological resources that can be produced in the aquatic environment.

Myself, I am convinced that this will be a reality in the future, and I definitely hope that this conference will make a valuable contribution to reaching the goal. I will repeat the words having been said before: We have just begun to scratch the surface. With these words I will wish you all a very prosperous conference and I am looking forward to all the interesting presentations that we have ahead of us.

**SUSTAINABLE AND MORE EFFICIENT
UTILIZATION OF RESOURCES**

FUTURE DEMAND FOR, AND SUPPLY OF, FISH AND SHELLFISH AS FOOD

U. Wijkstrom [1]
Fishery Department, Food and Agriculture Organization of the United Nations, Rome, Italy

Abstract

If real prices for fish remain at the levels they had attained by 1999 in the year 2050 demand for fish and shellfish as food could be of the order of 270 million tons (live weight equivalent) per year. If producers were able to supply these quantities consumption would rise by 176 % over this 50-year period. To meet the demand supply would have to expand at the rate of 2.1 % annually. But, a review of the pattern of population growth – and of historical patterns of increases in per capita consumption of fish – shows that annual growth in the volume of fish demanded is likely to be largest in the coming two decades, and then to taper off. Will producers be able to deliver?

It is clear that wild marine stocks at present harvested by capture fishermen can not support fisheries that would yield much more than 100 million tons per year and of this amount a significant proportion will continue to be used for fish meal and oil production. The question therefore narrows down to: can aquaculture, or non-traditional marine species, supply the required amounts?

The historical context of supply is considerably different from that in which prevailed during the last 30 years. At that time the growing demand in OECD countries was met partly through imports of fish produced in the seas and lakes of developing countries. During coming decades the increased demand in developing countries must be met essentially through their own resources. In fact, in poor countries it seems unlikely that supply will respond to demand unless they experience economic growth.

1. PRODUCTION AND CONSUMPTION OF FISH AT THE END OF THE 1990'S

1.1. Production

Let us first consider production of fish [2] from all sources. Production is now back at record high: an estimated 126 million metric tons in 1999. This means that production has recovered from the El Niño related decline in 1998. But, capture fisheries – fresh water and marine - has been stable since 1994, the combined production oscillating between 91 and 94 million tons, the only exception being 1998, when production was only 86 million tons.

1.1.1. *Capture fisheries*

China continues to be the most important fishing nation. The combined production of fresh water and marine capture fisheries in 1999 reached some 17.2 million metric tons, thus accounting for 19 % of world capture fisheries production. Peru and Chile followed China. Twelve countries accounted for about two thirds of all capture fisheries production in 1999. It is worth noting that the majority of these twelve countries exploit the Pacific. The significant exceptions are Norway and India. Of course, both USA and Russia also exploit other oceans.

1.1.2. *Aquaculture*

Production from culture continues to expand. The estimate for 1999 is that the world produced 33.3 million tons (excluding aquatic plants). The top nine producers are Asian economies. Together they accounted for 88 % of the volume of global aquaculture production in 1999.

1.2. Consumption

During the last decade the use of fish as food increased. The food-fish share of total production grew. The share peaked in 1998 – when El Nino severely restricted capture of pelagic species off the coasts of Peru and Chile. The result was a steep drop in fishmeal production. In spite of this drop, it is clear, that there is a tendency to use as food a larger share of the world's fish supplies. This reflects, on the one hand, an unchanged use of small pelagics for fishmeal, and, on the other, the growing share of aquaculture produce in world food fish production.

The main fish consumers are China, Japan, US, Europe (the EU), India, Indonesia. Together they accounted for about 62 % of consumption (and 55 % of world population). But even amongst these countries there are large differences. India, having a population about 4 times that of the US, in fact consumes less fish than the US. Japan on the other hand consumes as much fish as the EU, but with only one third the EU population.

1.2.1. *China*

The preceding analysis shows that China is important both as a producer and as a consumer. The Chinese supplies have grown rapidly during the last two decades. They have grown so fast as to make the Chinese population one of the most significant fish eaters of the world, with a per caput consumption rate of about 27-28 kgs per person and year in 1998. Also, most of China's fish production is used for food. It is only the last few years that fish meal and fish oil production has come to occupy a noticeable share of Chinese capture fishery production.

1.2.2. *World as a whole*

For the world as a whole apparent consumption of fish increased steadily from just below 9 kgs in 1961 to 16.3 kgs in 1999. But of course there are large variations; from less than half a kilo per person and year, to more than 100 kgs during the same period.

How does fish do against meat? FAO figures show that over the past four decades the average world consumer has increased the share of fish as a proportion of the total animal proteins that he/she consumes. In 1961 the proportion of fish proteins in animal proteins was 13.8 % in 1997 it had grown to 16.5 %. But, again here the variations are large. For some 5 % of the world's population fish proteins account for at least 50 % of all animal proteins consumed; for some 20 % fish provides at least 25 % of all animal proteins consumed. Given the importance of animal protein for development and growth, fish in these societies is an essential ingredient of the diet.

2. FISH CONSUMPTION: THE IMMEDIATE FUTURE.

Recent figures show that during the last few years the supply for human consumption has been growing at a steady rate, the recent exception being 1998. This is explained mainly by a very slow increase in capture fisheries – occurring against a background of oscillating production of small pelagics - and a steady growth in aquaculture output. The estimate of world fish consumption in 1999 is 97.2 million tons (live weight equivalent) [3].

2.1. Demand

At the beginning of the present decade the annual growth in demand is of the order of 2.9 million tons. This growth is caused by an expanding world population and an increasing economic well being. We estimate that about 1 million tons can be attributed to a world population increase of about 75 million people and 1.9 million tons attributed to a 2 % GDP/caput growth.

2.2. Supply

But how will supply respond? What will be the yearly increment in supply? Will it be more or less than the projected near-future increase in demand for fish and shellfish? Recent developments give an indication. The yearly increases in fish for human consumption during the five-year period 1995 – 1999 appear to hold steady at about 3 million tons. Breaking up the total supplies into those originating in capture fisheries and those originating in aquaculture shows that while aquaculture supplies increases annually at a rather steady absolute rate of 2 to 2.5 million tons per year, possibly with an upward trend. The contribution of capture fisheries (marine and inland fisheries combined) fluctuates, is low, and seems to have a declining trend.

Media have for some years reported that aquaculture is the fastest growing food production sector of all with an annual growth rate (in volume terms) of 10 % and more during the 1990s. While this is a mathematically correct affirmation it can be somewhat misleading. In fact a look at the growth rate for China on the one hand, and the rest of the world on the other, will show that for most of the world expansion has been slower than 10 % per year and that in China it is gradually slowing down, now being a bit below 10 % per year.

2.3. Consumption

It appears plausible the supply will match the incremental growth in demand during the next few years – particularly in view of the recent global economic slowdown. So prices, in the immediate future, are likely not to increase in real terms. The effects of that may be to place wild stocks – and the environments nearby aquaculture installations – under less pressure. This will be good for management.

3. FISH CONSUMPTION IN 2010 – A REGIONAL PERSPECTIVE

Let us consider fish consumption in 2010 not for the world as one unit, but separately for: Africa, North America, South America, Europe, including Russia, Oceania and Asia. Future fish consumption will depend on present state of consumption and on economic, technical and political developments.

While we can only speculate about future economic, technical and political developments, we have a fairly good understanding of how change sin population size and in economic growth will affect consumption. There is less agreement on how these factors will affect supplies of fish and fish products.

3.1. Fish Production and its use in 1997

Table 1 summarises, by continent, information on production, trade and consumption of fish products for the year 1997. A few observations on this table. Asia accounts for more than 50 % of production - 56 to be exact. South America is the dominant user of fish for purposes other than direct human food. Of their total production more than two thirds are converted into fish meal and oil, thus accounting for about one third of the world's production of fish meal and oil. Europe (here including Russia and countries in transition) is the biggest importer, but also the largest exporter of food fish. South America is the largest net exporter of food fish.

Table 1

Apparent consumption, or supply of fish for food, by region in 1997 (in thousand metric tons – live weight equivalent)

	Africa	North America	Caribbean, Central and South America	Europe including Asia	Russia	Oceania	World
Production	5822	6598	19806	69221	19527	1081	122055
Non-food uses	736	1278	12873	7469	6001	109	28466
Imports	1265	3155	931	7899	10971	317	24538
Exports	1164	2195	3133	6429	11126	714	24761
Total food fish supply	5187	6280	4731	63222	13371	575	93366
Population (in millions)	731.4	302.1	496.3	3536	739.2	29.3	5834.3
Per-caput consumption (kgs)	7.1	20.8	9.5	17.9	18.1	19.6	16

3.2. How much fish is needed in 2010?

Demand for fish will expand with population growth. This is illustrated in Table 2. The table is constructed as follows: the per-caput consumption figures of 1997 have been maintained unchanged. The population, as forecasted by the United Nations population Division in New York (median variant 1998) has been inserted for the year 2010 and the production needs calculated, assuming that non-food uses will remain constant. The total food fish needs increase by about 14 million tons, or about 15 %. The additional quantities needed are greatest in Asia – additional 11 million tons - because of population growth. But demand will grow fastest in Africa: an increase of about 1.9 million tons representing an expansion of 36 %, well above the world's average rate of growth. The needs for fish as food will decline in Europe.

I have already referred to the fact that economic growth has been shown to occur in parallel with increase in the volume of consumption per capita. This fact is incorporated in the exploratory calculation provided in Table 3. The assumptions used in that table are: a 1.5 % growth in

GDP/person world-wide per year from 1997 to 2010, and that each percentage increase in GDP/caput will lead to a two thirds of one percent increase in the volume of consumption.

The result is an increase in demand by a further 14 million tons, comparing the situation in 1997 with the projected situation in 2010. The fish as food needs for Asia grow by 9.7 million tons, for Africa by 0.9 million tons. The needs in Europe increase, but at a somewhat lower rate, by 1.2 million tons.

In summary, at the very least about 107 million tons of fish as food will be needed in 2010. That amount signifies that the population then inhabiting the earth will have the same per capita availability of fish as the world's population had in 1997. But needs are likely to be higher; possibly as high as 121 million tons. If that turns out to the be case 28

Table 2
Projected needs of fish as food by continent in 2010, at per/caput apparent consumption rates of 1997 (in thousand metric tons-live weight equivalent)

	Africa	North America	Caribbean Central & S America	Asia	Europe including Russia	Oceania	World
Total needs (Production + imports --exports)	7815	8184	18526	81665	18906	775	135870
Non-food uses (constant)	736	1278	12873	7469	6001	109	28466
Imports							24 600
Exports							24 600
Total food fish needs	7079	6906	5653	74196	12905	666	107404
Population (in millions)	997	332	595	4145	713	34	6816
Project Per-caput consumption (kgs)	7.1	20.8	9.5	17.9	18.1	19.6	16

Table 3
Projected needs of fish as food by continent in 2010, at per/caput apparent consumption rates of 1997 and assuming a yearly increase in GDP/caput of 1 % (in thousand metric tons − live weight equivalent)

	Africa	North America	Central & S. America	Asia	Europe including Russia	Oceania	World
Total needs (Production + imports − exports)	8735	9047	19180	91310	20584	862	149615
Non-food uses (constant)	736	1278	12873	7469	6001	109	28466
Imports*							24 600
Exports							24 600
Total food fish needs	7999	7769	6307	83841	14583	753	121149
Population (in millions)	997	332	595	4145	713	34	6816
Project Per-caput consumption (kgs)	8	23.4	10.6	20.2	20.5	22.1	17.8

*As world exports should equal world imports, trade directions are unimportant when estimating world needs for food fish.

million tons of fish for human food would be needed on top of the 93 million that were available in 1997. In this case, the average sustained increase in supplies over this period needs to be of the order of 2 million tons per year.

Where will this fish come from? For the world as a whole there does not seem to be major problem, as long as we can maintain increases in fish supplies at the level achieved during the 1990's. But is this equally true for all parts of the world?

3.2.1. *Africa*

Present consumption and recent trends

In Africa fish is a common and important part of the diet. Nevertheless, in Africa people eat less fish than in other continents. The average per caput consumption has now fallen to about 7 kgs per person and year. But, there are large differences within Africa. In West Africa the consumption is at about 9 kgs per year, in Central Africa about 7 kgs, being slightly below 5 in East Africa. In fact both Central Africa and West Africa depend on imports to keep consumption at present levels. Most of these imports come from outside Africa. Only minor quantities of fish are imported from South Africa and North West Africa, both of which are net exporters of fish.

Future demand

Demand for food and fish will grow rapidly in Africa simply because its population is expanding rapidly. Africa is expected to have just below 1 billion inhabitants in 2010. This represents more than a 35 % increase compared with the continent's estimated population in the year 1997. This increase alone − at present real prices and per caput incomes − would amount to an expansion of fish demand from 5.5 million tons to about

7.1 million tons. If economic growth (increasing GDP/capita) continues in Africa demand for fish will expand rapidly indeed, much faster than in any other continent. It may reach several percent per year. An average growth of 1.5 % per year in GDP/caput until 2010 would add about one million tons to the demand in 2010 [4].

Future supplies

At the beginning of the 21st Century fish supplies in Africa were characterised by fully to overexploited wild stocks and an incipient aquaculture, located essentially in West and Central Africa. Most likely supply from local sources will grow only slowly even in the face of expanding demand. This will mean, most likely, that commercial aquaculture for urban African markets will take off. But during the rest of this decade the volumes will be comparatively small. Thus, it seems likely that some of Africa's exports of fish will be directed at African markets. For a long time informed observers have maintained that the abundant small pelagic stocks off North West Africa should find a market in West – and possibly Central – Africa. Although some attempts have been made, most fishery entrepreneurs in the region have found high value, rather than high volume, species to be of interest and thus have not invested in the exploitation of small pelagic species. Nevertheless it seems quite plausible that inter-African trade in small pelagics from North West Africa could become economically interesting during the period. But, Africa is most likely to continue to be a net importer of cheap fish, and an exporter of expensive varieties.

Future consumption

Thus it is clear that the future fish consumption in Africa is closely linked to international developments and to the development of inter-African trade in fish products. In the worst scenario – that of stagnant or very low economic growth - there is no evident reason to expect anything else but a fall in per caput consumption throughout the region, and with the most important declines being in Central and West Africa. The basic reason is that in the absence of economic growth, Africa will find it difficult to forego the hard currency revenues generated by extra-continental exports while few consumers will be able to afford the price of imported fish products. The net result – in view of stagnating local supplies – will be decreasing consumption per caput.

3.2.2. North America

Present consumption and recent trends

In the US fish consumption is higher, on a per caput basis, than what it is on average for the world as a whole. It has been rather stable about 20 kgs (in live weight equivalent) since 1985 [5]. This is noteworthy, as the average consumption of meat (of all sources) has increased in the US during the same period. Most of the increase has been captured by poultry producers. Consumption of poultry has expanded faster than that of any other category and during the period achieved a partial replacement of red meats. In Canada fish consumption is slightly higher than in the US. But the consumption of fish has not been static in the US. During the last 15 to 20 years the consumption of small pelagic species has declined by half while the consumption of fish products derived from demersal species has grown rapidly. This has occurred in parallel with a growing share of fresh/frozen products, that has replaced cured products.

Future demand

In volume terms demand is going to expand at a modest rate over the next few decades. In value terms consumption may expand more rapidly. There are two main reasons for the slow growth in volume. On the one hand it is foreseen that population in North America (essentially the US and Canada) will grow from about 310 million in 2000 to some 342 million in 2010. This is an increase of about 10 %. On the other hand the population will become wealthier. This will expand consumption per head, but at a declining rate as an increasing share will arrive at an 'optimal' food basket from the point of view of health and variety.

Future supply

Capture fisheries production in North America will not exceed its present level of production. Aquaculture is an established industry and it will expand. But, it seems unlikely to expand much faster than the market demand. However, in the future the American market will remain open to fish imports, particularly of un-processed fish. And in fact the North American consumer will continue to depend on imports for some 10 % of his/her consumption, in particularly for supply of products such as shrimps, tuna and white fish blocks.

Future fish consumption

As local suppliers and fish importers are likely to be able to meet demand, consumption is going to expand at a rate that will not be much below that projected for demand. But national suppliers will have growing economic problems in attempting to compete with foreign suppliers and, as time goes by, American consumers most likely will increasingly depend on imported products, particularly imported aquaculture products.

3.2.3. *South America and the Caribbean*

Present consumption

Consumption of fish is higher than in Africa, but lower than in other continents. At the end of the 1990's apparent average consumption was about 9 kgs/person and year. But consumption of meat is high, the highest in the world. In Brazil, the largest country in the region, accounting for about one third of the population, consumption is below the average for the region at about 6 kgs/caput and year, about one third of which is imported. Also in the Caribbean the dependence on fish imports are high; about 50 % of the fish consumed come from outside the area.

Future Demand

Demand for fish will grow rather fast in Latin America. Population growth will account for close to 1. 5 % per year, as the population grows from some 520 million in 2000 to 594 million in 2010. However, as the continent experiences an increase in well being the wish for a more diversified diet will spread. Per capita demand for fish – at present real prices - is set to grow at 2 to 3 % per year.

Future Supply

Many of the food fisheries are overfished and wild stocks exploited by artisanal fisheries are under growing pressure. Most of the small pelagic species on the West Coast continue to be fished by industrial fisheries supplying the fish meal industry. There is a well-established aquaculture industry in Chile (Salmon) and Equador (Shrimps), and shrimp culture is now common the West Coast of South and Central America. Fresh water fish culture is expanding fast. The food-fish fisheries of Uruguay and Argentina are likely to cater to any growth in Brazilian demand for high quality food fish. But the volumes are likely to be comparatively small. The bulk of Argentinean and Uruguayan fish exports – over 1 million tons (live weight equivalent) per year – are likely to find markets outside Latin America.

Future Consumption

It seems likely, in view of the above supply and demand projections, that fish consumption will continue to grow in South America and in the Caribbean, and that this will take place without leading to any significant modification of Latin America's role as a net exporter of fish and fish products.

3.2.4. *Europe, Including Russia*

Present consumption

Fish consumption in Europe is far from uniform. At the end of the 1990s it ranged from high values in Northern Europe (average of 28 kgs), through a slightly lower value for the European Union countries, some 20 kgs/capita in the Russian Federation (after a low of 10 kgs in the mid 1990s) and about 7 kgs/caput in Eastern Europe. While Russia and Northern Europe are net exporters of fish, the rest of Europe is a net importer. In both the EU and Eastern Europe about a third of the fish consumed is imported.

Future demand

In volume terms demand is going to grow essentially as a function of economic growth, as the European population is projected to diminish from about 740 million in 2000 to some 713 million in 2010. Given the rather high levels of consumption already reached the demand increase will be marginal measured in terms of volume of fish.

Future supply

Also in Europe aquaculture represents the principle category of production that may achieve an increase in production, as wild stocks exploited by capture fishermen are fully used, and in some cases over-exploited. However, the growth rate of European aquaculture production has been below that for the rest of the world and it has depended also on non-European markets to sell some of its produce. Given that the bulk of the European aquaculture is made up of a rather small number of species – while European fish consumption is based on a vast array of species – European aquaculture producers will continue to depend on non-European markets for part of their sales. It seems unlikely that European aquaculture will be able to substitute for any significant share of present European fish imports. Thus, in all likelihood during the coming decades the average European will continue to eat a significant portion of non-European fish. That is, the dependency on imports will continue.

Future fish consumption

In Europe, consumers will continue to be supplied with as much fish as they desire – at present real prices. Growth in the total volume of consumption will be slow. During the coming decade or two growth in the volume of consumption will take place at a rate that is considerably below that of GDP/caput.

3.2.5. *Asia*

Present consumption

Asia accounts for about two thirds of production and about two thirds of consumption of fish in the world. There are of course large differences within Asia in the levels and composition of consumption. Consumption patterns spread from those resulting in a level of about 70 kgs per person and year (Japan) to those implying only 5 kgs – and less - in India and Pakistan.

Future consumption

A breakdown of the increase due to population growth (shown in Table 4) indicates that the increase in demand is less than what I have earlier stated. The reason is that population grows fastest where apparent consumption per caput is lowest, that is in the Indian sub-continent.

Table 4 shows that just below 40 % of the increase is accounted for by China; ASEAN member countries account for some 28 % and South Asia for about 17 %. Japan's consumption, in volume terms, will hardly change. But, as stated above the increase in demand caused by population growth is the minimum that can be expected. Economic growth will add to demand. A steady increase of 1.5 % in GDP/caput may add another 6 to 8 million tons over the same period.

Future Suppl

Imports and production will grow to meet this demand. Of Asian fish imports, some 40 % - measured in value – in 1995/97 came from other Asian countries, the remainder from other continents [6]. Japan has been the leading importer, lately followed by China. However, the growth in imports may slow as Japanese demand continues to be stable in volume terms. The most obvious exporting region is South America. Some fish will also come from Africa.

Asian capture fisheries are unlikely to contribute a significant share of the needed additional landings. Most inshore areas are fully exploited and tuna species will soon be. Towards the end of the period – that is by 2010 – it is feasible that effective management undertaken today, will lead to stock recoveries and some increase in the volumes landed. But, the additional volumes are not likely to be large.

Table 4
Additional demand for fish as food in Asia by 2010 resulting from population growth

	Population in 1996 (thousand)	Projected population in 2010 (thousand)	Population increase 1996-2010 (thousand)	Per capita consumption in 1996 (kgs)	Additional food fish requirements in 2010 (tons)
Brunei Darussalam	302.0	384.0	82.0	28.0	2.296
Cambodia	11.741	13.25	1.559	8.7	13.563
Indonesia	200.583	238.012	37.429	18.2	681.208
Lao PDR	4.801	6.965	2.164	8.7	18.827
Malaysia	20.46	25.919	5.459	55.6	303.52
Myanmar	45.079	50.903	5.824	16.8	97.843
Philippines	69.806	90.544	20.138	30.5	614.209
Singapore	3.585	3.885	300.000	31.8	9.54
Thailand	59.545	66.511	6.966	31.2	217.34
Vietnam	73.999	90.764	16.765	16.6	278.299
Bangladesh	136.336	167.926	31.59	10.1	319.059
China	1,231.068	1,366.215	135.147	24.2	3,270.557
India	943.561	1,164.020	220.459	4.7	1,036.157
Pakistan	126.867	181.385	54.518	2.0	109.036
Japan	125.826	128.22	2.394	69.0	164.186
Asia – others			129.715	(10.0)	1,297.150
Total	3,473.828	4,144.937	671.109	(12.6)	8,432.481

Aquaculture can provide long-term relief for the sector. China has continued to expand its aquaculture industries, in particular by bringing new species under culture. It seems possible that this pattern will be followed in other Asian countries.

Future consumption

It seems plausible that the Asian consumption of fish may evolve as described. But, as it does consumption may polarise, reflecting uneven economic growth. Aquaculture will be the principle source of additional supplies – and these will go towards all sectors of the market; those consisting of luxury products as well as staple. Imports from Latin America and Africa will cater to the upper part of the market.

In conclusion there seems to be only one major concern in terms of fish supply for the year 2010 and that concerns Africa. As already stated, needs for fish in Africa are likely to grow faster than what can be expected from traditional sources of supply. It seems important that concerned governments promote aquaculture.

4. WHERE WILL THE FISH COME FROM IN 2050?

Let us consider the situation in 2050. To develop an idea of what fish consumption may look like in 2050 it is of course important to consider both the likely demand and the possibilities of supply.

4.1. Demand

It seems likely that consumers who are economically well off will eventually stabilise their consumption of fish. There are several reasons for this. Amongst them are: the wish to have a diverse diet; the desire to eat healthy food – and fish is generally seen as healthy food; and a declining proportion of disposable income being spent on food. Thus the price of any individual food item will have a reduced importance in the purchase decision and other aspects – such as health considerations – will become more important. But, at what level will consumption stabilise? Will it be at 20 kgs/caput/year? This is the level where UK consumption has been for the last few decades. Or will it be at 70 kgs/per year like in Japan. A plausible, but not unduly pessimistic assumption would be, that consumption for most individuals would stabilise at about 40 to 50 kgs per person/year. Would they have attained this level by 2050? Most likely not. The reason is, as we will see, shortage of supply.

How much fish will be demanded by consumers in 2050? A rough, exploratory calculation has been made in Table 5. The per-capita consumption figures for 2050 has been taken as given. It differs from one continent to another but overall is well below the 40 to 50 kgs mentioned in the previous paragraph. The amount of fish needed to supply this per capita consumption has been calculated using the projected population by 2050. The amount of fish used for non-food purposes has been left unchanged.

The per capita consumption of fish in 2050 has thus been assumed to be 17.5 kgs in Africa. This estimate is the lowest of those made and may seem low. Nevertheless the amount is equivalent to a level that is 2.5 times the level achieved in Africa in 1997. As the African population is expected to grow rapidly, under this scenario the total needs of food fish will increase from 5.2 million tons per year in 1997 to 31 million tons in 2050, a six fold increase. It implies essentially that either Africa will become a major aquaculture producer, a net importer of fish, or - and this is likely - both.

The per capita figure projected for South America is also lower than that for the world as a whole. For this continent, however, the reason is not difficulty of access to fish – of

Table 5
Projected needs of fish as food by continent in 2050 (in thousand metric tons –liveweight equivalent)

	Africa	North America	Caribbean, Central & South America	Asia	Europe Russia	Oceania including	World
Production plus imports less exports	31641	13038	33098	191849	27981	1719	299326
Non-food uses	736	1278	12873	7469	6001	109	28466
Imports							100 000
Exports							100 000
Total food fish needs	30905	11760	20225	184380	21980	1610	270860
Population (in millions)	1766	392	809	5268	628	46	8909
Per-caput consumption (kgs)	17.5	30	25	35	35	35	30.4

which it has a lot - rather it is that the tendency to eat more meat than fish will change only slowly. The projected figure for Asia may appear low. However it should be remembered that it includes the Indian Sub-continent where very large numbers of its inhabitants each eat only small amounts of fish at present.

4.2. Supply

Thus, the projected need of fish as food in 2050 is about 270 million tons. This means that supply must increase by 176 % over the amount available in 1999. This is a sizeable increase. However, it could be achieved if an average annual growth of 2.1 % could be maintained. In purely arithmetic terms this would mean that today's availability should increase by about 2 million tons per year (which it did during the second half of the 1990s) ; while at the end of the 2040s the expansion would be of about 5.5 million tons. It would of course be preferable to inverse the availability. It is now that the large yearly increases are needed: a large number of consumers are at low levels of consumption and their number is increasing relatively rapidly. By 2050 population growth will have dwindled to very low levels and propensity to increase consumption of fish will be much lower on the average. But, apart from looking at growth rates – which may be reassuring – can they be achieved in a finite world? After all the present supplies of about 97 to 100 million tons per year are straining fishery resources and pushing aquaculturists. Will almost three times this amount be feasible?

In theory the possible sources of increased supplies are: presently exploited stocks; presently un-exploited stocks; less in discards; less fish used for fish meal; and an increase in aquaculture production. Most informed observers seem to concur that there is little if any substantial increase to be obtained from better use of presently exploited resources.

The major unexploited stocks are: krill, mesopelgics and oceanic squids. Krill is not strictly in this category as some fishing nations have taken small quantities for several decades. However, in spite of much work, a breakthrough has not been achieved – only relatively small amounts find their way to consumers. There does not seem to be any strong reason to believe that there will be such a breakthrough soon – in particular if the moratorium on harvesting of cetaceans continues. But, if fish meal continues to be an essential ingredient for fish feeds – which is likely - the day may still come when krill becomes a candidate ingredients in such feeds.

Mesopelagic species has also been the subject of studies and trial fishing intermittently over several decades. The objective has been to convert these mesopelagic species – of which there are large quantities e.g. off Yemen and the Arabian Peninsula – into fishmeal. Also here success has not come, partly because economics of production does not look promising and partly because the resulting fish meal is not of high-quality.

Ocean squids are closer to become a product for human consumption, although the consumption of squid today is limited to a relatively small number of countries (mostly Japan, and the countries around the Mediterranean). But, after the ban on drifting pelagic gill nets the fishery for oceanic squids has become difficult. Nevertheless, most likely there will be some increase in the exploitation of this resource. But, considered as a whole, un-exploited or under-exploited species do not seem to be a likely source of significant part of the additional 170 million tons needed by 2050.

What about fewer discards as source of additional supplies? Discards are believed to be of the magnitude of 20 million tons per year. A lot is being done to reduce discards through developing more selective fishing gear and methods. That solution of course means that the fish is not taken up at all. In the first instance it is not available to anyone. But the fish may become accessible to fishermen at a later moment or may become prey for fish that in turn are caught. So, it is far from certain that even if discards were reduced by 10 million tons in the course of the coming decades this would result in an increase of landings by the same amount in later years.

Most fishmeal manufacturers employ capital intensive technology and make use of low value fish. Generally raw material is not paid above US 100 per ton, often much below that. The fish is of low value not because it lacks nutritional value. It is of low value either because it is perishable and caught in very large volumes far from human population centres, or because the species lacks appeal to the average consumer.

Fishmeal is demanded mainly by livestock and aquaculture industries. The demand from these industries is set to grow strongly, while the demand for these species as human food will grow marginally, if at all in regions where they are easily available – that is where they are landed. Thus, it is unlikely that a decreasing share of the present capture fisheries production will be destined for fishmeal. And, this will be true even if the livestock and fishmeal industry manages to find replacements, which may be easier for fishmeal than for fish oil.

Thus, in summary it does not seem that the situation of world fish stocks, or of the use that present landings are put to, is such that we should expect a major increase in availability from world resources.

But, there is other – and equally valid reasons – not to expect that capture fisheries will be a major source of increased fish supplies. These reasons are social and economic in nature.

Professional fishers are engaged in one of the most dangerous occupations in the world – and/or an unusually uncomfortable one. For some fishermen it involves being away from home for long periods. For others, and particularly small scale fishermen, the exposure to the forces of nature is constant and dangerous. It should come as no surprise that where alternatives are available young men/women do not enter the profession. The average age of the active fisher population tend to increase, particularly in developed countries.

Also, often the profession is not economically attractive [7]. And, fishers have the cards stacked against them as they strive to keep up their real incomes. They are more restricted in the way they can do this than are farmers and foresters. In agriculture and forestry productivity is improved on the one hand through better machinery and methods, and, on the other, by improving the yields from plants and animals. In capture fisheries productivity increases relies almost exclusively [8] on improvements in equipment and techniques (boat and gear). Fishers have themselves little if any possibility to alter the characteristics, or abundance, of their prey. And, for the majority, access to the fishery is not controlled and the stocks may be dwindling in size making better effectiveness or efficiency more difficult to achieve.

Also, while the larger fisheries – those that fish large stocks tending to school – will continue, small-scale marine fisheries, exploiting local stocks with locally adapted gears, will find it increasingly difficult to remain economically competitive. Many of them, in the course of the next decades, will disappear as full-time commercial activities – as has already happened in many inland capture fisheries.

This places a major burden on aquaculture. Will it be able to produce all that which capture fisheries will not produce? We have no choice: aquaculture has to, and, in my view it will.

ANNOTATION

1. The views expressed in this paper are those of the author and not necessarily those of FAO.
2. In this paper the author includes in the term "fish" also crustaceans and molluscs, unless otherwise stated.
3. This is an overestimate as it represents the apparent supply for human consumption. These will have been looses in processing and trade.
4. These assumption of an income elasticity of demand of 0.67 is maintained. But, it may be low for Africa.
5. These data are FAO data. It is possible to encounter US estimates of per caput fish consumption that are higher than these.
6. FAO Yearbook of Fishery Statistics. Commodities. (1999) Vol.89, Appendix III-Trade Flow by Region.
7. The exceptions to this statement are those fishermen who benefited from a free entry into fisheries for which Individually Transferable Quotas (ITQ's) have been created.
8. Relatively few fisheries – possibly with the exception of the Japanese ones – benefit from stock enhancement and modification of fish habitats (artificial reefs).

BYCATCH AND DISCARD FISH IN THE JAPANESE FISHERIES

Yoh Watanabe
Fisheries Biology and Ecology Division, National Research Institute of Fisheries Science, Fisheries Research Agency, 2-12-4 Fukuura, Kanazawa, Yokohama 236-8648, Japan

Abstract

Discards of bycatch in fisheries is one of the most serious concerns in fisheries management. The types of discarded fish from Japanese fisheries can be mainly classified into four categories: 1) species with low value or no value in the market, 2) small sized species including young or juvenile, even if they have commercial values as adult, 3) species with substantially lower values in the market taken by fisheries targeting on highly valuable species such as tuna longline, and 4) species whose sales in the market are prohibited by the government and poisonous fish. Some methods to decrease the discarded fish are proposed: 1) effective utilization of bycatch species, 2) release to the sea in alive immediately after catch and 3) improvement on gears to avoid bycatch fishes. The improvement of utility value for the bycatch fish is considered to be the most effective solution on this issue. Preparing the circumstances to make the fishermen bring back the less value fishes without discards and the action to raise the value of the discard fish are most required at present.

1. INTRODUCTION

In the beginning, FAO focused on a large amount of discards by shrimp trawling in 1982 from the viewpoint of the waste of fishery resources. Later, the research of this subjects became popular all over the world, and it came to be considered one of the greatest problems in the management of the fisheries. In 1994, the FAO Fisheries Technical Paper 339 on Global Assessment of Fisheries Bycatch and Discards estimated the annual amount of discards in the world fishing from 1988 to 1990 as 27 million tons and showed clearly how it wastes the resources [1]. Now, it is recognized that discards of bycatch in fisheries is one of the most serious concerns in fisheries management and effective use on fishery resources.

2. DEFINITION OF BYCATCH AND DISCARD

A bycatch fish generally indicates fish except for the target fish; however, the definition is ambiguous. For example, bycatch fish in the bottom trawling which caught plural species has not clarified fully. In this report, I define bycatch fish as the fish other than the main target. A discard fish indicates a fish dumped into the sea from fishing boat after catch. There is a case fish is thrown away since the size does not suit, even though it is a target fish. Therefore, all discard fishes are not only bycatch fish. After landing in the fishing port, there is also a fish to dump in sorting due to less commercial value. It is thought that this should also be put into a category so-called discard fish. Generally, discard fishes consist of the following cases. Species with low value or no value in the market, small-size species including young or juvenile (in rare case, too large), species with low value in the market, compared to highly valuable target species and with no room to preserve on board, and species whose sales in the market are prohibited by the government and poisonous fishes. In the case when the fish length is below regulation size under length restrictions, and we back the fish in the water after caught immediately, we call this case as release. And this case is clearly distinguished from a discard as a release.

3. DISCARDS OF THE WORLD FISHERIES

As already mentioned, Alverson et al. [1] (1994) estimated the annual amount of discards in the world fisheries as 27 million tons in 1988 to 1990 at the request of FAO. According to the amount of discards in Northwest Pacific was also estimated as the 9.13 million tons, about one third of the global discards, followed by those in the Northeast Atlantic of 3.67 million tons, West Central Pacific of 2.78 million tons, Southeast Pacific of 2.60 million tons, West Central Atlantic of 1.60 million tons, West Indian Ocean of 1.47 million tons, and Northeast Pacific of 0.92 million tons. The amount of discards in the Northwest Pacific was overwhelming among the world in this estimation (Table 1). Although the amount of discards by the country was not described, it is thought that the amount of discards by the fishing of Japan occupies a large amount of 9.13 million tons.

However, in the meeting of Technical Consultation on Reduction of Wastage in Fisheries held in October 1996, jointly sponsored by FAO and the Fisheries Agency of Japan, Matsuoka [2] reported the amount of discards in the Japanese fisheries and pointed out that the estimation by the FAO Fishery Technical Paper 339 was excessive. He estimated that the annual discards of Japanese fisheries were 858 thousand tons in average from 1988 to 1990 and 793 thousand tons in 1994. He also estimated the discard in 1994 as 255 thousand tons for small trawl fishery, 215 thousand tons for boat seine fishery, 132 thousand tons for tuna longline fishery in distant waters, and 62 thousand tons for off-shore trawl fishery (Table 2). Since the total fish catch of Japan in 1994 was 6.59 million tons, this indicated that about 12 % of total catch by Japanese fleet in 1994 were discarded. Therefore, considering from the total fish catch of Northwest Pacific belonging to the Japan, Korea, North Korea, China, Taiwan and Russia, probably it should be concluded that at the most 3 million tons of fish are discarded in this area.

Table 1
Discard weight by major world region

Area	Discard weight (mt)
Northwest Pacific	9,131,752
Northeast Atlantic	3,671,346
West Central Pacific	2,776,726
Southeast Pacific	2,601,640
West Central Atlantic	1,600,897
West Indian Ocean	1,471,274
Northeast Pacific	924,783
Southwest Atlantic	802,884
East Indian Ocean	802,189
East Central Pacific	767,444
Northwest Atlantic	685,949
East Central Atlantic	594,232
Mediterranean and Black Sea	564,613
Southwest Pacific	293,394
Southeast Atlantic	277,730
Atlantic Antarctic	35,119
Indian Ocean Antarctic	10,018
Pacific Antarctic	109
Total	27,012,099

Alverson, D.L. et al. (1994) [1]

Furthermore, a fishery which produces most abundant discard fish in the world fisheries by the FAO Report was a shrimp trawl, and which had 9.51 million tons and occupied to 1/3 of total amount. The rate of a discard by the areas ranged 40 - 98%. The discard in the fishery performed very huge. In the estimation by Matsuoka [2], shrimp trawl fisheries in Japan was a little portion and the amount of discards was 13,9 thousand tons during the same period of FAO estimation.

4. BYCATCH AND DISCARDS OF EACH FISHERIES IN JAPAN

4. 1. Bottom trawl fishery

Hirakawa and Ono [3] described that in the small trawl fishery of Fukushima Prefecture, main species discarded were squid (*Lorigo japonica* :jindoika), shrimp (*Trachypenaus curvirostris* :saruebi), shrimp (*Metapeaeopsis dalei* :kishiebi) in except for fish, and moon dragonet (*Repomucenus lunatus*), sand lance (*Ammodytes personatus*), pinkgray goby (*Chaeturichthys hexanema*) in fishes. There were many species discarded since they were merely a small although they have commercial value in a market.

According to the research on the off-shore trawl of the Onahama fishing port by interview, Macrouridae such as bighand grenadier (*Abyssicola macrochir*) and threadfin hakeling (*Podonema longipes*) are discarded at present, even though they were landed at few yeas ago. Moreover, cloudy dogfish (*Scyliorhinus torazame*) which are occasionally caught more than one ton is also discarded. As other fish for a discard, there are snailfish

Table 2
Estimates of discards in fisheries in Japan (1988-90 and 1994)

Sectors	W/W discard ratio	1988-90 Products (mt)	Discards (mt)	1994 Products (mt)	Discards (mt)
Skipjack pole and line	0.0	228,773	0	168,870	0
Saury stick-held dip-net	0.0	279,348	0	249,950	0
Mackerel angling	0.0	3,114	0	2,882	0
Shellfish collection	0.0	106,833	0	73,836	0
Seaweed collection	0.0	195,133	0	134,458	0
L/M surrounding net for others	0.0	3,704,605	0	1,375,675	0
Purse seine	0.0	1,076,385	0	840,663	0
Small trawl (powered, shellfish portion)	0.0	220,372	0	298,125	0
L/M surrounding net for skipjack/tunas	0.000423	197,912	84	230,537	98
Tuna longline (distant waters)	0.672	185,084	124,376	196,725	132,199
(off-shore waters)	0.496	62,836	31,167	48,252	23,933
(coastal waters)	0.646	27,393	17,696	39,319	25,400
Distant water trawl in N Pacific (high seas, rounded)	0.0435	607,000	26,405	0	0
(other waters)	0.190	40,583	7,711	145,786	27,699
Distant water trawl (East China Sea)	0.618	88,957	54,975	45,420	28,070
(shrimp trawl)	10.0	1,385	13,853	687	6,870
Off-shore trawl	0.14	509,210	71,289	442,412	61,938
Small trawl (powered, other than shellfish)	1.53	209,811	321,011	166,584	254,874
(small sail trawl)	22.4	662	14,836	388	8,691
Large set-nets for salmon	0.014	110,523	1,547	146,118	2,046
Large set-net (others)	0.014	353,777	4,953	294,618	4,125
Small set-net	0.014	181,241	2,537	163,087	2,283
Boat seine	1.12	147,883	165,629	191,821	214,840
Squid angling (distant-waters)	0.0004	312,276	125	174,764	70
Squid angling (Other waters)	0.0004	74,609	30	265,835	106
Distant water trawl (S Pacific otter trawl)		385,547		179,061	
Other surrounding nets		8,387		11,823	
Mother-ship type salmon fishing		1,590		0	
Salmon drift gillnet		12,944		23,628	
Squid drift gillnet		138,963		0	
Swordfish drift gillnet		8,870		4,147	
Other gillnets		348,394		210,581	
Other lift-nets		105,042		81,981	
Beach seine		3,087		2,733	
Patch-ami type boat seine		119,773		63,719	
Salmon longline		753		198	
Other longlines		94,820		69,233	
Other anglings		86,744		79,976	
North Pacific longlines and gillnets		0		23,736	
Other fisheries		197,679		141,936	
Total		10,438,297	858,224	6,589,564	793,240
Corrected estimate			1,003,663		917,550

Matsuoka, T. (1997) [2]

Table 3
Annual catch of bycatch species in purse seining of Yaizu port, 1991-1993 [ton (%)]

Year	Bycatch species						
	Dorphin Fish	Rainbow runner	Mackerel scad	Blue marlin	Frigate mackerel	Wahoo	Total
1991	273	120	135	111	188	17	844
	(32.3)	(14.2)	(16.0)	(13.2)	(22.3)	(2.0)	(100.0)
1992	185	255	114	83	41	40	718
	(25.8)	(35.5)	(15.9)	(11.6)	(5.7)	(5.6)	(100.0)
1993	231	274	442	126	39	22	1134
	(20.4)	(24.2)	(39.0)	(11.1)	(3.4)	(1.9)	(100.0)

Takeuchi, S. (1995) [28]

(*Liparis tanakai*), pricklebacks (*Stichaeus grigorjewi*), headlightfish (*Diaphus coeruleus*), moon dragonet, Gray's cutthroatl (*Synaphobranchus kaupi*), etc. A fisherman told that the amount of discards is related to the number of labor for sorting catches onboard.

Takahashi and Nihira [4] mentioned that in small trawl fishery of Ibaragi Prefecture, the discard consisted of total of 93 species, including 69 species of fishes, 7 species of cuttlefish and octopus, 7 species of crab and shrimp, and 10 species of shellfish. There were discard fishes with high commercial value. The reasons why are small size compared to commercial size, for example, brown hakeling (*Physiculus maximowiczi*), crimson sea bream (*Evynnis japonica*), Aredwing searobin (*Lepidotigla microptera*) are comparatively large amount species for this reason. Another reason is the lack of labor in sorting on board, for example, greeneyes (*Chlorophthalmus albatrosis*), shell (*Buccinum isaotakii* :shiraitomakibai) are discarded for this reason. While, comparatively large catch but with no commercial value are cloudy dogfish (*Macrorhamposus sagifue*), etc.

Kobayashi et al. [5] described that in the trawl fishery of the Suruga Bay, main discards species were squid (*Rossia* spp. :yawarabouzuika), shrimp (*Plesionike martia* :jinkenebi), and greeneyes, and discards occupied 63% of the total weight of catch.

Kusakabe et al. [6] reported that in the small trawl fishery of the Osaka bay, there were 65 kinds of discards organisms including 20 species of fish. In number of fish species, pinkgray goby, cardinalfish (*Apogon lineatus*) and marbled sole (Limanda yokohamae) were abundant, and in particular, marbled sole is the most. It should be considered to control resources management of this species as discard considering that they occurs only in short term of autumn.

The Okayama Prefecture Fisheries Experimental Station has carried out the investigation of discard fish in small trawl fishery at eastern waters of Okayama Prefecture for a long time. Matumura and Fukuda [7-9] described species composition of discard fish. Ukida and Matsumura [10,11] described the discard fish compared with shrimp and squilla catch. Kamaki and Matsumura [12-14] studied composition and size of shrimp, 1989-1991, and they made a list of discarded species in those reports. Discarded organisms were 89, 83 and 117 species in each year, and species of fish in each year were 43, 35 and 44, and mainly contained croakers (*Sciaenidae*), cardinalfish, gobys

(*Gobiidae*), and moon dragonet. There were many landings from October to December in this area, but there were many discard fishes in June and August.

Ogawa and Shibata [15] surveyed experimentation of survival stages in discard species in small trawl fishery of Seto Inland Sea. A ratio of discard was 88% in number and 63% in weight. Survival rate in fish was very low as 13 % and that in squilla (*Oratosquilla oratoria*) was comparatively high as 66 %.

Kimura et al. [16-21] and Hiyama [22], Yamaguchi Prefecture Naikai Fisheries Experimental Station, studied the discards of a small trawl fishery at Suo-nada in Seto Inland Sea, 1984-1992 and described a composition of species and variation of specific diversity, survival rate, problem of discard rate, etc. They found that organisms of discards consisted of about 200 species. A number of species in decreasing order were crab (*Charybdis bimuaculata* :futaboshiishigani), shrimp (*Alpheus japonicus* :tenagateppoebi), shrimp (*Crangon affinis* :ebijako), saruebi shrimp, squilla in crustacian. Cardinalfish, streaked goby (*Acentrogobius pflaumi*), red eel goby (*Ctenotrypauchen microcephalus*) and Smithurst's ponyfish (*Leiognathus smithursti*) were large number in fishes. 84% out of total catch was landed in weight and remaining 16% was discarded.

Yokomatsu [23] described that discarded fish in the small trawl fishery in Buzen Sea, Ooita Prefecture, were cardinalfish, Richardson's dragonet (*Callionymus richardsoni*), gafftopsail goby (*Cryptocentrus filifer*), red eel goby in whole season. Also larvae of finespotted flounder (*Pleuronichthys cornutus*), marbled sole and anchovy (*Engraulis japonicus*) were discarded.

Nishinokubi (personal communication) indicated that as species of discard in the small trawl fishery at the Ariake Sea, small sized finespotted flounder, silver jewfish (*Pennahia argentata*), milkyspotted sole (*Aseraggodes kobensis*), mottled tonguefish (*Cynoglossus interruptus*), pigmy flounder (*Tarphous elegans*), grey goblinfish (*Minous monodactylus*), rays (*Rajiformes*), squid (*Euprymna morsei* :mimiika), etc. were discarded.

Fujiishi [24] described that in the investigation of small trawl fishery at Yuya Bay, Yamaguchi Prefecture in the Sea of Japan, weight ratio of discard fish was very high as 71.8%. This figure means there is large amount of meaningless fishery in the Bay. Main discard fishes were cardinalfish, Smithurst's ponyfish, ginpo (*Enedrias nebulosus*), gafftopsail goby and cardinalfish (*Apogon semilineatus*).

Kitazawa and Oaku [25] studied on discards by Danish seine fishery in western Wakasa Bay in 1980. Discarded fish ranged from 30 to 120 kg in weight per haul. In most sought after species of the fishery, the small, unmarketable size was discarded and this percentage ranged from 15 to 50 % of all discards in weight. Comparison between the catch number unmarketable and marketable size indicated that the former was nearly the same as or more than the other , and especially the discards of Korean flounder (*Glyptocephalus stelleri*), zuwai crab (*Chioneacetes opilio*) were remarkable.

4.2. Tuna longline fishery

In tuna longline fishery in distant waters, main target is tunas, and bycatch species are swordfishes, sharks, and other small fishes. Catches of sharks occupied about 16 % of the total catch in number [26], and since almost all sharks are discarded, the weight also seems to be considerable amount. Blue shark (*Prionace glauca*), sandbar shark (*Carcharhinus plumbeus*), oceanic whitetip shark (*C. longimanus*), shortfin maco (*Isurus* oxyrinchus), thresher sharks (*Alopiidae*), hammerhead sharks (*Sphyrnidae*) are main bycatch species of sharks. The sharks captured are often discarded after taking off the fins, because the commercial value of shark's fin is very high as material of Chinese food. Another bycatch species discarded are wahoo (*Acanthocybium solandri*), longnose lancetfish (*Alepisaurus ferox*), great barracuda (*Sphyraena picuda*), dolphin fish (*Coryphaena hippurus*), moonfish (*Lampris guttatus*), snake mackerel (*Gempylus serpens*), sickle pomfret (*Taractichthys steindachheri*), etc. However, recently it is difficult to fulfill the storage of the boat by reduction of tuna resources, then discard fishes has decreased largely. For example, moonfish and wahoo are become to be sold 150 yen and 100 yen per 1 kg. Shortfin mako shark has become one of the target fish in this fishery, and they are brought back and landed. Other sharks used as packing in the fish hold are also sold in the market without discards. Therefore, the estimation of discard amount by Matsuoka [2] in Japanese tuna longline in distant waters seems to be over estimation at present. On the other hand, in tuna longline fishery in adjacent waters, sharks mostly are landed and sold. Discard fishes are oilfish (*Ruvettus pretiosus*), escolar (*Lepidocybium flavobrunneum*), black grunt (*Lobotes surinamensis*), wahoo, etc. and the amount is not so large. As escolar has highly wax content in the meat, thus it is prohibited to sell in the market, and should be thrown away.

4.3. Purse seine fishery

In the purse seine fishery whose main target fishes are sardines, mackerels, horse mackerel, in off-shore and coastal waters, almost all fishes captured are taken into the cargo boat. Thus, there are little discard fish onboard. [27]. After sorting in the market, some fish of no value are discarded. On the other hand, in the purse seine fishery in tropical waters, main target fishes are skipjack (*Katsuwonus pelamis*), yellowfin tuna (*Thunnus albacares*) and some other discarded fishes. Main bycatch fishes in the seine are blue marlin (*Makaira mazara*), silky shark (*Carcharhinus falciformis*), mackerel scad (*Decapterus spp.*), rainbow runner (*Elagatis bipinnulata*), dolphin fish (*Coryphaena hippurus*), ocean triggerfish (*Canthidermis maculatus*), etc. The ratio of bycatch fish is very low compared to the target, and many of bycatch fishes are caught in flouting operation with driftwood, flouting objects, etc. Usually, most fish caught are preserved in a brain tank to freeze at first, then all fish are transferred from brain tank to refrigerate tank after frozen. In the latter process, most bycatch fish except blue marlin are selected from target fish and thrown away to the sea. However, at the time when the operation is ended at the port, fish kept in brain tank is not transferred to another tank, and whole fish loaded are landed and bycatch species are sold at very cheap price. Only ocean triggerfish is dumped due to no commercial value. Takeuchi [28] described the landing in purse seining of Yaize port in 1991-1993, and found that landing amount of six bycatch species, dolphin fish, rainbow runner, mackrel scad, blue marlin, frigate mackerel (*Auxis spp.*) and wahoo occupied 0.3 - 0.6% of a total fish landing and annual weight had reached in 718-1134 tons (Table 3). Since it seems to be a large amount of discard fish onboard, total amount of bycatch fish seems to reach awful quantity in this fishery.

4.4. Bottom longline fishery

In the small bottom longline fishery of the Sagami Bay located near the Tokyo Bay, silvergray seaperch (*Malakichthys griseus*), Japanese blue-fish (*Scombrops boops*), hilgendorf saucord (*Helicolenus hilgendorfi*) were main target fish, and bycatch fish were oilfish (*Ruvettus pretosus*), a brown hagfish (*Paramyxine atami*), sharks and pufferfishes [29]. Total amounts of the discards in this fishery are unknown, but seem to be not so large.

4.5. Gillnet fishery

By the internet in this web sit, Murakami [30] showed status about bycatch fishes in flounder gillnet fishery at Abashiri of Hokkaido and informed me bycatch species as personal letter, main target fish were brown sole (*Limanda herzensteii*) and cresthead flounder (*Limanda schrencki*). Bycatch fishes were starry flounder (Platichthys stellatus), longsnout flounder (*Limanda punctatissima*), sculpins (*Cottidae*), scorpion fishes (*Scorpaenidae*) and Pacific ribbed sculpin (*Pleurogrammus azonus*). Although small sized target fishes were landed because of their high commercial value, most of bycatch fish were likely to be discarded. In particular, amount of discards of the small-size (less than 18 cm TL) longsnout flounder was remarkably large due to the low price. As for the discard rate in number for each species, a longsnout flounder showed 48.4 % of the total amount, 10.3 % of brown sole and 6.5 % of cresthead flounder. A part of bycatch fish is sold to the meal factory at about 10 yen per kg. By personal communcation, other bycatch species are found. They were gobys including snowy sculpin (*Myoxocephalus brandti*), plain sculpin (*Myoxocephalus jaok*), frog sculpin (Myoxocephalus stelleri), black edged sculpin (*Gymnocanthus herzensteini*) and scorpion fishes including Schlegel's black rockfish (*Sebastes schlegeri*), white-edged rockfish (*Sebastes taczanowskii*), threestripe rockfish (*Sebastes trivittatus*).

Torisawa [31] described the bycatch species in the squilla gillnet fishery of the Ishikari Bay, Hokkiado. Many flounders with commercial value, such as a brown sole, a longsnout flounder, and pointhead flounder (*Cleisthenes pinetorum*) were caught as bycatch fish. Most of them are smaller than about 15cm SL, and are discarded due to no market value.

4.6. Set net fishery

Akiyama [32] described that discarded species in three types of set net fishery at Tateyama Bay, Chiba Prefecture. They were bigeye sadine (*Etrumeus micropus*), banded bluesprat (*Spratelloides japonicus*), horse mackerel (*Trachurus japonicus*) in small sized net, chub mackerel (*Scomber japonicus*), bigeye sardine, cardinalfishes (*Apogon semilineaus*) in bottom net and anchovy (*Engraulis japonicus*), darkblue headlightfish (*Diaphus watasei*), flathead silverside (*Hypoatherina bleekeri*) in large sized net.

Ishidoya and Ishizaki [33] described that seasonal discarded species in set net fishery at western part of Sagami Bay, Kanagawa Prefecture. They were red bulleye (*Priacanthus macracanthus*), anchovy, bigeye scad (*Selar crumenophthalmus*) in summer, chicken grunt (*Parapristipoma trilineatum*), Bermuda catfish (*Promethichthys prometheus*) in autumn, round scad (*Decapterus maruadsi*), Bermuda catfish, darkblue headlight fish, in winter and anchovy, darkblue headlightfish, Japanese sardine (*Sardinops melanostictus*) in spring. Unmarketable species are utilized as feed for aquaculture or discarded. Annual catch of unmarketable fish were 120 kg in maximum (39 % of total catch) and 1.4 % in average.

Takeno [34-36] discussed the suitable management system in set net fishery at Himi area of Toyama Prefecture in the Sea of Japan, and he described that in fish landed at Himi fishing port, discards during sorting were 70 species totally including 25 species of unutilized species and 49 utilized species. And before landing alive 29 species were released and discarded from boat in the sea.

Hamada [37] investigated composition of discard species in small set net fishery in the Buzen Sea of Seto Inland Sea of Japan. He mentioned that main discard species were Japanese scaled sardine (*Sardinella zunasi*), spotted sardine (*Konosirus punctatus*), keelback mullet (*Liza carinata carinata*) and Japanese sea bass (*Lateolabrax japonicus*) in fishes, and crab (*Charybdis japonica* :ishigani) and swimming crab (*Portunus trituberculatus*) in crabs. The reason of discard were unmarketable species, as Japanese scaled sardine, spotted sardine and keelback mullet, and small size species even though maketable, as Japanese sea bass, swimming crab and ishigani crab.

Murakami, Experimental Station of Hokkaido, infomed me as personal letter that main discarded fishes were fringed blenny (*Azuma emmnion*), agassiz's snailfish (*Liparis agassizii*), black edged scalpin (*Gymnocanthus herzensteini*), frog sculpin (*Myoxocephalus stelleri*), Pacific ribbed sculpin (*Pleurogrammus azonus*), greenrings (*Hexagrammos otakii*), Schlegel's black rockfish (*Sebastes schlegeri*), scorpionfishes (*Sebastes taczanowskii*), threestripe rockfish (*Sebastes trivittatus*), etc. in small bottom trap net, a kind of set net, fishery at Abashiri, Hokkaido.

4.7. Saury stick-held dip-net fishery

The number of saury fishing boats with recently developed size sorter machines are increasing in recent years. In fishing ground they sort by size of saury (*Cololabis saira*) using the machine and small sized saury with low market value are discarded. Although the company to catch saury has denied the situation, a few people have reported on a large number of dead saury in fishing ground and around. Quantity of discards in small sized saury remains unknown.

5. CONCLUSION

The types of discarded fish from Japanese fisheries can be mainly classified into four categories: 1) species with low value or no value in the market, e.g. snailfish, grenadiers, etc. 2) small sized species including young or juvenile, even if they have commercial value as adult, such as sea bream, flounder, etc. 3) species with substantially lower value in the market taken by fisheries targeting on highly valuable species such as tuna longline, and 4) species whose sales in the market are prohibited by the government such as escolar, and poisonous fish such as great barracuda, pufferfish. There are another cases of discarded in addition to these four categories. For example, in tuna longline fishery, damaged tunas by whale's or shark's biting are discarded. And in saury stick-held dip-net fishery, in order to control the assignment of TAC (Total Allowable Catch), small sized saury were discarded to maximize the income..

Some methods to decrease the discarded fish are proposed: 1) effective utilization of bycatch species, 2) release to the sea alive immediately after catch, and 3) improvement on gears to avoid bycatch fishes. But, the effectiveness of these methods depends on the characterization of fisheries as well as characterization of discarded species. According to the study on the release immediately after catch, survival rates of fish are heavily affected by the duration of sorting, and it may change by seasons. Thus, considering the operation on board, release does not seem so effective actually. As for escapement of the small fish by improvement of fishing gear, it is effective but still has a limit. There is no gear to catch only single target species and size. Therefore, the improvement of utility value for the bycatch fish is considered to be the most effective solution on this issue now. In other word, raising added value as food with machining technology. Of course, it needs a premise that the fishermen brings back fish which is discarded as usual. In the tuna longline fishery in distant waters, however, there is a case that sharks as bycatch is difficult to bring back due to their size and low value. Therefore, one promising solution of this problem is how to raise the commercial value of discard fish as a method for diminishing the discard fish. Preparing the circumstances to make the fishermen bring back the less value fishes without discards, and the action to raise the value of the discard fish are most required at present.

REFERENCES

1. Alverson, D.L., Freeberg, M.H., Murawski, S.A. and Pope, J.G. (1994) A global assessment of fisheries bycatch and discards. FAO Fisheries Technical Paper 339. Food and Agriculture Organization of the United Nations. Roma.
2. Matsuoka, T. (1997) Discards in Japanese marine capture fisheries and their estimation. FAO Fish. Rep. 547, Suppement. Food and Agriculture Organization of the United Nations. Roma, pp.309-329.
3. Hirakawa, H. and Ono, T. (1984) Investigation of discard fish in the small trawl fishery. Annual Rep. Fukusima Pref. Fish. Exp. St. 1983, 190-192.
4. Takahashi, M. and Nihira, A. (2000) Investigation of discards fish. Annual Rep. Ibaraki Pref. Fish. Exp. St. 1999, pp.129-132.
5. Kobayashi, S., Tanaka A. and Kosaka M. (1998) Composition of discard fish of trawl fishery in Suruga Bay. Abstracts for the Meeting of the Jap. Spc. Fish. Sci, Sep. 23-27, 1998, p.21.
6. Kusakabe, T., Tsujino, K. and Abe, T. (1990) Discards of small trawl fishery in Osaka Bay. Western Region Bottom Fish Research Bulletin, 22:74-81.
7. Matsumura, S. and Fukuda, T. (1980) Specific composition and actual condition of cast away of catch by a small trawl boat in eastern part of Okayama Prefecture. Annual Rep. Okayama Pref. Fish. Exp. St. 1979, pp.70-90.
8. Matsumura, S. and Fukuda, T. (1981) Specific composition and actual condition of cast away of catch by a small trawl boat in eastern part of Okayama Prefecture, fiscal year 1980. Annual Rep. Okayama Pref. Fish. Exp. St. 1980, pp.56-71.
9. Matsumura, S. and Fukuda T. (1982) Discards fish of small trawl fishery in eastern part of Okayama. Western Region Bottom Fish Research Bulletin, 14:17-32.
10. Ukida, K. and Matsumura, S. (1989) Specific composition and size of shrimp, mantis shrimp and soleidae by a small trawl boat in the eastern part of Okayama Prefecture, 1988. Bull. Okayama Pref. Fish. Exp. St., 4:127-135.

11. Ukida, K. and Matsumura S. (1989) Specific composition and size of shrimp and manta shrimp, and discard fishes by a small trawl boat in the eastern part of Okayama Prefecture. Western Region Bottom Fish Research Bulletin. 21:79-101.
12. Kamaki, A. and Matsumura, S. (1989) Specific composition and size of shrimp mants shrimp Soleoidei and actual condition of sea animals casted away by a small trawl boat in the eastern part of Okayama prefecture 1989. Bull. Okayama Fish. Ext. St., 5:93-108.
13. Kamaki, A. and Matsumura, S. (1991) Species composition and size of shrimp, mantis shrimp Oratosquilla oratoria and sole cought by a small trawler in the eastern waters of Okayama prefecture in 1990, and discarded species. Bull. Okayama Fish. Ext. St., 6:173-188.
14. Kamaki, A. and Matsumura, S. (1992) Species composition and size of shrimp, Mantis shrimp Oratosquilla oratoris, and sole caught by a small trawler in the eastern waters of Okayama Prefecture in 1991 and the discarded species. Bull. Okayama Fish. Ext. St., 7:104-120.
15. Ogawa, Y. and Shibata R. (1996) Survival experiments of discard fishes by a small trawl fishery in the Seto Inland Sea, Western Region Bottom fish Research Bulletin, 23: 13-37.
16. Kimura, H. (1994) Study on discards of a small trawl net − IV Survival rates of discarded fishes. Bull. Yamaguchi Pref. Naikai Fish. Exp. St., 23:1-8.
17. Kimura, H., Hiyama, S. and Yoshioka S. (1994) Study on discards of a small trawl net − V Problems of discarding fry of marketable fish from regard to "discard rate". Bull. Yamaguchi Pref. Naikai Fish. Exp. St., 23:9-13.
18. Kimura, H., Hiyama, S. and Yoshioka, S. (1994) Study on discards of a small trawl net − VI. Composition of species and variation of specific diversity in a long period. Bull. Yamaguchi Pref. Naikai Fish. Exp. St., 23:14-18.
19. Kimura, H., Hiyama, S., Yoshioka, S. and Okabe, S. (1993) Study on discards of a small trawl net − I Amount of discards. Bull. Yamaguchi Pref. Naikai Fish. Exp. St., 22:21-25.
20. Kimura, H., Hiyama, S., Yoshioka, S. and Okabe, S. (1993) Study on discards of a small trawl net − II Comparison of fisheries efficiency. Bull. Yamaguchi Pref. Naikai Fish. Exp. St., 22:26-35.
21. Kimua, H., Yoshioka, S. and Okabe S. (1993) Study on discards of a small trawl net − III. Length measured in each portion of discard fish. Bull. Yamaguchi Pref. Naikai Fish. Exp. St., 22:36-40.
22. Hiyama, S. (1987) A discard fish in small trawl fishery at Suo-nada of Yamaguchi Prefecture − I. Species of discard fish. Bull. Yamaguchi Pref. Naikai Fish. Exp. St., 15:28-38.
23. Yokomatsu, Y. (1984) Composition of discard organisms by small trawl fishery in Buzen waters in Ooita Prefecture, Proceedings of the 16th study meeting on the Nansei area inland sea fishery, pp.9-29.
24. Fujiishi, A. (1995) Small trawl fisheries. In: By-catch in Japanese Fisheries. Matsuda, K. (ed.), Koseishakoseikaku, Tokyo, pp. 30-42.
25. Kitazawa, H. and Oaku, T. (1982) Study on the discards by the Danish seine fishery in western Wakasa Bay. Nippon Suisan Gakkaishi, 48:1089-1093.

26. Kobayashi, H. and Yamaguchi, Y. (1978) The hooked ratio of longline-caught fish and shark damage. Bull. Faculty of Fish. Mie Univ., 5:117-128.
27. Hara, I. (1995) Coastal purse seine fisheries. In: By-catch in Japanese Fisheries. Matsuda, K. (ed.), Koseishakoseikaku, Tokyo, pp. 80-87.
28. Takeuchi, S. (1995) Skipjack purse seine fisheries. In: By-catch in Japanese Fisheries. Matsuda, K.(ed.), Koseishakoseikaku, Tokyo, pp. 71-79.
29. Arimoto, T., Ogura, M. and Inoue, Y. (1983) Catch variation with immersion time of gear in coast set-line. Nippon Suisan Gakkaisi, 49:705-709.
30. Murakami, O. (2000) Problem of bycatch and discard fishes (for preservation of small fish). Experimental study at present. Internet (http://www.fishexp.pref.hokkaido.jp/exp/).
31. Torisawa, M. (1995) Coastal gillnet fishey. In: By-catch in Japanese Fisheries. Matsuda, K. (ed.), Koseishakoseikaku, Tokyo, 62-70.
32. Akiyama, S. (1997) Discarded catch of set-net fisheries in Tateyama Bay. J. Tokyo Univ. Fish., 84:53-64.
33. Ishidoya, H. and Ishizaki, H. (1995) Setnet fisheries for horse mackerel, mackerel and sardine. In: By-catch in Japanese Fisheries. Matsuda, K. (ed.), Koseishakoseikaku, Tokyo, pp. 88-95.
34. Takeno, Y. (1995) Study of suitable management in set net fishery. Annual Rep. Toyama Pref. Fish. Exp. St., 1994, pp.29-33.
35. Takeno, Y. (1996) Study of suitable management in set net fishery. Annual Rep. Toyama Pref. Fish. Exp. St., 1995, pp.38-49.
36. Takeno, Y. (1997) Study of suitable management in set net fishery. Annual Rep. Toyama Pref. Fish. Exp. St., 1996, pp.31-37.
37. Hamada, H. (1997) Comparison of discarded species composition between bag-nets set at different position in a small set-net in the Buzen Sea, Inland Sea of Japan. Nippon Suisan Gakkaishi, 63: 43-49.

INTRINSIC AND EXTRINSIC FACTORS AFFECTING EFFICIENT UTILIZATION OF MARINE RESOURCES.

Michael T. Morrissey[a] and Gilbert Sylvia[b]
[a]Seafood Laboratory, Department of Food Science and Technology, Oregon State University, 2001 Marine Drive, Rm 253 Astoria, OR 97103, USA
[b]Resource Economics, Coastal Oregon Marine Experiment Station, 2030 Marine Science Drive, Newport, OR 97365, USA

Abstract

This paper will focus on various fisheries in which intrinsic and extrinsic factors are critical variables for improving utilization and economic benefits of aquatic products. Pacific whiting, an important US West Coast fishery, has endogenous protease enzymes that impact yield and quality in surimi production. Protease activity can be mitigated, in part, through onboard handling practices centering on harvest-to-process time/temperature parameters that reduce loss of myofibrillar proteins and improve gel strength. Seasonal changes in nutrient composition in the flesh will also affect final product quality and utilization. The timing of salmon harvest and improved handling practices through training and education can reduce postharvest fishery losses and provide additional opportunities for markets and product development. Changes in management regulations have created interesting challenges in the ability of the fishing industry to maintain high quality and improved yields. Intrinsic factors in fish must be considered along with extrinsic factors of the fishing and processing sectors for optimum fisheries management. To improve utilization of fishery resources there needs to be greater coordination of effort and cooperation among resource managers, food scientists, and the fishing industry.

1. INTRODUCTION

Optimum utilization of harvested fishery resources is an important issue for the seafood industry and researchers. There is a general consensus that marine fisheries harvests have peaked and that further increases in production must be accomplished through improved fishery and processing operations. Intrinsic and extrinsic factors have a profound influence on quality and yield parameters of the majority of fisheries. Intrinsic factors are inherent chemical/physiological characteristics in fish species that are important to quality and marketability of fish. Extrinsic factors describe harvesting and processing inputs that affect final product quality. Maximum utilization and increased economic benefits to the industry can occur through capturing fish at their peak physiological condition with respect to quality and through a reduction of postharvest quality losses by optimizing handling and processing conditions. Most commercial marine species undergo physiological changes due to maturation, spawning, migration and feeding which affect potential quality and processing yield. These intrinsic characteristics of fish include changes in gross composition such as protein, lipid and moisture content, as well as other characteristics that influence the eating and storage quality of different species. The pioneering work of Love [1] showed that groundfish as well as pelagic species undergo compositional changes depending on diet, growth and maturity. These changes affect fish quality, especially texture, sensory quality, and consumer acceptability. There are many examples of dramatic gross lipid changes in small pelagics. Fat content in the muscle tissue of sardines and herring may vary from 5 to 20% depending on maturation and spawning condition of the fish [2]. Lean fish (<5% fat) also vary in lipid and protein content depending on the season and maturation. Although these variations are not as dramatic, they can cause a significant change in quality and yield for both small-scale and large volume operations. Extrinsic factors, especially those that focus on onboard handling and storage conditions, are critical for maintaining prime quality in both fresh and frozen fish. Intrinsic and extrinsic factors affect harvest yields, processor recovery rates and the market value of harvested and processed seafood products. Work by Sylvia et al. [3] shows that changes in flesh composition as a result of normal seasonal variation can be incorporated in fisheries management to benefit both conservation of the stocks and obtain maximum economic value. This paper will briefly discuss intrinsic properties in three species (Pacific whiting, Alaska pollock and Pacific salmon) and discuss how intrinsic and extrinsic factors should influence fisheries management to maximize net return to the industry.

2. PACIFIC WHITING

Pacific whiting (*Merluccius productus*) is the largest groundfish biomass off the West Coast of the U.S. excluding Alaska. Until 1991 it was underutilized as a shoreside fishery despite its proximity to coastal catching and processing operations. Whiting was viewed as a problem species with soft flesh due to high proteolytic activity and poor storage qualities. There were few domestic markets for the species and the majority of the catch was processed at sea by foreign vessels for international markets. In 1991, due in part to the growing domestic at-sea processing industry increasing restrictions of other groundfish species, and the need to maintain viable coastal operations, an active fisheries was developed for at-sea and shoreside processing of Pacific whiting. By the mid 1990s, shoreside processing increased from less than 10,000 mt to more than 70,000 mt of which more than 80% was dedicated to surimi production [4]. The surimi process is primarily an extraction of myofibrillar proteins from fish flesh through a series of washing and dewatering steps [5]. Because of protease enzymes in the flesh, there can be significant loss of myofibrillar proteins if time and temperature parameters are not followed [6]. Scientists at Oregon State University began a program in the early 1990s to determine the

Table 1
Intrinsic and extrinsic variables for Pacific whiting surimi

Fish characteristics (Intrinsic)	Processing characteristics (Extrinsic)
Moisture content	Processing water temperature
Protein content	Mince recovery
Lipid content	Surimi yield
Ash content	Surimi grade
Percent parasitization	Mixing time in wash tanks
Protease concentration	Water:meat ratios for 1st wash
Average length and weight	Water:meat ratios for 2nd wash
Salinity of flesh	Temperature at start of processing
pH of flesh	Temperature after 1st wash, 2nd wash
	Temperature after refiner, screw press
Site/trip characteristics (Extrinsic)	Processing plant temperature
Longitude and Latitude	pH at start of processing
Tow size (mt)	pH after 1st wash, 2nd wash
Number of tows on board	pH after refiner, screw press
Refrigeration system	Salinity at start of processing
Julian date	Salinity after 1st wash, 2nd wash
Time of day	Salinity after refiner, screw press
Storage temperature in boat	Initial microbial count
Storage time prior to processing	Microbial count after 1st wash, 2nd wash
Rate at which the fish were cooled	Microbial count after refiner, screw press
Weather conditions	Refiner speed
Ocean temperature	Percent inhibitor added to surimi
Net depth	Percent cryoprotectants added
Percent by-catch	Time to complete processing

intrinsic and extrinsic factors important in the whiting fishery with the objective of improving quality and yield for the fishery. Cooperative work and data gathering in the field was undertaken with industry and state agencies as well as laboratory research with controlled experiments. Results showed that both intrinsic and extrinsic factors played a major role in processing yields and final product quality in the fishery [7]. Several types of variables, consisting of large data sets were collected over a three year period. These are classified into intrinsic and extrinsic variables and are shown in Table 1. Since the majority of the whiting was being used for surimi production the variables were correlated to surimi quality parameters such as gel strength.

For the Pacific whiting fishery, intrinsic factors include the seasonal changes in proximate composition of the flesh (protein, lipid, moisture), pH, protease levels, length-weight ratios and geographic availability. Extrinsic factors include distance to harvest grounds, length and size of tows, onboard storage among others. Extrinsic factors in the processing sector were focused on the surimi operation. Consequently, processing factors (extrinsic) include the efficiency of the heading and gutting operation, the water: mince ratios, numbers and time of wash, protease inhibitors, water temperature, processing time and a host of other factors. We were able to measure several of these factors in the laboratory as well as obtain actual fishing and processing information from the industry [7-11]. Analysis of the data by neural networks and multiple linear regression were used

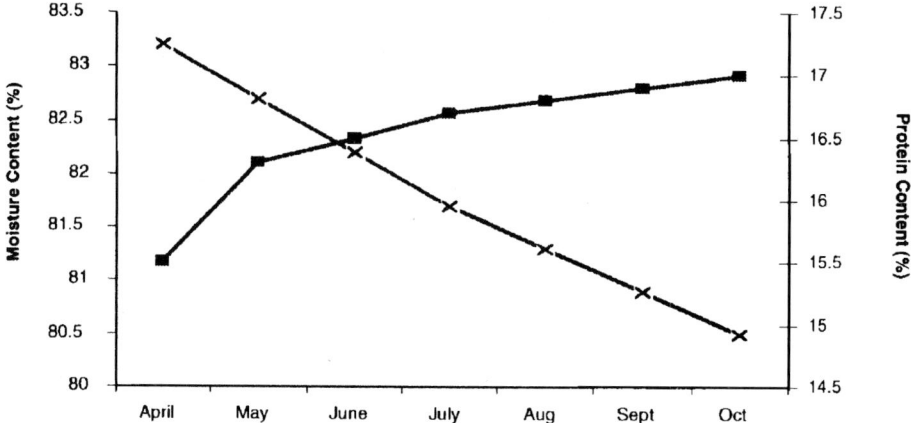

Figure 1. Seasonal variation of flesh composition of Pacific whiting.
×, moisture; ■, protein

to determine significant factors that affect the product quality and yield. Fishing and process data analyses showed that intrinsic factors such as moisture, protein, salinity and pH affected the final product quality. Extrinsic factors such as the time between capture and processing and fish storage temperature were important fishing variables while wash ratios were important processing factors [7]. Lin and Park [10,11] were able to collaborate several of these results in the laboratory and determine the specific effects of extrinsic processing variables on yield and quality. This information was transferred to the industry through workshops and publications so that they may improve both fishing and processing operations.

Data analysis showed that protein content and fat in whiting flesh increased during the summer months while relative moisture decreased (Fig. 1). The optimum condition of whiting flesh for surimi production and other products was during the months of July and August for both quality and yield. This data was used to develop a bioeconomic model that maximizes social benefits subject to dynamic biological and economic conditions that often occur in wild-caught fisheries. Relationships between product characteristics and prices, costs, recovery rates and production practices were determined using interseasonal data [12]. The model maximizes the value of the fishery over time and generated management options for the fishery. In the U.S., management decisions are made by the National Marine Fisheries Service in consultation with regional management councils (e.g. the Pacific Fisheries Management Council). In general fishing plans include fishing quotas and opening and closing dates. These decisions are made to satisfy three broadly defined goals including 1) conservation of the resources, 2) maximum economic value, and 3) efficient utilization. In the past, these decisions have ignored intrinsic factors in a fishery that could impact all three goals. The bioeconomic model showed that inseason timing of the harvests plays an important role in accomplishing these goals [3]. Previously the whiting fishery opened in April soon after spawning and moved into coastal waters off the US Pacific Northwest coast. During the summer months, fish rapidly increased in weight, protein content, and overall condition. The bioeconomic model demonstrated that a delayed opening would better serve the fishery. Conservation of the resource (Goal 1) was clearly met by delaying the harvest as the fish gained weight during the summer months. Under this management plan, fewer but heavier fish are needed to capture a specific quota. For Goal 2, the economic value to the fishery was greatly improved, by a delayed harvest, as both quality and yield tended to increase. The model indicated that the industry could double their net present value by fishing when the fish were at optimum condition later in the year. Efficient utilization (Goal 3) was realized as increased surimi yields demonstrate that the industry is maximizing production under specific harvest guidelines. When this information became available, the shore-based industry requested that the regional fisheries management council delay the opening of the fishery to coincide with the optimum physiological condition of the fish to optimize yield and quality. Under current management policy, the fishery now opens on June 15 and has shown steady increases in production yields for surimi and other fishery products.

3. PACIFIC SALMON AND ALASKAN POLLOCK

Pacific salmon begins to deteriorate in quality once it enters its natal river system and migrates for its spawning grounds. Most salmon will stop feeding and use its body lipids and then its muscle protein for energy for final migration and spawning. The eating quality of the flesh changes as it travels upstream depending on the length of migration and month of spawning. Ocean caught salmon or salmon that are harvested soon after entering the river system are often superior in quality to fish captured near hatcheries or spawning grounds. Wild and hatchery-raised Alaska pink salmon is a case in point. A study is currently underway to determine the quality of pink salmon captured in different areas in the Gulf of Alaska and Prince William Sound [13]. Preliminary results showed there are significant differences in the quality of the flesh depending on the time of the year and geographical location. Previously, fishery managers would not allow the harvesting of fish until a certain percentage were captured for artificial spawning in the hatcheries. Consequently, the majority of the fish had already reached areas near their spawning grounds and, although easy to capture, were often of poor quality. Fishery managers are now permitting early season harvest of pink salmon well before anticipated spawning dates. Although there are some management risks, as escapements can be variable, it has become critical for the industry to improve quality so that it can compete on the global market with other salmon products.

The Alaskan pollock fishery represents the largest groundfish fishery in the world as more than 2 million metric tons are harvested annually. The largest pollock fishery is in U.S. waters and is divided between shore-based trawlers and factory ships. Over the last five years there has been a dramatic increase in surimi yields from pollock. This is due to two major factors, 1) the use of high speed centrifuges to recover protein formally lost in the wash water and 2) the better management of fishery operations with the ending of Olympic-style fisheries and incorporation of quotas between and among the different sectors. Although the fisheries is still limited by specific seasons, more thought is given toward managing the fisheries for optimum yield. Regulations continue to play an important role as evidenced by recent rulings on fisheries and marine mammal interactions. Because of possible interactions with endangered Steller sea lions, recent rulings closed off several prime areas for pollock fisheries, especially for shore-based trawlers. Under normal fishing conditions, shorebased trawlers could capture, off-load and return to the fishing grounds within 24 hr. With the new regulations, most operations have been extended to 72 hr or longer. This has dramatically affected product quality, yield, and economic return to both the fisherman and processor.

4. SUMMARY

Fishery resources are limited and most scientists agree that we are approaching the maximum sustainable harvest levels from our oceans. To increase production we must improve our utilization of capture fisheries. This can be accomplished by better understanding the effects of intrinsic and extrinsic factors on seafood quality and yield. There has been ample documentation of compositional changes in many of the commercial harvested species. Seasonal changes in lipid, protein, moisture content, etc. affect final product yields, quality and economic return to the industry. There has also been a tremendous amount of research on how variables in capture, storage, processing and packaging impact yields and product acceptability. What has been missing is the linkage between our understanding of these factors and the way we manage our fisheries. We have seen how variations in intraseason intrinsic factors can be an important key for successful development and management of several wild-stock fisheries. However, for many species, management has disregarded the inseason timing of harvests in order to focus on other issues such as allocation of quota among competing sectors within the industry. Failure to consider intrinsic variability as well as extrinsic factors in the harvesting and processing sectors results in sub-optimal management of fast growing or rapidly-changing stocks. The results can decrease benefits to society in an increasingly demanding global marketplace and potentially the ecosystem. Seafood science can play an increasingly important role in how we manage our fisheries in the future. Incorporation of intrinsic and extrinsic variables in the development of fishery management plans will help managers meet their goals of improved conservation, efficiency and utilization of the resource.

REFERENCES

1. Love, R.M. (1988) The Food Fishes: Their Intrinsic Variation and Practical Implications. Farrand Press, London, AVI Books, NY.
2. Ackman, R.G. (1995) Composition and nutritive value of fish and shellfish lipids. In Fish and Fishery Products: Composition, Nutritive Properties and Stability. Ruiter, A. (ed.), CAB International, Oxford, pp. 117-156.
3. Sylvia, G., Larkin, S.L. and Morrissey, M.T. (1997) Intrinsic product quality and fisheries management: a bioeconomic model of inter- and intraseasonal quota

allocation in the Pacific whiting fishery. Proceedings of the 2nd World Fisheries Congress — Developing and Sustaining World Fisheries Resources: the State of Science and Management. Smith, D. and Beumer, J. (eds.) Brisbane, Australia, pp. 229-253.
4. Radtke, H. (1996) Windows on Pacific Whiting: and Economic Success Story for Oregon's Fishing Industry. Prepared for OCZMA, Newport, OR.
5. Park, J.W. (2000) Surimi and Surimi Seafoods. Marcel Dekker, Inc., NY.
6. Nelson, R.W., Barnett, H.J. and Kudo, J. (1985) Preservation and processing characteristics of Pacific whiting (*Merluccius productus*). Mar. Fish. Rev., 47: 60-67.
7. Peters, G., Morrissey, M.T., Sylvia, G. and Bolte, J. (1996) Linear regression, neural network, and induction analysis to determine harvesting and processing effects on surimi quality. J. Food Sci., 61: 876-880.
8. Morrissey, M.T., Wu, J.W., Lin, D. and An, H. (1993) Protease inhibitor effects on torsion measurements and autolysis of Pacific whiting surimi. J. Food Sci., 58: 1050-54.
9. Morrissey, M.T., Hartley, P. and An, H. (1995) Proteolytic effects in Pacific whiting and effect on surimi processing. J. Aquatic Food Prod. Technol., 4: 5-18.
10. Lin, J. and Park, J.W. (1996) Extraction of proteins from Pacific whiting at various washing conditions. J. Food Sci., 61: 432-38.
11. Lin, J. and Park, J.W. (1996) Protein solubility in Pacific whiting affected by proteolysis during storage. J. Food Sci., 61: 536-39.
12. Larkin, S. and Sylvia, G. (1999) Intrinsic fish characteristics and intraseason production efficiency: A management level bioeconomic analysis of a commercial fishery. Am. J. of Ag. Economics, 81: 29-43.
13. Crapo, C. (2001) Personal communication. Fisheries Industry Technical Center, University of Alaska, Kodiak, AK.

# UTILIZATION OF ANTARCTIC KRILL FOR FOOD AND FEED

Bunji Yoshitomi
Central Research Laboratory, Nippon Suisan Kaisha, Ltd., 559-6 Kitano-machi, Hachioji, Tokyo 192-0906, Japan

Abstract

The catchable stock of krill is estimated at more than 10 million tons per year. As for food shortage, because of an increase in the world population, krill is expected to be the last marine resource remained in the world. However, the annual krill catch amounts to only about 80 thousand tons. Many researchers have been studying on the utilization of krill for a long time, but even now the main use of krill is limited to fish feed for aquaculture and bait for leisure fishing. Unfortunately, we can not find any high-value-added krill products in market at present.

Considering the low utilization of krill, krill digestive protease was an obstacle for its food processing. Since the protease is highly intensive, it is easy to digest its muscle protein rapidly. Therefore, many researchers have thus tried to inhibit and to remove the protease in order to keep the muscle protein intact. But they have not found effective way keeping the muscle intact. On the other hand, for some medical uses, krill protease is going to play an important part; for example, it would be effective for bedsore remedy and for skin cancer. It would be another way to utilize krill.

1. BACKGROUND

Krill is shrimp-like zooplankton and it is normally found in all over the world. Krill is composed of 85 species and the most prominent among them is *Euphausia superba* Dana in the Antarctic Ocean. The length of *Euphausia superba* Dana varies from 1.5 to 6.0 cm, and mature animals have a liveweight of around 1.0 to 1.5 g. Their living space is in a water depth of approximately 50 m. Antarctic krill is considered to be a keystone species, an organism upon which very many Antarctic predators depend [1-3, 14-18].

Union of Soviet Socialist Republics (USSR) started to catch krill on 1961 and the international study for utilization of krill has also started at the same time. About ten years later, 1972, Japan started to catch and to study krill. Before now, the United States, Poland, Germany, French, Taiwan, Korea, Chile etc. have studied using research ships [4].

Table 1
Total amount of krill catching in the Antarctic Ocean

Year	91/92	92/93	93/94	94/95	95/96	96/97	97/98	98/99	99/00
Japan	74,325	59,272	62,322	60,303	60,546	58,798	63,233	71,318	67,188
Russia	151,725	4,249	965						
Chile	6,066	3,261	3,834						
Poland	8,607	15,909	7,915	9,384	20,610	19,156	15,312	18,554	20,721
Korea	519						1,621	1,228	5,444
Ukraine	61,719	6,083	8,852	48,886	20,056	4,246		5,694	985
Argentine								6,524	
Uruguay									6,948
Other nations		74	41	502	308	634			
Total amount (ton)	302,961	88,774	83,962	118,714	101,714	82,508	80,800	103,318	101,286

91/92, a fishing term from July 1st, 1991 to June 30th, 1992.

The main species of krill for commercial use is Antarctic krill, *Euphausia superba* Dana. Table 1 shows total amount of krill catching in the Antarctic Ocean from 1991 to 2000. From 1991 to 1992, USSR (Russia) caught about 150 thousand tons, therefore the world annual catch was about 0.3 million. But since then, Russia quit catching, so the world annual catch has being around 80 thousand tons.

The catching ground of Japanese trawlers is mainly around the Antarctic Peninsula in summer, usually from December to June. But in winter, normally from July to October, at such high latitude area heavy storms come so frequently that they move to work around the South Georgia Island, South Sandwich Islands or South Orkney Islands, etc.

2. MAIN PRODUCT OF KRILL ON BOARD

Krill products by Japanese trawlers are shown in Table 2. Frozen krill occupies about 80 % of total product and boiled krill occupies about 13 %. Peeled krill is only product for food and occupies about 3 %. In Europe, some frozen food used peeled krill can be found, but it is not so popular. Krill meal occupies about 4 % and it is mainly used for feed industry to improve feed palatability.

General utilization of krill is shown in Fig. 1 [4]. The main use of krill is for aquaculture feed and bait for leisure fishing. On the other hand, its use for human food is

Table 2
Krill products by Japanese trawlers

Year	91/92	92/93	93/94	94/95	95/96	96/97	97/98	98/99	99/00
Boiled	5,516	5,924	5,628	5,430	4,771	6,442	5,070	5,076	4,268
Frozen	20,511	16,848	22,066	27,385	28,319	27,743	24,265	27,519	23,904
Peeled	2,206	1,477	1,479	469	738	431	797	848	1,116
Meal and others	3,487	2,163	1,975	1,986	1,977	2,009	2,617	2,982	2,795
Total amount (ton)	31,720	26,412	31,148	35,270	35,805	36,605	32,749	36,425	32,083

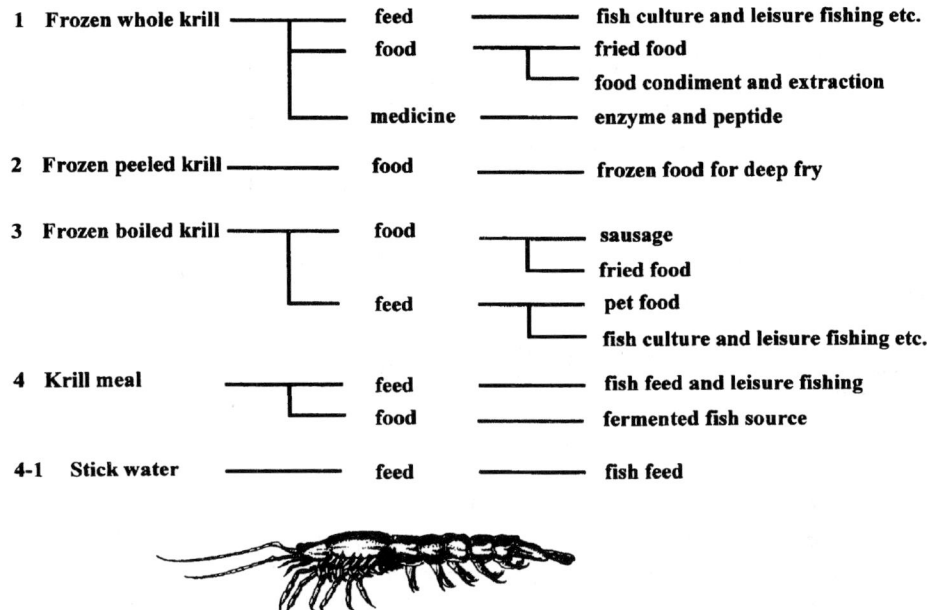

Figure 1. General utilization of krill in Japan.

minor. The reason is that krill has intensive proteases that work at low temperature and its muscle is very fragile [7-10], so it is very difficult to keep the muscle intact after catching. Concerning krill protease for medical use, some studies have started several years ago and recently, some remarkable results described below are found [11-13].

3. GENERAL COMPOSITION OF KRILL

3.1. Seasonal variation of general composition of krill

There are many reports on general composition of krill [19-21], but few of them have referred to the seasonal variation, especially that of in winter. Our latest result is shown in Fig. 2. The frozen krill samples were obtained by Nippon Suisan Kaisha, Ltd. the trawler Niitaka-maru, in the area of the Antarctic Ocean from December, 1999 to August, 2000. The fat content varied remarkably as many researchers had reported so far [22-25]. The crude fat content was low in mid-summer (December and January), but it increased about 4 times in mid-winter (May and June). This would be an important variation for krill to pass through a winter season in the Antarctic Ocean.

The seasonal variation of lipid composition is shown in Fig. 3. In mid-summer, phospholipid occupied about 75% of the total lipid. But in mid-winter, the phospholipid content was about 65%, the lowest in the year. On the other hand, triglyceride content was low in mid-summer, but it was at the peak in mid-winter. These phenomena would have some relation with feeding on phytoplankton for the krill.

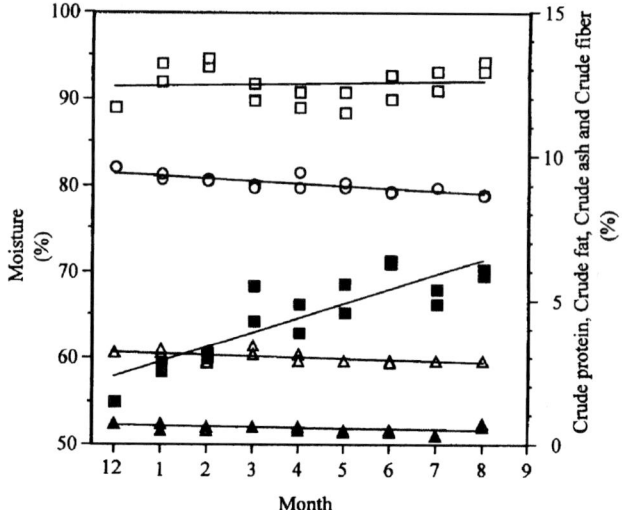

Figure 2. Seasonal variation of general composition of krill caught in the Antarctic Ocean from December 1999 to August 2000. ○, moisture; □, crude protein; ■, crude fat; △, crude ash; ▲, crude fiber.

3.2. Seasonal variation of crude krill protease activity

Crude krill protease (CKP) activity has measured on board from December 1998 to August, 1999 and from December 1999 to August, 2000. The method to measure crude krill protease activity was as follows. Fresh krill (50 g) was homogenized with cold water (100 mL) and centrifuged at 3,000 rpm at 10°C for 10 min. The supernatant (0.05 mL) was added into 1.0 mL of 1.0 % azocasein (Sigma) solubilized in 0.1 M Tris-HCl (pH 7.5). The enzyme reaction was performed at 37°C for 60 min. Two mL of 5 % of trichloroacetic acid was added to the mixture to stop the reaction, followed by centrifugation at 3,500 rpm for 10 min. Optical density of the supernatant was measured at 440 nm.

Figure 4 shows the seasonal variation of crude krill protease activity. The enzyme activity was very high in summer, but it went down in winter. In addition, Fig. 4 shows the relation between the protease activity and the feed intake index. In summer, there is a lot of phytoplankton in the Antarctic Ocean and krill can thus feed on them and have to digest them intensively to accumulate energy. Therefore, their protease activity have to keep high, but in winter there are not any phytoplankton available and they do not have to keep the activity high, so it would be better to keep the activity lower for saving energy [26].

4. AQUACULTURE FEEDS

Because of recent expansion of the global aquaculture market, krill has been required

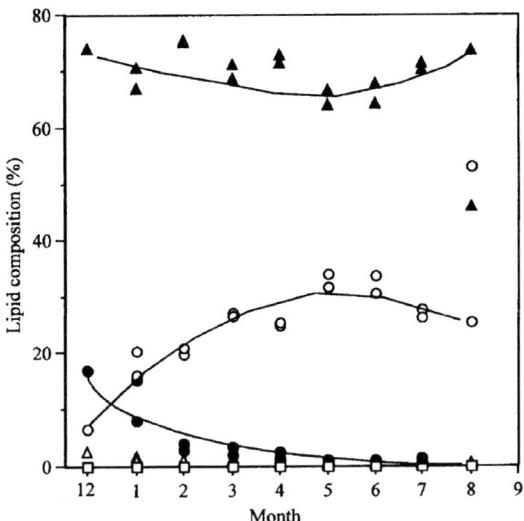

Figure 3. Seasonal variation of lipid composition of krill catched in Antarctic Ocean from December 1999 to August 2000. ○, triglyceride; ●, free fatty acid; □, hydrocarbon; ■, waxester; △, cholesterol; ▲, phospholipid.

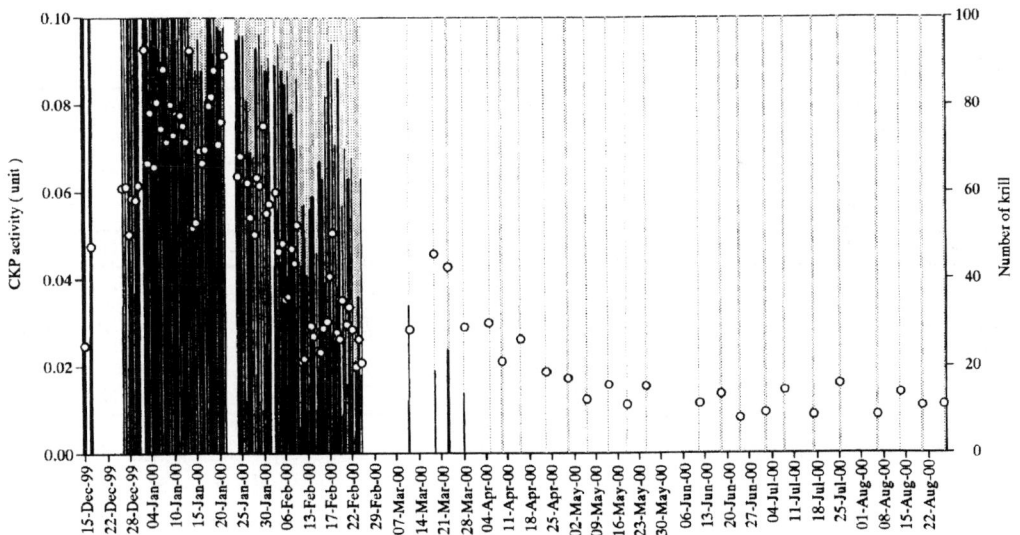

Figure 4. Seasonal variation of crude krill protease activity. ○, crude krill protease activity; ■, feed intake index 3; ▨, feed intake index 2; ▢, feed intake index 1. Feed intake index is a visual observation of feed intake condition for 100 krill. When krill feeds a lot of phytoplankton, its digestive organs turns into dark green. The feed intake index is 2 and when krill feeds nothing, their digestive organs have no color; feed intake index is 1. The optical density of crude krill protease (CKP) activity by the azocasein method at 440 nm was converted to activity unit as follows: 1 unit means increase in 1 optical density at 440 nm per 10 min.

Table 3
Dietary ingredients of the experimental (%)

Ingredient	Diet 1	Diet 2	Diet 3	Diet 4	Diet 5
Fishmeal	63.5	58.5	54.5	59.5	55.0
Krill meal	0.0	5.0	9.9	0.0	0.0
Hydrolyzed krill	0.0	0.0	0.0	5.0	9.9
Rice bran	8.0	8.0	7.1	7.0	6.6
Wheatfeed flour	12.9	12.9	12.9	12.9	12.9
Soybean meal	6.9	6.9	6.9	6.9	6.9
Vitamin mixture	1.0	1.0	1.0	1.0	1.0
Mineral mixture	0.8	0.8	0.8	0.8	0.8
Fish oil	6.9	6.9	6.9	6.9	6.9
Total	100.0	100.0	100.0	100.0	100.0

Hydolyzed krill meal was produced by Nippon Suisan Kaisha Ltd. Central Research Laboratory.
Experimental feed was extruded by twin-screw extruder (Suehiro EPM Co. Ltd.). Proximate feed composition: crude protein 51.6%, crude fat13.5%, crude fiber 1.0%.

Table 4
Growth of red sea bream *Pagrus major* fed feed with krill meal and hydolized krill

	Diet 1	Diet 2	Diet 3	Diet 4	Diet 5
Initial body weight (g)	144.4	144.3	144.4	144.4	144.4
Final body weight (g)	218.9	219.6	220.8	220.8	220.8
Daily feed intake (%)	1.95	1.97	1.87	1.93	1.87
Feed efficiency (%)	65.7	65.6	69.9	75.1	80.0

Thirty red sea bream were fed for 40days in a 500 L tank.

Table 5
Proximate composition of each material (%)

Proximate composition	Fish meal	Krill meal	Hydrolyzed krill
Crude protein	70.2	65.9	66.6
Crude fat	10.7	6.7	8.5
Crude ash	15.8	18.3	16.2
Crude fiber	0.4	1.9	2.9
Free amino acid	2.0	7.5	17.1

Fish meal was made of Chilean horse mackerel.
Free amino acid was analyzed by HPLC.

not only as a protein source but also a material with a high palatability. Especially, krill meal is one of the most important materials for fish feed.

A growth experiment of red sea bream *Pagrus major* was conducted to clarify the character of krill meal and hydrolyzed krill [27]. The dietary ingredients and proximate composition are shown in Table 3. The proximate composition of each experimental feed was almost the same. The result is in Table 4. The feed containing krill meal showed better feed efficiencies than the control feed. In addition, the addition of the krill meal at 10 % in the feed showed a better result than at 5%. Moreover, the feed containig hydrolyzed krill showed the best result of all.

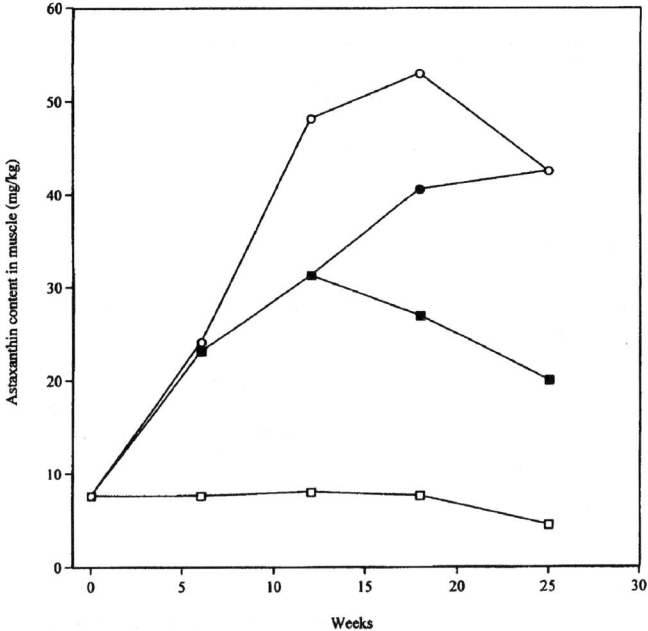

Figure 5 Astaxanthin content in Rainbow trout Oncorhynchus mykiss muscle fed krill oil. ○, 4.5 % krill oil (8.3mg/100 g astaxanthin in feed); ●, 2.2 % krill oil (4.1mg/100g astaxanthin in feed); ■, 2.2% krill oil for 12 weeks; □, non-pigment. Krill oil was extracted from krill meal by solvent. Experimental fish were fed dry feed with krill oil for 12 weeks or 25 weeks.

Table 5 shows proximate composition of each material. Hydrolyzed krill has more free amino acid than krill meal and fishmeal, so it would be effective in the palatability and growth. The krill product as aquaculture feed materials is meaningful to improve feed quality.

Krill was formerly investigated as a rich source of a natural pigment as carotenoids [28-30]. Figure 5 shows the change of astaxanthin content in rainbow trout muscle that was bred the feed with krill oil. Krill oil containing astaxanthin was effective in coloring the muscle. However, a synthetic pigment has been replaced krill meal due to lower price. In the field of poultry feed, krill meal is sometimes used for natural colorants of egg yolk. The only problem of krill meal is its high price. If the price could be reduced to the level of fishmeal, the potential market would be huge.

5. FOOD

The utilization of krill has not been very popular in the field of food processing because its protease activity is extremely high and its muscle protein is rapidly hydrolyzed after catching [5,6,31]. Moreover, the structure of the muscle protein itself is markedly fragile. The reason why krill has an intense protease activity and a delicate muscle would be attributable to the water temperature that it inhabits. Therefore, it is difficult to use frozen krill for food materials, because we thaw the frozen krill, the potential protease is rapidly activated and the muscle protein is hydrolyzed. Only immediate boiling of krill after catching on board, we can solve the problem. However, the water-soluble substances in krill body are lost into the boiling water, so the flavor turns weaker.

Another problem is the texture of krill shell, which is not suitable for food and its fluorine content also would be a problem. Peeled krill has developed to solve these problems using a revolutionary peeling machine, but the product is not tasty because it is washed several times in fresh water. In addition, the yield is very low.

Meanwhile, making great use of its potent protease, a kind of krill extract similar to that made from shrimp is on the market. A fish sauce made from krill has started to be marketed. The fermented krill sauce rich in free amino acid is more profound in its taste than the soy bean sauce. Fermented krill sauce has a characteristic flavor of krill, so it can not replace as soybean sauce. On the other hand, the unique function of the fermented krill sauce has been reported that chicken meat pickled in the fermented krill sauce turned to be juicier than non pickled one. The fermentation would thus be a new way to utilize krill [32].

6. MEDICAL USE

6.1. Peptide
Some water-soluble peptides have been studied for anti-hypotensive drug, but the intense effects have not been obtained [14].

6.2. Protease
The utilization of krill protease has been studied for a long time. Recently the study of some drugs for skin cancer and bedsore has developed rapidly in Europe and its achievements have become a center of attraction [11-13]. In this field, it is important to check the seasonal variation of krill proteolytic activity.

7. ACKNOWLEDGMENT

The author wishes to thank Dr. Kazuhiro Yoshikawa and Dr. Ruben, O. Bustos Cerda for their helpful advice.

REFERENCES

1. Ichii, T. (2001) The world trend of Antarctic krill as a marine resource. Kaiyo Monthly, 272: 244-251.
2. Sayed, Z. (1997) South Ocean Ecology: the Biomass Perspective. El-Sayed,

Cambridge Unversity Press, pp. 129-187.
3. Kawaguchi, S. (2001) Study of Antarctic krill. Kaiyo Monthly, 272: 147-155.
4. Kawaguchi, S. (2001) The fishing grounds of krill in the Antarctic Ocean (18). National Reseach Institute of Far Seas Fisheries, Shimizu, pp. 18-19.
5. Ooizumi, T., Usuki, E., Yabe, K. and Arai, K. (1983) Digestibility of actomyosin of Antarctic krill *Euphausia superba* by its digestive protease. Bull. Japan. Soc. Sci. Fish., 49: 113-121.
6. Seki, N., Sakatani, H. and Onozawa, T. (1977) Studies on proteases from Antarctic krill. Bull. Japan. Soc. Sci. Fish., 43: 955-962.
7. Hashimoto, A. and Arai, K. (1979) Thermo-stability of myofibriller Ca-ATPase of Antarctic krill. Bull. Japan. Soc. Sci. Fish., 45: 1453-1460.
8. Hashimoto, A., Kobayashi, A. and Arai, K. (1982) Thermostability of fish myofibrillar Ca-ATPase and adaptation to enviornmental temperature. Bull. Japan. Soc. Sci. Fish., 48: 671-684.
9. Bustos, R. R., Romo, C. R. and Healy, M. G. (1996) Stabilisation of trypsin-like enzymes from Antarctic krill: Effect of polyols, polysaccharides and proteins. J. Chem. Tech. Biotechnol., 65: 193-199.
10. Bustos, R. R., Romo, C. R. and Healy, M. G. (1999) Purification of trypsin-like enzymes from Antarctic krill processing wastewater. Process Biochemistry, 35: 327-333.
11. Bucht, A. and Karlstam, B. (1991) Isolation and immunological characterization of three highly purified serine proteinases from Antarctic krill (*Euphausia superba*). Polar Biology, 11: 495-500.
12. Karlstam, B., Johansson, B. and Bryno, C. (1991) Identification of proteolytic isozymes from Antarctic krill (*Euphausia superba*) in an emzymatic debrider. Comp. Biochem. Physiol., 100B: 817-820.
13. Karlstam, B. (1991) Crossed immunoelectrophoretic analysis of proteins from Antarctic krill (*Euphausia superba*) with special reference to serin proteinases. Polar Biology, 11: 489-493.
14. Dohmoto, N., Wang, K. C., Mori, T., Kimura, I., Koriyama, T. and Abe, H. (2001) Development of a new type fish sauce using the soy sauce fermentation method. Nippon Suisan Gakkaishi, 67: 1103-1109.
15. Siegel, V., de la Mare, W. K. and Loeb, V. (1997) Long-term monitoring of krill recruitment and abundance indices in the Elephant Island area (Antarctic Peninsula). CCAMLR Science, 4: 19-35.
16. Siegel, V. (1987) Age and growth of Antarctic Euphausiacea (crustacea) under natural conditions. Marine Biology, 96: 483-495.
17. Kawaguchi, S., Ichii, T. and Naganobu, M. (1997) Catch per unit effort and proportional recruitment indices from Japanese krill fishery data in subarea 48. 1. CCAMLR Science, 4: 47-632.
18. Endou, Y. (1990) Ecology of Antarctic krill. Kaiyo Monthly, 22: 10.
19. Clarke, A. (1980) The biochemical composition of krill, *Euphausia superba* Dana, from South Georgia. J. Exp. Mar. Biol. Ecol., 43: 221-236.
20. Kinumaki, H. (1973) Krill. Food Science Journal, 11: 74-86.
21. Toyama, K. (1982) Krill. Food Science Journal, 67: 60-66.

22. Vander Vee, J., Medwadowski, B. and Olcott, H.S. (1971) The lipids of krill (*Euphausia species*) and red crab (*Pleuroncodes Planipes*). Lipids, 6: 481-485.
23. Bottino, N. (1975) Lipid composition of two species of Antarctic krill *Euphausia superba* and *E. crystallorophias*. Comp. Biochem. Physiol., 50B: 479-484.
24. Ellingsen, T. E. and Mohr, V. (1982) Lipids in Antarctic krill. Composition and post mortem changes. Proceedings Scandinavian Symposium on Lipids 11th 1981, pp. 110-117.
25. Fricke, H., Gercken, G., Schereiber, W. and Oehle Schlager, J. (1984) Lipid, sterol and fatty acid composition of Antarctic krill (*Euphausia superba* Dana). Lipids, 19: 821-827.
26. Miyazaki, I. (1978) Plankton fishery. Food Science Journal, 44: 91-103.
27. Hertrampf, J. W. and Piedad-Pascual, F. (2000) Krill meal. Handbook on Ingredients for Aquaculture Feeds. Kluwer Academic Publishers, London., pp. 221-228.
28. Yamaguchi, K. (1983) The composition of carotenoid pigments in Antarctic krill *Euphausia superba*. Bull. Japan. Soc. Sci. Fish., 49: 1411-1415.
29. Fujita, T. (1983) Pigmentation of cultured red sea bream with astaxanthin diester purified from krill oil. Bull. Japan. Soc. Sci. Fish., 49: 1855-1861.
30. Miki, W. (1985) Origin of tunaxanthin in the integument of yellowtail (*Seriola quinqueradiata*). Comp. Biochem. Physiol., Vol. 80B: 195-201.
31. Kolkovsi, S., Czesny S. and Dabrowski, K. (2000) Use of krill hydrolysate as a feed attractant for fish larvae and juveniles. Journal of the world aquaculture society, 31: 81-88.
32. Nakamura, H., Mohri, Y., Muraoka, I. and Ito, K. (1979) Studies on brewing food containing Antarctic krill-1 Production of soysauce with cryo-ground Antarctic krill, *Euphausia superba*. Bull. Japan. Soc. Sci. Fish., 45: 1389-1393.

TOTAL UTILIZATION OF SQUIDS: ITS ZERO EMISSIONS APPROACH

Teisuke Miura
Laboratory of Marine Environment and Resources Sensing, Division of Marine Environment and Resources, Graduate School of Fisheries Science, Hokkaido University, 3-1-1 Minato-cho, Hakodate, Hokkaido 041-8611, Japan

Abstract

Zero emission was advocated as ZERI (Zero Emission Research Initiative) by Gunter Pauli in United Nations University in 1994. Afterwards, this concept has been rapidly generalized. In the fishery, this idea is also beginning to spread for solution of problems. A large amount of raw squid material is used in our country, and, as a result, the squids-processing industry has generated a large amount of wastes. Thinking of this point, we have begun the research on utilization of squid wastes. One is the development of the dyestuff extracted from squid ink. Another is the artificial bait made from squid liver. The former is a trial of clarifying the industrial character in region by the reason why squids are used not only for food but also for culture and art. The latter was developed for the tuna long line fishery. This fishery is currently done in the sea all over the world. Because the nitrogen included in the bait is carried out to the sea area in foreign countries, it would contribute to the world circulation of nitrogen. In this respect, it is thought that the artificial bait needs some incentives in price competition. Moreover, it should be recognized as an ecological commodity in the future.

1. BACKGROUND

Zero emission was advocated as ZERI (Zero Emission Research Initiative) by Gunter Pauli in United Nations University in 1994 followed by rapid generalization of this concept [1,2]. UNU's Second Medium-Term Perspective (1990-1995) has articulated such five areas of concentration (UNU 1996) as (1) Universal human values and global responsibilities, (2) New directions for the world economy, (3) Sustaining global life-support systems, (4) Advances in science and technology, (5) Population dynamics and human welfare.

The third theme contains five sub-programs: (i) "Eco-restructuring" for sustainable development, (ii) Integrated studies of ecosystems, (iii) Information systems for environmental management, (iv) Natural resources in Africa, and (v) Environmental law and governance. The first subprogram contains six themes, and the fourth theme is a zero emission research. "Eco-restructuring" is strongly related to the concept of "deep ecology" and this concept was generated from "living system theory". This was deeply influenced from "integrated biology", "gestalt psychology", "ecology", "general system theory" and "cybernetics". It is important to understand the idea of such a background about zero emission.

Japan experienced the environmental pollution problems in the rapid growth age and tried to achieve the solution of them in the 70s by the method of processing the pollutant exhausted from the industrial process in the end of the outlet. It is called "end-of-pipe solution". However, the antipollution appliance did not bring any economic effects. As a result, a lot of concerns began to be paid to "cleaner production (the low pollution type of the production technology)" in the latter half of 80's. However, because these technologies were missed to a basic common principle, a big result was not achieved. On the other hand, zero emission shows a basic idea of efficient utilization of the resources and the decrease in the environmental burden. It becomes possible by the networking. The construction of "industrial cluster" becomes important in the achievement of zero emission. And, this composition should approach the one which looked like ecosystem. Therefore, this idea is obviously different from "end-of-pipe solution". Zero emission goes back on the upstream side of the process by thinking about the whole of the material circulation and advances the optimization of the entire systems [3].

Let me introduce zero emission technologies here as an example of the processing industry. In general, when raw material is processed, by-product is abandoned in the factory. However, in case of the industrial process which supported by zero emission technologies, wastes (material or energy) are converted again into profitable one by the proper method. It is difficult to achieve to zero emission in an only single enterprise. Therefore, to decrease the environmental burden, some different types of business make the industrial network as a general approach and thus the emission approaches gradually zero. Figure 1 shows this concept [1]. If the meaning of zero emission is simply explained, it basically contains the following three items:

1. Development of the industrial process in which waste is not generated.
2. When generated once, waste is effectively used by recycling.
3. We decrease the waste generated from the society and industry until it approaches zero as much as possible.

Moreover, it can be said that it is an ideal to satisfy three bases under the condition that a current life level is not dropped.

2. MATERIAL FLOW AND NITROGEN BALANCE OF SQUIDS IN JAPAN

The squid-jigging fisheries are done in the offshore as follows; Southwest Pacific Ocean, New Zealand, Peru and Japan. It is also done in the coastal waters of Japan. The total of these catch amounts to 450,463 t, additionally, pelagic trawling (4,000 t) is also done in the offshore of New Zealand, North America, South America and Africa. The remainder is 181,000 t caught by other fishing methods. The import of squid in Japan

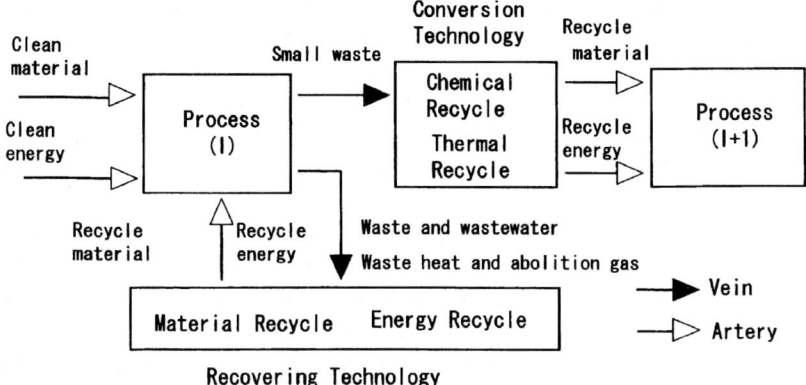

Figure 1. Core technologies which support zeroemission.
Adapted from Fujimura, Y. (1996) [1].

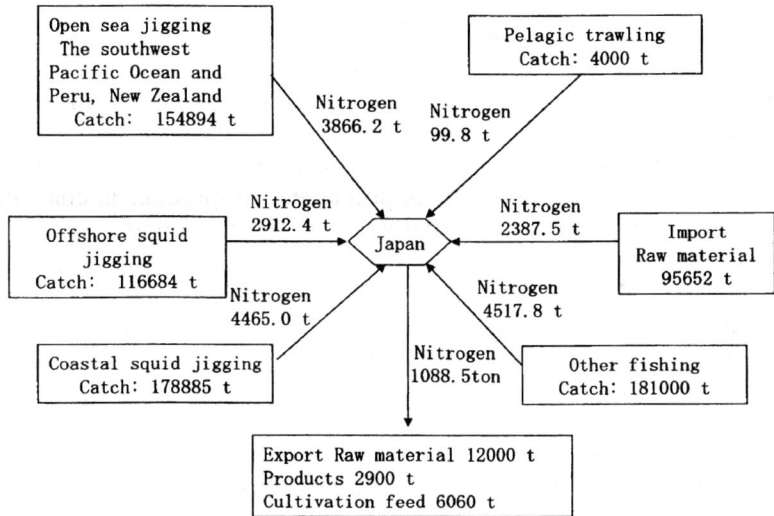

Figure 2. Material and nitrogen balance of squids in Japan (1997).
Miura, T. et al. (2000) [6].

(1997) is 95,652 t a year. The main export countries are Thailand (23,000 t), China (14,000 t), Morocco (9,000 t), Vietnam (6,000 t) and others. The raw material of squid exported from Japan is 12,000 t and the products are 2,900 t according to Japanese preliminary trade statistics (1997). These are only artery system products. On the other hand, some routes of the vein system are located in Hakodate and Hachinohe where the feed for cultivation of prawn (*Penaeus monodon*) is made from squid waste. The products of 6,060 t were already exported to the Philippines and other countries in 1997 [4,5].

Nitrogen balance of squid (1997) in Japan is shown in Fig. 2. The total of input is

18,200 t, that is, the total of output amounts to 1,100 t and the input exceeds the output by 17,100 t. The nitrogen content of squids was obtained as follows. First, the protein content was referred from "Standard Table of Food Composition in Japan" and next, the nitrogen content was calculated from the weight of protein by using conversion factor (6.25).

3. CURRENT STATE OF THE SQUID WASTE GENERATED IN HAKODATE

The squids of about 110,000 t enter the city of Hakodate a year. Hakodate-special-food-industry-union (1997) used the 98,000 t of them. As a result, the waste of 6,500 t is generated from the squid processing factories. Nihon Kagaku Shiryo Co., Ltd converts the waste into the prawn's feed and exports a year about 1,500 t. However, the processing cost of the waste is demanded by the waste processor by now. It is a problem that additional expenditure is necessary for the utilization of the squid waste. The Nihon Kagaku Shiryo Co., Ltd is undertaking it for 12,000 yen/a ton and has the processing performance of 80 t a day. Processing is done well in the current state. However, this load of cost begins to burden the management of the squid processor. The waste is processed as cultured prawn's feed for Southeast Asia. However, it is a fact to have actualized the problem from the instability of export by the economic stagnation and fluctuation of Asia. It seems that the idea of the zero emission technology is very necessary for the squid processing industry as the superannuating of facilities makes the problem serious further. Just now should be the chance of thinking about the processing method by the whole persons who concern that [7].

4. ZERO EMISSION TECHNOLOGIES

Table 1 shows the possibility of effective use of squid waste. We have begun the trial of how squid wastes are used as artificial bait. One is the investigation for the long line fisheries executed by cooperation with Kagoshima University, and it has begun in 1999. Because the price such as "alfonsin (*Beryx splendens*)" is comparatively high, there is a

Table 1
Possibility of deployment of squid wastes

Material	Products	Situation
Squid liver	Liquid crystal, Fishing bait, and Omega 3 (EPA,DHA), Feed of cultured fish and Domestic animals	Practical use (Hakodate)
Squid ink	Dyestuff, Health food and Anti-tumorous material	Special products (Hakodate and Aomori)
Squid skin	Dyestuff	Under experiment (Aomori)
Squid glade	Natural fiber (Artificial skin)	Under experiment (Sapporo)

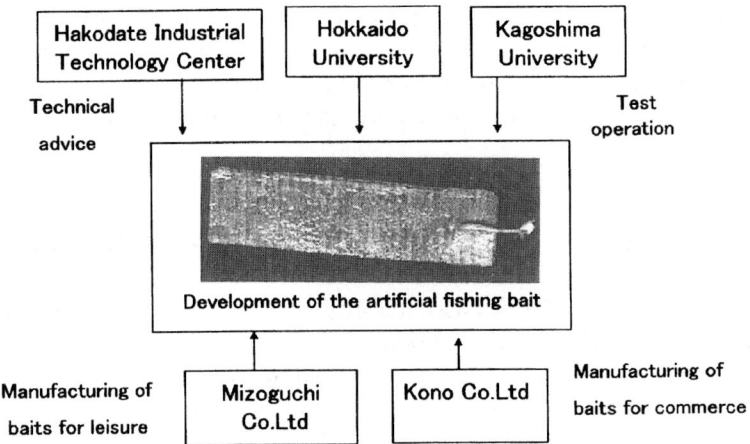

Figure 3. Organization for development of the artificial fishing baits.

big expectation for the artificial bait from the regional fishery. Another investigation is executed by Hokkaido University. We carried out the test operations of the tuna long line by using the artificial bait in the offshore of Hawaii. Figure 3 shows this overall plan. It can be called that the tuna long line fishery is the representative of the pelagic fisheries. Tuna fishing boats now are operated with large-scale gears reaching for as much as 100 km in the total extension through a year. Amount of money of tuna landing is about 120million yen per ship. The natural bait consumed per year is about 60,000 t as a whole in this fishery. It begins to have some serious problems: e.g. a decrease of the tuna's stock and the price-change of the natural baits such as squid and saury (*Cololabis saira*). It is said that 45 yen is a price of the profit turning point in the current state. Therefore, stabilizing the price and moreover securing the demand of amount of the baits become important. If the natural bait is substituted by the artificial one made from the squid waste in such a situation, it is stably offered by the price in the latter half of almost 30 yen level. Moreover, because the squid waste is generated by about 18,000 t in year in the southern part of Hokkaido, about 30% of the entire natural baits can be quantitatively substituted as a result.

The cadmium included in the product such as "shiokara (salted and fermented seafood)" does not become a problem in the amount taken with a usual meal because of the extremely small amount. However, when the squid waste is used for cultured fish's feed or farm fertilizer, it is necessary to do the removal processing of cadmium is required. When the squid waste is used for a specific object for a long term, the bio-concentration might be caused. However, there is not worry of the bio-concentration because artificial bait and tuna's encountering is only one or two times and cadmium is distributed to the entire sea area by the tuna long line fisheries widely and thinly. Moreover, because the level of the cadmium contained in the bait is almost the same as the one of the internal organs of the squid which lives in the sea, the artificial bait made from the squid waste is advantageous also in the point of the environmental burden.

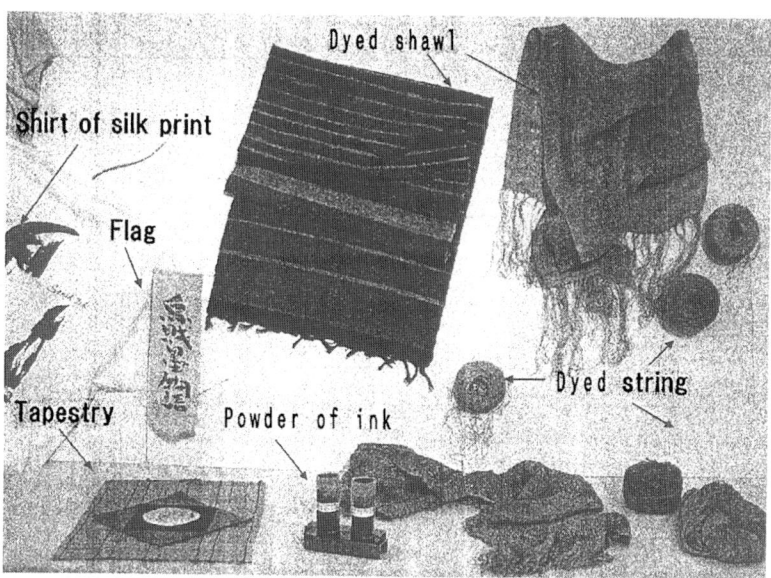

Figure 4. Some artistic works made by using squid ink.

The utilization of the squid ink is currently progressing. The squid is able to disturb the enemy using the ink as a defense means and can escape well. Since fish becomes difficult to breathe when the ink adheres to his gills, it is an undesirable material for the fish. Therefore, it is necessary to remove the ink sac completely to produce the artificial bait. The sepia coloring matter is extracted from the abandoned ink sac and the commercialization of the squid ink-picture and the silk print has started. The commodity registration "Ikasumizome (dyed goods)" is completed and the establishment of the company is done now. Some artistic works made by using squid ink are shown in Fig. 4. This plan is trial of clarifying the industrial character in region by the reason why squids are used not only for food but also for culture and art. Consequentially, it gives a very big added value to the developed products. Technically, there are some problems of a deodorization from the dyestuff, a color fixation, and a color and tone quantitative assessment. We are now trying to make the solution of these problems. The administration assistance of Hakodate is obtained from the view point of the sightseeing policy now; 5,500,000 tourists in year are expected in the future. A big business can be expected in this respect.

5. HOW MUCH IS THE RAW MATERIAL CONVERTED FROM SQUID WASTE ?

The raw material of squid includes such tissues as liver, ink sac, gradii, skin, eyes, etc., all of which are processed as tissues are abandoned as the wastes when they are mixed, because they are worthless wastes to some products in factories. If it is possible to separate to each part, they become some effective materials for various enterprises, and bear some big added values. We obtained 60 squids for the inspection; they are generally used as raw material for the processing in Hakodate (Table 1). Weight ratios of each part to the whole body were obtained from the squid samples. The squid of 766,000 t is consumed in Japan, and each part of weight was presumed by using these values and the total amount. The amount of liver was the largest (89,000 t), followed by that of sexual organs and digestive organs (57,000 t). Moreover, the amount of skins, eyeballs, buccal mass, gradii and ink sacs were 19,000, 20,000, 10,000, 2,000 and 2,000 t, respectively.

6. CONCLUSION

Mass balances of squid are considerably biased in the total of Japan. The squid processing industry imports a large amount of raw material as well as other food industries, and the most of them is consumed domestically. The import of the raw material was 730,000 t in 1997, while the export of the Squid raw material was 21,000 t. As the result, 94% of the entire nitrogen stays domestically. It is necessary to develop the product consumed in foreign areas like the bait of tuna fishery for improving this state. The feed for prawn cultivation has already been exported to Southeast Asia. It is one of products to improve this situation; the feed changes the appearance into the prawn, and lands to Japan again. Naturally, the cadmium included in squid liver is contained in the body of the prawn. Even if the heavy metal is harmless in amount for human health, it will be necessary to improve this problem in the future.

It should be important to think of input-output preconditions, when we will construct the system aiming at zero emission. The system should be most efficient in every respect such as the resource consumption, economy, and environmental burden. Moreover, the system should circulate smoothly not only material but also energy. Gunter Pauli says, "Industry mimics nature's sustainable cycles and humanity".

On the other hand, the artificial bait of tuna long line distributes widely the nitrogen to the sea, resulting in contribution to the world circulation of nitrogen. Under environmental condition of Japan, it will especially contribute to the prevention of the eutrophication of the hydrosphere. It is necessary to give the artificial bait some incentives by price competition in this respect because the artificial bait will be recognized as an ecological commodity in the future. Therefore, the developmental supports of the university, institute and national policy responses will also be needed.

7. ACKNOWLEDGEMENTS

I express my sincere thanks to the president Mr. Toshio Abe of KOHNO Co. Ltd., and the executive director Mr. Hiroyasu Mizoguchi of MIZOGUCHI Co. Ltd. for their generous support of my work.

REFERENCES

1. Fujimura, H. (1996) Zero emission and corporate management. United nations university and Diamond Ltd. co-sponsoring forum, Tokyo.
2. Capra, F. and Pauli, G. (1998) Steering business toward sustainability. United Nations Network UNU Conference, Tokyo.
3. Suzuki, M. (1999) Achieving an Eco-Industrial Society through Zero Emissions. International Conference on Zero Emissions in Industrialized Society, Tokyo.
4. Miura, T. (1998) Approach to Zero Emissions in Squid Processing. The 3rd Zero Emissions Network UNU Conference, Tokyo.
5. Fuwa, F., Ishizaki, M., Tanaka, A, Ebata, K, Miura, T. and Abe, T. (2000) The efficiency of the processed bait for long line made by the leftovers of Squid. Nippon Suisan Gakkaishi, 66:888-889.
6. Miura, T., Watanabe, K., Shiode, D., Fujimori, Y. and Shimizu, S. (2000) Squid material flow, zeroemission technology and the nitrogen balance in Japan. Kankyo Kagakukaishi, 13:612-617.
7. Miura, T., Takahashi, K. and Kanehiro, H. (2001) Present state and task ahead of zero emission on fisheries. Nippon Suisan Gakkaishi, 67:308-318.

UTILIZATION OF THE RESOURCES OF LANTERN FISH AS FISHERIES PRODUCTS

Satoshi F. Noguchi
Central Research Institute, MARUHA Corporation, 16-2, Wadai, Tsukuba, Ibaraki 300-4295, Japan

Abstract

Present report is concerned with the utilization of resources of Myctophiformes or lantern fish as fisheries products. Myctophiformes are a typical underutilized deep-sea fish, and are estimated to have rich resources. The present studies research on the fishing area, fishing net for deep-sea fish, chemical analysis of the components of each fish species, and developments of various kinds of fisheries products.

Ten species of lantern fish caught in the offshore of Sanriku located in Pacific Ocean near northern part of Japan were analyzed. It is special characteristic that some species of Myctophiformes hold a wax. The overall utilization of whole body of fish without waste is the concept of our present research work. We tackled with the development of surrying of the whole body to meet the above concept. The manufacturing of seasoning, feed for cultured fish and cosmetics were investigated using Myctophiformes in the commercial or test plants.

The largest fraction is investigated as a seasoning, and is evaluated good and similar to Japanese natural seasoning. The light fraction by centrifuge is consists of wax, and these qualities of refined fractions are excellent as a wax material and equal the quality of the wax of Orange Roughy that is used as a raw material of cosmetics.

1. INTRODUCTION

Developments of underutilized fish resources have resulted in new fisheries products for a long time. These developments have been possible, because of the recent remarkable progress of fishing gear, confirmation of the school of fish and navigation techniques. The present report is concerned with the utilization of resources of Myctophiformes or lantern fish as fisheries products. Lantern fish is a typical underutilized deep-sea fish and is at a relatively low stage in a food chain.

I have hopes that the active feed supplies by fisheries products should be investigate in the present time and on the base of idea of food chain. However, research on underutilized fish species has been limited due to two major problems: lack of funding for research and lack of data on the underutilized fish species.

For example, study was carried out from 1994 to 1997 by unique Japanese organization named Marino-forum 21 (MF-21). The research funds of MF-21 are provided by Japanese government and by some private firms such as Japan Marine Fishery Resources Research Center, Cooperative Society of Utilization of Big-catch Fishes in Choshi, Kyowa Technos Inc., Nippon Suisan Kaisha and our MARUHA Corporation.

Second, specific data lacking for lantern fish [1,2], such as the stock of resources, chemical components and biochemical information of these species for the development of processing. Consequently, the present study researches fishing area, fishing net for deep-sea fish, chemical analysis of components of each fish species, and the development of various kinds of fisheries products.

2. DEVELOPMENT OF METHODS FOR LANTERN FISH CATCH

Lantern fish inhabit archibenthic waters and cover a wide area of the world. The rich resources of these fish species are estimated to be like the pelagic fishes, squids and krill. However, the information about the formation of fish schools with the spawning, and the diurnal variation of vertical distribution of these species was lacking. Moreover, since the almost present Japanese trawlers are using the bottom trawling net, the information about the midwater trawl, which is the indispensable technique for the fishing of lantern fish was lacking. New midwater trawling net for the deep-sea fish were designed and manufactured for the present experimental trawling.

In the present study, the offshore of Sanriku located in Pacific Ocean near northern part of Japan was selected as the present experimental fishing area by the investigation of oceanography. Figure 1 shows the present experimental fishing area. The experimental trawling in summer time was repeated in both night and day times at each research point shown the figure. The depths of fishing water were set up in seven layers from 0 to 700m to know the diurnal variation of vertical distribution of these species. As a result, it was known that the school of *Stenobrachius leucopsarus* is distributed in the depths of 500–600m in the day time and in the depths of 20–100m in the night time. In the case of *Stenoburachius nannochir* and *Lampanyctus regalis*, these schools are distributed in the depths of 500–700m in the both of day and night times. Incidentally, the above-mentioned three species of lantern fish are the typical wax-rich species as shown later. In the final year of the present project, the experimental towing was repeated during day time in the depths of 500–700m based on these data.

The selection of incidental catches and sorting of each species of lantern fish were complicated and difficult. In the present experiment, the selection and sorting were carried out with research workers hands on the deck of boat. Since the works of selection and sorting of each species of lantern fish required much time and effort, the mechanization of these processes should be required for the practical processing plant.

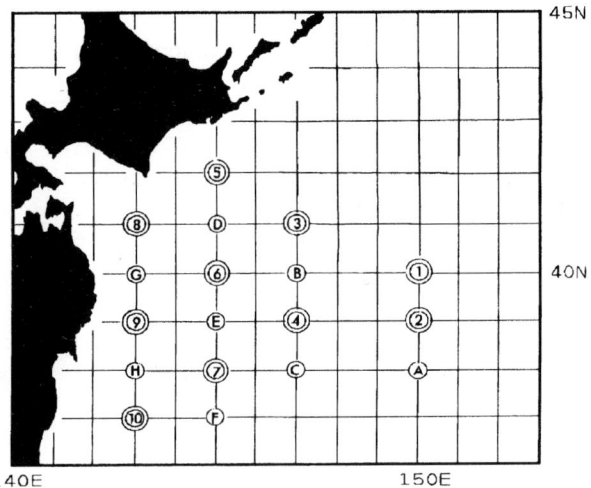

Figure 1. The present experimental fishing area and research points.◎, Towing point at every layer; ○, Towing point by slanting towing.

3. MATERIALS

Ten species of lantern fish were caught in summer season, from July to August, and in the present experimental fishing area. Caught lantern fish were selected according to the fish species by hand and were frozen on the boat. Frozen samples were transported to our Tsukuba laboratory, and chemical components of each species were analyzed. Namely, ten fish of each species were homogenized, and water, protein, ash and lipid were analyzed in accordance with the standard methods. The concentration of wax in the lipid fraction was analyzed using IATROSCAN TH-10 analyzer. Table 1 shows the chemical components of each species.

It is special characteristic that some species of lantern fish hold a wax. Caught lantern fishes were divided into three groups based on the concentrations of lipid, protein and wax of each species. The classified three groups are shown as follows:

Group A (low lipid and low wax)
- *Diaphus gigas* (suitouhadaka)
- *Diaphus theta* (todohadaka)
- *Lampanyctus jordani* (mamehadaka)

Group B (high lipid and low wax)
- *Ceratoscoplus warmingii* (gokouhadaka)
- *Lampadena luminosa* (kagamihadaka)
- *Notoscopelus japonius* (ookuchiiwashi)
- *Symbolophrus californiensis* (nagahadaka)

Group C (high lipid and high wax)
- *Stenobrachius leucopsarus* (kohirehadaka)
- *Lampanyctus regalis* (mikadohadaka)
- *Stenobrachius nannochir* (sekkihadaka)

Table 1
The chemical components of the caught species of lantern fish in the present experimental fishing area

Name of fish species	Water (%)	Ash (%)	Protein (%)	Lipid (%)	Wax (%)*
Diaphus gigas (suitouhadaka)	73.2	2.9	15.7	7.3	undetected
Diaphus theta (todohadaka)	79.2	3.1	13.3	4.9	0.4
Lampanyctus jordani (mamehadaka)	75.2	2.4	13.4	9.3	traces
Ceratoscoplus warmingii (gokouhadaka)	64.9	2.3	11.5	20.6	2.4
Lampadena luminosa (kagamihadaka)	66.6	1.8	11.3	19.8	traces
Notoscopelus japonius (ookuchiiwashi)	58.4	1.9	13.5	26.2	undetected
Symbolophru californiensis (nagahadaka)	57.5	2.0	12.3	28.5	traces
Stenobrachius leucopsarus (kohirehadaka)	65.9	2.0	12.8	19.5	92.1
Lampanyctus regalis (mikadohadaka)	73.1	1.9	12.1	13.4	82.6
Stenobrachius nannochir (sekkihadaka)	67.2	1.9	12.5	18.0	94.6

Wax (%)*, Wax/Lipid×100; Fish names in the bracket show the standard Japanese name.

In the present study, the manufacturing of fisheries products were investigated based on the above-mentioned biochemical characters of each species, especially lipid and wax contents. The investigated utilization or manufacturing techniques of lantern fishes in the present Marino-forum 21 project were shown as follows:
① Feed production for the cultured fish used with the whole body of the fish.
② Surimi production used with the fish meat of the fish.
③ Cosmetics and lubricating oil refining used with the fat tissue of the fish.
④ Production of the seasoning, feed, natural apatite and cosmetics used with the whole body of the fish for the overall utilization.

The present study is concerned with the overall utilization of the whole body of lantern fish as show in ④ which were carried out by the members of our MARUHA Central Research Institute.

4. METHODS

4.1. Slurrying of the whole fish body

The overall utilization of whole body of fish without waste is the concept of our present research work, and it is the special characteristic that some species of Myctophiformes or lantern fish hold a wax. First of all, we tackled with the development

Figure 2. Figure of *Stenobrachius nannochir* (sekkihadaka).

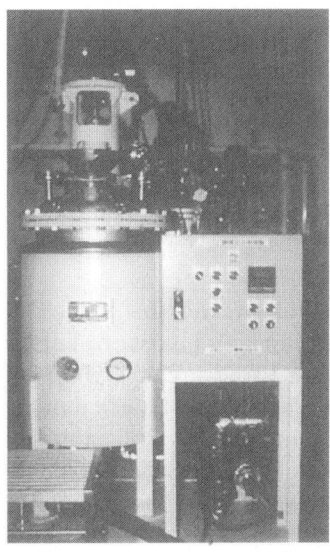

Figure 3. The digestion tank that was designed and manufactured for the slurrying of whole body of lantern fish.

of surrying of the whole body of the lantern fish that have high lipid and high wax content to meet the above concept. The fish species used for the present experiments were *S. leucopsarus* and *S. nannochir* which have the high lipid and high wax content. The body lengths of these fishes were about 10−15cm, and weights of these fishes were about 7−13g. Figure 2 shows the figure of *S. nannochir*.

The several kinds of protease on the market were investigated with the small-scale experimental fermenter. Alkalase 2.4L (Novo Nordisk) was selected as the suitable protease for slurrying these fishes. Figure 3 shows the digestion tank that was designed and manufactured for the slurrying of whole body of lantern fish. Especially the structural considerations of the tank were needed for the smooth discharge of bone and scales.

The whole body (100 kg) of lantern fishes that have the high lipid and high wax content and 27 ml of Alkalase 2.4L were put into the digestion tank, and the temperature

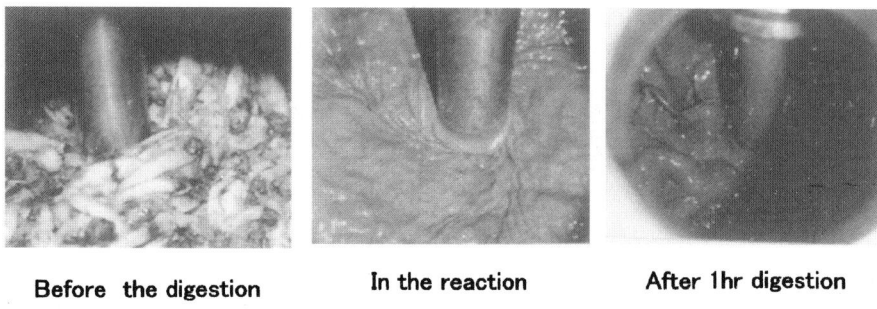

Before the digestion   In the reaction   After 1hr digestion

Figure 4. Changes in the appearance of body of *S. nannochir* during the digestion.

of the mixture were kept at 50−55°C for 1 hour. After 1 hour of digestion, the temperature was raised up to 80°C and kept for 10 min for the deactivation of enzymes. Figure 4 shows the changes in appearance of the body of *S. nannochir* during the digestion in the digestion tank. The completed features of lantern fishes are liquid and smooth as shown in Fig.4.

4.2. Fractionation of the slurry of lantern fish.

The completed slurry of lantern fishes were separated into the five kinds of intermediate materials using the three-layer (450, 104 and 75μm) vibrating mesh separator (Kyowa, KF-400-2W) and three-layer centrifuge (Alfa-Laval, WSPX303). Figure 5 shows the manufacturing process of each intermediate raw material and chemical components in the case of *S. nannochir*. Fractions of paste and bone that were fractionated with mesh separator were used for the intermediates of feed or chemicals,

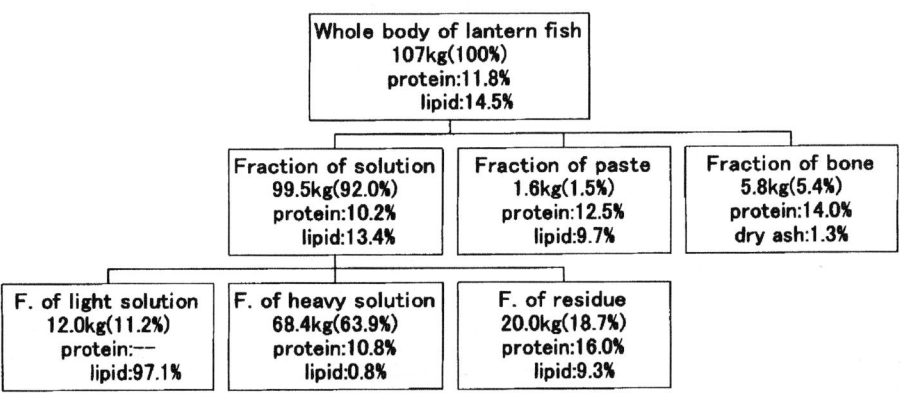

Figure 5. Fractionation of lantern fish (*S. nannochir*) for the intermediate raw materials. The chemical components of each materials and the yield rate of each process were shown in the boxes.

Table 2
The yields of refined wax and the acid values of waxes with the present refining plant

Name of fish species	Crude wax		Refiened wax		Collection
	Weight (kg)	Acid value	Weight (kg)	Acid value	(%)
Stenobrachius nannochir (sekkihadaka)	11.1	2.2	7.6	0.1	68.5
Stenobrachius nannochir (sekkihadaka)	10.5	2.2	7.0	0.1	66.7
Stenobrachius leucopsarus (kohirehadaka)	13.6	2.6	9.0	0.1	66.2

and the fraction of solution was transported to the centrifuge for following fractionations. Fractions of heavy solution and fractions of residue with centrifugal separator were used for the intermediates of seasoning or feed, and the fraction of light solution was transported to the refining plant of oil.

4.3. Refining of the fraction of light solution (a crude wax)

Since high lipid and high wax species of lantern fish (*S. leucopsarus* and *S. nannochir*) were used as a raw fish material in the present experiments, the fraction of light solution with the centrifugal separation is a crude wax.

About 10 kg of the fraction of light solution, or crude wax, was refined with the refining plant of oil that was designed and manufactured for the present experiment. The procedures of degumming, acid treatment, decolorizing and deodorization were according to the rule in general. However, the set up temperature of degumming, acid treatment and decolorizing were kept at 60°C, and that of deodorization was kept at 190°C to avoid the chemical changes. Table 2 shows the data of the yield rate and acid value during the process. The mean value of yield rate was 67%, and is low as compared with the value of the commercial edible oil processing. The present low value is assumed to caused by the presence of intermediate materials residues in the laying pipes and the filter press. However, the commercial scale continuous refining plant should not have this problem and therefore be able to improve the yield rate.

5. RESULTS

5.1. Evaluation of the fraction of heavy solution with centrifugal separator.

The fraction of heavy solution is the largest intermediate materials (about 64%) in the present overall utilization process of whole body of lantern fish. Since the fraction of heavy solution is digested fish proteins, application to natural seasoning was investigated. In the present general processing, the mixture of whole body of lantern fish and protease was kept at 50–55°C for 1 hour. Table 3 shows the free amino acids of the fraction of heavy solution of *S. nannochir* and the commercial extract of saury pike (sanma). The concentration of free amino acids related with good taste such as glutamic acid and alanine were high.

Table 3
The results of organoleptic test on the fraction of heavy solution of S. nannochir (sekkihadaka) as a natural seasoning

	Umami taste 1hr digestion	Umami taste 2hrs digestion	Bitter taste 1hr digestion	Bitter taste 2hrs digestion	Comments on the character
A	++	+++	++	+	squid or shellfish like taste, bitter taste noticed
B	+	++	+++	0	refined and Japanese seasoning like taste, bitter taste noticed
C	++	++	+	−	chicken broth like taste
D	−	−	−	−	simple taste
E	++	++	+	+	refined good taste, but too simple

Organoleptic evaluation, Favor ←+++,++,+,−,→ Disfavor

Table 4
Free amino acids compositions of the fraction of heavy solution of S. nannochir

Amino acids (parameter)	Fish species S.nannochir	S. leucopsarus	Colorabis saira
Total free amino acids/Total amino acids	11.0%	13.9%	13.9%
Glutamic acid/all free amino acids	7.1%	5.0%	3.6%
Glycine/all free amino acids	1.6%	1.7%	2.4%
Alanine/all free amino acids	8.0%	7.6%	2.4%

In the present organoleptic test, the commercial values of the fraction of heavy solution as natural seasoning were evaluated with five panels of trained experts who work in the industry of seasoning. The concentration of tested sample solutions was fixed at Brix 30 with pure water. Two kinds of digested fish solutions in different digestion time (at 50−55°C for 1 hour, 2 hours) were supplied for the present organoleptic test. Table 4 shows the results of organoleptic test on the fraction of heavy solution of S. nannochir.

Sensory testing showed that the character of the heavy solution of S. nannochir is good, and is similar to the Japanese traditional natural seasoning, but it is weak, bitter taste. Differences of flavor and taste were not remarkable between two kinds of digested fish solutions in different digestion time.

Results of amino acids analysis and organoleptic test suggested that there is every possibility of producing the natural seasoning with the fraction of heavy solution of lantern fish.

5.2. Evaluation of the residual fraction with centrifugal separator

The residual fraction with centrifugal separator accounted for about 20 percent of the digested whole body of lantern fish. Since the fraction contains the digestible protein, the application to the feed for animals was investigated. Table 5 shows the concentration of amino acids of the residual fraction. The data of the commercial fish Soluble are shown in the same table to make comparative study. The balance of amino acids of the tested fraction is better than that of the commercial fish Soluble. However, since the

Table 5
Concentration of free amino acids for feed analysis of the fraction of residue with centrifugal separator of *S. nannochir* and the commercial fish soluble

Amino acids	Fish soluble*	Fraction of residue S. nannochir
Threonine	0.77% (w/w)	0.45% (w/w)
Serine	1.30	0.44
Glycine	4.36	0.56
Valine	1.71	0.38
Methionine	0.40	0.29
Cystine	0.14	0.11
Isoleucine	0.47	0.30
Leucine	1.12	0.81
Tryptophan	0.13	0.66
Tyrosine	0.29	0.33
Phenylalanine	0.59	0.42
Lysine	1.12	0.88
Histidine	0.54	0.20
Arginine	1.65	0.52
Water content	51.0%	73.8%
Lipid	5.2	13.5
Ash	7.0	1.9
Total Nitorgen	5.89	1.58

*Standard tables of feed composition in Japan (1995)

concentration of water and lipid of the tested fraction are higher than that of the commercial fish Soluble, the hygienic investigations at room temperature are needed, such as the study on the countermeasure for the transportation at normal temperature.

5.3. Evaluation of the fraction of light solution with a centrifugal separator

The fraction of light solution with centrifugal separator (crude wax) accounted for about 11 percent of the digested whole body of lantern fish. The application of the wax of lantern fish was investigated with the refined wax that was shown in section 3-3. Since wax consists of alcohol and fatty acid, the construction or molecular composition of the wax of lantern fish was analyzed with gas chromatography. Table 6 shows the molecular composition of the refined wax of lantern fish (*S. leucopsarus* and *S. nannochir*). Since the character of wax is a very important index for evaluating the raw material of chemical products, the special analyses of oil chemistry on the present materials were carried with the standard methods for fats, oil and related materials (Japan Oil Chemists' Society) [3], and results are shown in Table 7.

The concentrations of wax in the present refined samples are very high (*S. nannochir* (sekkihadaka): 97.2%, and *S. leucopsarus* (kohirehadaka): 94.6%). The parameters of color quality are good (light or faint yellow color) and the smells are faint. These qualities are excellent as a wax material, and equal the quality of the wax of Orange Roughy (*Hoplostethous atlanticus*) that is in use as a raw material of cosmetics by Kyowa Technos Inc. that is a Japanese specialty company of wax oil. Finally, the present refined

Table 6
Molecular composition of the refined wax of lantern fishes (*S. leucopsarus* and *S. nannochir*)

Composition (Alcohol)			Composition (Fatty acids)		
Molecular structure	S. leucopsarus	S.nannochir	Molecular structure	S. leucopsarus	S.nannochir
14 : 0	9.6%	12.8%	14 : 0	0.9%	2.8%
15 : 1	0.7	0.7			
16 : 0	54.8	29.9	16 : 0	3.0	2.1
16 : 1	3.3	2.6	16 : In-7	12.6	14.8
17 : 0	0.5	0.4	17 : 0	2.9	1.1
17 : 1	0.3	0.2			
18 : 0	2.8	1.4	18 : In-9+n-7	44.6	28.6
18 : 1	8.2	3.4	18 : 2n-6	2.1	1.9
			18 : 3n-3	1.0	0.8
			18 : 4n-3	1.7	1.3
20 : 1	5.5	15.8	20 : In-9+n-7	9.8	19.9
			20 : 2n-6	0.2	0.2
			20 : 4n-6	0.3	0.3
			20 : 5n-3	4.7	2.5
22 : 1	11.5	28.7	22 : In-11+n-9+n+7	6.0	14.2
			20 : 6n-3	3.3	2.0
24 : 1	0.3	0.8	24 : In-91		0.7
Others	2.5	3.3	Others	6.9	6.8

wax was refined again with the hydrogenation and the bleaching by adsorption with activated clay, and the character of high degree refined wax is shown in Table 7. The color of the present high degree refined wax is transparent and colorless, and equal the quality of the commercial purified wax of Orange Roughy that is in use as a ingredient of cosmetic oil by Iwase-Cosuha Inc. that is a Japanese company of cosmetics.

6. DISCUSSION

The completed surry of lantern fishes were separated into the five kinds of intermediate materials, and the potential of seasoning, feed and cosmetic oil of these intermediate materials were investigated and explained in detail in the above sections. Moreover, the fraction of bone and scale as the remaining materials were investigated on the raw materials of calcium phosphate and natural apatite. Since the proteins in the fraction of bone and scale are digested during the surrying process, the purity of calcium is high. The quality of the present intermediate materials of bone and scale is good as raw materials of calcium phosphate and natural apatite.

Other members in the present organization (MF-21) developed other fishery products, and the outlines of each development were shown as follows.

Cooperative Society of Utilization of Big-catch Fishers in Choshi developed the potential of surimi production with the fish meat of lantern fishes. Since the bodies of

Table 7
Character of refined wax samples of lantern fishes (*S. nannochir* and *S. leucopsasu*)

Parameters and composition*	Refined Wax S. leucopsarus	Refined Wax S.nannochir	High degree refined Wax
Acid value	0.1% (w/w)	0.1% (w/w)	0.1% (w/w)
Peroxiside value (meq/kg)	3.6	2.7	0.9
Saponification value	112.9	106.6	76.6
Iodine value	97.3	93.3	101.8
Cloud point	8.0	9.5	16.0
Specific gravity (d^{40})	0.8580	0.8560	0.8510
Refractive index	1.4622	1.4618	1.4618
Color (Lovibond)	Y:20, R:10	Y:8, R:7	Y:0.2, R:0.1
Color (Gardner)	3	1	—
Color (APHA)	—	—	5~10
Wax ester	94.6%	97.2%	—
Triacylglycerol	5.4	2.8	—
Phosphatides	tr.	tr.	—
Free cholesterol	tr.	tr.	—
Hydrocarbon	n.d.	n.d.	—
Free fatty acid	n.d.	n.d.	—
Unsaponifiable matter (%)	44.2%	52.9%	—
Mixed free fatty acids (%)	50.2%	42.4%	—

*Standard methods for fats, oil and related materials (Japan Oil Chemists' Society)

lantern fishes are very small and the used raw fishes were frozen, lantern fish surimi is not as excellent as Alaska pollock surimi. However, since the lantern fishes meat is white and has not any disagreeable flavor, there is possibility of a born mixed surimi products. For example, in Japan, the born mixed surimi products named Jakoten and Honeku that are made with small fishes like *Malakichthys wakiyai*.

The potential of the feed production for the cultured fish used with the whole body of the lantern fish were developed by Nippon Suisan Kaisha Ltd. The juvenile fish of flounder (*Paralichthys olivaceus*) had been reared for two months with the expansion pellet feed samples containing *Diaphus theta* (todohadaka) and *Stenobrachius nannochir* (sekkihadaka). Though the used lantern fish pellet feed samples hold 2% wax, the quality of these feed samples were evaluated as equal to the quality of the commercial feed pellet.

Kyowa Technos Inc. investigated the potential of the chemical oil products. The quality of the present high degree refined wax is evaluated as equal in quality to commercial purified wax of Orange Roughy. These are explained in detail in the above section. Moreover sulfarized lantern fish wax was synthesized and the potential of the lubricating and cutting oils were examined. The quality of the present sulfarizend lantern fish wax was evaluated as equal in quality to commercial lubricating and cutting oils. Alkyl vinyl ether of lantern fish wax alcohol was synthesized and the potential of the plasticizer were examined. The quality of the present alkyl vinyl ether was evaluated as equal in quality to commercial plasticizer.

During the present study, some scientists in the universities and the national laboratory studied on the basic research of Myctophiformes as cooperative scientists, and reported their results in the meeting of our meeting. Namely Dr. Koichi Kawaguchi in Tokyo University reported on the resources of Myctophiformes, and Dr. Kenichiro Fujimoto in Tohoku University and Dr. Hiroaki Saito in the National Research Institute of Fisheries Science reported on the oil chemistry of Myctophiformes [4-6].

In the conclusion of the present research works, we think that the potential of lantern fish as fishery products are high in general. However, we have a little information for the commercial operation of lantern fish at the same time. I think that reliable and abundant data is indispensable for the sustainable development of the natural resources, and I hope that many fishery scientists will continue research works on the accumulation of the related data with lantern fish.

REFERENCES:

1. Haque, A., Pettersen, H.A, Larsen, J.T. and Opstvedt, J. (1981) Fish meal and oil from lantern fish. J. Sci. Food. Agric., 32: 61-70.
2. Neighbors, M. A. and Nafpaktitis, B. G. (1982) Lipid compositions, water contents, swimblader morphologies and buoyancies of nineteen species of midwater fishes (18 Myctophids and 1 Neoscopelid). Mar. Biol., 66:207-215.
3. Nippon Yukagaku Kyoukai (1990) Yushikagakubinnran, Maruzen, Tokyo.
4. Seo, H. S., Endo, Y., Fujimoto, K., Watanabe, H., Moku, M. and Kawaguchi, K. (1996) Characterization of lipids in Myctophid fish in the subarctic and tropical ocean. Fisheries Sci., 62: 447-453.
5) Seo, H. S., Endo, Y., Fujimoto, K., Watanabe, H., Moku, M. and Kawaguchi, K. (1998) Denaturation of myofibrillar characterization of lipids in Myctophid fish in the subarctic and tropical ocean. Fisheries Sci., 64: 423-427.
6) Seo, H. S., Endo, Y., Fujimoto, K., Watanabe, H., Moku, M. and Kawaguchi, K. (1998) Amino acid composition of proteins in Myctophid fishes in the subarctic and tropical ocean. Fisheries Sci., 64: 652-653.

PACU (*PIARACTUS MESOPOTAMICUS*) A NEW FISH SPECIES IN ISRAELI AQUACULTURE: POSSIBILITY OF UTILIZATION

A. Gelman[a], V. Drabkin[a], O. Sachs[b], K. Chechic[a], I. Gabay[a] and L. Glatman[a]
[a]Kimron Veterinary Institute, P.O.Box 12, Bet Dagan 50250, Israel
[b]Dor Aquaculture Experimental Station, Dor, Israel

Abstract

Pacu (*Piaractus mesopotamicus*) is an exotic species, endemic to South America, and newly introduced into Israeli aquaculture. The high potential of this species for aquaculture arises from its rapid growth under a variety of conditions. However, some problems have arisen in industrial processing and marketing of the fish, because its high fat content makes its taste unacceptable to many people. The purpose of this study was to develop a method of cold smoking pacu that would enhance its market appeal by modifying its properties to meet the demands of the modern food market better. Fish were first eviscerated and salted by soaking in brine, because of the high fat content. Fish weighing less than 2,100 g were immersed whole; those weighing more than 2,100 g were cut longitudinally into two symmetrical halves. Fish halves were smoked for 8 h, and whole fish for 10 h. Samples of fish were taken for organoleptic, chemical and bacteriological analyses. Both types of cold-smoked pacu a pleasant, delicious smell and taste, and that their fat content was much less perceptible acquired; however, the halved fish possessed the better sensory properties. All samples of smoked pacu showed high stability during 3 months of cold storage.

1. INTRODUCTION

Worldwide, fish breeding in natural habitats is already limited by various factors, and this has led to an intensified development of aquaculture in fresh, brackish and salt water. Aquaculture in Israel is very dynamic and is characterized by the introduction of exotic new species. Pacu (*Piaractus mesopotamicus*), endemic to South America, is one of these species. The water temperature suitable for pacu breeding varies from 15 to 35°C with an optimum range of 23-28°C, but the fish have survived down to 7°C [1,2]. This tropical freshwater fish shows good tolerance of changes in the oxygen content and pH of the water. The high potential of this species for aquaculture arises from its rapid growth under a variety of conditions (2.2-2.6 times faster than carp) and the possible use of cheap dried feeds based on fruit wastes and other vegetable sources. Pacu reaches sexual maturity at

3-4 years old, when its body weight has usually reached 5 kg. Their breeding is profitable and should provide a cheap fish product for human consumption. However, some problems have arisen in processing and marketing the fish [3,4], because its high fat content make its taste unacceptable to many people.

The purpose of the present study was to develop a method of cold-smoking pacu, that would enhance its market appeal by modifying the properties of the fish, to the better meet the demands of the modern food market. The study also aimed to evaluate the quality characteristics of the resulting product.

2. MATERIALS AND METHODS

2.1. Fish

The pacu fish used in this investigation were descendants of fish that were supplied to Israel in 1996. Fish were harvested from the pond in the Dor Aquaculture Experimental Station at a temperature of 20°C, and were immediately placed in sealed plastic bags containing ice; they were transported to the laboratory in sealed Styrofoam boxes with ice beneath and on top of the bag. The time span from the packing fish to their arrival at the laboratory was 3 h. The fish ranged from 840 to 5,190 g in weight. A sample of four fish was taken for organoleptic, bacteriological and chemical analyses immediately after arrival.

2.2. Gutting and brining

Fish were first eviscerated and salted by soaking in brine, because of their high fat content. Two methods of salting and smoking were used: 1. fish weighing less than 2,100 g were immersed whole in 10% salt solution for 5 days; and 2. fish weighing more than 2,100 g were cut longitudinally into two symmetrical halves and were immersed in 3.5% salt solution for 2 days. Weight changes were determined after each operation.

2.3. Smoking

Cold smoking was performed at 25°C with the apparatus "Afos" as follows: fish halves were smoked for 8 h and whole fish for 10 h in smoke of oak sawdust (Gunter Springer Spanholz GmbH). The smoked fish were placed in Styrofoam boxes and cold-stored at 2-3°C. Samples of eviscerated whole and halved fish were taken for organoleptic, chemical and bacteriological analyses after each step of the technological process.

2.4. Organoleptic evaluation

Organoleptic analysis was performed by evaluating four characteristics (odor, color, texture, and taste) according to a 10-point scale. The higher scores denoted better quality. A trained taster panel comprising 32 experienced members performed the tests; they included women and men, 25-60 years old. Each panelist evaluated two samples: fillet of whole fish and fillet of halved fish. The samples of fish were presented individually in covered Petri dishes (diameter 100 mm) at room temperature. The cover was removed and the sample was examined, smelt and tasted. The allocated scores were averaged over all tasters.

2.5. Chemical analyses

Total volatile basic nitrogen (TVB-N) was determined in a Kjeltec System 1026 (Foss Tecator, Sweden). Bases from the weighed sample were distilled in the presence of magnesium oxide. The distillate was collected in a solution of excess boric acid in the presence of indicators (methyl red mixed with bromocresol green) and titrated with 0.01 N sulfuric acid; the TVB-N value was expressed as milligrams of N per 100 g of fish.

Protein, lipid, minerals, water, dry content, and salt were determined according to the Association of Official Analytical Chemists [5].

2.6. Bacteriological analysis

Total aerobic bacterial counts were done both on the surface and in the flesh of the fish. Surface counts were obtained by plating a (5 × 5 cm) sterile template on the fish surface, at the middle region along the lateral line, and swabbing the area enclosed by the template. The swabs were transferred into 10 ml of sterile peptone (0.1% w/v) as diluting medium and shaken well before plating. Prior to taking the flesh samples of, the skin was removed aseptically. Then underlying muscles (10 g) were removed and homogenized in a Stomacher 400 (Model BA6021, Seward Laboratory, UAC House UK) for 1 min in 90 ml of sterile peptone (0.1 % w/v). Tenfold serial dilutions of the same diluting medium were plated on the plate count agar (PCA, Difco), which were then incubated at 30°C for 48 h. Each 20 dominant colonies from the PCA were re-streaked onto nutrient agar and stored for identification. The following tests were used for determination of the purified strains: Gram-staining, KOH-test, production of oxidase and catalase (3% H_2O_2), and growth on appropriate media (Difco) (MacConkey agar, Baird-Parker agar, Pseudomonas agar, and Triple Sugar Iron agar, respectively). Final biochemical identification was carried out with API NE, API 20E and API Staph (BioMerieux Vitek, Inc., France).

3. RESULTS AND DISCUSSION

3.1. Gutting

The yield of fish after gutting was rather high - between 78.2 and 87.2% - and did not depend on the initial fish weights (Table 1). In Israel pacu reach sexual maturity at 2 years of age. The fish weighing less than 2600 g were mostly nonmature. The fish weighing from 2,850 to 3,950 g had undeveloped gonads; the only fish with fully developed gonads weighed 5,190 g, and its gonads weighed 580 g. Over all the fish, the gonads weight varied between 0.9 and 11.2% of the fish weight; that of the intestines between 0.9 and 11.2%; and that of the blood and mucus between 0.6 and 2.3%.

3.2. Yield of final product

The yields of the eviscerated fish after cold smoking and after all the intermediate steps of the process are shown in Table 2. During salting the fish swelled and their weights increased by approximately 2%. Though the fish weights decreased significantly during smoking, the final yield was quite a high percentage of the initial fish weight, ranging from 66.7 to 71.6%. The smoking process lasted 10 h., i.e., it was rather prolonged due to the presence of the scales and the thick skin.

Table 1
Yield of the eviscerated pacu (g)

No	Whole fish	%	Intestine +fat	%	Gonads	%	Slime and blood	%	Eviscerated fish	%
1	2,630	100	370	14.1	-	-	40	1.5	2,220	84.4
2	3,380	100	430	12.7	30	0.9	32	0.9	2,888	85.4
3	172	100	200	11.6	-	-	20	1.2	1,500	87.2
4	5,190	100	430	8.3	580	11.2	120	2.3	4,060	78.2
5	3,950	100	480	12.2	84	2.1	40	1	3,346	84.7
6	2,080	100	340	16.3	-	-	30	1.4	1,710	82.7
7	2,230	100	390	17.5	-	-	40	1.8	1,800	80.7
8	2,850	100	310	10.9	60	2.1	20	0.7	2,460	86.3
9	1,700	100	250	14.7	-	-	10	0.6	1,440	84.7
10	1,500	100	270	18	-	-	30	2	1,200	80
Av.	4,254	100	347	13.6	189	4.1	38	1.2	2,262	83.4
SD	4,617	0	90	3.1	262	4.8	30	0.6	928	2.9

Table 2
Yield of the eviscerated pacu after smoking (g)

No	Whole	%	Eviscerated	%	Salted	%	Smoked	%
1	2,230	100	1,800	80.7	1,870	83.8	1,500	67.3
2	2,080	100	1,740	82.7	1,780	85.1	1,490	71.6
3	1,700	100	1,400	82.3	1,420	83.5	1,160	68.2
4	1,500	100	1,200	80.0	1,220	81.3	1,000	66.7
Av.	1,878	100	1,535	81.7	1,573	83.4	1,288	68.5
SD	336	0	284	1.6	305	1.6	248	2.2

Table 3
Yield of the halved pacu after smoking (g)

No	Whole	%	Eviscerated	%	Salted	%	Smoked	%
1	5,190	100	4,070	78.4	4,220	81.3	3,180	61.3
2	3,950	100	3,430	86.8	3,470	87.8	2,800	70.9
3	2,850	100	2,460	86.3	2,500	87.7	2,060	62.3
4	2,860	100	2,450	85.6	2,480	86.7	1,940	67.8
Av.	3,712	100	3,103	84.3	3,168	85.9	2,495	65.6
SD	1,112	0	792	3.9	840	3.1	594	4.5

Table 3 records the yield after smoking of the halved fish. Swelling of the halved fish was around 1.5%, i.e., slightly less than in whole fish, probably because the latter was in brine longer (5 days). The yield was also high, and ranging from 61.3 to 70.9%, even though the weight loss was likely to have been higher, because of the greater area of evaporation. This was probably because of the shorter time used for smoking the halved fish (8 h). Thus, the final yields of fish smoked in the two ways were quite similar: 68.5±2.2% for whole eviscerated fish and 65.6±4.55% for halved fish.

3.3. Organoleptic evaluation

The organoleptic analysis of the two types of smoked pacu (Fig. 1) resulted in high evaluation scores for both types, for all the parameters used. Those of the halved pacu were higher than those of the whole eviscerated fish, especially for taste (8.7±0.42 and 7.5±0.69, respectively) and odor (8.83±0.46 and 7.91±0.62, respectively) and total evaluation (9.0±0.41 and 7.68±0.51, respectively). Both types of cold-smoked pacu acquired a pleasant goldish color and a pleasant, delicious smell and taste, reminiscent of traditional cold-smoked balyks. Their fat content was much less perceptible and the flesh consistency was homogeneous. All samples of smoked pacu showed high stability and freedom from spoilage during 3 months of storage at 2-3°C, probably because of their low bacterial counts and low water contents, and their evaluation scores hardly changed (around 7.5-8).

3.4. Chemical composition

Figures 2 and 3 record the chemical composition changes in the halved and the whole eviscerated fish during salting and smoking. Since it was the larger fish which had been cut in half the halved fish initially contained a high fat percent (34 vs 23%) and a relatively low percentage of protein (around 12% compared with around 14%) and accordingly, their water content was lower (54 vs 61%). After smoking, the protein content of the halved fish arose relatively and reached approximately 16%, and their fat content decreased slightly, (to around 31%), probably because of its flowing out of the larger area evaporation during the process; the water content decreased to 48%, and the salt content ranged from 2.5 to 3.0% (Fig. 2).

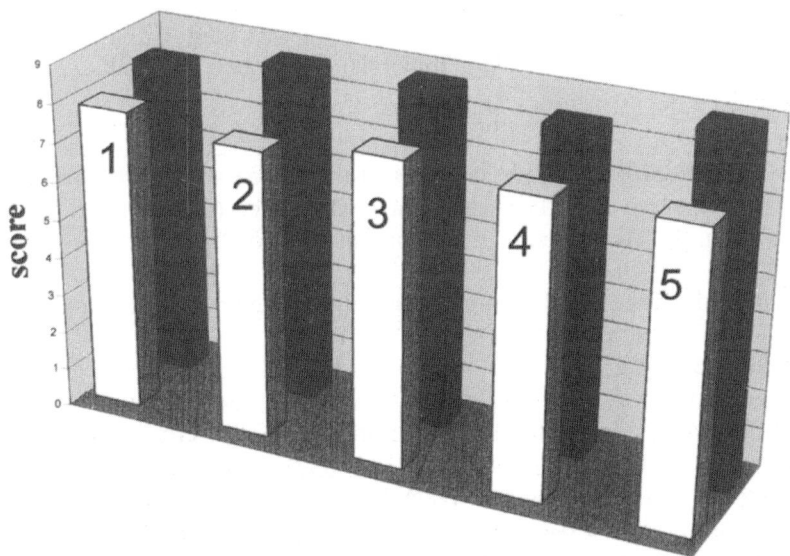

Figure 1. Organoleptic evaluation of halved and eviscerated pacu. ■, halved; □, eviscerated. 1, appearance; 2, taste; 3, odor; 4, texture; 5, total evaluation.

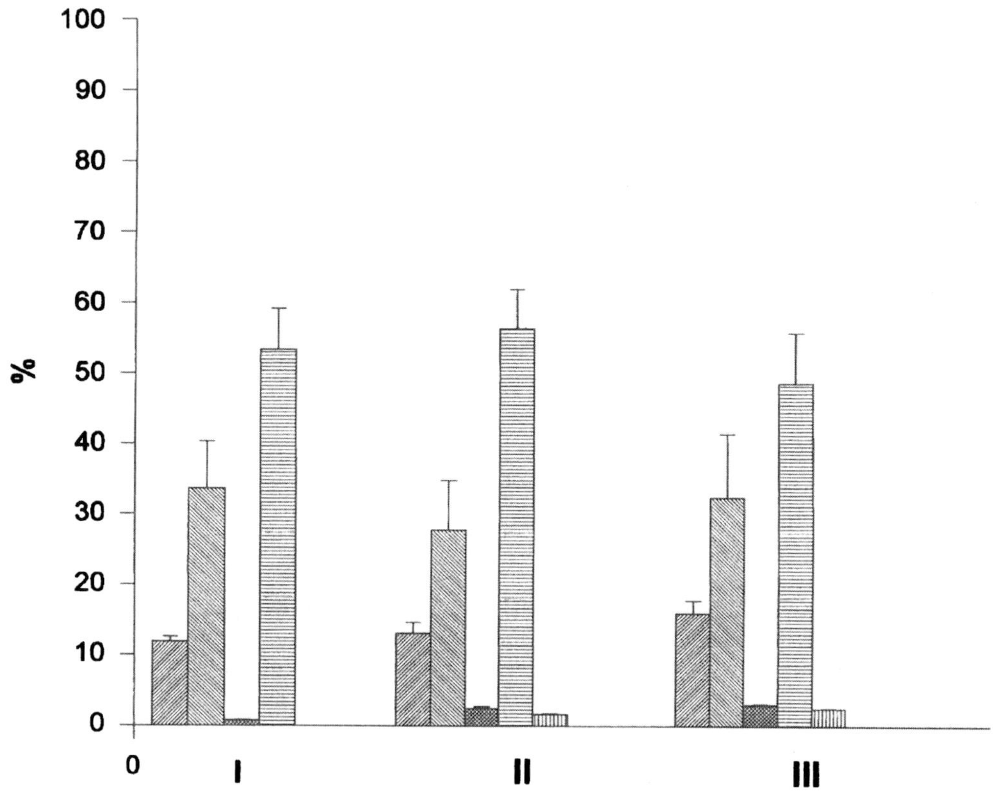

Figure 2. Chemical composition changes in fresh (I), salted (II) and smoked (III) halved pacu. ▨, protein; ▩, fat; ▦, minerals; ≡, water; ▥, salt.

After smoking the protein and fat contents of the eviscerated whole fish arose slightly (to 16 and 28%, respectively), the water content decreased (to 53%) and the salt content was 2.0-2.5% (Fig. 3). Thus, the chemical compositions of the two types of smoked pacu were similar.

TVB-N values did not exceed 20 mg N per 100 g fish in any of the samples tested.

3.5. Bacteriological changes

The change in total aerobic bacterial counts from the fish surface and flesh are presented in Table 4. The initial total counts of the fresh fish were 1.3×10^3 CFU/cm^2 for the fish surface and $<10^2$ CFU/g for the fish flesh. These values are in agreement with the results reported earlier for tropical and freshwater fish [6-9]. Salting of the halved fish for two days led to a drop in the bacterial counts isolated from the fish surface (to 3.8×10^2 CFU/cm^2), whereas salting of the whole eviscerated fish for 5 days resulted in increases in the counts (to 1.0×10^5 CFU/cm^2 for the fish surface and 1.2×10^2 CFU/g for the fish

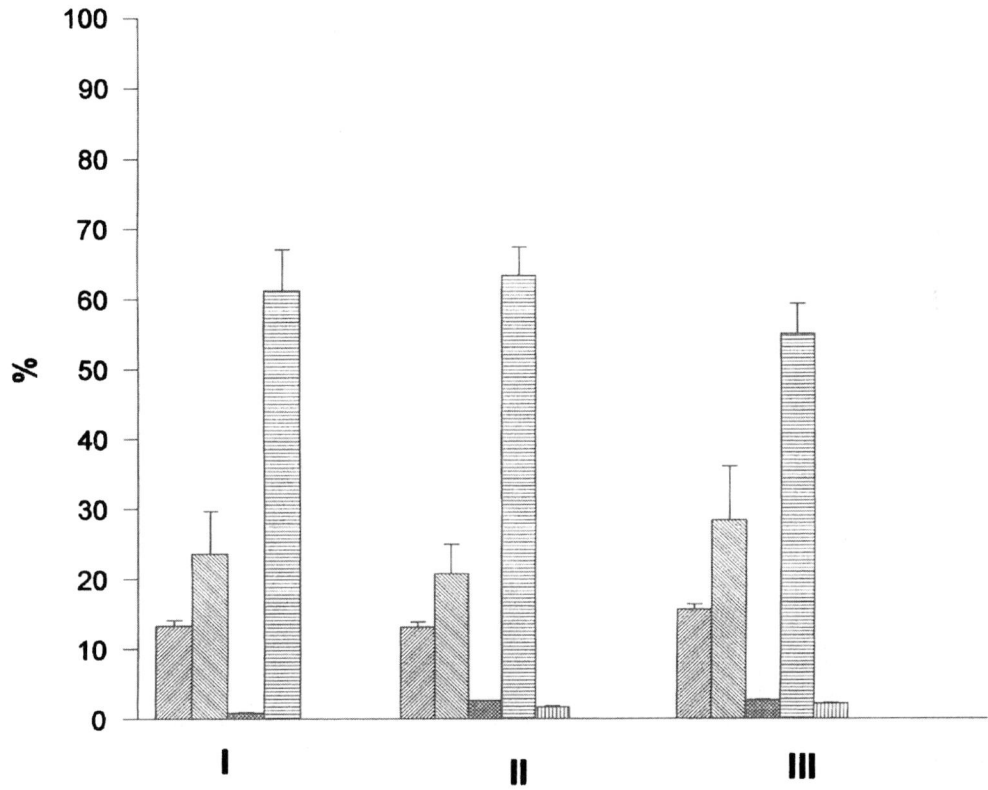

Figure 3. Chemical composition changes in fresh (I) , salted (II) and smoked (III) eviscerated pacu. ▨, protein; ▨, fat; ▨, minerals; ≡, water; ▥, salt.

flesh). After both types of cold-smoking the total bacterial count for the fish surface and for the fish flesh were < 10^2 CFU/ cm^2 and <10^2 CFU/g, respectively.

Composition changes in bacterial flora were also examined. The initial flora on fresh fish surface was predominantly mesophylic and Gram-positive, and consisted mostly (80%) of *Micrococcus*, *Bacillus* and *Corinebacterium* as well as Gram-negative *Moraxella* spp. (20%). Two days of salting of the halved fish changed the bacterial flora composition in the fish surface insignificantly: Gram-positive bacteria remained predominant, but, in addition to an increase in *Moraxella* spp. to 35%, representatives of *Aeromonas hydrophila* occurred sporadically on two of the four samples. After 5 days of salting of the whole eviscerated fish, there were significant changes in the composition of bacterial flora on the fish surface: the proportion of Gram-positive bacteria dropped to around 8%, while that of Gram-negative ones rose to approximately 90%, comprising of *Moraxella* spp., *Aeromonas hydrophila*, *Aeromonas sobria* and *Shewanella putrefaciens*. The data obtained are consistent with the results obtained elsewhere, for other fish species

Table 4
Total bacterial counts of fresh, salted and smoked pacu

Fish	Surface (CFU/cm^2)	Flesh (CFU/g)
Fresh (whole)	$1.3 \times 10^3 \pm 7.5 \times 10^1$	$< 10^2$
Eviscerated and Salted (whole)	$1.0 \times 10^5 \pm 4.10 \times 10^2$	$1.2 \times 10^2 \pm 1.5 \times 10^1$
Salted (halved)	$3.8 \times 10^2 \pm 4.2 \times 10^1$	$< 10^2$
Eviscerated and Smoked (whole)	$< 10^2$	$< 10^2$
Smoked (halved)	$< 10^2$	$< 10^2$

Counts are expressed as mean±SD.

[11-12]. Thus, the prolonged salting time of the whole eviscerated fish resulted in an increase in spoilage bacteria.

4. CONCLUSION

Pacu has a poor market, especially in Israel, because of its high fat content; its flesh is rather tasteless and its fat gives it a waxy consistency. The cold smoking process was chosen in order to improve the appearance, taste and smell of the fish, in order to meet the demands of the modern food market better. Moreover, its rapid growth under a variety of conditions makes its breeding very profitable. The results obtained showed that both types of cold-smoked pacu acquired a pleasant, delicious smell and taste reminiscent of traditional cold-smoked balyks, and their fat content became much less perceptible. However, the halved fish exhibited the better sensory properties and the better bacteriological quality; also the smoking process in this case would be more profitable technologically (less time, materials, etc.). The yield of smoked fish was high. All samples of smoked pacu showed high stability and freedom from spoilage during 3 months of storage at 2-3°C. The 32 trained and experienced panelists of various ages and sexes, who evaluated the organoleptic properties of the products, represented a variety of social and educational groups. Probably cold smoking, which modifies the properties of the fish product, would enhance the market appeal of pacu, not only in Israel but also in other countries.

REFERENCES

1. Milstein, A., Gomelsky, B., Zoran, M., Cherfas, N. and Peretz, Y. (1996) A new candidate for the Israeli aqua culture, *Piaractus mesopotamicus*. A report for the chief scientist of the Ministry of Agriculture, pp.6.
2. Ferraz de Lima, J.A., de Souza, J.H., Bustamante, A. and Gaspar, L.A. (1992) Recomendacoes tecnicas para a criacao de pacu *Piaractus mesopotamicus* em viverios. Boletim Red Acuiculture, 6 : 8-11.

3. Pullela, S.V., Fernandes, C.F., Flick, G.J., Libey, G.S., Smith, S. A. and Coale, C.W. (2000) Quality composition of aquacultured pacu (*Piaractus mesopotamicus*) fillets with the other aquacultured fish fillets using subjective sensorial traits. J. Aquat. Food Prod. Technol., 9: 65-76.
4. Sant'-Ana, L.S. (2000) Influence of the addition of antioxidants in vivo on the fatty acid composition of fish fillets. Food Chem., 68: 175-178.
5. Cunniff, P. (ed.) (1995) Official Methods of Analysis, 16th edition, Association of Official Analytical Chemists (AOAC), Virginia, U.S.A.
6. Estrada, M., Olymoia, M., Mateo, R., Milla, A., dala Cruz and Embuscado, M. (1985) Mesophylic spoilage of whiting (*Sillago maculata*), and tilapia (*Oreochromis niloticus*). In: Spoilage of Tropical Fish and Product Development. Proceedings of a Symposium held in conjunction with the sixth session of the Indo-Pacific Fishery Commission Working Party on Fish Technology and Marketing. Melbourne, Australia, October 1984. Really A. (ed.), FAO Fish. Rep., 317.; 133-145.
7. Gorczyka, E. and Len, L.P.(1985) Mesophylic spoilage of bay trout (*Arripis trutta*), bream (*Acanthropagrus butcheri*) and mullet (*Aldrichetta forsteri*), In: Spoilage of Tropical Fish and Product Development. Really, A. (ed.), Proceedings of a Symposium held in conjunction with the sixth session of the Indo-Pacific Fishery Commission Working Party on Fish Technology and Marketing. Melbourne, Australia, October 1984. FAO Fish. Rep., 317: 123-132.
8. Putro, S., Saleh, M. and Bandol, U.D.C. (1985) Storage life of rabbit fish (*Siganus* sp.) during icing, In: Spoilage of Tropical Fish and Product Development. Really, A. (ed.), Proceedings of a Symposium held in conjunction with the sixth session of the Indo-Pacific Fishery Commission Working Party on Fish Technology and Marketing. Melbourne, Australia, October 1984. FAO Fish. Rep., 317: 54-61.
9. Pullela, S., Fernandes, C.F., Flick, G.J., Libey, G.S., Smith, S.A. and Coale, C.W. (1998) Indicative and pathogenic microbial quality of aquacultured finfish grown in different production systems. J. Food Prot., 61: 205-210.
10. Deng, J., Toledo, R.T. and Lillard, D.A. (1974) Effect of smoking temperature on acceptability and storage stability of smoked spanish mackerel. J. Food Sci., 39: 596-601.
11. Hansen, L.T., Røntved, S.D. and Huss, H.H. (1998) Microbiological quality and shelf life of cold-smoked salmon from three different processing plants. Food Microbiol., 15: 137-150.
12. Leroi, F., Joffraud, J.J., Chevalier, F. and Cardinal, M. (1998) Study of microbial ecology of cold-smoked salmon during storage at 8°C. Int. J. Food Microbiol., 39: 111-121.

BIOACTIVE AND FUNCTIONAL COMPONENTS

MEDICAL APPLICATIONS OF FISHERIES BY-PRODUCTS

Koretaro Takahashi
Division of Marine Biosciences, Graduate School of Fisheries Sciences, Hokkaido University, 3-1-1 Minato, Hakodate, Hokkaido 041-8611, Japan

Abstract

The *sn*-2 position docosahexaenoic acid inserted phospholipids (*sn*-2 DHA-PL) are beginning to receive attentions because of their beneficial functions. There are two alternatives to obtain *sn*-2 DHA-PL. Squid skin, muscle and testis of late run salmon, fish roe are by-products or wastes that can be subjected to extract crude *sn*-2 DHA-PL. High purity *sn*-2 DHA-PL can be obtained through phospholipase A_2 mediated esterification of highly purified DHA into soy lysolecithin. The obtained high purity *sn*-2 DHA-PL showed promotional effects on leukemia cell differentiation when retinoic acid or dibutyryl cyclic adenosine monophosphate was used as a differentiator. When *sn*-2 DHA-PL, a mixture of *sn*-2 DHA-phosphatidylcholine and *sn*-2 DHA-phosphatidylserine were made into liposomes, they showed antitumor activity against Meth-A fibrosarcoma both *in vitro* and *in vivo*. Highly branched glycogen obtained from scallop and a novel peptide glycan from squid ink, also showed antitumor activities against Meth-A fibrosarcoma. Squid pen β-chitin laminated with salmon skin collagen was borne out to be a favorable artificial human skin. DNA from salmon testis could apply to antibacterial film with silver ion. All materials presented here are no doubt highly bioavailable, and should thus become applicable to varieties of medical uses.

1. INTRODUCTION

To decrease by-products and wastes from bioresources, various ways of utilizations from highly value added products to a fairly low value added but with large mass demand must be developed. At present, the only available utilization of by-products and wastes from marine sources in a large scale has been the production of feeds. However, markets for feeds are limited. For this reason, developing a more effective way to utilize fisheries related by-products and wastes are becoming more and more important.

In this article, components contained in the by-products and wastes from salmon, squid and scallop that may be applicable to medical uses are demonstrated, since these three species are consumed in an enormous amount in the world. And vast excess by-products and wastes from these three species have been a burden to seafood processing companies. Equipment investment for reducing by-products and wastes might become possible if we could develop value added products. Medical uses should have the potential to be the most highly value added, and expected to be consumed in large amounts, though it is very hard to realize for the time being.

The purpose of this article is to show the potential benefits of utilizing phospholipids, glycogen, peptide glycan, chitin and DNA contained in the by-products and wastes from salmon, squid and scallop.

2. THE sn-2 DOCOSAHEXAENOIC ACID INSERTED PHOSPHOLIPIDS

Benefits of taking docosahexaenoic acid (DHA) are well recognized. Hara [1] pointed out that incorporated DHA must be converted to phospholipids before exerting its biological functions. In the last decade, it has been shown through many studies, both in vivo and in vivo, that DHA inserted phospholipids, especially the sn-2 DHA inserted phospholipids (sn-2 DHA-PL) have many functional benefits other than the health benefits of DHA itself. For example, the sn-2 DHA-PL is known to enhance survivals of tumor bearing mouse [2] and to promote cell differentiation of erythroleukemia cancer cells [3]. We also demonstrated that sn-2 DHA-PL liposomes with 70% phosphatidylcholine (PC) and 30% phosphatidylserine decreases tumor size of the Meth-A fibrosarcoma bearing mice without any immunomodulator encapsulated (Fig. 1) [4]. In vitro study done at the same time designated that those liposomes decrease the viability of the Meth-A fibrosarcoma (Fig. 2) and increase phagocytic effect of macrophage like J774-1 cells and also slightly increase the number of the phagocytic activated cells (Table 1). Another functional benefit of sn-2 DHA-PL is the promotional effect of cell differentiation. Suzuki [3] and his coworkers showed that sn-2 DHA-PL induces cell differentiation of embryo. We have been bearing out that the sn-2 DHA-PL may be beneficial for therapy of leukemia, because it promotes the effect of HL-60 cell (the cell line of leukemia) differentiators such as all *trans* retinoic acid [5]. Differentiated HL-60 cells become mortal and soon they disappear.

Compared with the sn-2 DHA-PC, sn-2 DHA inserted phosphatidylethanolamine (sn-2

Table 1
Phagocytic effect of macrophage like J774-1 cells induced by sn-2 DHA PC/PS* liposomes

Concentration (μg/mL)	Phagocytic activity			
	(A)	(%)	(B)	(%)
0	47.4	100.0	8.5	100.0
10	64.4	135.9	9.5	112.3
25	82.1	173.2	10.1	119.3

(A), Number of phagocytic cells/ 300cells ; (B), Number of beads/Individual phagocytic cells.
*Same abbreviation as in Fig. 1.

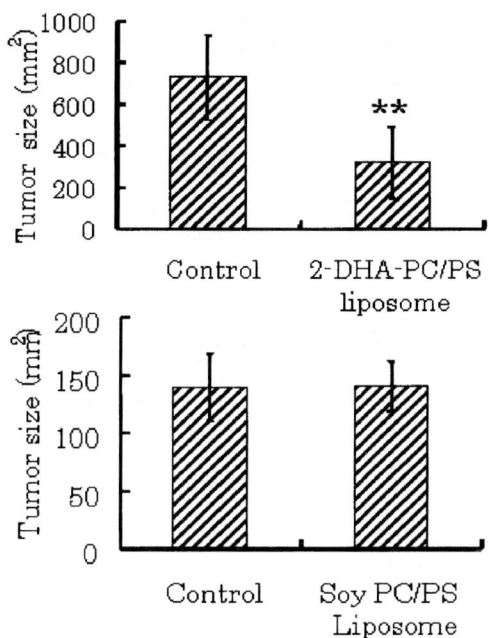

Figure 1. Antitumor effect of sn-2-DHA—PC/PS* liposome on Meth-A fibrosarcoma bearing BALB/c mice. *sn-2 DHA inserted phospholipids with 70% phosphatidylcholine (PC) and 30% phosphatidylserine (PS) in mol%. **$p<0.01$ vs. control. Fujimoto, A. et al. (2001) [4].

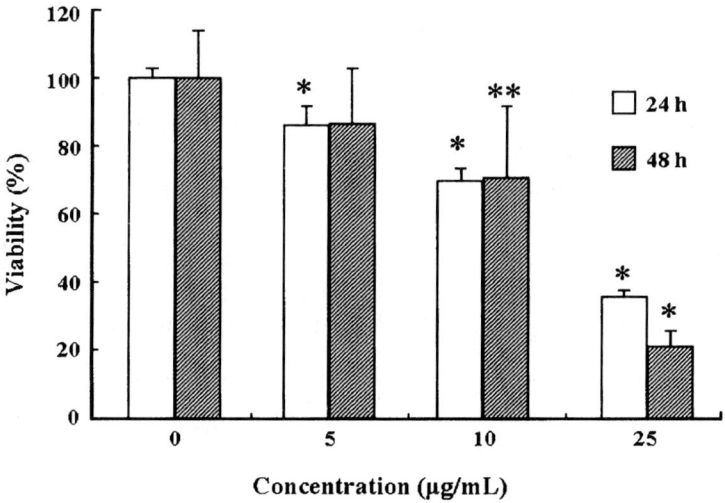

Figure 2. Cytotoxic effect of sn-2 DHA–PC/PS liposomes on Meth-A fibrosarcoma. Data are shown as means ± S.D. (n=6). *$p<0.01$, **$p<0.05$ vs. 0 μg/mL. Fujimoto, A. et al. (2001) [4].

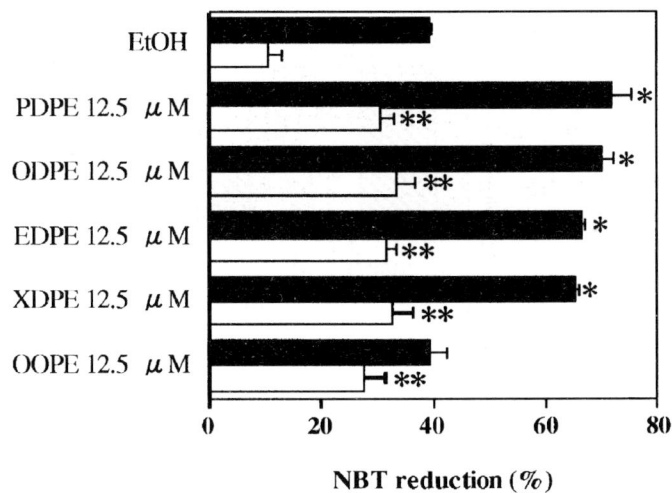

Figure 3. Effect of individual sn-2 DHA-PE molecular species on differentiation of HL-60 cells. Values represent means ± S.D. (n=3). *$p<0.01$ vs. control (retinoic acid (RA) added, dark bar); **$p<0.01$ vs. control (RA free, white bar). Tochizawa, K. et al. (1997) [5].

DHA-PE) was more effective on differentiating HL-60 cells. And as shown in Fig. 3, HL-60 cells coincubated with 12.5μM sn-2 DHA-PE and 100nM retinoic acid doubled the nitro blue tetrazolium reduction-positive cell number (an indicator of HL-60 cell differentiation) compared to those with 100nM retinoic acid alone [5]. We also demonstrated that a mixture of sn-2 DHA-PE and sn-2 EPA-PE obtained from natural sources such as late run chum salmon muscle, testis, and roe have promotional effects on retinoic acid induced (data not shown) or dibutyryl cyclic adenosine monophosphate (Bt^2-cAMP) induced HL-60 cell differentiations (Fig. 4) [6]. Squid skin is known to contain an extremely DHA enriched sn-2 DHA-PL (Table 2) [7]. There is no doubt that this PL can promote retinoic acid or Bt^2-cAMP induced cell differentiation. The reasons why sn-2 DHA-PL can promote cell differentiation are under investigations by the authors and other groups.

Recently, Inoue [8] reported that among the supplemented oils which are squid skin PL, egg yolk PL, and fish triacylglycerol, squid skin PL supplemented diet was the only available lipid to prevent apoplexy of SHR-SP rats as illustrated in Fig. 5. It was suggested that DHA should be bound with glycerophospholipid backbone, but not with glycerol backbone to exert this noticeable functionality.

3. HIGHLY BRANCHED GLYCOGEN

The existence of an antitumor fraction in a hot water extract from scallop was first reported by Sasaki et al. in 1978 [9]. And in 1998, Takaya and his coworkers [10] bore out that highly branched glycogen was the one that cured the Meth-A fibrosarcoma

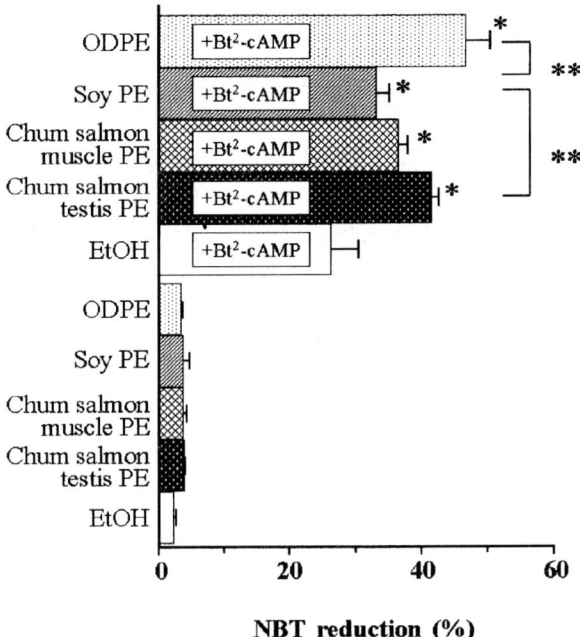

Figure 4. Comparison of various PEs on differentiation of HL-60 cells induced by Bt^2-cAMP. HL-60 cells (5×10^4 cells/mL) were incubated with 100 μM Bt^2-cAMP and/or 50 μMP PEs for 72 h after preincubation for 24 h. Data are shown as means ± S.D. (n=3). *$p<0.01$ vs. Bt^2-cAMP-treated cells; **$p<0.01$ vs. soy PE + Bt^2-cAMP-treated cells. Hosokawa, M. et al. (2001) [6].

implanted mice. Plant for isolating this glycogen has been designed by Aomori Advanced Industrial Technology Center in Japan.

4. A NOVEL ANTITUMOR PEPTIDOGLYCAN

Matsue et al. [11] discovered that squid ink from *Illex argentinus* has an antitumor activity. They found that a novel peptidoglycan with fucose as shown in Fig. 6 was the one that exerts the antitumor activity through modifying the biological response of the phagocytic activity of macrophages [12]. As shown in Figs. 7 and 8, both the unheated and heated novel peptidoglycans show antitumor effect against Meth-A tumor cells [12].

5. SQUID PEN β-CHITIN LAMINATED WITH SALMON SKIN COLLAGEN AS AN ARTIFICIAL HUMAN SKIN

Compared to a well known α-chitin from crab, squid pen β-chitin is much softer even

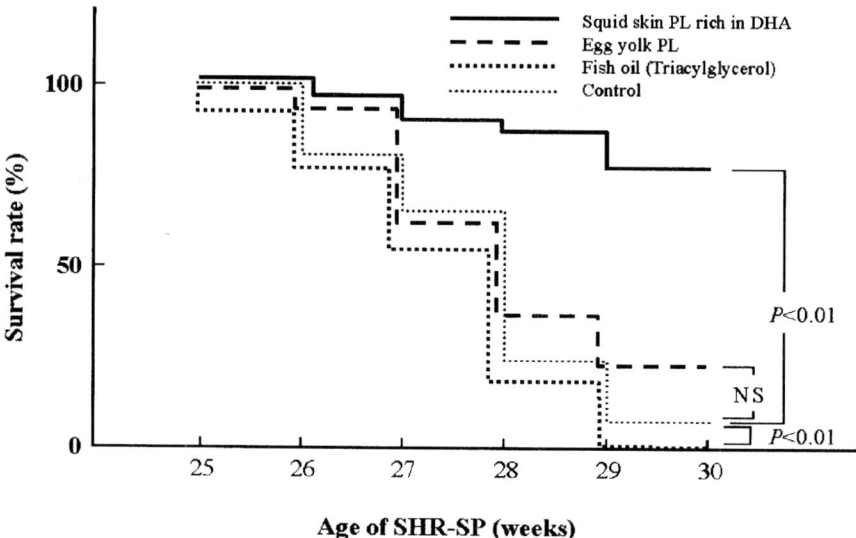

Figure 5. Effect of feeding diet containing various types of lipids on survival of SHR-SP rats. PL, phospholipids. Inoue, Y. (2001) [8].

Table 2
Fatty acid composition of egg yolk phospholipid (PL), DHA-enriched egg yolk PL obtained from fish oil-fed hens, and squid skin PL

	16:0	18:0	18:1	20:4	22:6
Egg yolk PL	31.2	15.3	26.6	4.9	3.5
DHA yolk PL	35.2	10.7	24.5	1.5	10.9
Squid PL	27.0	7.8	2.7	2.7	33.3

Ono, M. et al. (1997) [7]

though strength against tearing is comparable to each other (Table 3) [13]. For this reason, squid pen β-chitin laminated with salmon skin collagen must be a favorable artificial human skin. In fact, after culturing fibroblasts on that laminated sheet, proliferation of those cells immediately occurred and covered the whole sheet within 8 days as shown in Fig. 9 [14]. DNA labeling index which is a marker for the occurrence of cancer was low throughout the incubation period of the fibroblasts proliferation on the salmon skin collagen laminated squid β-chitin sheet (data not shown) [13].

6. SALMON TESTIS DNA-ALGINIC ACID FILM

Iwata and his coworkers [15] invented a DNA-alginic acid film by mixing up DNA sodium salt and sodium alginate solution, then cast in a sheet form, dry and treated with

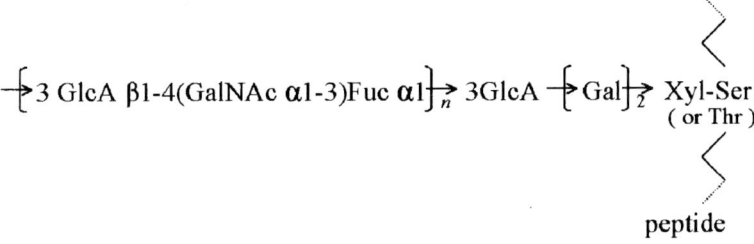

Figure 6. Antitumor peptidoglycan with novel carbohydrate chain obtained from squid (*Illex argentinus*) ink. Fuc, fucose; GlcA, glucuronic acid; GalNAc, N-acetylgalactosamine. Matsue, H. et al. (1997) [11].

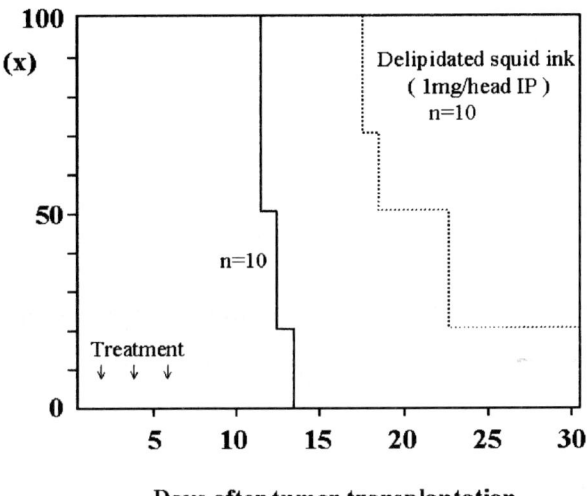

Figure 7. Anti-tumor activity of delipidated squid ink. Meth-A tumor cells (2×10^4) were intraperitoneally (IP) transplanted into mice. Mice were IP treated with 1mg/mL/head of delipidated ink, three times on days 2, 4 and 6 after tumor transplantation. Sasaki, J. et al. (1997) [12].

calcium chloride. The sheet rapidly absorbed ethidium bromide, a well-known carcinogenic compound. Therefore, this filter might be applicable not only to the medical uses, but also to variety of fields.

Kitamura and his coworkers [16] developed a salmon testis DNA-alginic acid film to impregnate silver ion. This was successfully done merely by mixing up silver nitrate before casting. As shown in Table 4, the silver ion impregnation efficiency of the DNA-alginic acid film was about five times superior than the DNA free alginic acid film. And for this reason, this novel silver-DNA-alginic acid sheet may be useful for antibacterial purposes.

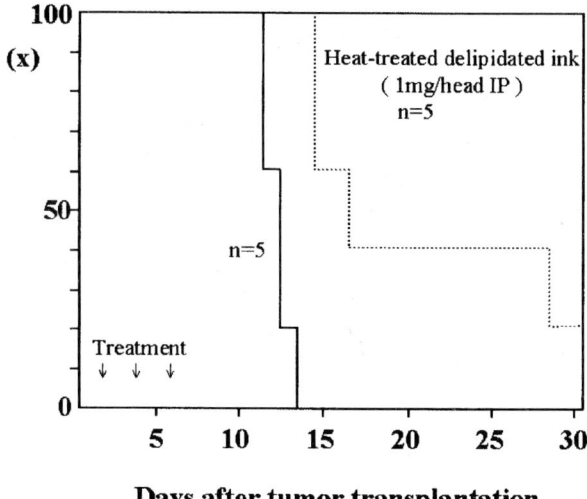

Figure 8. Anti-tumor activity of delipidated squid ink after heat treatment at 100°C for 10min. Meth-A tumor cells (2×10^4) were intraperitoneally (IP) transplanted into mice. Mice were IP treated with 1mg/mL/head of heat-treated delipidated ink, three times on days 2, 4 and 6 after tumor transplantation. Sasaki, J. et al. (1997) [12].

Table 3
Comparison of properties of chitin sheets between squid and crab

	Stiffness	Burst strength (kP·m²/g)	Braking strain (km)	Expansibility (g/m²)
Squid chitin sheet	12	6.9	6.9	21.9
Crab chitin sheet	66	3.8	7.1	21.9

Takai, M. et al. (1995) [13]

7. FUTURE VIEW

So far, medically beneficial components from wastes of abundantly supplied species e.g. salmon, squid and scallop, have been introduced. Unfortunately, most of the studies other than squid β-chitin-salmon collagen sheet, an artificial skin, still remain in the low phases of development for practically medical uses. However, there is no doubt that phospholipids, glycogen, peptidoglycan, chitin, DNA and alginic acid have high bioavailability.

Antitumor active compounds depicted here are so called biological response modifiers (BRM). BRM is known to make immune systems more active. Side effects of antitumor drugs are expected to be minimum when using these BRM. The *sn*-2 DHA-PL is considered to be another BRM. It may make the cancer cells more sensitive to drugs. For this reason, if the *sn*-2 DHA-PL is preloaded prior to the anti cancer drugs, it may be

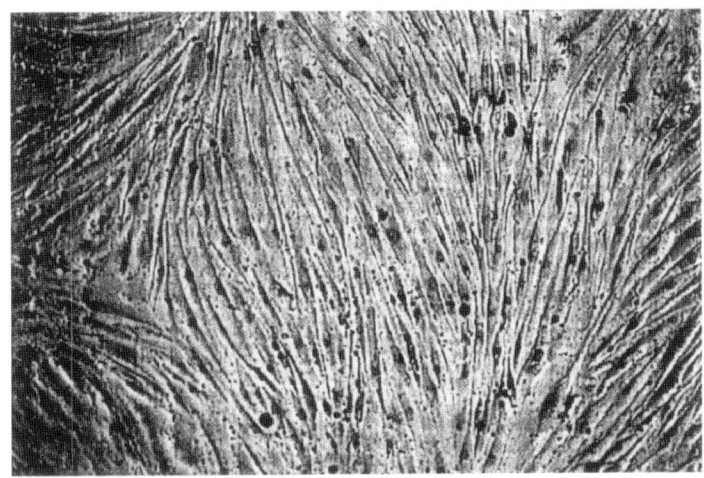

Figure 9. Proliferation of fibroblast on salmon skin collagen laminated squid β-chitin sheet. Takai, M. (1996) [14].

Table 4
Antibacterial activity of alginate (AL) film with and without Salmon testis DNA as a carrier of silver ion (Ag) in relation to the Ag impregnation

	AL-Ag	DNA-AL-Ag
Ag(μg)	4.21	21.7
E. coli	89*	226*
S. aureus	387*	708*

*Inhibition area (mm^2) of cell-growth. Kitamura, H. et al., (1997) [16]

expected to reduce the side effects of the drugs by reducing their dose amount. To increase the quality of life of patients, and to give intensive to reduce wastes, or to consume byproducts, it should be very important to study the BRM contained in those sources.

REFERENCES

1. Hara, K. (1993) Biochemistry and applications of physiologically functional lipids., Saiwai-shoboh, Tokyo.
2. Jenski, L. J., Nanda, P. K., Jiricko, P. and Stillwell, W. (1995) ω-3 Fatty acid-containing liposomes in cancer therapy. Pro. Soc. Exp. Biol. Med., 210:227-233.
3. Suzuki, M., Asahi, K., Isono, K., Sakurai, A. and Takahashi, N. (1992) Differentiation inducing phosphatidylcholine(s) from the embryos rainbow trout (*Salmo gairdneri*): isolation and structured elucidation. Develop. Growth & Differ.,

34:301-307.
4. Fujimoto, A., Sasaki, J., Hosokawa, M. and Takahashi, K. (2001) Antitumor activity of highly unsaturated phospholipid liposomes. Proceedings of the Japanese Conference on the Biochemistry of Lipids, 43:318-321.
5. Tochizawa, K., Hosokawa, M., Kurihara, H., Kohno, H., Odashima, S. and Takahashi, K. (1997) Effect of phospholipids containing docosahexaenoic acid on differentiation and growth of HL-60 human promyelocytic leukemia cells. J. Jpn. Oil Chem. Soc. (J. Oleo. Sci.), 46:382-390.
6. Hosokawa, M., Sato, A., Ishigamori, H., Kohno, H., Tanaka, T. and Takahashi, K. (2001) Synergistic effects of highly unsaturated fatty acid-containing phosphatidylethanolamine on differentiation of human leukemia HL-60 cells by dibutyryl cyclic adenosine monophosphate. Jpn. J. Cancer Res., 92:666-672.
7. Ono, M., Hosokawa, M., Inoue, Y. and Takahashi, K. (1997) Water activity-adjusted enzymatic partial hydrolysis of phospholipids to concentrate polyunsaturated fatty acids. J. Am. Oil Chem. Soc., 74:1415-1417.
8. Inoue, Y. (2001) Recent developments of polyunsaturated fatty acids and their derivatives related researches. New Food Industry, 43:22-26.
9. Sasaki, T., Takasuka, N. and Abiko, N. (1978) Anlitumor, activity of a boiled scallop extract. J. Natl. Cancer Inst., 60:1499-1500.
10. Takaya, Y., Uchisawa, H., Ichinohe, H., Sasaki, J., Ishida, K. and Matsue, H. (1998) Antitumor glycogen from scallops and the interrelationship of structure and antitumor activity. J. Mar. Biotechnol., 6:208-213.
11. Matsue, H., Takaya, Y., Uchisawa, H., Naraoka, T., Okuzaki, B., Narumi, F., Ishida, K. and Sasaki, J. (1997) Antitumor peptidoglycan with new carbohydrate structure from squid ink. In Food Factors for Cancer Prevention., Ohigashi, H., Osawa, T., Terao, J., Watanabe, S. and Yoshikawa, T. (eds.), Springer-Verlag, Tokyo, pp. 331-336.
12. Sasaki, J., Ishita, K., Takaya, Y., Uchisawa, H. and Matsue, H. (1997) Anti-tumor activity of squid ink. J. Nutr. Sci. Vitaminol., 43:455-461.
13. Takai, M. and Shimizu, Y. (1995) Squid β-chitin sheet laminated with salmon skin collagen as an artificial human skin. Report of the cooperative study on "Development of functional membrane using marine bioresource collagen". Supported by grant in aid of Ministry of International Trade and Industry of Japan for Industrial-Academe cooporation, pp. 20-35.
14. Takai, M. (1996) Forming squid chitin-salmon skin collagen complex suitable for human fibroblast adhesion and proliferation. Report of the cooperative study on "Development of functional membrane using marine bioresource collagen". Supported by grant in aid of Ministry of International Trade and Industry of Japan for Industrial-Academe cooperation, pp. 20-31.
15. Iwata, K., Sawadaishi, T., Nishimura, S-I., Tokura, S. and Nishi, N. (1996) Utilization of DNA as functional materials: preparation of filters containing DNA insolubilized with alginic acid gel. Int. J. Biol. Macromol., 18:149-150.
16. Kitamura, H., Kondo, Y., Sakairi, N. and Nishi, N. (1997) Preparation and characterization of antibacterial alginate film containing DNA as a carrier of silver ion. Nucleic Acids (Symposium Series), 37:273-274.

SEPARATION AND PHYSIOLOGICAL FUNCTIONS OF ANSERINE FROM FISH EXTRACT

Kazuaki Kikuchi, Yoshiharu Matahira and Kazuo Sakai
Yaizu Suisankagaku Industry Co., Ltd.,5-8-13 Kogawashinmachi, Yaizu, Shizuoka 425-8570, Japan

Abstract

A separation process that produces anserine powder with a 10.6 % purity and a 9 % yield from the raw material bonito extract was established. This powder was further purified to anserine hydrochloride with a purity of 98 %. In tests of its physiological effect, anserine that had been administered orally to mice was quickly absorbed into the blood stream and was found to increase the swimming time and the chinning exercise time of the mice. In addition, the content of lactic acid in plasma was lower in mice given anserine than in mice that had not received the dipeptide. These results suggest that anserine has an anti-fatigue effect. The anti-oxidant effect of anserine in vitro by ESR was investigated: anserine, as well as the other imidazole dipeptide carnosine, showed a strong ability to eliminate hydroxyl radicals and singlet oxygens. On the basis of these interesting properties, anserine was considered to have a potential for use in a wide range of applications, including as a component of functional foods and cosmetics.

1. INTRODUCTION

Low molecular mass nitrogenous components, such as free amino acids and peptides that are used as seasoning additives to improve the taste of our food, have been extracted from various fish species. Free imidazole dipeptides include, for example, anserine (β-alanyl-1-methyl-L-histidine), carnosine (β-alanyl-L-histidine) and balenine (β-alanyl-3-methyl-L-histidine), the structures of all of which have been elucidated. Because these dipeptides contain an imidazole residue, they are reported to have a buffering capacity at physiological pH [1].

The white muscle of migratory marine fish such as tuna and bonito contains a large amount of free anserine [2], and for some time it has been thought that the migratory performance of such fish may be related to their remarkably high content of anserine [3, 4]. It is thought that the physiological function of anserine may be a buffering action against proton produced by anaerobic exercise. In vitro studies have shown that anserine and carnosine are activators of the myofibrillar-ATPase activity of skeletal muscle [5] and inhibitors of a lipid peroxidation that depends on iron ions [6]. Although anserine is potentially a very interesting substance, its advanced use as a functional food additive has not been well investigated. Furthermore, there is little information on the metabolic fate of anserine that has been orally administered to humans, even though it is a common component of our diet.

Since anserine is a free and water-soluble component, it is found in by-products of the manufacture of canned tuna or dried bonito, as fish broth. The fish broth generally contains a variety of components such as proteins, peptides, amino acids, nucleotides, and minerals. Due to the low concentration of anserine in the by-products and to the occurrence of other components with very similar molecular masses, the costs of recovering and purifying anserine are high, resulting in very limited utilization. Here, we have used a loose reverse osmosis (LRO) membrane to elevate the concentrations of anserine. In order to establish a separation system for anserine, it was first necessary to determine the membrane characteristics, in particular, the species of molecular size cut-off.

In this work, we have studied the separation methods and the physiologies of anserine extracted from bonito extract to determine whether fish broth and other by-products are potentially useful in other applications.

2. MATERIALS AND METHODS

2.1. Materials

Bonito extract was prepared using fish broth (by-product) from manufacturing process of canning. In practice, after the beheaded and eviscerated bonitos were boiled for 1 h, the broth was collected. The broth was filtered and concentrated in vacuo until the water content become to be below 25 %. As a result, we obtained the bonito extract, which consisted of 23.8 % water, 7.2 % free amino acid, 14 % NaCl, and 55 % protein. This concentrated extract was the starting raw material for the following membrane separation process.

2.2. Membrane separation process

A two-stage membrane separation technique was used to elevate the anserine content using reverse osmosis apparatus equipped with spiral modules operated under a diafiltration process. First, high molecular components such as proteins were removed from the bonito extract by an LRO membrane, NTR-7410HG (Nitto Electric Industrial Co., Japan), which has a 10 % rejection coefficient for sodium chloride. Solutes with a relative molecular mass below about 600 can permeate an NTR-7410HG membrane. Initially, the feed reservoir held an aqueous solution (1000 L) containing 100 kg of bonito extract, which was equilibrated by circulating the solution at 25 ± 2°C at a flow rate of 6.0 L/min, corresponding to a operating pressure of 3.5 MPa. During the diafiltration process, deionized water (1000 L) was added continuously to the feed reservoir at a rate equal to the membrane permeation rate, thus maintaining a constant volume of feed solution. The permeate (1000 L), comprising the low molecular components such as amino acids, peptides, nucleotides, minerals, was collected. Second, this permeate was concentrated 10-fold with an NTR-7250HG membrane (Nitto Electric Industrial Co., Japan), which has a 60 % rejection coefficient for sodium chloride. Finally, the concentrate in the feed reservoir was collected and powdered. Nine kg of powder (unwrought product) was obtained. The content of anserine, carnosine, and histidine in this product was analyzed by a Hitachi amino acid analyzer (L-8500A).

2.3. Ion exchange chromatography

The unwrought product was purified by ion exchange chromatography. An aqueous solution (200 L) containing 9 kg of unwrought product was loaded on a column (1000 L) containing a weak acidic resin, DIAION WK-20 (H^+ type). The resin was washed with 5000 L of deionized water, and subsequently the adsorbed basic components were eluted using 5000 L of 1 M NaCl. The eluate was desalted by electrodialysis and concentrated in vacuo. This basic fraction was further purified on a column (25 L) containing a strong acidic resin, Dowex 50W × 4 (200-400 mesh, Na^+ type), by isocratic elution with a 0.38 M sodium citrate buffer (pH 4.0) system according to the method of Suyama et al. [7]. Every 2 L of the eluent was collected and the absorbance was measured at 570 nm (ninhydrin reaction). The purity of anserine in every fraction was also analyzed by a Hitachi amino acid analyzer. Fractions containing anserine were recovered and desalted. The solution was adjusted to pH 6.0 with 1 N HCl to convert anserine to its hydrochloride form. After lyophilization, a highly purified product (anserine hydrochloride) was obtained.

2.4. Effects of anserine on exercise performance for mice

We used the purified anserine hydrochloride in experiments to examine the effects of anserine on the exercise performance of twenty 6-week-old male mice (Slc : ddY). The mice were housed individually in stainless-steel wire cages in a room maintained at 22 ± 3°C under a controlled 12 h light:dark cycle. The mice were assigned randomly to two groups with 10 mice in each group. The mice were fasted for 4 h, and then an aqueous solution of anserine hydrochloride was orally administered to the test group at a dose of 200 mg per kg of body weight, while the same volume of distilled water was orally administered to the control group. Exactly 1 h after the administration of anserine, the mice were placed in a compulsory swimming environment in which they were loaded with a sinker weight (corresponding to 10 % of the average body weight in each group) and the length of time until head remained under water surface for 7 s was measured as the 'swimming time'. Three and thirty minutes after completing the swimming task, the mice were subjected to chinning tasks (first run and second run, respectively) in which the time that they could hang from a wire was measured as the 'chinning exercise time'. Exactly 2 h after the administration of anserine, blood was drawn from the vena cava of each mouse into a heparinized tube. The plasma was separated from the blood cells and stored at −70°C until analysis. The lactic acid in plasma was determined by an automatic analyzer (BM-12, JEOL, Tokyo), whereas the anserine was converted into a fluorescent derivative and quantified by HPLC [8]. All data are presented as means ± S.E. The statistical significance of the differences between the means of each group was calculated by Student's t-test.

2.5. Anserine transfer to human blood

We gave a man a draft of an aqueous solution containing 10g of anserine hydrochloride, and then collected a sample of venous blood every 5 min for 1 h. Samples of both whole blood and separated plasma were stored at −70°C until analysis. The anserine in these samples was converted to a fluorescent derivative and quantified by HPLC [8].

2.6. Radical-scavenging activity

The effect of anserine on free radicals was examined in vitro using ESR spectroscopy. To measure hydroxyl radicals (•OH) and singlet oxygens ($^1O_2$), we used the Free Radical Monitor (JES-FR 30, JOEL, Tokyo). •OH radicals generated by ultraviolet photolysis of an aqueous H_2O_2 solution were analyzed as a spin adduct of 5,5-dimethyl-1-pyrroline-N-oxide (DMPO) [9]. $^1O_2$ was produced from oxygen in the presence of a photosensitizer (riboflavin) and detected as a spin adduct of 2,2,6,6-tetramethyl-4-piperidone hydrochloride (TMPD). The ESR spectrometry conditions used to estimate the free radicals were as follows: magnetic field, 335.7 mT; power, 4 mW; modulation frequency, 9.4 GHz; modulation amplitude, 0.1 mT; time constant, 0.1 s; amplitude, 25; and sweep time, 1 min. The radical-scavenging activities of anserine, carnosine, and histidine were evaluated to estimate the concentrations required to reduce the relative peak height of the DMPO-OH or TMPD-$^1O_2$ by 50% (ID_{50}). Finally, the radical-scavenging activity (mM^{-1}) was defined by the multiplicative inverse of ID_{50}. All experiments were carried out in phosphate buffered saline (pH 7.4).

3. RESULTS AND DISCUSSION

3.1. Industrial production of anserine

Figure 1 shows a flowchart of the separation procedure for anserine. The membrane separation method produced 9.0 kg of unwrought product from 100 kg of the raw starting material, bonito extract. We validated that an LRO membrane, NTR-7410HG, is a sufficient tool by which to remove proteins from the raw material because proteins with a molecular mass higher than ca. 1000 were hardly detected in the unwrought product (data not shown). As shown in Table 1, the amount of anserine, carnosine, and histidine in the unwrought product was 10.6 %, 0.8 %, and 8.6 %, respectively. In contrast, bonito extract contained 1.38 % anserine, 0.11 % carnosine, and 1.58 % histidine. Thus, the membrane separation process (NTR-7250HG) increased the anserine content from 1.38 % to 10.6 %. The content of histidine also increased from 1.58 % to 8.6 %; however, the ratios of histidine content to dipeptide content (i.e., histidine/anserine + carnosine) were 1.06 and 0.75 for bonito extract and unwrought product, respectively. Thus, NTR-7250HG membrane treatment appeared to alter this ratio by preferentially allowing the permeation of histidine, which has a lower molecular mass than anserine or carnosine. Nevertheless, we believe that this membrane separation technique is suitable for elevating the anserine content and is cost-effective for industrial processes. The unwrought product, containing 10.6 % anserine, may be applicable for use mainly as a functional food additive.

Furthermore, 0.63 kg of a highly purified product was obtained after two-step ion

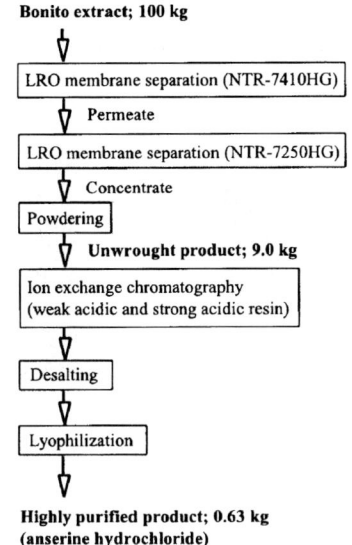

Figure 1. Separation process of anserine from bonito extract.

Table 1
The contents of anserine, carnosine and histidine in the unwrought product and the highly purified product (%)

	Anserine	Carnosine	Histidine
Bonito extract (starting material)	1.38	0.11	1.58
Unwrought product	10.6	0.8	8.6
Highly purified product	98*	2.0*	0

*Content as a hydrochloride

exchange chromatography of 9.0 kg of the unwrought product. The content of anserine, present as the hydrochloride, was found to be 98.0 % in the highly purified product (Table 1).

3.2. Effects of anserine on exercise performance for mice

The effects of oral anserine intake are summarized in Fig. 2. The test group (anserine-administered mice) showed a tendency towards longer swimming times and longer chinning exercise (first run) times. The chinning exercise (second run) time of the test group was increased significantly ($p < 0.01$) by 27.1 % as compared with the control group. After all exercises were completed (2 h after anserine intake), the content of plasma lactic acid in the mice was 32.6 ± 5.4 mg/dL and 24.8 ± 3.4 mg/dL for the control group and the test group, respectively. That is, the content of plasma lactic acid (a product of fatigue) was significantly lower ($p < 0.05$) in the mice that had received anserine. The intake of oral anserine increased the plasma anserine content from a trace level to 27.5 ±

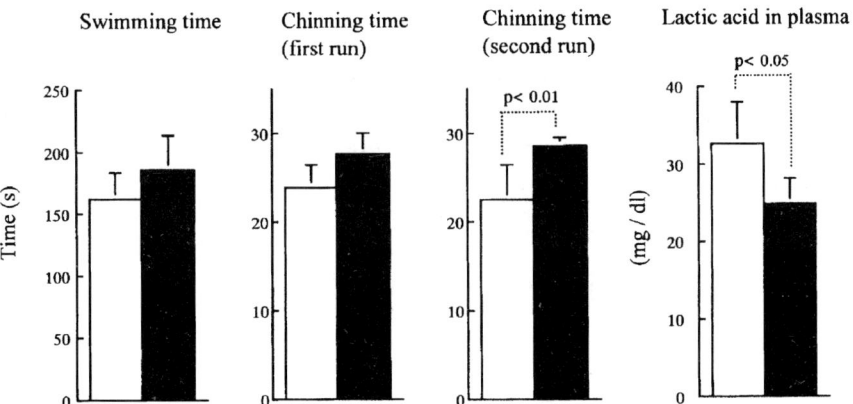

Figure 2. Effect of oral anserine intake on exercise performance for mice. Prior to the exercises, an aqueous solution of anserine hydrochloride was orally administered to the test group at a dose of 200 mg per kg of body weight. To the control group, a distilled water was orally administered. The plasma samples for the measurement of lactic acid were collected at exactly 2 h after administration of anserine. Values are means ± S.E. of 10 mice. □, Control group; ■, Test group.

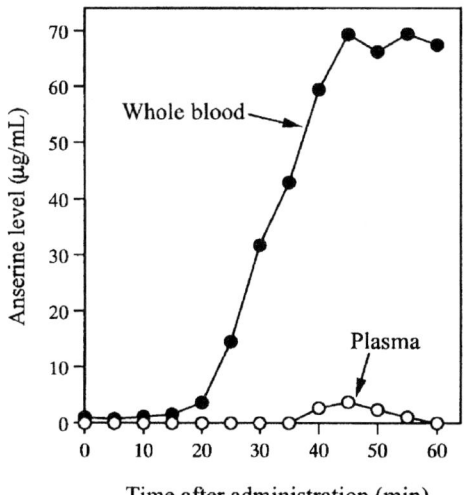

Figure 3. Time-dependent changes in blood anserine level after administration of anserine hydrochloride. The blood samples were collected from a man that had ingested a draft of an aqueous solution containing 10g of anserine hydrochloride.

2.8 mg/mL at 2 h post anserine intake. Taking these results together, orally administered anserine was transferred readily to the blood, where it effected an improvement in exercise performance, seemingly through an anti-fatigue mechanism.

3.3. Anserine transfer to human blood

Figure 3 shows the time course of anserine levels in human blood after oral administration of anserine hydrochloride (10g). The plasma anserine level began to increase after 35 min, reached the maximum concentration 45 min after the oral administration, and disappeared by 60 min. In contrast, the anserine level in whole blood started to increase after 20 min, reached the maximum concentration of 70 mg per mL of blood at 45 min, and then remained at this maximum level until at least 60 min after the administration. The concentration of anserine in whole blood was about 20 times greater than in plasma at 45 min after anserine administration. We therefore presume that anserine is incorporated preferentially into blood cells. The difference of anserine level between in whole blood and in plasma may be related to the presence of a serum dipeptidase that has hydrolyzing activities towards anserine and carnosine [10].

3.4. Radical-scavenging activity of anserine

The ESR signals derived from •OH radicals and $^1O_2$ decreased dose-dependently with added anserine (Fig. 4). The signal of DMPO-OH disappeared completely at a final concentration of 15 mM anserine, whereas the signal of TMPD-$^1O_2$ was almost undetectable at a final concentration of 0.5 mM anserine. Table 2 shows the radical-scavenging activity (1/ID$_{50}$) of anserine compared with that of other compounds. Among imidazole dipeptides, the activity of anserine for quenching $^1O_2$ was about 80 % of that of carnosine, which exhibited the highest activity. In regard to •OH radicals, carnosine and histidine have scavenging activities comparable to that of gallic acid, which is well known as an antioxidant. In contrast, the •OH radical scavenging activity of anserine was about 50 % less than that of gallic acid. The radical-scavenging activity of carnosine, as determined by spin traps, has been reported so far [11]; there is little information on that of anserine. Our study suggests that anserine, like carnosine, is an efficient scavenger of both •OH and $^1O_2$. As antioxidants, the imidazole dipeptides might be of potential use in the cosmetic and pharmaceutical industry to protect against lipid peroxidation. Moreover, as free radicals generated from neutrophils and mast cells are involved in the development of inflammation and tissue injury [12], the imidazole dipeptides may be potentially useful as natural anti-inflammatory agents.

The interesting properties of anserine, such as its anti-fatigue effect, its radical-scavenging activity and its good absorbability from the digestive system, suggest that this dipeptide has potential in a wide range of applications, for example, in functional foods, cosmetics, and pharmaceuticals. In addition, our study has defined a protocol for

Table 2
The radical-scavenging activity (mM^{-1}) of anserine, carnosine and histidine

	Anserine	Carnosine	Histidine	Gallic acid
Singlet oxygen ($^1O_2$)	6.8	8.6	5.6	-
Hydroxyl radical (•OH)	0.44	0.73	0.71	0.83

The radical-scavenging activity is defined by the multiplicative inverse of final sample concentration (ID$_{50}$) at 50% inhibition of DMPO-OH or TMPD-$^1O_2$ signal strength.

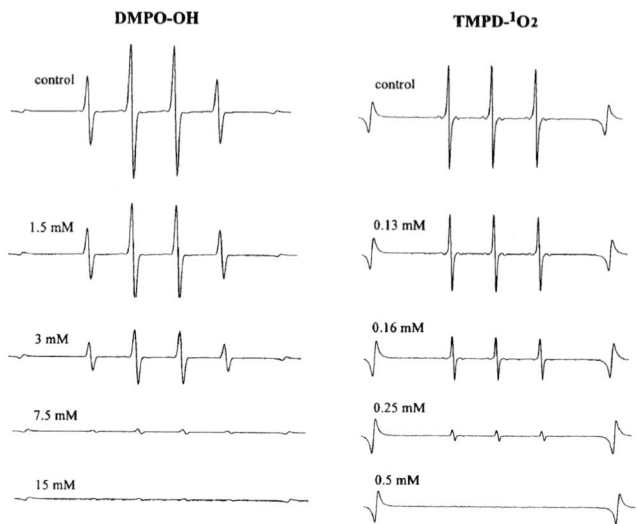

Figure 4. The ESR signals of DMPO-OH and TMPD-$^1O_2$ observed upon the addition of anserine. Values (mM) are the final concentrations of anserine in the reaction solutions for the measurement of free radicals.

industrially producing from bonito extract a 10.6 % anserine powder, which will be useful mainly for functional foods, and a highly purified anserine powder.

4. ACKNOWLEDGEMENTS

This work was supported in part by a grant from Japan Food Industry Center (JAFIC).

REFERENCES

1. Smith, E.C.B. (1938) The buffering of muscle in rigor; protein, phosphate and carnosine. J. Physiol., 92: 336-343.
2. Suzuki, T., Hirano, T. and Suyama, M. (1986) Free imidazole compounds in white and dark muscles of migratory marine fish. Comp. Biochem. Physiol., 87B: 615-619.
3. Suyama, M., Hirano, T. and Suzuki, T. (1986) Buffering capacity of free histidine and its related dipeptides in white and dark muscles of yellowfin tuna. Bull. Japan. Soc. Sci. Fish., 52: 2171-2175.
4. Abe, H. and Okuma, E. (1991) Effect of temperature on the buffering capacities of histidine-related compounds and fish skeletal muscle. Nippon Suisan Gakkaishi, 57: 2101-2107.
5. Parker, C. J. Jr. and Ring, E. (1970) A comparative study of the effect of carnosine on myofibrillar-ATPase activity of vertebrate and invertebrate muscles. Comp. Biochem.

Physiol., 37: 413-419.
6. Boldyrev, A.A., Dupin, A.M., Pindel, E.V. and Severin, S.E. (1988) Antioxidative properties of histidine-containing dipeptides from skeletal muscles of vertebrates. Comp. Biochem. Physiol., 89B: 245-250.
7. Suyama, M., Suzuki, T. and Nonaka, J. (1967) Chromatographic determination of imidazole compounds in the whale meat. Bull. Japan. Soc. Scient. Fish., 33: 141-146.
8. Kasziba, E., Flancbaum, L., Fitzpatrick, J.C., Schneiderman, J. and Fisher, H. (1988) Simaltaneous determination of histidine-containing dipeptides, histamine, methylhistamine and histidine by high-performance liquid chromatography. J. Chromatogr., 432: 315-320.
9. Harbour, J.R., Chow, V. and Bolton, J.R.(1974) An electron spin resonance study of the spin adducts of OH and HO_2 radicals with nitrones in the ultraviolet photolysis of aqueous hydrogen peroxide solution. Can. J. Chem., 52: 3549-3556.
10. Jackson, M.C., Kucera, C.M. and Lenney, J.F. (1991) Purification and properties of human serum carnosinase. Clin. Chim. Acta, 196: 193-205.
11. Rubtsov, A.M., Schara, M., Sentjurc, M. and Boldtrev, A.A. (1991) Hydroxyl radical-scavenging activity of carnosine: a spin trapping study. Acta Pharm. Jugosl., 41: 401-407.
12. Kuehl, F.A., Humes, J.L., Torchiana, M.L., Ham E.A. and Egan, R.W. (1979) Oxygen-centered radicals in inflammatory processes. In: Advances in Inflammation Research., Weissmann G. (ed), Raven Press, New York., pp. 419-430.

BONE AND LIPID METABOLISM IN THE BONE MODELING OF RATS ADMINISTERED WITH THE BONITO DOCOSA-HEXAENOIC ACID, THE VISCERA VITAMIN D AND THE CUTTLEFISH CALCIUM

Keiko Yoshioka[a], Azusa Seki[b], Ai Yamada[a], Mamoru Fujita[a], Yasuhiko Sakaguchi[c], Koichi Nagafusa[d] and Shun Wada[c]

[a]Department of Food and Nutrition, Nakamura Gakuen University, 5-7-1 Befu, Johnan-ku, Fukuoka 814-0198, Japan

[b]Mitsubishi Chemical Safety Institute Ltd., 2-1-30 Shiba, Minato-ku, Tokyo 105-0014, Japan

[c]Department of Food Science and Technology, Tokyo University of Fisheries, 4-5-7 Konan, Minato-ku, Tokyo 108-8477, Japan

[d]Yaizu Meal Coop., 1001-1 Nakazato, Yaizu, Shizuoka 425-0014, Japan

Abstract

The effects of the bone and lipid metabolism were investigated in rats administered the diets prepared from cuttlefish shells (C-Ca), bonito viscera extract (VD) and bonito fish oil docosahexaenoic acid (DHA) from the viewpoint of the utility of marine resources. Six-week old male Wistar rats after being provided with the standard diet (AIN-93G) for 10 days were distributed into 7 groups (A~G) and fed by each diet for about 5 weeks. A:standard diet; B:control diet (Ca,VD and DHA had been removed from the standard diet); C:control diet+Ca; D:control diet+C-Ca; E:control diet +C-Ca+VD, Groups A to E were administered with soybean oil. F:control diet +Ca+DHA; G:control diet+Ca+VD+DHA. The lipid metabolism markers (TG, PL and T-Cho) in the plasma were lower for F and G. Regarding the effect of the addition of vitamin D on the soybean oil administered groups (A, C) and the DHA administered groups (F, G), there was no significant difference between C and F, but A and G had significant differences ($p<0.05$) in the Ca content in the femoral bone. In the bone metabolism the total BMD in the metaphysis region were 1.51, 1.51, 1.65, 1.53 and 1.60 times higher in C, D, E, F and G than B, respectively. These results revealed that the trabecular BMD level is maintained by the addition of DHA, the total and cortical BMD levels are maintained by the addition of vitamin D and the whole BMD is maintained by taking cuttlefish calcium together with vitamin D.

1. INTRODUCTION

The calcium intake of the Japanese people is low based on the current required intake of calcium [1]. To maintain this requirement, it is necessary to eat foods rich in calcium. Due to the aging Japanese population, the patients with osteoporosis have increased and calcium fortified foods are in great demand. It is reported that n-3 polyunsaturated fatty acid has an effect on bone modeling or preventing bone loss during aging. As a supplement for the lack of calcium intake, we paid attention to the absorbable calcium from fishery products [2]. In this study the effects of the bone and lipid metabolisms were investigated in rats administered with diets prepared from cuttlefish shells (C-Ca), bonito viscera extract vitamin D (VD) and bonito fish oil docosahexaenoic acid (DHA) from the viewpoint of utilization of marine resources.

2. MATERIALS AND METHODS

Six-week-old male Wistar rats after being provided with the standard diet [3] for 10 days were divided into 7 groups (A-G) and the experimental diet of each group is shown in Table 1 and 2. The number of animals in each group was six. The mixed diet in each group was given and deionized water was freely drunk. The animal rooms were automatically maintained at a room temperature of $22\pm1°C$, a relative humidity of $55\pm5\%$ and lighted for 12 hours per day.

All the rats were dissected, and their livers were immediately removed and the lipid contents, the composition of fatty acids and the lipids class were examined. The liver lipid was extracted using the method of Bligh and Dyer [4], and the composition was examined by thin layer chromatography (TLC) and gas liquid chromatography (GLC).

The biochemical parameters in the plasma were measured using a bioautoanalyzer. The calcium and the inorganic phosphate contents in the femoral bone were determined by the o-CPC method[5] and the enzymatic method, respectively.

Table 1
The experimental diet of each group

Group A	Standard diet (AIN-93G diet)
Group B	Control diet (Ca and VD are removed from Diet A)
Group C	Control diet + Ca
Group D	Control diet + C-Ca
Group E	Control diet + C-Ca + B-VD
Group F	Control diet + Ca + DHA
Group G	Control diet + Ca + VD + DHA

Control diet : Calcium, VD and soybean oil are removed from standard diet (AIN-93G diet, Oriental Yeast Industry Co., Ltd.). Soybean oil was administered for 70.0 g/kg diet to groups B to E. DHA prepared from bonito fish oil (Yaizu Meal Coop). Ca was administered for 5,000 mg/kg diet to all groups except Group B. Ca, AIN-93G calcium; C-Ca, cuttlefish shell calcium (Ca extracted from cuttlefish shell powder, Yaizu Meal Coop.). Vitamin D_3 was administered for 1,000 I.U/kg diet to all groups except Group B. B-VD, bonito viscera extract vitamin D (Yaizu Meal Coop.); VD, AIN-93G vitamin D.

Table 2
The composition of AIN-93G diet (g/kg diet)

Cornstarch	397.486
Casein ($\geq$ 85% protein)	200.000
Dextrinized cornstarch	132.000
Sucrose	100.000
Soybean oil (no additives)	70.000
Fiber	50.000
Mineral mix (AIN-93G-MX)	35.000
Vitamin mix (AIN-93G-VX)	10.000
L-Cystine	3.000
Choline bitartrate	2.500
Tert-butylhydroquinone	0.014

The femoral bone was examined for the total and trabecular bone mineral density (BMD) and stress strain index (SSI) using pQCT (XCT-960A, Norland Stratec Inc.).

Furthermore, the bone tissue of their femoral bones was histologically observed. The femoral bones were fixed with 10% neutral buffered formalin, decalcificated with 10% EDTA, embedded in paraffin, and thin sections were stained and observed under a light microscope.

Statistical analysis of the data obtained for each test was performed by the least significant difference (LSD) method. Values were expressed as the means±SD.

3. RESULTS AND DISCUSSION

The changes in the body weights of the rats in each group are shown in Fig. 1. There was no significant difference in the food consumption rate of all groups. However, the body weight gain was inhibited in Group B.

The lipid metabolism markers, triacylglycerol (TG), phospholipid (PL) and total cholesterol (T-Cho) in the plasma and the fatty acid composition of the rat liver lipid were lower for Groups F and G. The ratios of DHA and TG in the liver lipid were higher in Groups F and G, whose TG ratios were less than that in the standard diet.

The calcium and inorganic phosphate contents in the plasma of the rats are shown in Fig. 2. The calcium level in the plasma of Group B was the lowest, but those in the other groups were normal. As shown in Fig. 3, groups A, D and E showed lower than the others for alkaline phosphatase activity (ALP) in the plasma of rats.

The calcium and inorganic phosphate contents in the femoral bone showed the same tendency. Group A, groups D and E, and groups F and G except Group B showed almost the same value, respectively. Regarding the effect of the addition of vitamin D to the soybean oil administered groups (A and C) and the DHA administered groups (F and G), there was no significant difference between groups C and F, but groups A and G had significant differences ($p<0.05$) in the Ca content of the femoral bone. These interactive effects are shown in Fig. 4. With the addition of vitamin D, an interaction was seen in groups A and C ($p<0.01$), although the addition of vitamin D had little effect on the

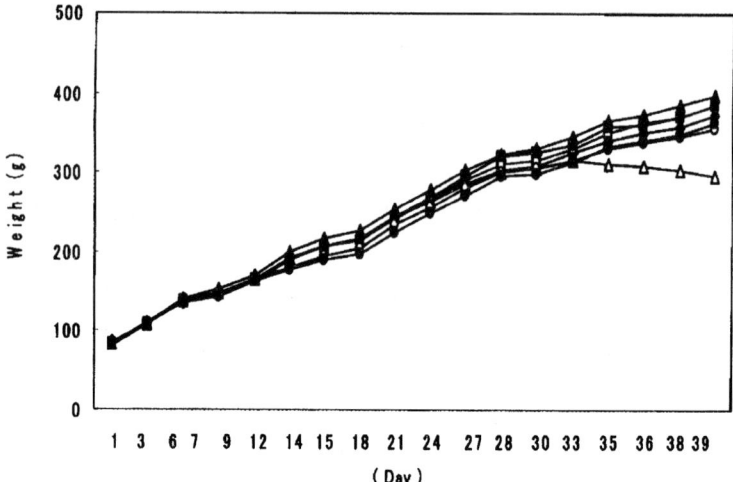

Figure 1. Changes in the body weight of rats in each group. The body weight in each group is the mean value of 6 rats. ◆A, Standard diet (AIN-93G diet); △B, Control diet (Ca and VD removed from AIN-93G diet) ; ▲C, Control diet + Ca; □D, Control diet +C-Ca; ■E, Control diet + C-Ca + B-VD; ○F, Control diet + Ca + DHA; ●G, Control diet + Ca + VD + DHA.

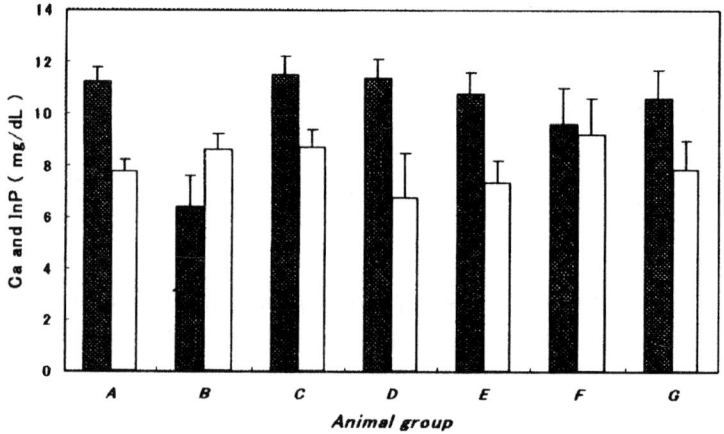

Figure 2. Content of calcium and inorganic phosphate in plasma of rats. Values are shown as the mean±SD of 6 rats per group. A, Standard diet (AIN-93G diet); B, Control diet (Ca and VD removed from AIN-93G diet); C, Control diet + Ca; D, Control diet + C-Ca; E, Control diet + C-Ca + VD; F, Control diet + Ca + DHA; G, Control diet + Ca + VD+ DHA; ■, Ca; □, InP.

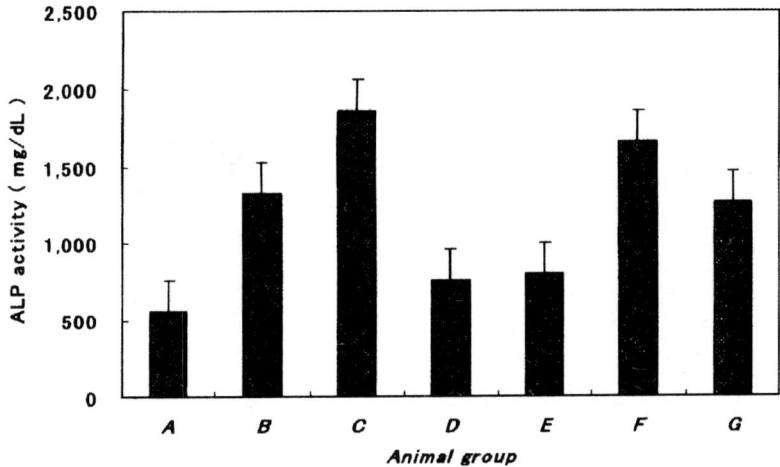

Figure 3. Alkaline phosphatase (ALP) activity in plasma of rats. Values are shown as the mean±SD of 6 rats per group. See footnotes in Fig. 2 about experimental animal groups.

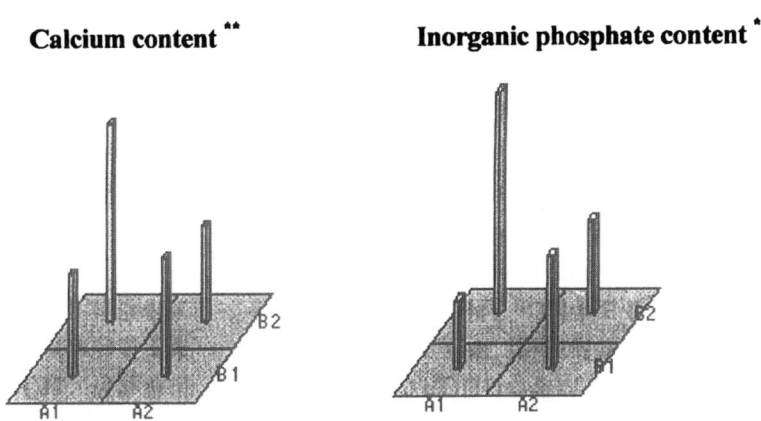

Figure 4. Interaction of the combination with soybean oil and DHA with and without VD. A1, soybean oil; A2, DHA (bonito fish oil); B1, without vitamin D; B2, with vitamin D. **Interactions were effective in calcium content and inorganic phosphate content ($p<0.01$).

interaction in groups F and G. It is suggested that this is due to the difference in the lipid composition between the soybean oil and DHA.

For the bone metabolism, the pQCT image is shown in Fig. 5. Trabecular and total BMD in the region of 3mm of metaphysis and cortical BMD and SSI in the region of 12mm of diaphysis from the distal growth cartilage were measured by pQCT. The total BMD of groups C, D, E, F and G in the metaphysis region were higher by 1.51, 1.51,

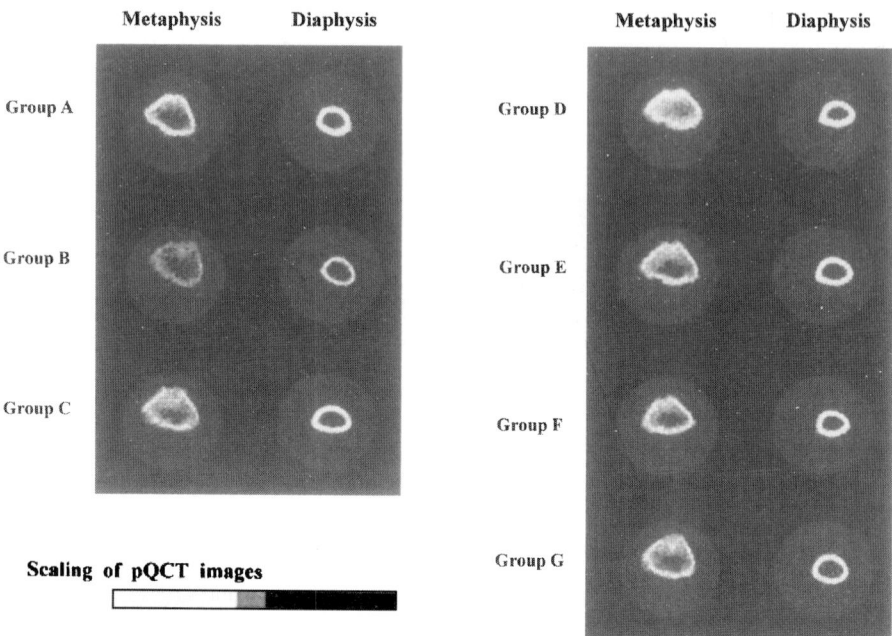

Figure 5 Bone mineral density of rat femur pQCT image. Trabecular and total BMD are metaphysis at the region of 3mm and cortical BMD and SSI are diaphysis at the region of 12mm from the distal growth cartilage. See footnotes in Fig. 2 about experimental animal groups.

1.65, 1.53 and 1.60 times than that of Group B, respectively. The trabecular BMD was higher in groups D, E, F, and G than in Group B. The cortical BMD and thickness in the diaphysis region were markedly decreased in Group B, but maintained in groups D and E, and to a lesser degree in groups F and G. As a stress strain index, the polar values among the test substance groups were almost the same level except in Group B, and were almost twice the level recorded in Group B.

The BMD and calcium content in each group were almost coincident with the observations by microscopy [6].

These results revealed that cuttlefish calcium and fish oil-DHA would be effective in the bone modeling, and the utility of these fish residues is thought to be available for the environmental conservation.

REFERENCES

1. Ministry of Health and Welfare (2001) Present condition of national nutrition-results of the national nutrition survey in 1999. Daiichi Publication, Tokyo, pp.29-31.
2. Kato, K., Toba, Y., Takada, Y. and Aoe, S. (1997) Mineral bioabilability of fortified food based on skim milk powder in rats. The Japanese J. Nutr., 55: 189-196.

3. Reeves, P., Nielsen, F. and Fahey, G. (1993) AIN-93G purified diets for laboratory Rodents: Final report of the American Institute of Nutrition Ad Hoc Writing Committee on the reformulation of the AIN-76A rodent diet. J. Nutr., 123: 1939-1951.
4. Bligh, E. G. and Dyer, W. J. (1959) A rapid method of total lipid extraction and purification. Can. J. Biochem. Physiol. , 37: 911-917.
5. Gitelman, H.J. (1967) An improved automated procedure for the determination of calcium in biological speciments. Anal. Biochem., 18: 521-531.
6. Yamada, A., Baba, R., Tanaka, R., Fujita, M. and Yoshioka, K. (2002) Effect of calcium extracted from cuttlefish shells on the rat bone tissue. Bull. Nakamura Gakuen Univ. and Nakamura Gakuen Juni. Coll., 34: 295-301

VALUE-ADDING TO AUSTRALIAN MARINE OILS

Peter D. Nichols, Ben D. Mooney and Nicholas G. Elliott
CSIRO Marine Research, Hobart, Tasmania 7000, AUSTRALIA

Abstract

Marine biotechnology in Australia is a relatively new field. Australia has one of the largest exclusive economic zones (EEZ) globally, an area known to be rich in marine biodiversity from polar, temperate and tropical waters. It has been acknowledged for some time that although Australia is surrounded by a rich marine resource, we are not taking advantage of the potential benefits from the abundant biodiversity. Future research efforts will address this shortcoming. An overview on aspects of current Australian research and development in marine biotechnology, focusing on marine omega-3 oils, will be presented. A comprehensive data-base has been established on the nutritional composition (oils emphasis) of principal Australian seafood. Results are available for strategic marketing of Australian fish species as well as for use by medical and consumer groups. Most Australian fish contain high levels of the nutritionally important omega-3 polyunsaturated fatty acids (PUFA) - eicosapentaenoic acid (EPA, 20:5ω3) and docosahexaenoic acid (DHA, 22:6ω3). Australian fish generally have higher relative levels of DHA compared to fish and oils from northern hemisphere waters. Tuna are a particularly good source of DHA-rich oil, with the manufacture of Australian-made, value-added tuna oil products commenced in 1998. The tuna oil products contain in excess of 25% DHA, a DHA to EPA ratio of >3, and a lower proportion of cholesterol than is found in flesh from tuna and some other seafood. Species specific differences and regional variation in lipid and fatty acid profiles of seafood highlight the ecological significance of this class of biochemical compounds; in addition changes in oil profiles with cooking and aquaculture are examined. Microbial sources of PUFA have also been developed, with a range of applications possible. Lipid class profiles of wax ester rich oil fishes enabled a species associated with consumer illness to be identified.

1. INTRODUCTION

Early Commonwealth Scientific and Industrial Research Organization (CSIRO) research in the 1930s suggested that the oil and vitamin content of a range of shallow and mid water Australian fish had potential to be value-added [1]. After 1950, the availability of synthetically produced vitamins saw the use of fish-derived oils diminish. Limited research and development then occurred in Australia, with purified omega-3 polyunsaturated fatty acid (PUFA) containing oils being imported over the past few decades. However, crude omega-3 PUFA fish oils were being exported or used in lower value products such as feeds for the aquaculture industry.

There is increasing nutritional interest in the omega-3 long-chain (C_{20} and C_{22}) polyunsaturated fatty acids (LC-PUFA). In particular, eicosapentaenoic acid (EPA, 20:5ω3) and docosahexaenoic acid (DHA, 22:6ω3) in seafood and marine oil products receive special attention. This is the result of the well-documented nutritional benefits of these unique PUFA [2-6]. They help against coronary heart disease, high blood pressure, rheumatoid arthritis, and may also be beneficial against other disorders, including some forms of cancer, depression and other neural illnesses. Australian research over the past decade on omega-3 PUFA oils has included characterisation of these oils from a range of edible seafood species [7,8].

Development of new or refined processing conditions suitable for use with Australian fish oils has occurred, with transfer of know-how and technology to industry through research and licensing agreements. Uses for the omega-3 PUFA oils include health and nutritional products, infant formula and, in the case of lower value material, aquaculture and other feeds. More recent marine oils research has focused on specific nutritional requirements of new aquaculture species. Research is also performed on shark liver and wax ester rich oils, with new Australian products reaching the national and international markets during the 1990s (e.g. nutraceuticals, degreasers, hand cleaners, cutting oils) [9]. The research has established strong ties with local industry, giving an increased return for both the fishermen and oil processors, without an increase in catching effort.

In this paper, we highlight our research on characterizing the oils of Australian seafood, examining the effects of processing and aquaculture on the oil content and composition of fillets, demonstrate commercial uptake of the research, and describe novel sources of oils and uses of oil profiles in ecological studies.

2. MATERIALS AND METHODS

2.1. Samples
Seafood, by-product and oil samples were provided by CSIRO colleagues, industry collaborators and others. Unless otherwise stated, samples analysed were consistently taken from the right shoulder region of fish, and the tail of crustaceans. Effects of processing or cooking involved sampling the same region of individual fillets before and after treatment. Results presented are generally the mean of three samples. All samples were stored frozen prior to analysis. Microbial samples referred to were prepared as described in [10-12].

2.2. Oil analyses
Full details on oil (lipid) analyses are provided elsewhere [13-15]. Briefly, samples were extracted using a single phase Bligh and Dyer procedure [16]. Oil yield was determined gravimetrically. An aliquot of the oil was analysed by TLC-FID to determine

lipid class composition. Fatty acid profiles were obtained by capillary GC and GC-MS analysis following transmethylation of an aliquot of the extracted oil.

3. RESULTS AND DISCUSSION

3.1. Oil composition of Australian seafood

A comprehensive database has been established on the content and composition of the oils from nearly 300 species of Australian fish, shellfish and crustaceans [7] (and Nichols et al., unpublished data, 2001). Results have been published in the Guide "Seafood the Good Food" [8], and are available for marketing of Australian fish and for use by various medical, nutritional, consumer and research groups. The data also has been incorporated within nutritional data-bases.

For the Australian species analysed during our study, summary findings were:
- Local fish had lower levels of oil than species from northern hemisphere waters
- Most Australian fish have high levels of omega-3 PUFA and low levels of cholesterol.
- Australian seafood contain attractive levels of omega-3 PUFA, typically tenfold or more greater than other food groups (Table 1). The key omega-3 PUFA are EPA and DHA. Representative results are provided in Fig. 1.
- The relative level of PUFA in fish (expressed as % of total fatty acids) generally increased with decreasing oil content, suggesting that better oil quality occurs for the low oil species, and that Australian fish are an excellent source of the essential omega-3 PUFA

Table 1
Summary of average content of LC omega-3 PUFA in wild-caught Australian seafood, with comparison to representative farmed species and other food groups

Food group	LC omega-3 PUFA mg/100 g
Australian seafood (wild)	
Fish	235
Oysters	150
Prawns	130
Lobster	105
Farmed Australian fish	
Atlantic salmon	1930
Striped perch	2480
Silver perch	790
Barramundi	1970
Other food groups	
Turkey	35
Beef	22
Chicken	19
Lamb	18
Pork	0
Veal	0

Data for Australian seafood [7,14] and non-seafood items [32-34]

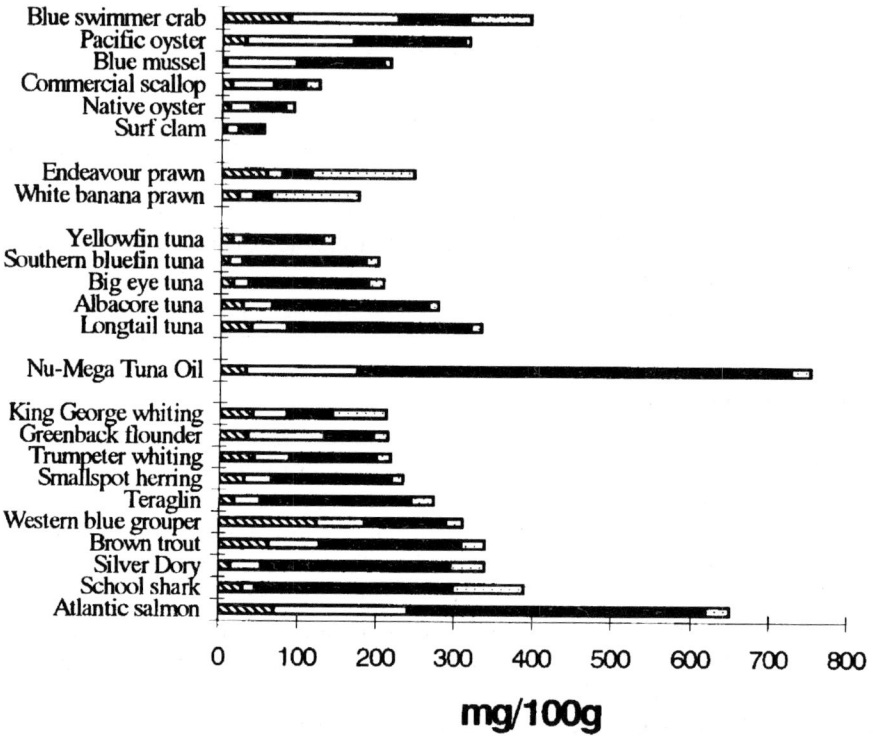

Figure 1. AA, EPA, DHA and cholesterol levels of selected Australian seafood (mg per 100 g serving, wet mass) and for purified tuna oil (mg per two 1 g capsules). ◨,AA; ☐,EPA; ■,DHA; ▦,Cholesterol.

- Prawns (shrimps) have lower levels of omega-3 PUFA and higher levels of cholesterol compared with fish
- Australian fish have higher relative levels of DHA compared with fish, and nutritional supplements containing fish oil, from northern hemisphere waters
- fishes from warmer waters and those with specific diets have lower omega-3 / omega-6 ratios compared with fishes from temperate waters, due largely to higher relative levels of arachidonic acid (AA)
- Season, diet and other factors can influence the oil and fatty acid content and composition of seafood.

Our findings, together with results from earlier Australian reports, provide oil compositional information for use in the selection of fish or fish oils for nutritional studies and for the marketing and communication of the health benefits of seafood. These studies indicate that seafood is clearly the best source of omega-3 LC-PUFA from the common food groups (Table 1). Standardised procedures were adopted during the study, however, intra-species variation may occur. The results serve as a guide for medical practitioners, nutritionists and other user groups. Comparative nutritional studies with seafood and/or specific marine oils of known oil composition and content will provide additional information on the health benefits of omega-3 LC-PUFA.

Intake of the omega-3 PUFA is well documented to aid in decreasing the incidence of coronary heart disease and stroke in humans and also play a role against a range of other disorders, including arthritis [2-6]. Recent nutritional studies have also indicated that DHA may be more active than EPA for certain disorders. Australian seafood, and oils derived from various species, contain generally higher relative levels of DHA and may well prove to be nutritionally more beneficial than those from northern hemisphere waters. Species-specific differences and regional variation in lipid and fatty acid profiles were also observed (Fig. 1), highlighting the ecological significance of this class of biochemical compounds.

3.2. Cooking seafood – effect on oil composition

A frequently asked question is whether different cooking methods affect the nutritional value of seafood. A popular local fish species - blue eye trevalla (*Hyperoglyphe antarctica*) – was chosen to examine the effect of cooking on oil composition. Fresh blue eye contains 1.4% oil, with polar lipid and triglyceride as the main oil classes (84% and 14% respectively, Table 2). The omega-3 LC-PUFA, predominantly DHA and EPA, accounted for approximately 35% of the total fatty acids, with total PUFA at 41%. DHA is the dominant component at 27% (Table 2). On an absolute basis, the average omega-3 LC-PUFA content was 244 mg/100 g (Fig. 2), which is slightly higher than the average for Australian fish (235 mg/100 g; see Table 1).

Specimens of the blue eye were cooked in five ways: grill, steam, microwave, pan and deep fry. Oil content varied depending on the method of cooking used. A peanut oil was used for deep frying, while a cottonseed based oil was used in the pan frying and grilling. Cooking by microwave or steaming did not affect oil and omega-3 LC-PUFA content and composition relative to fresh fish (Table 2, Fig. 2). Oil content of blue eye increased from 1.4% in fresh fish, to 2.3% with pan frying and 5.3% with deep frying. The higher oil content observed in the deep-fried and pan-fried fish is consistent with uptake of cooking oil by the fillet during frying. This is also reflected in the oil class results; the deep fried sample contained 84% triglyceride and 14% polar lipid, the complete opposite to the fresh sample at 14% triglyceride and 84% polar lipid. Similarly, pan frying and grilling also increased the triglyceride at the expense of polar lipid. This was not seen with microwaving or steaming.

The content of the main omega-3 LC-PUFA - DHA and EPA - was largely not affected by cooking (Fig. 2). The differences that were observed in the absolute amount of these beneficial PUFA (e.g. grilling increase to 324 mg/100 g) can be attributed to within-sample variation, and also due to loss of water during cooking.

Examination of the fatty acid profiles expressed in percent form shows that the deep-fried and pan-fried and grilled fish contained higher relative levels of vegetable oil derived components. As noted with the higher oil content observed with these cooking methods, this feature is due to incorporation of cooking oil by the fillet. The deep-fried fish contained higher levels of the omega-6 PUFA linoleic acid (LA, 18:2ω6, 43% cf 1% in fresh samples, Table 2) consistent with the use of peanut oil. Oleic acid (OA, 18:1ω9c) was elevated in the pan-fried blue eye (41%). OA is the major fatty acid (62%) in the

Table 2
Lipid class and fatty acid composition (%) of fresh and cooked blue eye, and soup by-product

	Fresh	Pan fry	Deep fry	Steam	Grill	Microwave	Soup
Total lipid (%)	1.4±0.5	2.3±0.7	5.3±1.8	1.1±0.2	1.9±0.8	1.5±0.5	1.1±0.1
Lipid class (%)							
Wax ester	0.0±0.0	0.2±0.1	0.3±0.1	0.1±0.1	0.1±0.1	0.0±0.1	0.0±0.0
Triglyceride	14.1±8.6	58.1±11.5	84.4±2.9	5.7±4.8	26.8±14.4	8.5±9.0	94.1±0.5
Free fatty	0.3±0.1	0.2±0.1	0.3±0.1	0.3±0.1	0.2±0.0	0.2±0.1	0.5±0.1
Cholesterol	1.7±0.2	1.1±0.4	1.0±0.1	2.0±0.7	1.5±0.4	1.7±0.6	0.5±0.1
Phospholipid	83.9±8.4	40.4±11.0	14.1±2.8	92.0±4.2	71.4±14.7	89.5±8.5	5.0±0.3
Fatty acid (%)							
14:0	1.5±0.6	0.5±0.2	0.7±0.1	0.6±0.3	0.4±0.2	0.8±0.4	8.8±0.1
16:1ω7	0.8±0.1	0.5±0.1	0.6±0.1	0.8±0.4	0.5±0.2	0.9±0.7	2.3±0.0
16:0	18.3±1.8	10.2±1.5	21.5±0.2	18.5±1.4	11.4±0.7	18.2±1.1	27.0±0.1
18:2ω6 LA	0.9±0.1	12.8±1.1	43.3±3.2	2.3±0.9	10.7±1.1	0.9±0.1	3.4±0.0
18:1ω9 OA	13.6±2.3	41.1±4.5	18.8±0.1	12.1±3.1	35.7±2.6	11.3±5.0	27.3±0.1
18:1ω7	1.9±0.2	2.9±0.4	1.2±0.1	2.1±0.3	2.7±0.2	2.1±0.6	1.5±0.0
18:0	6.9±0.1	3.9±0.6	3.4±0.3	6.1±0.7	4.3±0.4	6.0±0.7	12.7±0.0
20:4ω6 AA	2.8±0.6	1.2±0.4	0.5±0.1	2.9±0.6	1.6±0.4	3.0±0.8	0.4±0.0
20:5ω3 EPA	4.3±0.8	2.2±0.3	0.8±0.6	5.0±1.6	2.8±0.5	5.2±1.5	0.7±0.0
20:1ω9	8.6±2.2	2.5±1.2	0.6±0.2	3.2±0.8	2.5±0.9	4.0±1.4	1.6±0.0
22:6ω3 DHA	27.2±3.7	13.9±3.0	5.0±2.1	36.0±5.3	19.4±3.3	36.6±6.2	2.4±0.0
22:5ω3	2.0±0.2	1.1±0.1	0.3±0.2	2.5±0.5	1.3±0.1	2.6±0.2	0.4±0.0
22:1ω11	2.7±1.0	0.7±1.0	0.1±0.1	0.6±0.3	0.4±0.4	0.9±0.8	0.6±0.0
Others	8.6	6.4	3.1	7.1	6.1	7.6	10.7
Total SFA	27.9±1.5	15.8±1.9	26.5±0.3	26.5±1.7	17.5±1.1	26.3±1.4	53.4±0.1
Total MUFA	31.1±6.0	51.3±4.4	22.0±0.5	21.7±4.1	44.9±3.0	22.6±7.2	36.0±0.0
Total PUFA	41.0±5.0	32.8±2.6	51.6±0.8	51.8±4.1	37.5±2.5	51.2±5.7	10.6±0.0
Totalω3	34.9±4.3	17.7±2.8	6.3±2.7	44.4±4.8	24.0±2.9	45.2±4.7	3.7±0.0
Totalω6	5.6±0.7	15.0±0.5	44.1±3.0	7.1±1.2	13.4±0.7	5.8±1.3	4.1±0.0

Others include 12:0,i14:0,14:1,i15:0,a15:0,15:0,C16PUFA,i16:0,16:1ω9,16:2,16:1ω5,i17:0, 17:1ω8/a17:0,17:0,18:3ω6,18:4ω3,18:3ω3,18:2,18:1ω5,20:3ω6,20:4ω3,20:2ω6,20:1ω11, 20:1ω7,20:0,21PUFA,22:5ω6,22:4ω6,22:3ω3,22:1ω9,22:1ω7,22:0,C23PUFA,C24PUFA,24:1,24:0.

cottonseed oil used for pan-frying, with LA dominant (53%) in peanut oil used for deep frying. Steaming and microwaving had no observable effect on oil content and composition in blue eye.

This study represents, to our knowledge, one of the first detailed comparative studies of the effect of various cooking processes on the oil content and composition of an Australian species. Importantly, no loss of the beneficial omega-3 LC-PUFA was observed for any of the forms of cooking examined. An increase in oil content and levels of specific components was seen with frying and grilling, reflecting uptake of vegetable oil components. In terms of nutritional value, whilst the content of omega-3 LC-PUFA

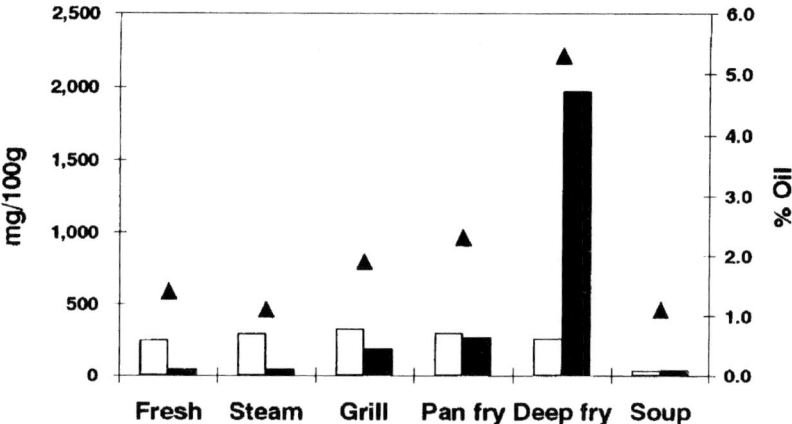

Figure 2. The effect of cooking on the content (mg/100 g, wet mass) of omega-6 PUFA and omega-3 PUFA and oil (%, wet mass) in blue eye trevalla fillets, also showing fish soup. ■, Omega-6 PUFA; □, Omega-3 PUFA; ▲, Oil.

did not decrease, the ratio of omega-3 to omega-6 fatty acids did decrease markedly with frying and grilling. Such changes in oil, in particular fatty acid, profiles will need further nutritional evaluation. The use of cooking oils containing lower levels of omega-6 PUFA may be worthy of consideration.

Soup products derived partly from seafood by-products can provide a further source of omega-3 LC-PUFA (Fig. 2, Table 2), although in this instance levels in the soup were lower than observed for the fresh and cooked fish. A variety of seafood by-products can contain high concentrations of the beneficial omega-3 LC-PUFA (e.g. salmon frames and belly flap contain higher levels omega-3 LC-PUFA than occurs in fillet). In this study, only low amounts of omega-3 PUFA were found for the fish-derived soup product (Fig. 2), therefore scope exists to further increase the level of beneficial oils in soups or stocks.

3.3. Farmed fish — the effect on oil composition

Barramundi (*Lates calcarifer*) is a popular wild-caught table species harvested by commercial and recreational fishers from northern Australian waters. Wild-harvested specimens contained on average 0.4% oil (in saltwater specimens) and 0.8% (freshwater specimens) [7]. In comparison, cultured barramundi contained 10% oil (wet weight basis), with triglyceride as the main oil class (97%, Table 3). In wild-harvested specimens, polar lipid was the dominant oil class. The main fatty acids in cultured barramundi in decreasing order of abundance were: $18:1\omega9$, $16:0$, DHA, $16:1\omega7$, EPA, $14:0$, $18:2\omega6$ and $18:0$; these eight components accounted for 77% of the total fatty acids.

The omega-3 LC-PUFA, predominantly DHA and EPA, accounted for 22% of the total fatty acids in cultured barramundi, with total PUFA at 29%. DHA was the dominant

Table 3
Lipid content and composition of cultured barramundi

LipidClass	Percentage composition
Triglyceride	97.4±0.7
Free fatty acid	0.2±0.0
Cholesterol	0.2±0.0
Phospholipid	2.2±0.6
Oil content (wet weight basis)	10.0±1.2

PUFA at 10%. The relative (percent) level of omega-3 LC-PUFA in cultured barramundi was similar to wild-caught freshwater barramundi (26%), but lower than in wild-caught saltwater barramundi (43%). The ratio omega-3 PUFA / omega-6 PUFA was higher (3.0) in cultured fish compared to the wild fish (0.8-1.9).

On an absolute basis, the omega-3 LC-PUFA content of cultured barramundi was 1970 mg/100 g (Fig. 3); this content is markedly higher than in most wild-caught seafood. In comparison to wild-caught fish, cultured fish fed marine oil based diets generally contain higher levels of the beneficial oils.

Under the current feeding practices cultured barramundi is an excellent source of the

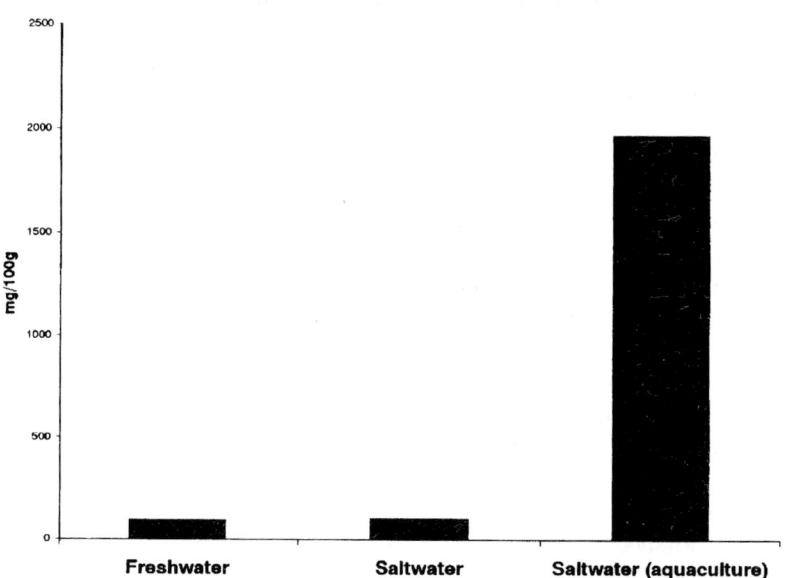

Figure 3. Content (mg/100 g, wet mass) of omega-3 LC-PUFA in wild-caught (freshwater and saltwater specimens) and cultured barramundi.

beneficial omega-3 LC-PUFA. Should fish meal and fish oil in current aquaculture diets be replaced with other protein and oil sources, the current high levels of omega-3 LC-PUFA may decrease, as may the product and nutritional value. Scope to further increase the level of beneficial oils in cultured barramundi through manipulation of diet also may exist.

3.4. By-product oils

In addition to research devoted to examination of the oil composition of edible species, collaborative research and development has occurred to better utilize marine resources [9].

3.5. Tuna oil

Of the Australian fishes, tuna were found to be a particularly good source of DHA-rich oil. Tuna oil is derived as a by-product from canning operations. Collaborative research and development by CSIRO and Clover Corporation, presented an opportunity to exploit the waste derived from processing Pacific tuna. The joint venture company Nu-Mega Lipids commenced manufacture of Australian-made, value-added tuna oil products in early 1998.

The tuna oil products contain around 25% DHA, and a DHA to EPA ratio of >3. It is these two parameters in particular that make the oil suitable, after the addition of AA, for nutraceutical application in infant formulas. When used as a nutritional supplement, consumption of two capsules (2×1 g) per day of tuna oil would provide higher levels of DHA (approximately 500 mg, Fig. 1) than is present in an average serve of Australian seafood (average LC-omega-3 PUFA, 235 mg/100 g; Table 1). In comparison, for most Australians the average intake of EPA and DHA is about 100 mg/day, which is less than half that recommended by the UK Department of Health, and only one-tenth of the average population intake recommended by a British Nutrition Foundation Task Force [4]. Of additional interest was the observation that in analyses of purified tuna oil performed to date, we have noted a lower proportion of cholesterol (typically 0.1% or less of the oil) than is found in flesh from tuna and other seafood (average cholesterol in fish, 28 mg/100 g, Fig. 1, [7,14]).

In comparison to the tuna oil supplement (500 mg DHA per 2 g oil), 2 g of a traditional fish oil supplement (e.g. MaxEPA capsules) contains 240 mg of DHA. Similarly, 2 g of newly developed plant-derived (microalgae) supplements available to the US market provide 400 mg of DHA (Nichols, P. and Mooney, B., unpublished data, 1997). The tuna oil supplement therefore generally contains higher levels of DHA than other supplements currently available.

3.6. Other oils

Liver oil profiles have been obtained for southern and northern Australian sharks, together with by-product oils from other fisheries. Examination of the fatty acid profiles indicates that new sources are potentially available for omega-3 (Fig. 4) and also diacylglyceryl ether containing oils. These two types of oils have been manufactured over the last 5 years in Australia from by-catch and by-products of other fisheries. The scope

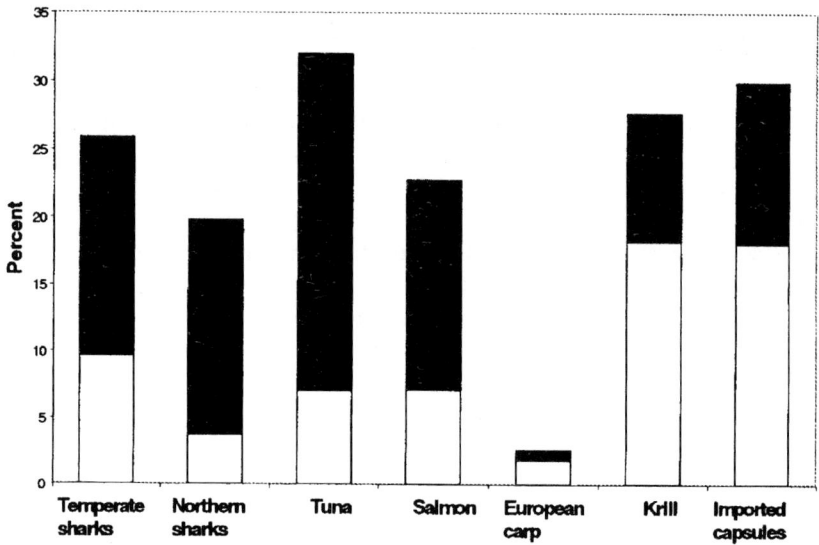

Figure 4. EPA and DHA composition (as % of total fatty acids) of selected by-product oils from Australian species compared with that in imported fish oil capsules. Northern shark data represents mean data for 41 species (unpublished data). □, EPA; ■, DHA.

may exist to utilize by-products of the northern shark fishery and other fisheries as new scources for these valuable oils.

3.7. Infant formula and functional food products

The use of fish oil supplements has gained considerable interest in a variety of applications, including as an additive in infant formulas. It has been demonstrated that with unsupplemented bottle feeding, a deficiency of DHA content (up to 50%) exists in erythrocyte lipid, phosphatidylcholine and phosphatidylethanolamine compared to breast-fed infants [17,18]. DHA supplemented (fish oil source) formula-fed infants exhibited a more rapid rate of development of visual acuity compared to control formula-fed babies [19,20]. The AA : EPA : DHA ratio in Northern Hemisphere fish oil (commonly 0.2 : 2.1 : 1.0 [14]) differs markedly from human milk (2 : 0.2 : 1). The high levels of EPA in fish oils from northern hemisphere species may act as an antagonist or an inhibitor of the infant's own endogenous AA synthesis, therefore infant formula may require AA co-supplementation [20].

Flesh from a select few Australian fishes contains the three essential fatty acids at a ratio more similar to that observed in breast milk [14]. However, to our knowledge, by-product oils containing the three essential fatty acids at the ratio found in breast milk are not available from Australian fishes. As noted above, tuna oil contains a high DHA to EPA ratio. After microencapsulation using a process developed by Food Science Australia and Clover Corporation, tuna oil is now being added to infant formula with either AA or gamma-linolenic acid (GLA, 18:3ω6) (see www.clovercorp.com.au). Tuna oil is also being utilized in new Australian bread products, and will be incorporated into other functional foods in the future. These examples highlight that the Australian fishing and associated industries do have the capacity to better utilise existing resources.

A perceived issue for the use of single cell oils (SCOs) from microalgae in infant formula is the possibility of toxicity problems. Noteworthy in this area is the recent reporting of the arresting of embryonic development in copepods by inhibitory compounds in diatom cells, with the possibility existing that algal oil constituents may be the toxic components [22]. The safety of unusual lipids derived from microalgal oils used for infant formulae production needs to be well established as noted at the International Meeting on Infant Nutrition held in Barcelona during late 1996 (INFORM, February 1997). As oil derived from fish generally does not contain unusual components, including e.g. unknown sterols, use of fish oil in infant formula would overcome this potential problem.

3.8. Microbial oils

Marine microorganisms contain an array of bioactive molecules, including oils, that have benefits in aquaculture, nutriceuticals, in new pharmaceuticals or as lead compounds. The possibility of using bacteria as aquaculture feeds has been previously considered, but their perceived lack of PUFA was thought to be a major drawback. It is evident that certain strains of Antarctic bacteria do produce high levels of PUFA [23]. The ability to produce PUFA can allow these bacteria to be considered as a valuable addition or alternative to current aquaculture feeds. Bacterial production of PUFA represents a renewable resource, in comparison to the variable nature of fish catches that are currently the most common source of omega-3 PUFA. PUFA incorporation into live feeds (rotifers) has been demonstrated using both Antarctic bacteria [10,11,23] and several recently isolated novel Australian thraustochytrids [12]. The latter group of microheterotrophs represent an attractive source of PUFA-rich oils, with new strains isolated that are rich in DHA and AA (Fig. 5) [24].

3.9. Species identification

Health complaints have occurred recently in Australia associated with the consumption of fillets sold as "rudderfish". The marketing group rudderfish consists of species from three trevalla (family *Centrolophidae*) genera, *Centrolophus*, *Schedophilus* and *Tubbia*, with several undescribed species and uncertain distribution in Australian waters. Consumers had purchased "rudderfish" fillets and, after cooking and eating, some had suffered severe diarrhoeal effects not usually associated with these fish, but reported for others such as escolar (or oil fish) [8]. Concern existed over the species identity of the fillets and the cause of the health effects, particularly as both rudderfish and escolar are both caught as long-line by-catch and anedoctal evidence suggested that escolar species were being sold as rudderfish. The name "escolar" in Australian fisheries includes two known gemfish (family *Gempylidae*) species, *Lepidocybium flavobrunneum* and *Ruvettus pretiosus*. The latter species has also been referred to as "castor oil fish [25]. Both species have reported purgative properties [26].

We compared the oil content and composition profiles, with particular emphasis on the non-saponifiable lipids, of two fish fillets associated with consumer illness, with those

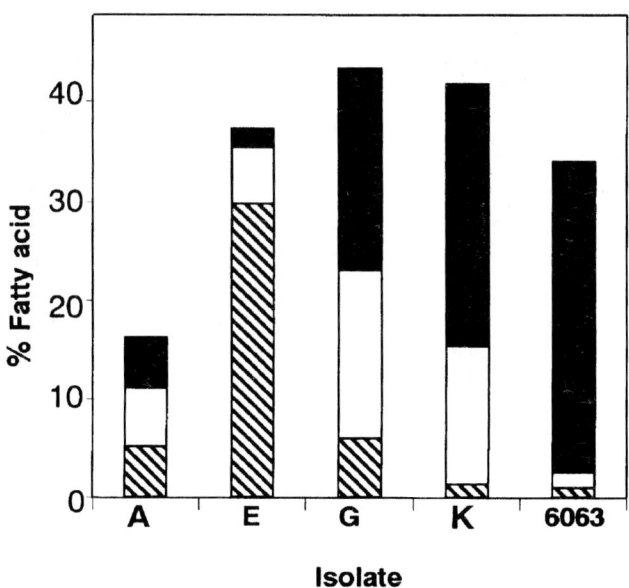

Figure 5. Composition of main PUFA (% of total fatty acids) in the oils of novel Australian thraustochytrids. Uauy, R. D. (1990) [19]; Miralto, A. et al. (1997) [22]. ◨, AA; ☐, EPA; ■, DHA. Lewis, T. et al. (1998) [24]; Lewis, T. et al. (2001) [12].

obtained for reference samples from the escolar and rudderfish groups. The analyses, supported by general protein fingerprinting (Elliott N., unpublished data, 1999) of the samples, highlight the potential for lipid profiles to be used for identification of selected escolar and rudderfish samples to at least group level.

Both of the unknown fillet samples that had been associated with consumer illness had a very high oil content (>22%, as % wet weight) and the dominant lipid class was wax ester (97%, as % of total oil, Fig. 6). Such a high oil content and unusual composition were similar to that found in the reference escolar samples and very different to rudderfish specimens (Fig. 6).

Although the lipid class and fatty alcohol profiles for *L. flavobrunneum* and *R. pretiosus* are very similar, higher levels of 18:1ω9 and 16:1ω9, and lower levels of 18:0 were observed in R. pretiosus [27]. Based on comparison of the fatty alcohol profiles, the unknown samples grouped more closely with *L. flavobrunneum* than with *R. pretiosus*. This finding is consistent with results obtained from general protein fingerprinting of the two unknown (Elliott, N., unpublished data, 1999) and reference samples [8].

Members of the escolar and rudderfish groups are unusual in containing very high levels of oil (14-25%, as % of total oil) and having oil composition profiles different to most seafood. Oils from seafood, with some exceptions, are generally rich in triacylglycerol and / or polar lipid. The unusual oil profiles of specific members of the escolar and rudderfish groups are not consistent within their families. Members of both families (*Gempylidae* and *Centrolopidae*) are known to have more conventional triacylglycerol or polar lipids [7,14].

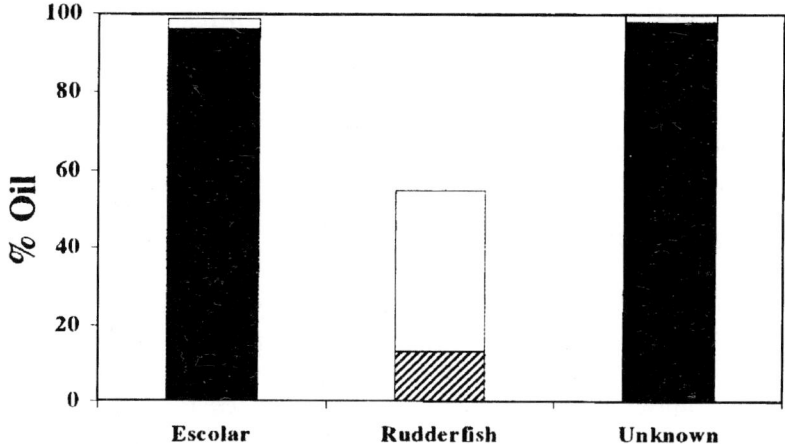

Figure 6. Lipid class composition (% of oil) of oil fishes (escolar and rudderfish) from Australian waters, including an unknown specimen associated with consumer illness. ☐, Polar Oil; ■, Wax Ester; ▨, Triglyceride.

The first report on the purgative properties of *R. pretiosus* occurred in 1841 [28]. More recently, and based on the high wax ester content in Japanese specimens of *L. flavobrunneum* and *R. pretiosus* [25], the diarrhoea and seborrhoea-producing activity of these fishes was investigated [29]. Based on the results for feeding trials in rats, the flesh and acetone-derived oil of both species were deemed not suitable for human food.

High levels of wax ester rich oil in orange roughy has been reported (fillets 7-10% oil; oil composition 95% wax ester) [30]. Extrapolating results obtained from feeding growing rats and pigs with orange roughy, it was proposed that "normal consumption" of orange roughy by humans was unlikely to cause serious health problems [31]. However, the level of wax ester oil in escolar (18-24%) is nearly three times greater than in orange roughy. Wax ester oils derived from orange roughy and oreo dories have been incorporated into several industrial cleaning, degreasing and other products in the past decade [9]. Based on the findings of this study, escolar may represent an additional source of wax ester oils for consideration by industry. The results, and the possible incorrect naming of the fillets, suggest that consumers should be made aware of the oil type in these two groups and that strict use be made of recommended marketing names to avoid similar health issues.

4. CONCLUSION

In summary, research and developments with omega-3 and other marine oils have occurred in Australia over the past decade and an exciting era is being entered where the potential benefits from the abundant marine biodiversity will be both increasingly understood and utilised. The Australian Marine Oils industry has progressed rapidly over the past 5 years and opportunities are now available to consolidate developments and commence new initiatives. Basic research opportunities also exist for using oil composition profiles in food-web studies, including as part of ecosystem studies. Ongoing oil characterization and process-oriented research will complement and strengthen existing industry initiatives and allow the Marine Oils industry to maximise returns on the present fish catch.

5. ACKNOWLEDGEMENT

The authors are grateful to MEUFFP organizers, in particular Drs Takashi Hirata and Allan Bremner, for their encouragement to both attend the Symposium and submission of the manuscript. The research was largely funded by the Australian Fisheries Research and Development Corporation. Clover Corporation is thanked for ongoing support to the project. Chris Strauss of CSIRO Molecular Science is kindly thanked for ongoing input to the project, as are colleagues in the Fish taxonomy group at CSIRO Marine. Mures Fish Centre kindly provided fish and soup analysed in this study, and Bluewater Barramundi provided fresh specimens. Collaboration on novel microbial oils has been performed with Drs David Nichols and Tom Lewis of the University of Tasmania. We thank Jill and George Mure and Paul Lynch for their help with the cooking performed in the Mures Upper Deck Restaurant, and for their ongoing interest in and support of the project. Danny Holdsworth managed the CSIRO GC-MS facility.

REFERENCES

1. Jowett, W. G. and Davies, W. (1938) A chemical study of some Australian fish. CSIRO pamphlet No. 85.
2. Kinsella, J.E. (1986) Food components with potential therapeutic benefits: the n-3 polyunsaturated fatty acids of fish oils, Food Technology. Feb; 89-97.
3. Kinsella, J.E. (1987) Seafoods and fish oils in human health and disease, Marcel Dekker Inc., New York. pp. 317.
4. Howe, P. R. C. (1998) Omega-3 fatty acids – an Australian perspective. World Rev. Nutr. Diet., 83: 215-218.
5. Howe, P. R. C. and Nestel, P. J. (1992) Antihypertensive effects of fish oil combined with a low sodium diet. Heartbeat, 3: 3-4.
6. Connor, W. E. (1997) The beneficial effects of omega-3 fatty acids: cardiovascular disease and neurodevelopment. Current Opinion in Lipidology, 8: 1-3.
7. Nichols, P. D, Virtue, P., Mooney, B. D., Elliott, N. G. and Yearsley, G. K. (1998) Seafood the Good Food. The oil content and composition of Australian commercial fishes, shellfishes and crustaceans. FRDC Project 95/122. Guide prepared for the Fisheries Research and Development Corporation.
8. Yearsley, G. K., Last, P. R. and Ward, R. D. (1999) Australian Seafood Handbook. CSIRO Marine Research.
9. Nichols, P., Elliott, N., Bakes, M. and Mooney, B. (1997) Marine oils from Australian fish: characterisation and value-added products. FRDC Project 94/115. Final report

prepared for the Fisheries Research and Development Corporation.
10. Lewis, T., Nichols, P. D., Hart, P. R., Nichols, D. S. and McMeekin, T. A. (1998) Enrichment of rotifers *Brachionus plicatilis* with eicosapentaenoic acid and docosahexaenoic acid produced by bacteria. J. World Aquaculture Society 29, 313-318.
11. Nichols, P. D., Nichols, D. S., Lewis, T. Bowman, J. P., Brown, J., Skerratt, J. H. and McMeekin, T. A. (1998) Novel bacteria as alternate sources of polyunsaturated fatty acids for use in aquaculture and other industries. In: Proceedings Marine Microorganisms for Industry Conference. LeGal, Y. and Muller-Feuga, A. (eds.), Editions IFREMER, Plouzane, pp. 26-32.
12. Lewis, T. (2001) Characterization and application of thraustochytrids. PhD thesis, University of Tasmania.
13. Bakes, M. J., Elliott, N. G., Green, G. J. and Nichols, P. (1995) Within and between species variation in lipid composition of Australian and North Atlantic oreo and orange roughy: regional specific differences. Comp. Biochem. Physiol. 111B: 633-642.
14. Nichols, P. D., Mooney, B., Virtue, P. and Elliott, N. (1998) Nutritional value of Australian Fish: oil, fatty acid and cholesterol composition of edible species. FRDC Project 95/122. Final report prepared for the Fisheries Research and Development Corporation.
15. Nichols, P. D., Bakes, M. J., and Elliott, N. G. (1998) Docosahexaenoic acid-rich liver oils from temperate Australian sharks. Marine and Freshwater Research., 49: 763-767.
16. Bligh, E. G. and Dyer, W. J. (1959) A rapid method of total lipid extraction and purification. Can. J. Biochem. Physiol., 35: 911-917.
17. Putnam, J. C., Carlson, S. E., DeVoe, P. W. and Harness, L. A. (1982) The effect of variations in dietary fatty acids on the fatty acid composition of erythrocyte phosphatidylcholine and phosphatidylethanolamine in human infants. Am. J. Clin. Nutr., 36: 106-110.
18. Simopoulos, A. P. (1989) Summary of the NATO advanced research workshop on dietary ω3 and ω6 fatty acids: biological effects and nutritional essentiality. J. Nutr. 119, 521-528.
19. Uauy, R. D. (1990) Are ω3 fatty acids required for normal eye and brain development in the human? J. Pediatr. Gastroenterol. Nutr. ,11: 296-302.
20. Uauy, R. D., Birch, D. G., Birch, E. E., Tyson, J. E. and Hoffman, D. R. (1990) Effect of dietary omega-3 fatty acids on retinal function of very-low-birth-weight neonates. Pediat. Res., 28: 485-492.
21. Singh, A. and Ward, O. P. (1997) Microbial production of docosahexaenoic acid (DHA, C22:6). Advances in Applied Microbiology, 45: 271-312.
22. Miralto, A., Ianora, A., Poulet, S. A., Romano, G., Buttino, I. and Scala, S. (1997) Embryonic development in copepods is arrested by inhibitory compounds in diatom cells. 4th International Marine Biotechnology Conference, Italy, September 1997, abstracts, p.79.
23. Nichols, D. S., Hart, P., Nichols, P. D. and McMeekin, T. A. (1996) Enrichment of the rotifer *Brachionus plicatilis* fed an Antarctic bacterium containing

polyunsaturated fatty acids. Aquaculture, 147: 115-125.
24. Lewis, T., Mooney, B., McMeekin, T. and Nichols, P. (1998) New Australian microbial sources of polyunsaturated fatty acids. Chemistry in Australia, 65: 37-39.
25. Nevenzel, J. C., Rodegker, W. A. and Mead, J. F. (1965) The lipids of *Rivettus pretiosus* muscle and liver. Biochem., 4: 1589-1594.
26. Berman, P., Harley, E. H. and Spark, A. A. (1981) Keriorrhoea - the passage of oil *per rectum* — after ingestion of marine wax esters. South African Medical Journal, 59: 791-792.
27. Nichols, P. D, Mooney, B. D. and Elliott, N. G. (2001) Unusually high levels of non-saponifible lipids in the fish, escolar and rudderfish: a tool for identification. J. Chromatog. A., 936: 183-191.
28. Lowe, R. T. (1841) A synopsis of the fishes of Madeira, etc. Tran. Zoological Society, London, 2: 180-181.
29. Mori, M., Saiko, T., Nakanishi, Y., Miyazawa, K. and Hashimoto, Y. (1966) The composition and toxicity of wax in the flesh of castor oil fishes. Bull. Japan. Soc. Sci. Fish., 32: 137-145.
30. Bakes, M. J., Elliott, N. G., Green, G. J. and Nichols, P. D. (1995) Within and between species variation in lipid composition of Australian and North Atlantic oreo and orange roughy: regional specific differences. Comp. Biochem. Physiol., 111B: 633-642.
31. James, K. A. C., Body, D. R. and Smith, W. C. (1986) A nutritional evaluation of orange roughy (*Hoplostethus atlanticus*) using growing pigs. NZ Journal of Technology, 2: 219-223.
32. Cashel, K., English, R. and Lewis, J. (1989) Composition of Foods Australia. Volume 1. Australian Government Publishing Service, Canberra.
33. Mann, N., Johnson, L. G., Warrick, G. E. and Sinclair, A. J. (1995) The arachidonic acid content of the Australian diet is lower than previously estimated. Journal Nutrition, 125: 1-8.
34. Sinclair, A. J., O'Dea, K. and Naughton, J. M. (1983) Elevated levels of arachidonic acid in fish from northern Australian coastal waters. Lipids, 18: 877-881.

UTILIZATION OF MARINE INVERTEBRATES AS RESOURCE FOR BIOACTIVE METABOLITES: ISOLATION OF NEW MYCALOLIDES AND CALYCULINS

Shigeki Matsunaga and Nobuhiro Fusetani
Laboratory of Aquatic Natural Products Chemistry, Graduate School of Agricultural and Life Sciences, The University of Tokyo, 1-1-1 Yayoi, Bunkyo-ku, Tokyo 113-8657, Japan

Abstract

As a result of our study toward exploration of Japanese marine invertebrates as a source of secondary metabolites of biomedical importance, we have discovered mycalolides from marine sponges of the genus *Mycale* and calyculins from the marine sponge *Discodermia calyx*. Mycalolides were discovered as potent cytotoxins against KB cell line and later shown to be actin-depolymerizing agents. Calyculins were isolated as inhibitors against the development of fertilized sea urchin embryo and turned out to be potent inhibitors of protein phosphatases 1 and 2A. The discovery and structure elucidation of several novel metabolites in these classes are described.

1. INTRODUCTION

Research efforts for the past 30 years by marine natural product chemists worldwide have disclosed that marine organisms are a rich source of secondary metabolites, whose structural features are significantly different from those of the terrestrial counterparts [1]. We have focused on the discovery of biologically active compounds with biomedical importance from marine invertebrates collected along the coast of Japan. Bioassays such as antimicrobial, antifungal, cytotoxic, enzyme inhibitory, and receptor-binding inhibitory activities were employed. We have isolated a variety of new metabolites with novel structural features. Among them are mycalolides and calyculins both of which are now commercialized as biochemical reagents to inhibit specific enzymes. We will describe the discovery and structure elucidation of several novel metabolites in these classes [2,3].

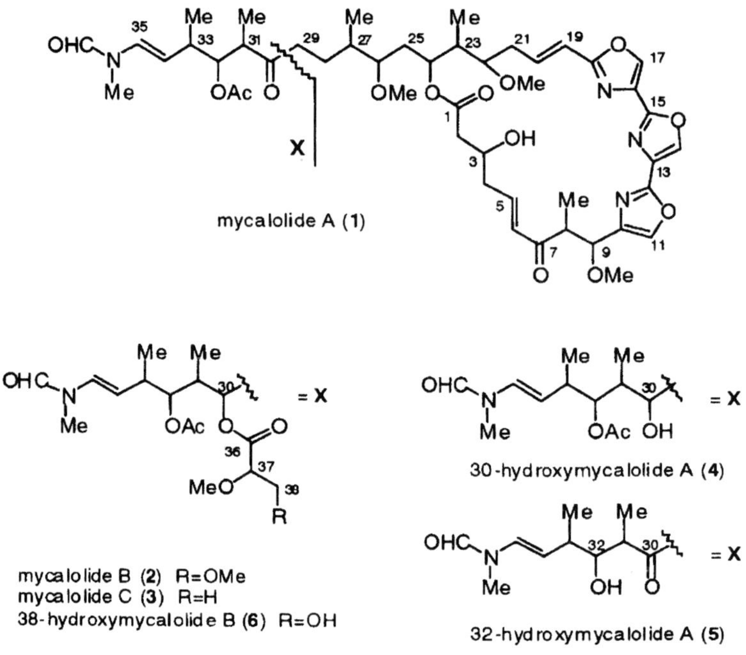

Figure 1. Structures of natural mycalolides 1-6.

2. MATERIALS AND METHODS

2.1. A representative scheme for extraction of *Mycale magellanica* and isolation of mycalolides

The sponge was collected at a depth of 15 m off Kumomi on the Izu Peninsula, Shizuoka Prefecture, Japan, and identified by Professor P. Bergquist (University of Auckland) as *Mycale magellanica* Thiele. The frozen sample (1 kg) was extracted with EtOH and the combined extract was evaporated to yield an aqueous suspension which was partitioned between water and ether. The ether phase was partitioned between *n*-hexane and MeOH/H_2O (9:1), and the latter phase, which was antifungal against *Mortierella ramanniana* was applied to a column of ODS and eluted with aqueous MeOH by increasing the concentration of MeOH. The fractions eluted with MeOH/H_2O (8:2) and MeOH/H_2O (9:1) were combined and further chromatographed on a column of Sephadex LH-20 (MeOH). The antifungal fractions were further purified by ODS HPLC with MeOH/H_2O (7:3) followed by ODS HPLC with MeCN/H_2O (1:1) to yield 30-hydroxymycalolide A (**4**, 7.5 mg), 32-hydroxymycalolide A (**5**, 12.3 mg), and 38-hydroxymycalolide B (**6**, 14.6 mg) together with mycalolide A (**1**, 8.5 mg) and mycalolide B (**2**, 32.3 mg) (Fig. 1).

2.2. NaBH$_4$ reduction of 30-hydroxymycalolide A (4) and mycalolide A (1)

To a solution of 30-hydroxymycalolide A (1.7 mg) in MeOH (0.5 mL) was added

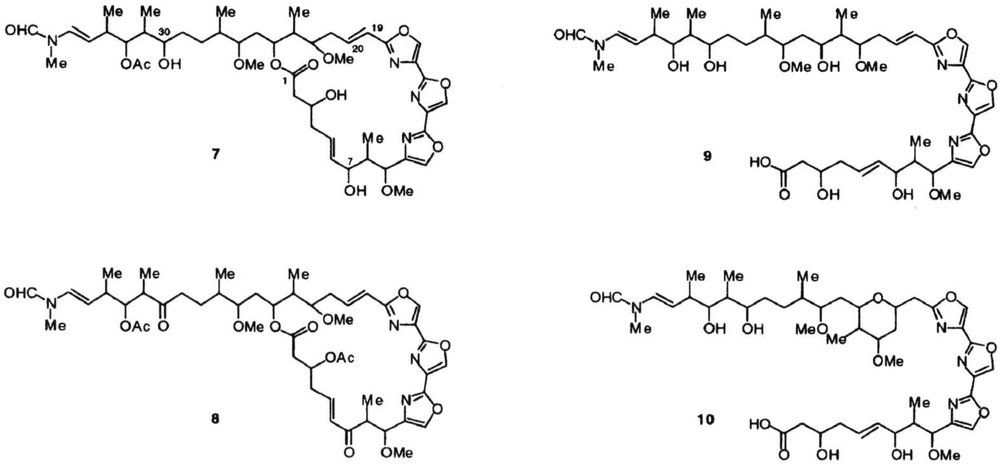

Figure 2. Structures of compounds **7-10**.

NaBH₄ (2.2 mg) and the mixture stirred for 30 min at 0°C. To the reaction mixture was added 5% AcOH in H₂O (1 mL) and the solution was subjected to ODS-HPLC with MeOH/H₂O (8:2) to yield **7** (1.6 mg). Mycalolide A (1.7 mg) was similarly treated to afford **7** (0.9 mg) and the C-30 epimer of **7** (0.6 mg) (Fig. 2).

2.3. Acetylation of 32-hydroxymycalolide A (5) and mycalolide A (1)

Compound **5** (1 mg) was dissolved in a 1:1 mixture of acetic anhydride and pyridine (1 mL) and the mixture stirred overnight at room temperature (rt). After removal of the solvent by lyophilization, the reaction product was applied to a silica gel column (1 × 2 cm) and eluted with CHCl₃ followed by a mixture of CHCl₃/MeOH (98:2) to furnish the diacetate **8** in quantitative yield. Mycalolide A was treated in the same way to afford **8**.

2.4. Base hydrolysis of mycalolide B (2), mycalolide C (3), 30-hydroxymycalolide A (4), and 38-hydroxymycalolide B (6)

Mycalolide B (0.8 mg) was reduced with NaBH₄ as described above and the major peak in the reversed-phase HPLC was collected. The reduction product was dissolved in a mixture of MeOH/1N LiOH in H₂O (2:1, 1 mL) and the mixture left standing at rt overnight. To the reaction mixture was added AcOH (100 mL) and after removal of MeOH by evaporation, the aqueous solution was subjected to ODS HPLC with MeOH/H₂O (7:3) to furnish **10** (0.6 mg). Mycalolide C, 30-hydroxymycalolide A, and 38-hydroxymycalolide B were treated in the same way to yield **10** whose ¹H NMR spectrum was indistinguishable from the spectrum of the compound prepared from **2**.

2.5. A representative scheme for the extraction of *D. calyx* and isolation of calyculins

The sponge *D. calyx* was collected by SCUBA at depths of 15-20 m off the Izu Peninsula. Specimens were immediately frozen and kept frozen at -20°C until processed.

After epibionts were removed, the frozen sponge (1.7 kg) was homogenized and extracted with ethanol (3 × 3 L). The combined extracts were concentrated and partitioned between CH_2Cl_2 and H_2O. The organic phase (5.2 g) was subjected to silica gel flash chromatography with CH_2Cl_2/MeOH solvent pairs. Fractions were monitored by silica gel TLC ($CHCl_3$-MeOH, 9:1). The CH_2Cl_2/MeOH (95:5) fraction was subjected to ODS flash chromatography with aqueous methanol. The 80% and 100% MeOH eluates were further fractionated by silica gel column chromatography, eluting with mixtures of $CHCl_3$/MeOH. Final purification of the new calyculins was accomplished by ODS-HPLC with 82% MeOH and then with 70% MeCN to yield calyculin J (**12**) (1.9 mg), calyculinamide A (**13**) (0.1 mg), calyculinamide F (**4**) (0.2 mg), and des-*N*-methylcalyculin A (**15**) (0.4 mg).

2.6. Preparation of calyculin J from calyculin A

To a solution of calyculin A (1 mg) in THF was added *N*-bromosuccinimide (10 μL of a 0.1 M solution in THF), and the mixture stirred at rt for 90 min. The reaction mixture was separated by HPLC (ODS column, 82%MeOH) to afford calyculin J (0.4 mg).

2.7. Preparation of calyculinamide A from calyculin A

Calyculin A (2 mg) was treated with a mixture of H_2O_2 (20 μL) and 25% NH_4OH (100 μL), and the mixture stirred at rt for 15 h. The reaction mixture was evaporated to dryness to yield a white powder which was subjected to ODS column chromatography (0.5 × 3 cm) with 40% MeOH (2 mL) and MeOH (2 mL). The MeOH eluate was dissolved in MeOH and reduced with Ph_3P (3 mg) at rt for 5 days. The mixture was dried and separated by SiO_2 column chromatography with $CHCl_3$, $CHCl_3$/MeOH(95:5), $CHCl_3$/MeOH (9:1) and MeOH, followed by HPLC on an ODS column with 70% MeCN to obtain calyculinamide A (0.8 mg).

3. RESULTS AND DISCUSSION

The EtOH extract of the sponge *Mycale magellanica* afforded six mycalolides, including three new derivatives, 30-hydroxymycalolide A (**4**), 32-hydroxymycalolide A (**5**), and 38-hydroxymycalolide B (**6**) together with mycalolides A-C (**1-3**), whose structures were assigned on the basis of spectral data.

30-Hydroxymycalolide A (**4**) has a molecular formula of $C_{47}H_{66}N_4O_{14}$ as determined by HRFABMS. The ^1H NMR spectrum immediately revealed the presence of three singlets arising from the tris-oxazole moiety and the 1:2 doublet of the formamide signals, which are characteristic of the mycalolide/kabiramide class of compounds. Subsequent analysis of 2D-NMR data indicated **4** was very similar to mycalolide A (**1**) except that the C-30 ketone was replaced by a secondary alcohol in **4**. This was supported by considerable up-field shifts of H_2-29 and H-31 as well as the appearance of an oxygenated methine proton at 3.51 ppm (H-30). Therefore, compound **4** was 30-hydroxymycalolide A. However, it was not concluded whether **1** and **4** have the same stereochemistry at the corresponding chiral centers. In order to clarify this issue, both mycalolide A and 30-hydroxymycalolide A were subjected to reduction with $NaBH_4$; 30-

hydroxymycalolide A furnished the allylic alcohol **7** as the major product, while mycalolide A gave two compounds, one of which co-eluted with **7** in the reversed-phase HPLC and exhibited the ^{1}H NMR spectrum superimposable on that of **7**. Therefore, **1** and **4** have the same stereochemistry at all the chiral centers except for C-30.

32-Hydroxymycalolide A (**5**) had a molecular formula of $C_{45}H_{62}N_4O_{13}$, which was smaller than mycalolide A by C_2H_2O. Comparison of ^{1}H and ^{13}C NMR data of **5** with those of **1** revealed the absence of the acetyl signal and an up-field shift of H-32 in **5**, thereby indicating that **5** had a secondary alcohol at C-32 instead of the acetoxy in **1**. Acetylation of either mycalolide A or of compound **5** furnished the diacetate **8**, allowing us to confirm that the stereochemistry of all the chiral carbons of **5** is identical with that of mycalolide A.

38-Hydroxymycalolide B (**6**) had NMR spectral features similar to those of mycalolide B (**2**). However, there was a difference in the number of *O*-methyl groups; one of the five *O*-methyl signals of mycalolide B was missing in **6**. Therefore, **6** was likely to be the de-*O*-methyl derivative of **2**, which was supported by HRFABMS data. Comparison of NMR data for **6** and **2** readily revealed that the C-38 carbon signal in **6** experienced an up-field shift of 9 ppm, thus suggesting that the C-38 methoxyl group in **2** was replaced by a hydroxyl group in **6**. In order to correlate the stereochemistry of mycalolide B with that of compound **6**, each was reduced with $NaBH_4$ followed by hydrolysis with LiOH. Although the expected pentol **9** was not obtained, compound **10** was obtained in each case, which indicated that both **6** and **2** had identical stereochemistry except for the unidentified chiral center C-37 in the glycerate residue. Furthermore, mycalolide C (**3**) and 30-hydroxymycalolide A (**4**) were treated in the same way to yield **10**, thereby disclosing that compounds **1-6** had the same stereochemistry at the corresponding chiral centers. Compounds **1-6** exhibited cytotoxic activity against L1210 cells with IC_{50} values between 0.01 μg/mL and 0.02 μg/mL.

Frozen specimens (1.7 kg, wet weight) of the sponge *D. calyx* were extracted with EtOH; the extract was partitioned between water and CH_2Cl_2. The organic phase was purified by silica gel and ODS chromatographies followed by ODS-HPLC to afford calyculin J (**12**; 1.9 mg, 1.1×10^{-4} % wet weight), calyculinamide A (**13**; 0.1 mg, 5.8×10^{-6}% wet weight), calyculinamide F (**14**; 0.2 mg, 1.2×10^{-5} % wet weight), and des-*N*-methyl-calyculin A (**15**; 0.4 mg, 2.3×10^{-5} % wet weight) (Fig. 3).

The presence of one bromine atom in calyculin J (**12**) was readily inferred from the 1:1 intensity of the $(M+H)^+$ ion peaks at *m/z* 1088 and 1090 in FABMS. The molecular formula was assigned as $C_{50}H_{80}BrN_4O_{15}P$ on the basis of HRFABMS and NMR data. The ^{1}H NMR spectrum was similar to that of calyculin A. Interpretation of the 2D NMR data including HMBC spectrum led us to propose the gross structure of **12**.

The structure of calyculin J was confirmed by chemical transformation from calyculin A. Treatment of calyculin A with *N*-bromosuccinimide furnished a major product whose ^{1}H NMR spectrum and optical rotation were identical with those of **12**. Therefore, the absolute stereochemistry of calyculin J was assigned as shown. Formation of calyculin J was rationalized as follows. Conformation of calyculin A in solution was almost identical with that in the solid state, which was disclosed by X-ray crystallography. With respect to the C-10/C-11 bond, C-9 and C-12 are in a gauche relationship reflecting the steric repulsion of Me-46 and Me-45. The electrophilic bromine atom attacks the $\Delta^{8(9)}$ olefin

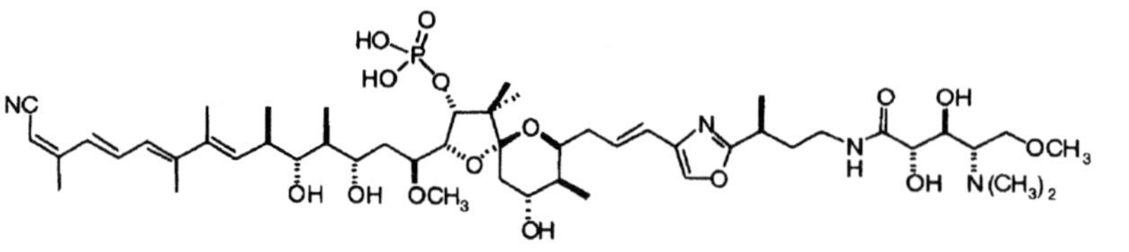

Calyculin A (**11**)

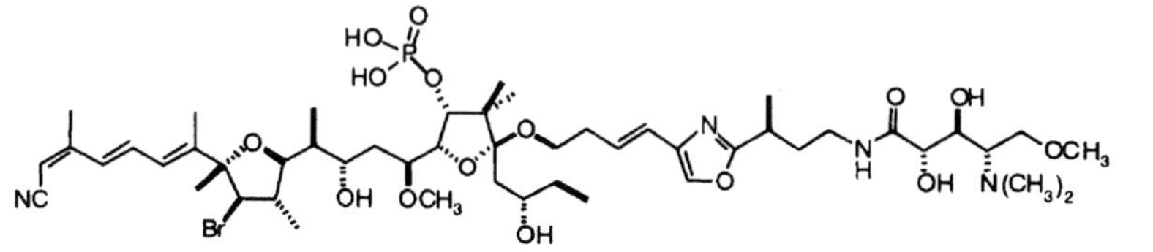

Calyculin J (**12**)

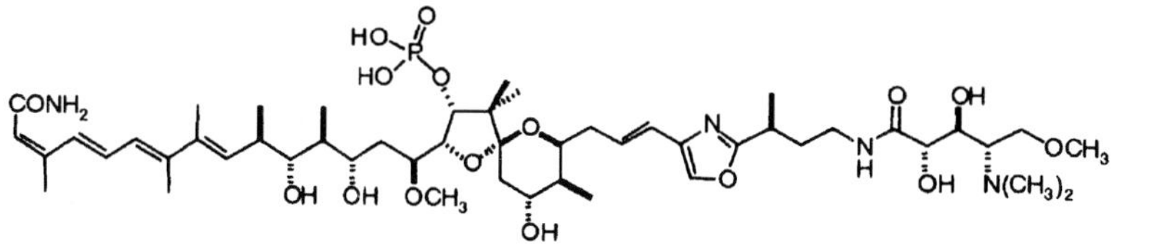

Calyculinamide A (**13**)

Figure 3. Structures of calyculin derivatives **11-13**.

from the less hindered side, i.e., the face opposite C-11, to form an intermediate bromonium ion, which is opened by the attack of 11-OH, preferentially forming a tetrahydrofuran ring.

The second new compound **13** exhibited a pseudomolecular ion 18 units larger than that of **11**. HRFABMS revealed that compound **13** was larger by the elements of H_2O than calyculin A, suggesting a hydrated **11**. The ^{1}H NMR spectrum of **13** was superimposable on that of **11** except for H-2 and H-4 signals, both of which experienced considerable down-field shifts. The C-2 signal, whose chemical shift value was determined by the HMQC spectrum, was shifted down-field by 24 ppm. Although further information such as ^{13}C NMR or HMBC data, was not obtained due to the paucity of material, compound **13** was likely to have a terminal amide instead of the nitrile in **11**. Therefore, attempts were made to hydrate the nitrile group in calyculin A without affecting other parts of the molecule. Hydrolysis with KOH/*t*-BuOH or with 2-mercaptoethanol afforded complex products, whereas treatment of **11** with H_2O_2 in 25% NH_4OH furnished a product with the molecular weight of 1043, which was 16 units larger than **13**, thus indicating that in addition to hydration of the nitrile the tertiary amine was oxidized to an amine oxide. Therefore, the amine oxide was reduced with PPh_3 and separated by HPLC to yield a compound which exhibited a ^{1}H NMR spectrum identical with that of **13**. Compound **13** was named calyculinamide A.

Compounds **13** and **14** had the same Rf values on silica gel TLC, eluted closely in reversed phase HPLC, and had the same molecular formula (Fig. 4). The ^{1}H NMR spectra

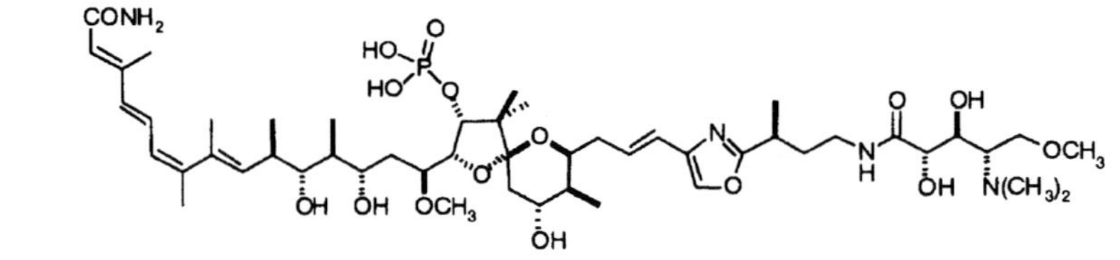

Calyculinamide F (**14**)

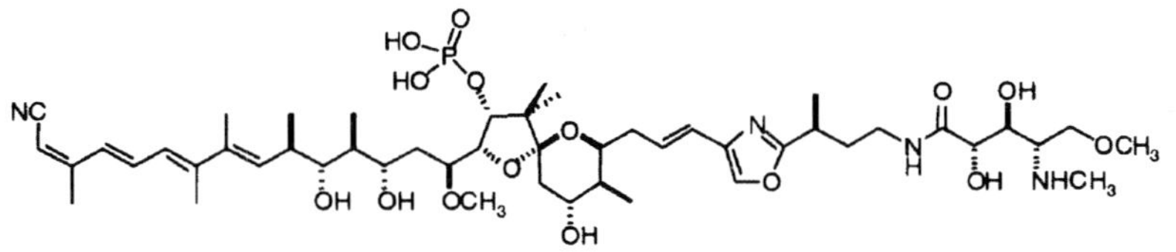

Des-*N*-methylcalyculin A (**15**)

Figure 4. Structures of calyculin derivatives **14** and **15**.

of these two compounds were similar, except for signals of the terminal tetraene portion, thus indicating that they are isomers differing in the geometry of one or more double bonds. ^{13}C chemical shift values of C-48 (15.0 ppm), C-49 (23.8 ppm), and C-50 (14.0 ppm) in **14** indicated 2*E*, 4*E*, 6Z-geometry, while the HMBC spectrum placed a carbonyl group (δ168.0) at C-1. Therefore, compound **14** was calyculinamide F, the C-1 amide of calyculin F.

The last compound **15** had a molecular formula of $C_{49}H_{79}N_4O_{15}P$ as determined by HRFABMS and NMR data, differing from calyculin A by a loss of CH_2. Interpretation of the COSY spectrum measured in benzene-d_6 led to the carbon framework identical with that of calyculin A, except for the lack of both *N*-dimethyl signals. The C-36 was shifted down-field by 6 ppm, consistent with *N*-mono-methylation of the nitrogen on this carbon. Although the *N*-methyl signal was not observed in benzene-d_6 due to broadening, it appeared at δ_H 2.76/δ_C 30.2 in CD_3OD; thus **15** is des-*N*-methylcalyculin A.

Compounds **11-15** inhibited protein phosphatase 2A with IC_{50} values of 1.0, 20, 1.5, 1.2, and 1.0 nM, respectively. Conversion of the terminal nitrile to the amide did not affect the activity as observed in compounds **13** and **14**, whereas loss of the C-11-hydroxyl group in calyculin J (**12**) resulted in reduction of activity. Replacement of dimethylamino group with a methylamino group in des-*N*-methylcalyclun A (**15**) did not affect the activity.

REFERENCES

1. Faulkner, D.J. (2000) Highlights of marine natural products chemistry (1972-1999).

Nat. Prod. Rep., 17: 1-6.
2. Matsunaga, S., Liu P., Celatka, C. A., Panek, J. S. and Fusetani, N. (1999) Relative and absolute stereochemistry of mycalolides, bioactive macrolides from the marine sponge *Mycale magellanica*. J. Am. Chem. Soc., 121: 5605-5606.
3. Wakimoto, T., Matunaga, S., Takai, A. and Fusetani, N. (2002) Insight into binding of calyculin A to protein phosphatase 1: Isolation of hemicalyculin A and chemical transformation of calyculin A. Chem. Biol., 9, 309-319.

ANTIOXIDATIVE ACTIVITIES OF LOW MOLECULAR FUCOIDANS FROM KELP *LAMINARIA JAPONICA*

Changhu Xue, Lei Chen, Zhaojie Li, Yuepiao Cai, Hong Lin and Yu Fang
Department of Food Science and Technology, Faculty of Fisheries, Ocean University of Qingdao, 5 Yushan, Qingdao, Shandong Province, China

Abstract
The chemical characters of three fucoidan fractions from *Laminaria japonica* were determined. Their inhibition activities on PC peroxidation by HPLC and the free radical scavenging effects on superoxide radical and hydroxyl radical of them by chemiluminescence method were studied. The results showed that all three fucoidan fractions had significant scavenging effects on free radicals. Among them fraction I with high glucuronic acid and low sulfate contents displayed the highest scavenging effects on free radicals. Its IC_{50} on $O_2^{\cdot-}$ was 0.044 mg/ml, and its IC_{50} on $^{\bullet}OH$ was 0.062 mg/ml. The free radical scavenging effects of three fucoidan fractions decreased after active oxygen damage. All three fucoidan fractions exhibited inhibitory activity on PC peroxidation, but they had no clear induction period.

1. INTRODUCTION

Oxygen-derived free radicals are potent agents causing many pathological effects and aging [1]. Recent studies showed that polysaccharides from seaweed had antioxidative property [2] and could scavenge oxygen-derived free radicals effectively [3-5]. Fucoidans are known to contain L-fucose residues as the main sugar constituent and sulfate esters, but their composition varies from species to species. The structures and activities of fucoidans have been studied in a range of brown seaweeds such as *Fucus versiculosus, Ascophyllum nodosom, Eisenia bicyclis, Sargassum thumbergii, Ecklonia kurome, Chorda filum* etc. *Laminaria japonica* is widely cultivated along the north coast of China and the annual production reached 200,000 tons. The main products extracted from *L. japonica* are iodine, mannitol, and alginate. Fucoidan extracted from *L. japonica* has many biological properties, including anticoagulant, antithrombosis, antiinflammatory, antitumor and antioxdative activities etc. Its biological activities varied depending on the molecular weights and sulfate contents of fucoidan from *L. japonica*. In this paper we describle the antioxidantive properties of the three low molecular mass fucoidan fractions (LMF) from *L. japonica* by means of chemiluminescence and HPLC analysis.

2. MATERIALS AND METHODS

2.1. Materials
Brown seaweed *Laminaria japonica* was collected in Rongcheng, Shandong, China. Egg yolk phosphatidylcholine (PC) was purchased from Sigma Chemical Co., 2,2'-Azobis(2-amidino propane) hydrochloride (AAPH) was purchased from Wako Pure Chemicals.

2.2. Methods
The crude fucoidan was prepared by water extraction and alcohol fractionation. Low molecular mass fucoidan (LMF) A was prepared from crude fucoidan by 0.01 mol/L HCl degradation at 80°C for 4h. One gram of A was fractionated on a 1.8×21cm column packed with DEAE-Sephadex-A25 with 0.5−2 mol/L NaCl and 0.01 mol/L HCl gradient elution. Fraction I and IV were collected, dialyzed and lyophilized.

Sulfate content was measured as described by Dodgson [6]. Uronic acid content was determined according to Bitter [7]. Neutral sugar components were analyzed by the method of Li TL [8].

The scavenging effects of fucoidans on superoxide free radical were measured as follows [3]:50μL of phosphate buffered saline (PBS) (0.05mol/L, pH 7.8), 50μL of pyrogallol (0.625mol/L) and 900μL of luminol (1 mmol/L) (Merck, Darmstadt, FRG) were added into plastic tube (55×10 mm) and the intensity of chemiluminescence was measured for 20 s. When LMF was added into the tube, the volume of the buffer was decreased subsequently, and the total volume was 1 mL.

The scavenging effects of fucoidans on hydroxyl free radical were meausured as follows [3]: 0.2 mL of ascorbic acid(2mmol/L), 0.4 mL of $CuSO_4$(2mmol/L), 50μL of luminol (0.1mmol/L), 0.1 mL of yeast (75 g/L), 0.65 mL of PBS (50 mmol/L, pH 7.8) were added into plastic tube(55×10mm) and incubated at 37°C for 0.5 hr. Then 0.6 mL of H_2O_2 (68 mmol/L) was added into the tube and the intensity of chemiluminescence was measured for 10 s. When LMF was added into the tube, the volume of the buffer was decreased subsequently, and the total volume was 2 mL.

The scavenging effects of fucoidan fractions on free radical after exposed to active oxygen were evaluated as follows [9]. Three samples of A (0.1g each) were dissolved in three Fe^{2+} and EDTA solution (final concentration :1mmol/L and 2mmol/L, respectively) and the H_2O_2 concentration in them were 0.6g/kg, 1.5 g/kg, 6 g/kg, respectively. Then they were incubated at 37°C for 5 h before dialyzation and lyophilization. The three damaged LMF (A1, A2 and A3) were prepared. Their scavenging effects were also evaluated by both of the chemiluminescence systems.

The antioxidative activity of LMF on PC system was meausured as follows [10]: 0.1mL of PC (50 mmol/L) was placed into a tube and dried by N_2 then 1mL of Tris-HCl (10 mmol/L, pH7.4) was added. The mixture was sonicated for 1 minute and incubated at 37°C for 5 min, and then 0.1mL of free radical initiator AAPH (0.1 mol/L) (Wako,

Table 1
The Chemical composition of 3 fucoidan fractions (g/kg)

Fractions	Content of total carbohydrates	Uronic acids	Organic sulfate
A	468.5	116.3	223.0
I	498.1	153.5	190.4
IV	324.1	65.5	469.8

Table 2
Molar ratio of neutral sugar components of the three LMF

	Fucose	Arabinose	Xylose	Mannose	Galactose	Glucose
A	1	0.01	0.06	0.23	0.71	0.12
I	1	0.11	0.33	0.33	3.30	0.44
IV	1	0.08	0.08	0.08	0.26	0.05

Osaka, Japan) were added to the mixture. The reaction mixture was incubated at 37°C in the dark. After regular intervals, a 20μL portion was injected into the HPLC system. HPLC was run on a Spherisorb-C8 (60×150mm). The column was eluted with methanol/water (95:5) at a flow rate of 1.0 mL/min. The eluate was monitored with a UV detector at 235nm.

The scavenging effects were evaluated by calculating the inhibitory rate of the intensity of chemiluminescence.

$$\text{Inhibitory rate (\%)} = \frac{\text{Intensity without LMF} - \text{intensity with LMF}}{\text{Intensity without LMF}} \times 100$$

3. RESULTS AND DISCUSSION

The chemical composition and sugar component of fraction A, I, IV was shown in Table 1 and Table 2. After the ion exchange fractionation, fraction I had high uronic acid and low sulfate ester. The neutral sugar composition was also different from that of the others.

Figure 1 showed that all of the three fucoidan fractions had scavenging effects on $O_2^{\cdot -}$. The concentration that sample had the inhibitory rates of 50% (IC_{50}) of A, I and IV were 0.115 mg/mL, 0.044mg/mL and 0.046mg/mL, respectively. Although I and IV had different sulfate and uronic acid contents, they had similar effects on the scavenging $O_2^{\cdot -}$. At the inhibitory rate of 50%, fucoidan (g^{-1}) had about the same scavenging effect of 40 units of superoxide dismutase SOD (3,800u/mg).

Figure 2 showed that all of the three fucoidans fractions had scavenging effects on $\cdot OH$. The IC_{50} of A, I and IV were 0.2 mg/mL, 0.062 mg/mL and 0.098 mg/mL, respectively. They all displayed much stronger effects than mannitol. Although the scavenging effects decreased greatly after exposed to active oxygen, LMF still retained the scavenging effects (Fig. 3). The fucoidan Fractions also showed their inhibitory activities on PC oxidation induced by AAPH (Fig. 4).

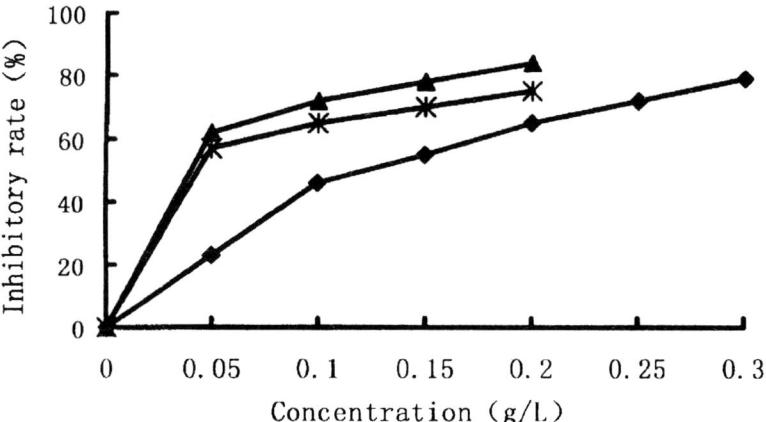

Figure 1. The scavenging effects on $O_2^{·-}$ of fucoidan fractions from *Laminaria japonica*. ▲, ✕ and ◆ show the inhibitory rate in the presence of I, IV and A, respectively.

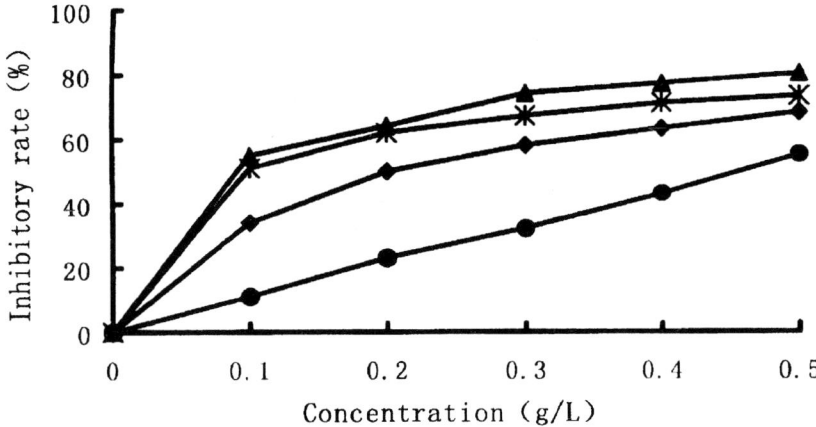

Figure 2. The scavenging effects on ·OH of fucoidan fractions. ▲, ✕, ◆ and ● show the inhibitory rate in the presence of I, IV, A and mannitol, respectively.

Many sulfated polysaccharides in algae had anti-tumor activity [11,12] and protective effects on myocardial ischemia-reperfusion injury [13]. It was found that these functions were associated with their antioxidative abilities. The scavenging effect of antioxidant could be determined by chemiluminescence analysis indirectly. Both SOD and mannitol were known for their ability to scavenge $O_2^{·-}$ and ·OH. After SOD and mannitol were added into the two systems, the chemiluminescence intensity decreased markedly. This inferred that both of these two systems produced oxygen free radical. All three LMF fractions had free radical scavenging effects and lipid oxidation inhibitory property. And their abilities to prevent oxidation increased with the increase in the fraction concentration. And Fraction I had the strongest scavenging effect. The main difference

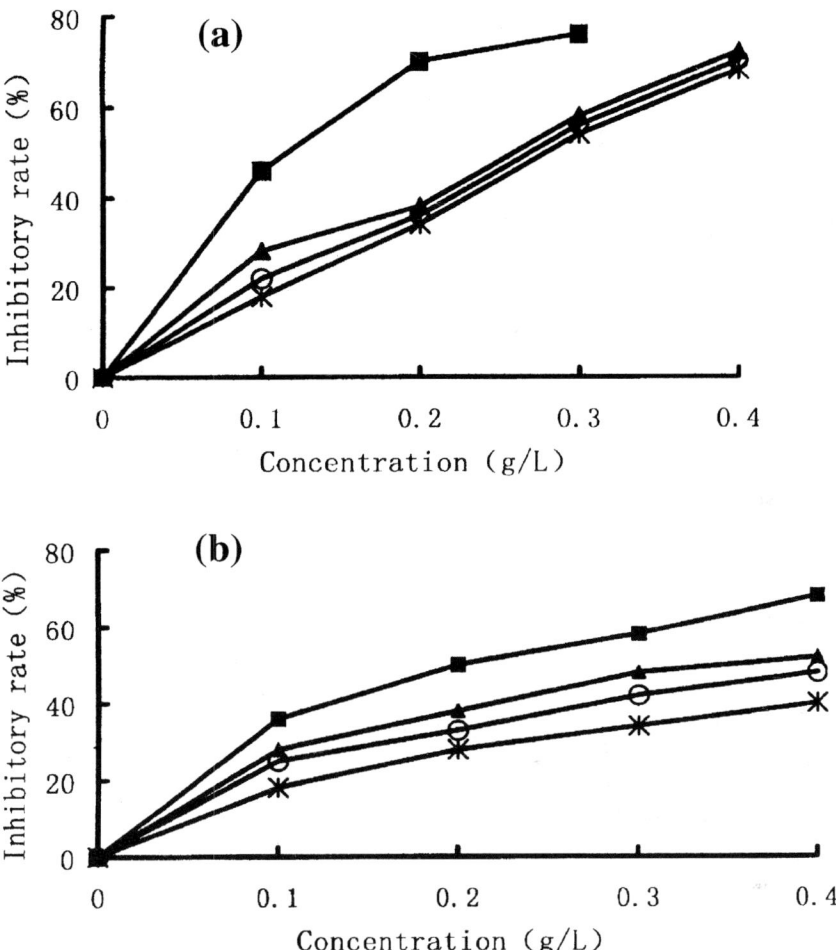

Figure 3. The scavenging effects on active oxygen radicals of fucoidan fractions after damaged by active oxygen, (a) on $O_2^{·-}$ and (b) on ·OH. ■,▲,○ and ✶ show the inhibitory rate in the presence of A, A1, A2 and A3, respectively.

between it and the other two was that it had higher uronic acid but lower sulfate content. Active oxygen free radical could degrade polysaccharides and change their conformation [14,15]. After exposed to oxygen free radical, the scavenging effect of LMF decreased. HPLC was another method to measure the scavenging effect of different polysaccharides indirectly. PC produced PC-OOH when incubated with AAPH. We compared the effects of LMF to BHT by checking the amount of PC-OOH during the oxidation. Figure 4 showed that LMF could inhibit lipid peroxidation although their effects were not stronger than those butylated hydroxytoluene (BHT). Probably, this was due to their water-soluble nature, so their activities were limited in the emulsion.

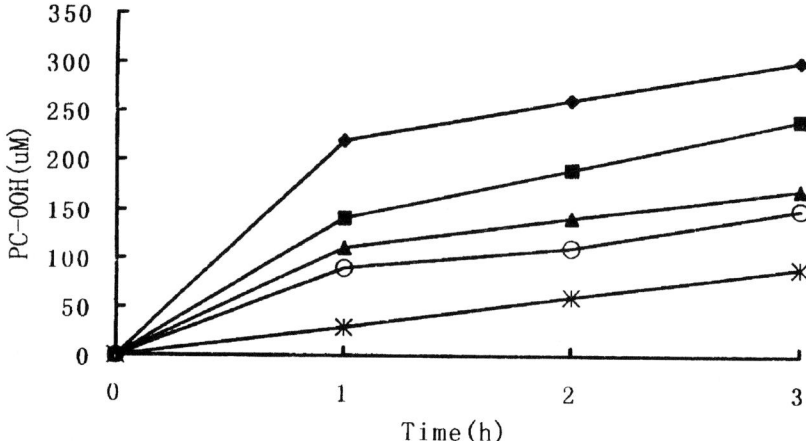

Figure 4. Effects of fucoidan fractions on PC liposome peroxidation. ◆,■,▲,○ and ✳ show the PC-OOH produced in the absence(control) and presence of A, IV, I (4g/kg) and BHT (0.2g/kg), respectively.

REFERENCES

1. Ames, B.N., Shigensaga, M.K. and Hagen, T.M. (1993) Oxidants, antioxidants, and the degenerative diseases of aging. Proc. Natl. Acad. Sci. USA, 90:7915-7922
2. Teng, X., Cong, J.B., Tian, X.H., Shi, D.J. (1998) Anti-oxidation and anti-tumor activities of sulfated polysaccharides from *Laminaria japonica*. Acta Nutrimenta Sinica, 20: 48-51.
3. Zhang, E.X., Yu, J.L. and Yu, X. (1995) Studies on oxygen free radical-scavenging effect of polysaccharides from *Sargassum thunbergii*. Chinese Journal of Marine Drug, 53: 1-4.
4. Zhang, E.X. and Yu, J.L. (1997) Studies on polysaccharide from *Sargassum thunbergii* for its ability to scavenge active oxygen species. Chinese Journal of Marine Drug, 63: 1-4.
5. Tian, X.H., Cong, J.B., Shi, D.J. and Teng, X. (1997) An in vitro ESR study of the free radical scavenging effects of the sulfated polysaccharides from brown seaweed and their kinetics. Acta Nutrimenta Sinica, 19: 32-37.
6. Dodgson, K.S. and Price, R.G. (1962) A note on the determination of the ester sulfate content of sulfated polysaccharides. J. Biochem., 84: 106-110.
7. Bitter, T. and Muir, H.M. (1962) A modified uronic acid carbazole reaction. Anal. Chem., 28:350-356.
8. Li, T.L., Wu, C.X. and Zhang, Y.X. (1982) Studies of gas chromatographic analysis of saccharides and Alditols: some improvements in analysis of acetylated aldononitriles by gas chromatograpgy. Chinese J. Anal. Chem., 10: 272-276.
9. Li, H.Q. and Xu, S.J.(1994) Damage of proteoglycan by reactive oxygen and the mineralization in the presence of proteoglycan. Acta Biochimica Biophysica, 26: 303-305.

10. Xue, C.H., Yu, G.L., Hirata, T., Terao, J. and Lin, H. (1998) Anti-oxidative activities of several marine polysaccharides evaluated in a phosphatidylcholine-liposomal suspension and organic solvents. Biosci. Biotechnol. Biochem., 62: 206-209.
11. Yamamoto, I., Takahashi, M., Tamura, E. and Maruyama, K. (1982) Antitumor activity of crude extracts from edible marine algae against L-1210 Leukemia Botancia Marina, 25: 455-457.
12. Zhuang, C., Itoh H. and Mizuno, T. (1995) Antitumor active fucoidan from the brown seaweed, *Umitoranoo(Sargassum thunbergii)*. Biosci. Biotech. Biochem., 59: 563-567.
13. Omata, M., Matsui, N., Inomata, N. and Ohno, T. (1997) Protective effects of polysaccharide fucoidan on myocardial ischemia-reperfusion injury in rats. J. Cardiovasc. Pharmacol., 30: 717-724.
14. Praest, B.M., Greilng, H. and Kock, R. (1997) Effects of oxygen-derived free radicals on the molecular weight and the polydispersity of hyaluronan solution. Carbohydr. Res., 303: 153-157.
15. Christensen, BE, Mildrid, H and Smidsrod, O. (1996) Degradation of double-stranded xanthan by hydrogen peroxide in the presence of ferrous ions: comparison to acid hydrosis. Carbohydr. Res., 280: 85-99.

PROPERTIES AND UTILIZTION OF SHARK COLLAGEN

Yoshihiro Nomura
Applied Protein Chemistry, Faculty of Agriculture, Tokyo University of Agriculture and Technology, Fuchu, Tokyo 183-8509, Japan

Abstract

The biochemical properties and the utility of shark collagen are explained in this review. Shark skin collagen is easy to prepare, and represents a possible resource for use on an industrial scale. Anti-shark skin collagen antiserum reacts with collagen from eel muscle. Fibroblasts from fetal rat skin could be cultured on shark collagen gel at 30°C, and shrinking of the collagen gel matrixes was apparent. However, it was not possible to culture the cells at 37°C because of the low denaturation temperature. Shark gelatin is useful as a substrate of matrix metalloprotease. The bone mineral density of bone epiphysis was increased by giving shark gelatin to ovariectomized rats, indicating that, shark gelatin would be useful for as a food supplement for treating osteoporosis.

1. BACKGROUND

Most blue sharks are landed as a secondary catch from long-line tuna fishing. About 80% of shark caught near Japan are landed at Kesennuma, a fishing port in the northern part of the main island of Honshu. The average haul of sharks is stable at about 12,000 t / year. The sharks are landed after removing the head and internal organs (Fig. 1) and sold on markets.

The species most abundantly available in Japan is the blue shark (*Prionace glauca*) belonging to the *Carcharhiniformes* genus, which inhabits the temperate zone and subtropical ocean regions around the world. The average length of a specimen is 2.5–3 m and the shark reaches adulthood in five years. A female produces 25 young sharks after a pregnancy of one year. Bony fish, cephalopods, dead whales, and turtles constitute the main diet, and other ecological aspects of the shark have already been described by Taniuchi [1], Yano [2], and Ferrari and Ferrari [3].

The principal purpose of catching shark is to obtain shark fin for high-class Chinese cuisine. The demand for shark fin has increased in hand with the economic growth of China and Southeast Asia. In most cases, after the fin has been removed, the remaining

Figure 1. Dressed blue shark.

carcass is dumped toat sea, which has raised an international problem about protecting shark resources.

Figure 2 shows the major useful components derived from shark tissues.

The deep-sea shark does not carry an ammoniacal smell and is marketed in the form of steaks or thin slices which are cooked in hot water. Shark meat is also used as a surimi material for making kamaboko. The relatively low propagating activity and over-fishing of the spiny dogfish (*Squalus acanthias*) has decimated its resources in the seas around Japan. The meat of the blue shark, which is the major species caught at present, is used for making hanpen.

Shark cartilage is used to produce chondroitin sulphate, which is taken as a food supplement for treating osteoarthritis, osteoporosis and cancer. Palmieri et al. [4] have shown that radiolabeled chondroitin sulfate (CS) fed to rats and dogs could be detected in the synovial fluid, and Conte et al. [5] have reported the presence of enzyme-digested CS fragments in the blood. The alleviating effect of CS on pain results from the accelerated synthesis by CS fragments of hyaluronan with a high molecular weight [6, 7]. The cartilage has application as an anticancer drug [8]. Lane [9] has reported the example of administering shark cartilage to cancer patients, and Lane and Contreras [10] and Conte et al. [11] have reported the inhibitory effect on cancer growth of administering to terminal cancer patients. Shark cartilage powder is sold as a food supplement to combat cancer.

Tsujimoto et al. [12] have isolated squalene from shark liver. Squalane, a reduced form of squalene, has been used as a low-temperature lubricating oil. The liver from the spiny dogfish has a high content of vitamin A and was used in the postwar period as liver oil to improve the nutritional state. However, liver oil manufacture declined with the

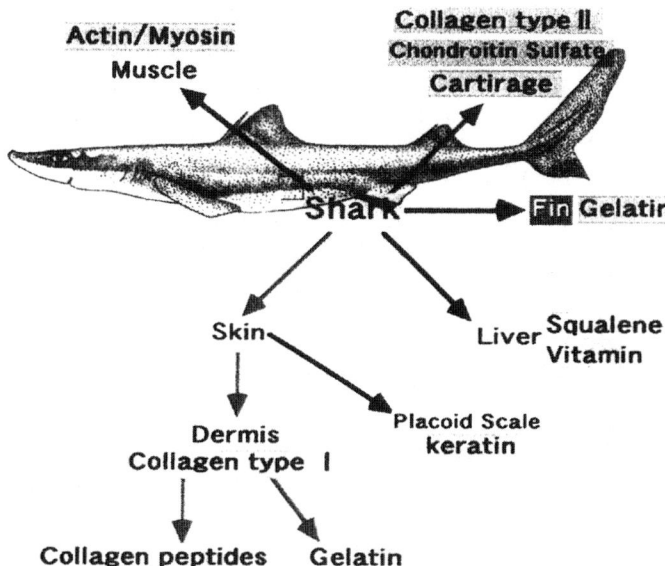

Figure 2. Major useful components derived from shark tissues.

availability of vitamin A from cheaper sources. The application of squalene for skin care has recently been studied for atopic dermatitis patients [13], and squalene from shark liver is now sold as a high-grade cosmetic [12, 14].

Fresh shark eggs have been used as an ingredient for sushi [15]. The heart from the salmon shark (*Lamna ditropis*), which is referred to as "Star of Moka." is eaten in the Sanriku area as a tonic in anticipation of its therapeutic effect.

The use of shark skin is mainly for shoes and handbags. During the 1940–1950 period, studies on the industrial application of fish leather were active, and these studies have been reviewed by Takahashi [16]. There is now relatively little use of shark skin and it is mainly discarded as industrial waste.

Almost all of shark skin, which constitutes about 11% of the body weight, are discarded as industrial waste. The purpose of this article is to describe a way to utilize this wasted part of the shark's body. Some examples of the utilization and properties of shark collagen are also described.

2. SHARK COLLAGEN

2.1. Biochemical properties

The skin and cartilage of the blue shark are respectively rich in types I and II collagen. Cartilage is the main source of type II collagen that is used as a food supplement, although there has been little biochemical study on it. A unique type of collagen is distributed in the transparent fibrous tissue between the skin and cartilage of the shark fin.

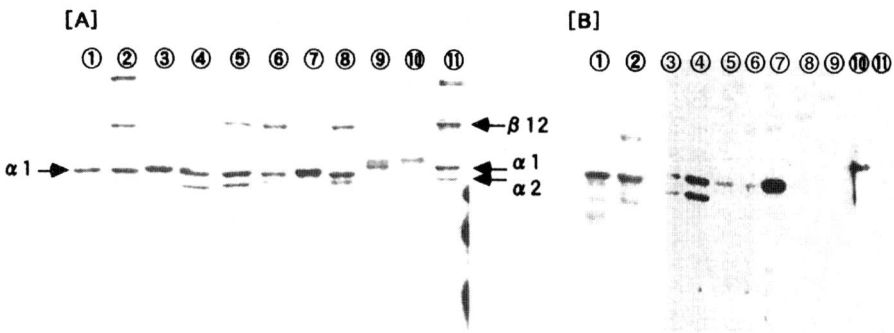

Figure 3. SDS-PAGE (A) and Western blotting (B) patterns of several collagens from aquatic animals.
The antibody used for western blotting was the anti-$\alpha 1$ chain from shark skin collagen.
①, $\alpha 1$ chain from blue shark skin collagen; ②, blue shark skin collagen; ③, salmon shark skin collagen; ④, salmon shark cartilage collagen; ⑤, horse mackerel skin collagen; ⑥, sardine skin collagen; ⑦, eel muscle collagen; ⑧, eel skin collagen; ⑨, sea anemone collagen; ⑩, jellyfish collagen; ⑪, bovine skin collagen.

This shark fin is referred to as elastoidin. The elastoidin from the blue shark can be easily rendered soluble by limited digestion with pepsin [17, 18]. Kimura et al. [19] have shown the collagen molecule of elastoidin to be an $[\alpha_1(E)]_3$ homotrimer which is different from types I and II collagen. On the otherhand, the vertebral cartilage of the shark, *Carcharhinus acutus*, contains types I and II collagen[20]. Subsequent work has been conducted on the value of shark cartilage as a food supplement.

Most of the studies on shark skin collagen have been related to leather manufacture. Takahashi [16] has described the structure, composition, thermal resistance and swelling ability of shark skin collagen, and the method for removing the placoid scale from the sandbar shark (*Carcharhinus plumbeus*), spiny dogfish, blue shark and shortfin mako (*Isurus oxyrinchus*). Kubota and Kimura [21] have shown that acid-soluble collagen from the skin of the blue shark had high solubility in an acid solution and a low denaturation temperature. This acid-soluble collagen is composed of α_1 and β_{12} chains and is a dimmer of α_1 and α_2 chains [22]. The ability for self-assembly and the acceptability as mammalian-type collagenase have been suggested to be comparable with those of domestic animal collagen [23].

Kimura [24] and other workers have studied the comparative biochemistry of lower vertebrate fish collagens. Nomura et al. [25] have shown that shark skin collagen had immunological homogeneity with pig skin collagen. This immunological homogeneity has more recently been studied. The reactivity of the anti-shark skin collagen α_1 chain (PAb-α) to lower vertebrate collagens is shown in Fig. 3. PAb-α reacts with collagen from the blue shark, salmon shark, horse mackerel, sardine and eel, but not with that of the sea anemone and jelly fish or with bovine collagen. In particular, eel muscle collagen strongly reacted with PAb-α; while the reason for this was not clear, it was considered that the strong reactivity might have been related to the structure of eel muscle collagen being composed of three α chains [26]. These results suggest that the structure of the collagen α chain is preserved within a certain range.

It is important to remove minerals with EDTA or diluted HCl when preparing the collagen from cartilage. Collagen from the skin is prepared after excising the epithelium and removing the placoid scale in an acidic or basic solution. Collagen can easily be extracted from the cartilage or skin with an acidic solvent or by pepsin digestion under acidic conditions.

Kubota et al. [27-29] have shown that acid-soluble collagen from the blue shark can be rendered insoluble by γ-ray irradiation, and Yoshimura et al. [30-33] have reported the properties of shark skin collagen. Compared with the collagen from domestic animal skins, shark skin collagen from the blue shark is easy to swell, has good reactivity with crosslinking agents, and shows lower viscosity at a high concentration. Nomura et al. [23, 34] have reported that it was possible to reconstruct the collagen fibrils in shark skin collagen within a limited temperature range and that these reconstructed collagen fibrils were not easily dissolved at a low temperature. The mechanical strength of the reconstructed shark collagen fibrils is higher than that of reconstructed pig collagen fibrils [34], and it is possible to prepare a hard gel from shark skin collagen.

Shark collagen is most strongly characterized by its low denaturation temperature; this characteristic provides the greatest advantage, but is also a problem. Therefore, increasing the denaturation temperature has been approached by two methods. The first method was to mix shark collagen with collagen from a domestic animal; a mixed collagen gel from pig and shark increased the denaturation temperature, and the gel did not melt at 37°C [35]. The second method was to apply a crosslinking agent that is acceptable for food use; microbial transglutaminase (MTGase) was used to modify the denaturation temperature of shark collagen [36, 37]. Yoshimura [33] has reviewed the properties of shark skin collagen.

2.2. Utilization

There have been several reports about the utilization of aquatic collagen and gelatin [38-41]. Examples of the use of shark skin collagen as the material for cell culture matrix, an matrix metalloprotease (MMP) substrate and food supplement are presented as follows.

2.2.1. *Cell culture matrix*

Collagen is an important material for maintaining the tissue structure and preparing the cell environment. Bell et al. [42] have presented the possibility of in vivo cell culture to grow fibroblasts in a collagen gel matrix, and several types of cell were subsequently cultured in or on such a collagen gel matrix. Collagen of cow origin has mainly been used, but that of salmon and sole skin is also known to be effective. To examine the effectiveness of shark collagen gel for use as a cell culture matrix, rat fetal skin fibroblasts were cultured on the collagen gel at 30°C (Fig. 4). The cell culture matrix was prepared from three types of collagen: from shark, pig and a mixture of both. A 0.3% collagen solution and twice the concentration of a culture medium containing 20% fetal bovine serum were mixed in culture dishes. After 18 hours, rat fetal skin fibroblasts and

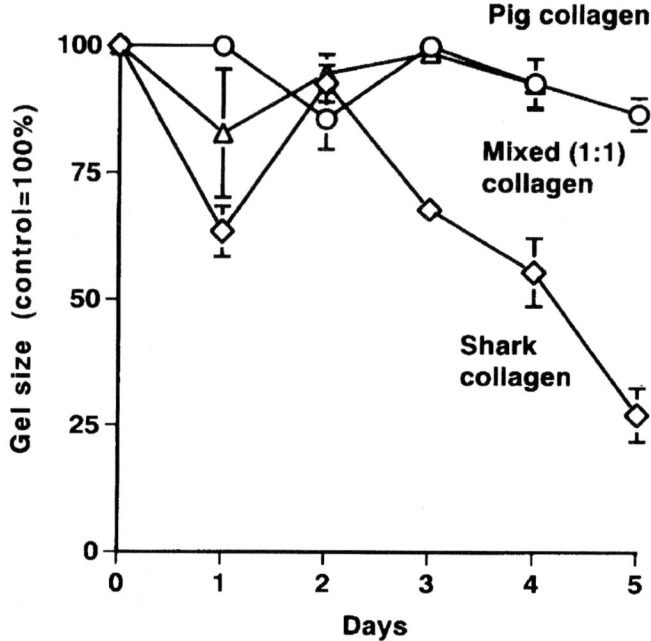

Figure 4. Changes in gel size during the culture of rat skin fibroblasts in collagen gel matrix. Conditions: medium, DMEM containing 10% Fetal calf serum; atmosphere, 5% CO_2 / 95% air; temp., 30°C; collagen conc., 0.15%.

the usual medium were added to the collagen gel matrix. The cells were cultured at 30°C, because the shark collagen gel melted when the culture was conducted at 37°C. The shark collagen gel progressively contracted to about 25% of its original size after five days of culture. The mixed shark and pig collagen gel contracted very little in comparison. Although it is usually difficult to observe cultured cells in contracted collagen gel, the shark collagen gel was highly transparent, so that the cells could be visualized. Fibroblasts interacted and aggregated with each other on the shark collagen gel, and the gel contraction could have occurred as a result of this cell aggregation.

2.2.2. Substrate for matrix metalloprotease (MMP)

Gelatin is now recognized as the best substrate for MMP involved in tissue metabolism. Since publishing the method of the gelatin zymogram [43], this technique has been found indispensable for studying MMP. Gelatin from bovine or pig skin has generally been used. We can demonstrate the use of gelatin prepared from shark collagen as the zymogram substrate. Electrophoretic gel samples were prepared to mix with denatured shark collagen by heating at 100°C for 5 minutes in a 10% acrylamide solution. MMP was obtained from the culture medium of rat fetal fibroblasts. Shark gelatin zymography showed the active and potential forms of MMP-2, and showed clearer patterns of MMP digestion than those from pig gelatin zymography (Fig. 5). A lower

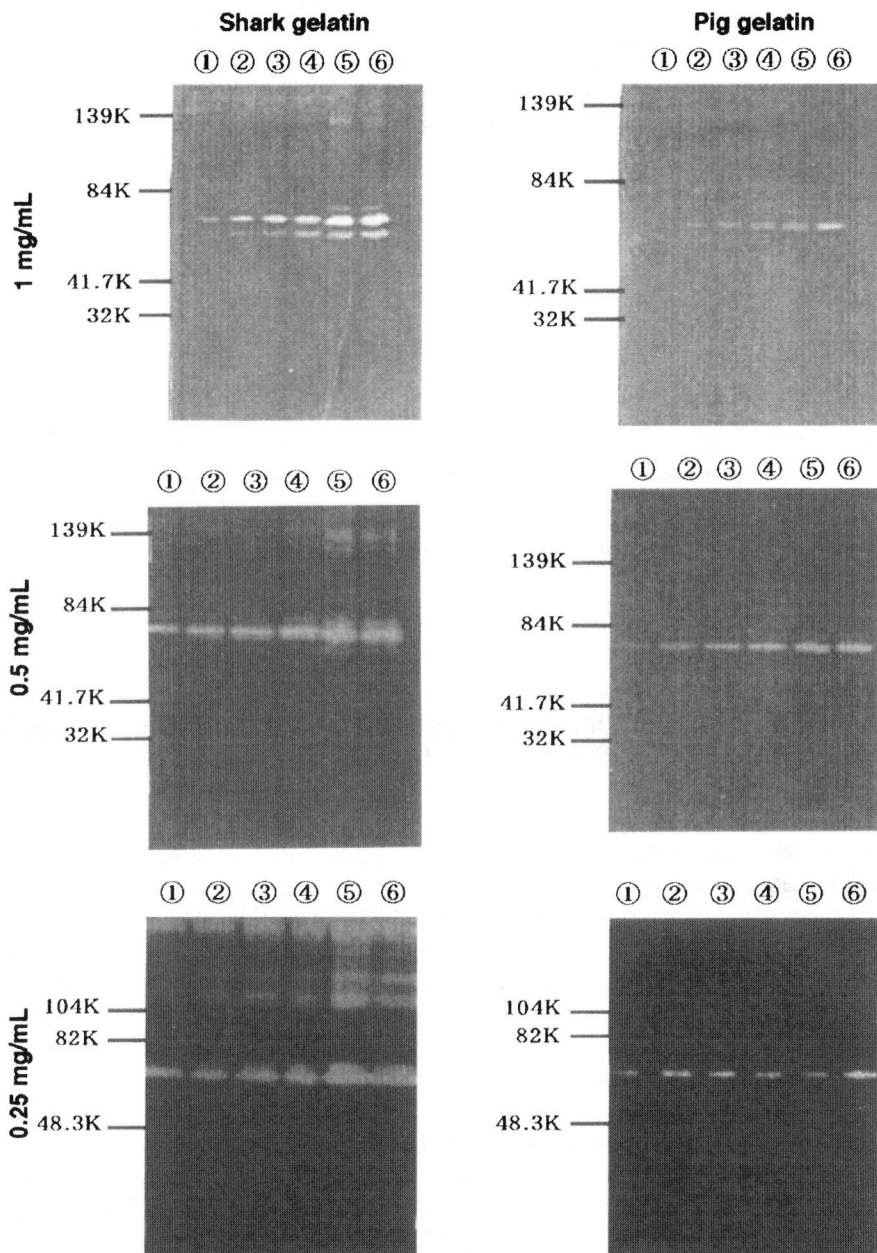

Figure 5. Gelatin zymographic patterns of the fibroblast culture medium. The concentrations of gelatin were 0.25, 0.5 and 1.0 mg/mL. All samples were the rat fibroblast culture medium after confluent growth at 2 (①), 4 (②), 8 (③), 12 (④), 24 (⑤) and 48h (⑥).

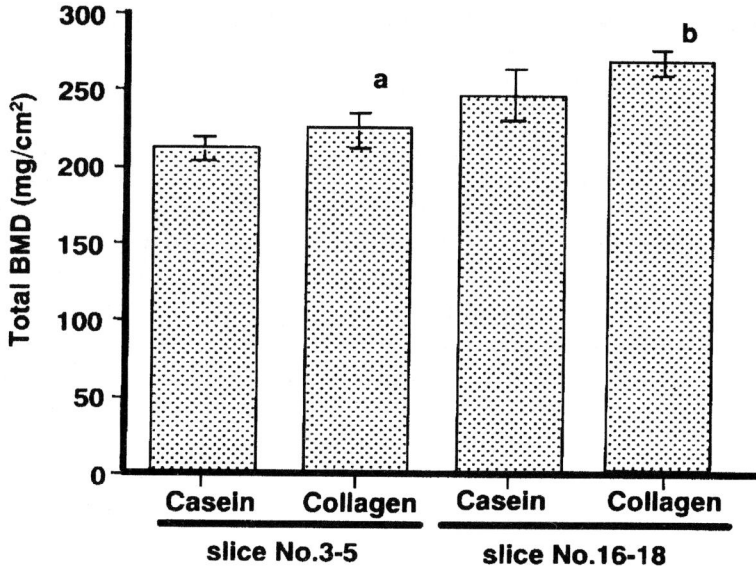

Figure 6. Effect of casein and shark gelatin on the bone mineral density (BMD) of both distal ends of the femoral bone from ovariectomised (ovx) rats. From the proximal end to distal end, the femur was divided into 20 sections, and BMD of each section was measured. The added BMD value of the slice number 3 to 5 and number 16 to 18 showed (n=10). a, significantly different from the ovx-casein group ($p<0.05$); b, significantly different from the ovx-casein group ($p<0.05$).

concentration of shark gelatin in the zymographic gel would enable bands to appear with lower mobility, perhaps for the form of MMP-9.

2.2.3. *Food supplement to treat osteoporosis*

Gelatin and collagen hydrolysates are now commercially available in anticipation of reducing the pain of osteoarthritis and improving the bone mineral density. The effectiveness of this supplement has been confirmed by administering to human volunteers [44, 45]. Koyama et al. [46] have reported that supplementation of gelatin to mice with a low-protein condition increased the bone mineral density (BMD). The use of shark gelatin to improve the bone mineral density has been studied with female Wistar rats which were four weeks old at the start of the experiment. After six days of normalization, the rats were fed on a low-casein diet. Nine days later, ten rats were ovariectomized and ten rats were given a sham operation. The ovariectomized and sham-operated rats were given shark gelatin or casein at a dosage of 40 mg /100 g of body weight /day. There was no difference in the body weight, bone length, and bone weight among the groups. The femur was divided into 20 sections from the proximal end to distal end, and BMD of each section was measured. Figure 6 shows the enhanced BMD value of the proximal end (nos. 3-5) and distal end (nos. 16-18). The group of ovariectomized rats fed on the shark gelatin showed an increase in BMD at epiphysis. The intake of shark collagen is likely to have been responsible for this improvement in bone mineral density.

The use of domestic animal collagens in cosmetics, functional foods and drinks is currently restricted by the prevalence of bovine spongiform encephalopathy and foot-and-mouth disease. This restriction has been increasing the attention on collagen from aquatic animals. A aquatic animal collagen has so far been derived from the skin of sole [47], yellowfin tuna [48]), salmon and trout [49] in Japan. There is also a plan to prepare a collagen hydrolysate by the process used to prepare chondroitin sulfate from shark fins. The main problem is the fishy smell; if this problem can be solved, fish collagen will become a material that is very easy to use. Shark collagen is biologically safe with new functions, and seems to have great potential for further applications as a biomaterial and food supplement.

It has been suggested that collagen can be the allergen responsible for fish allergy [50], so this problem must be taken into consideration when using aquatic animal collagen.

REFERENCES

1. Taniuchi, T. (1997) In: Natural history of the shark., Tokyo Daigaku Shuppankai, Japan.
2. Yano, K. (1998) In: Shark., Tokai Daigaku Shuppankai, Japan.
3. Ferrari, A. and Ferrari, A. (2001) In: Sharks and rays of the world., TBS Britannica, Japan.
4. Palmieri, L., Conte, A., Giovannini, L., Lualdi, P. and Ronca, G. (1990) Metabolic fate of exogenous chondroitin sulfate in the experimental animal. Arzeim.-Forsch./Drug Res., 40: 319-323.
5. Conte, A., de Bernardi, M., Palmieri, L., Lualdi, P., Mautone, G. and Ronca, G. (1991) Metabolic fate of exogenous chondroitin sulfate in man. Arzeim.-Forsch./Drug
6. Res., 41:768-772.
6. McAlindon, T.E., LaValley, M.P., Gulin, J.P. and Felson, D.T. (2000) Glucosamine and chondroitin for treatment of osteoarthritis. JAMA, 283: 1469-1468.
7. McCarty, M.F., Russell, A.L. and Seed, M.P. (2000) Sulfated glycosaminoglycans and glucosamine may synthesize in promoting synovial hyaluronic acid synthesis. Medical Hypotheses, 54: 798-802.
8. Lee, A. and Langer, R. (1983) Shark cartilage contains inhibitors of tumore angiogenesis. Science, 16: 1185-1187.
9. Lane, I.W. (1991) Shark cartilage: Its potential medical applications. J. Advancement in Medicine, 4: 263-271.
10. Lane, I.W. and Contreras Jr., E. (1992) High rate of bioactivity (reduction in gross tumor size) observed in advanced cancer patients treated with shark cartilage material. J. Naturopathic Medicine, 3: 86-88.
11. Conte, A., Volpi, N., Palmieri, L., Bahous, I., and Ronca, G. (1995) Biochemical and

pharmacokinetic aspects of oral treatment with chondroitin sulfate. Arzeim.-Forsch./Drug Res., 45, 918-925.
12. Kaiya A. (1999) Natural moisturizing agent "squalane." Kako Gijyutsu, 34: 322-325.
13. Tanii, T., Kato, J., Yashiro, N., Shinto, R., Satino, K. and Hamada, T. (1999) Clinical evaluations of squalane in patients with xerotic dermatoses and the results of patch test of squalane and its moisturizing effects. Skin, 33: 155-163.
14. Nippon Yushi Shinposya (1999) Cosmetics and raw materials, changing with deregulation. Squalene, revalued in cosmetics. Long-term prospect for new applications. Oil and Fat, 52: 22-23.
15. Iwase A. 3/7~10/31 (2000) Cultural history of shark. Kesennuma Kahoku, Miyagi, Japan.
16. Takahashi, T. (1957) Some fundamental knowledge on shark leather manufacture, Nippon Hikaku Gijyutsu Kyokaishi, 3: 59-78.
17. Kimura, S. and Kubota, M. (1966) Studies on elastoidin - I. Some chemical and physical properties of elastoidin and its components. J. Biochem., 60: 615-621.
18. Kimura, S. and Kubota, M. (1967) Studies on elastoidin - II. Dialysable peptides released by pepsin digestion of elastoidin. Bull. Japan. Soc. Scient. Fish., 33: 430-437.
19. Kimura, S., Uematsu, Y. and Miyauchi, Y. (1986) Shark (*Prionace glauca*) elastoidin: characterization of its collagen as $[\alpha_1(E)]_3$ homotrimers. Comp. Biochem. Physiol., 84B: 305-308.
20. Rama, S. and Chandrakasan, G. (1984) Distribution of different molecular species of collagen in the vertebral cartilage of shark (*Carcharius acutus*). Connective Tissue Res., 12: 111-118.
21. Kubota, M. and Kimura, S. (1967) Skin collagen of the great blue shark. Nippon Suisan Gakkaishi, 33: 338-342.
22. Kimura, S., Kamimura, T., Takema, Y. and Kubota M. (1981) Lower vertebrate collagen evidence for type I-like collagen in the skin of lamprey and shark. Bioch. Biophy. Acta, 669: 251-257.
23. Nomura, Y., Yamano, M., Hayakawa, C., Ishii, Y., and Shirai, K. (1997) Structural property and in vitro self-assembly of shark type I collagen. Biosci. Biotechnol. Biochem., 61: 1919-1923.
24. Kimura, S. (1985) The interstitial collagens of the fishes. In: Biology of Invertebrate and Lower Vertebrate Collagens. Bairati, A. and Garrone, R. (eds.), Plenum Press, New York, pp. 397-408.
25. Nomura, Y., Sasaki, Y., Arai, K., Ishii, Y. and Shirai, K. (1996) Separation of anti-shark type I collagen antibody from anti-pig type I collagen antiserum and its partial characterization. Biosci. Biotechnol. Biochem., 60: 697-698.
26. Kelly, J., Tanaka, S. Hardt, T., Eikenberry, E.F. and Brodsky, B. (1988) Fibril-forming collagen in lamprey, J. Biol. Chem., 263: 980-987.
27. Kubota, M., Kimura, S., and Ohasi, T. (1968) Influence of the γ-ray irradiation exerted on collagen -I. Hikakukagaku, 14: 105-111.
28. Kubota, M., Kimura, S. and Ohasi, T. (1969) Influence of the γ-ray irradiation exerted on collagen -II. Hikakukagaku, 15: 1-7.

29. Kubota, M., Kimura, S. and Ohasi, T. (1969) Influence of the γ-ray irradiation exerted on collagen -III. Hikakukagaku, 15: 8-13.
30. Yoshimura, K., Chonan, Y. and Shirai, K. (1997) Reactivity of shark, pig, and bovine skin collagens with formaldehyde and basic chromium sulfate. Anim. Sci. Technol., 68: 285-292.
31. Yoshimura, K., Chonan, Y. and Shirai, K. (1999) Preparation and dynamic viscoelastic characterization of pepsin-solubilized collagen from shark skin compared with pig skin. Anim. Sci. J., 70: 227-234.
32. Yoshimura, K., Terashima, M., Hozan, D. and Shirai, K. (2000) Preparation and dynamic viscoelasticity characterization of alkali-solubilized collagen from shark skin. J. Agric. Food Chem., 48: 685-690.
33. Yoshimura, K. (2000) Properties of shark skin collagen. Hikaku Kagaku, 46: 159-171.
34. Nomura, Y., Toki, S., Ishii, Y. and Shirai, K. (2000) The physicochemical property of shark type I collagen gel and membrane. J. Agric. Food Chem., 48: 2028-2032.
35. Nomura, Y., Toki, S., Ishii, Y. and Shirai, K. (2000) Improvement of the material property of shark type I collagen by composing with pig type I collagen. J. Agric. Food Chem., 48: 6332-6336.
36. Nomura, Y., Toki, S., Ishii, Y. and Shirai, K. (2001) Effect of transglutaminase on reconstruction and physicochemical properties of collagen gel from shark type I collagen. Biomacromolecules, 2: 105-110.
37. Nomura, Y., Toki, S., Ishii, Y. and Shirai, K. (2001) Improvement of shark type I collagen with microbial transglutaminase in urea. Biosci. Biotechnol. Biochem., 65: 982-985.
38. Berg, R.A., Silver F.H., Watt, W.R. and Norland, R.E. (1985) Fish gelatin in coating applications. Image Technology, 106-109.
39. Fraga, A.N. and Williams, R.J.J. (1985) Thermal properties of gelatin films. Polymer, 26: 113-118.
40. Hamada, M. (1990) Effects of the preparation conditions on the physical properties of shark gelatin gels. Nippon Suisan Gakkaishi, 56: 671-677.
41. Leuenberger, B.H. (1991) Investigation of viscosity and gelation properties of different mammalian and fish gelatins. Food Hydrocolloids, 5: 353-361.
42. Bell, E., Ivarsson, B. and Merrill, C. (1979) Production of tissue-like structure by contraction of collagen lattices by human fibroblasts of different proliferative potential in vitro. Proc. Natl. Acad. Sci. USA, 76: 1274-1278.
43. Makowski, G.S. and Ramsby, M.L. (1996) Calibrating gelatin zymograms with human gelatinase standards. Anal. Biochem., 236: 353-356.
44. Adam, M. (1991) Welche wirkung haben gelatinepraparate? Therapiewoche, 38: 2456-2461.
45. Moskowitz, R.W. (2000) Role of collagen hydrolysate in bone and joint disease. Seminars in Arthritis and Rheumatism, 30: 87-99.
46. Koyama, Y., Hirota, A., Mori, H., Takahara, H., Kuwaba, K., Kusubata, M., Matsubara, Y., Kasugai, S., Itoh, M. and Irie, S. (2001) Ingestion of gelatin has differential effect on bone mineral density and body weight. Protein Undernutrition,

47: 84-86.
47. Katakurachikkarin (2000) A marine pure collagen. Fragrance J., 8: 102-103.
48. Ichimarufarukosu (2000) A marine collagen. Fragrance J., 2: 76-77.
49. Shimizu, S. and Shimizu, H. (2000) Japan Kokai Tokkyo Koho, 50811 (Feb. 22, 2000).
50. Hamada, Y., Nagashima, Y, and Shiomi, K. (2001) Identification of collagen as a new fish allergen. Biosci. Biotechnol. Biochem., 65: 285-291.

MODERATION OF CHEMO–INDUCED CANCER BY WATER EXTRACT OF DRIED SHARK FIN : ANTI–CANCER EFFECT OF SHARK CARTILAGE

Kenji Sato[a], Naho Murata[b], Masahiro Tsutsumi[b], Masami Shimizu-Suganuma[c], Kazuhiro Shichinohe[c], Tsukasa Kitahashi[a], Kazunari Nishimura[a], Yasushi Nakamura[a] and Kozo Ohtsuki[a]

[a]Department of Food Sciences and Nutritional Health, Kyoto Prefectural University, Shimogamo, Kyoto 606-8522, Japan

[b]Department of Oncological Pathology, Nara Medical University, Kashihara, Nara 634-8521, Japan

[c]Department of Laboratory Animal Science, Nippon Medical School, 1-1-5 Sendagi, Tokyo 113-8602, Japan

Abstract

It is now well recognized that matrix metalloproteinase (MMP) plays significant roles in cancer progression. Shark cartilage contains the MMP inhibitors. On the basis of these findings, oral administration of the shark cartilage has been proposed to suppress cancer progression and improve quality of life of cancer patient. Now, the "shark cartilage therapy" is popular in USA, Japan and other countries. However, there are conflicting data on the effectiveness of the therapy. In addition, knowledge on the active component responsible for the potential anti-cancer effect is limited. To facilitate the further study on the anti-cancer effect of shark cartilage and utilization of shark fin cartilage, a by-product from fisheries industry, moderation of chemo-induced cancer by the water extract of dried shark fin is discussed.

1. BACKGROUND

Cancer is one of the major life-threatening diseases in many countries. Chemo- and radio-therapies based on cyto-toxic effect on cancer cell have been developed and used successfully to reduce tumor before surgery and suppress reoccurrence of cancer after surgery. However, these approaches are usually ineffective for the inoperative cancer, because cancer cell frequently acquires resistance to these therapies. Therefore, alternative approach targeting non-cancer cell has been proposed.

Folkman and his associates [1,2] have demonstrated that angiogenesis (neovascularization) is required for solid tumor growth. They also successfully demonstrated that inhibition of the tumor-induced angiogenesis can suppress growth and metastasis of cancer by using animal model [3,4]. Matrix metalloproteinase (MMP), a family of zinc dependent endoproteinase, is involved in the tumor-induced angiogenesis. The MMP also plays critical role in invasion and metastasis of cancer. Then the MMP may be a good target for cancer therapy. For contribution of the MMP on the cancer progression, refer to recent reviews [5-8].

The MMP forms a molecular family consists of more than 20 different gene products. Two groups of the MMP can degrade triple helical structure of collagen, a major extracellular component. MMP-1 (formally collagenase) degrades type I collagen, an interstitial fibrilar collagen. On the other hand, MMPs-2 and 9 (formally gelatinase) can degrade type IV and V collagens. These collagens (IV and V) are main constituents of the basal lamina and pericellular connective tissue, respectively [9]. The MMPs-2 and 9 are secreted as inactive pro-enzymes and converted to active forms by proteolytic cleavage. The active forms of the MMPs-2 and 9 are distributed preferentially on the tumor cells with high malignancy (refer to the reviews cited above). An example is shown in Fig. 1, the inactive and active forms of MMPs-2 and 9 are observed in the cancer lesion of pancreas of hamster, while only small levels of the inactive forms are observed in the normal pancreas. Therefore, these MMPs have been speculated to play a critical role in degradation of the extracellular matrix for growth, metastasis and invasion of pancreatic cancer [10]. Recently, several MMP inhibitors have been synthesized and examined their anti-cancer effect by human trial [11-14].

Cartilage, avascular tissue, has anti-angiogenesis activity under physiological condition. The implanted bovine and shark cartilage or their extract near tumor can

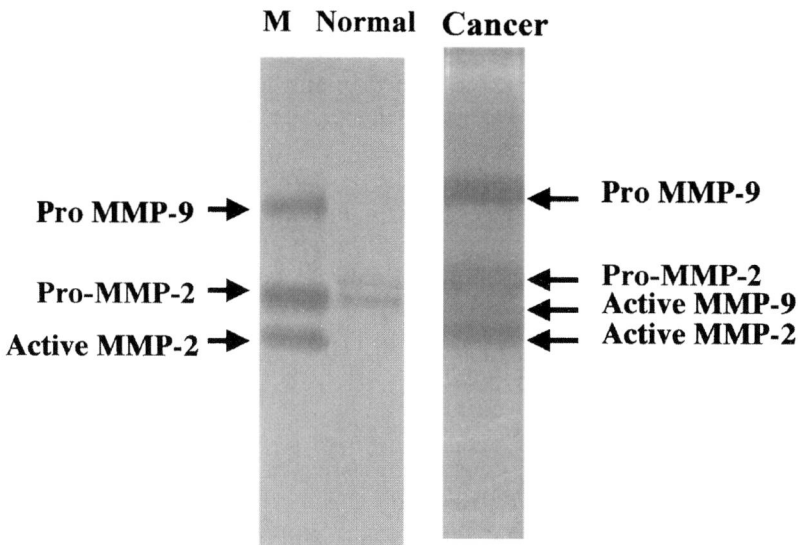

Figure 1. Gelatin zymographs of normal and cancer pancreas. M, maker MMPs.

inhibit the tumor-induced angiogenesis and growth of tumor [15-18]. The cartilage also contains the MMP inhibitors, such as tissue inhibitor of metalloproteinases (TIMPs)-1 and 2. Then, it has been suggested that TIMP-like proteins might be responsible for the anti-angiogenesis activity [19-21]. As shark has larger quantity of cartilage than cattle, pig and chicken, shark cartilage might be a good potential source for the anti-angiogenesis and MMPs inhibitor. On the basis of these findings, Lane and his coworkers have proposed to use the shark cartilage for the cancer therapy by oral administration. In spite of the limited number of subjects and short-term follow-up, they reported that oral and enteral administration of the shark cartilage powder could improve the quality of life of the patient suffering from inoperative advanced cancer [22]. TV, popular magazine, pamphlet, book and so on have reported these results and "the shark cartilage therapy" is currently popular in USA and other countries. However, there are discouraging data on the efficiency of shark cartilage powder [23]. In addition, the suggested dose of shark cartilage powder for the oral administration is too large (1g /kg) to continue this treatment, which has been a stumbling block for human trial. Recently, a water extract of the shark cartilage has been prepared and examined its therapeutic effect against cancer [24-26]. Some positive results are reported, although the active component responsible for the therapeutic effect has not been identified.

Shark is traditionally used as food ingredient in the forms of fillet, surimi and so on. Among them, dried shark fin prepared for Chinese cuisine is traded in high price. Preparation procedure of the dried shark fin is as follow. Shark fin was washed with water and solar-dried. The dried fin is rehydrated in water at 50-60°C and skinned. Collagenous transparent fiber, named elastoidin, on the both side of cartilaginous tissue is collected and dried. The dried elastoidin is used for high-grade Chinese cuisine, while the fin cartilage is low value product. Recently, we found that water extract of the dried shark fin cartilage has the in vitro inhibitory activity against MMPs-2 and 9 and also anti-cancer effect by oral administration in animal model. In the present article, we discribe our recent findings on the anti-cancer effect of the shark cartilage and discuss possible candidate components responsible for the anti-cancer effect.

2. EXTRACTION OF MMP INHIBITOR FROM THE DRIED SHARK CARTILAGE

In the early studies, the inhibitory activities against the tumor-induced angiogenesis and MMP have been extracted from small pieces of shark cartilage with guanidine chloride solution and then fractionated by ultrafiltration and chromatographic techniques [18,19]. Tissue inhibitor of metalloproteinase (TIMP)-like proteins have been identified in the extract and speculated to play a significant role in inhibiting the angiogenesis and MMP [19-21]. However, guanidine chloride solution can not be used for preparation of food ingredient. Therefore, we selected water for solvent to extract the active components from shark cartilage. To facilitate the extraction, a fine powder (20-50 mm) was prepared by the freeze-milling technique from the dried shark cartilage. We found that the inhibitory activity against the MMPs-2, 9 can be extracted with water from the powder. The extract was fractionated by a preparative isoelectrofocusing without added

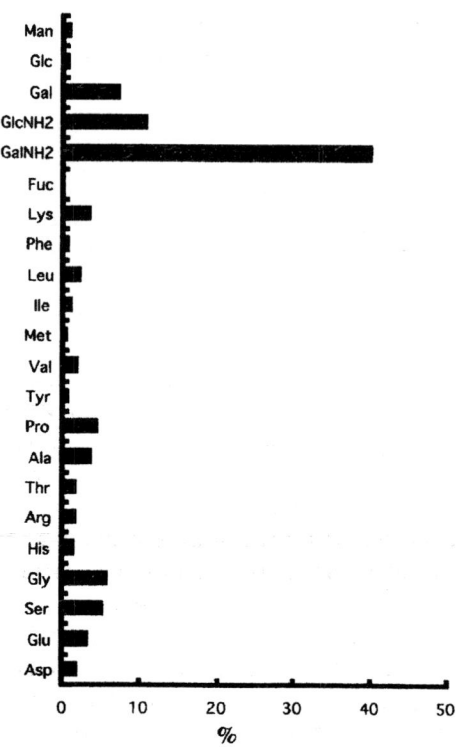

Figure 2. Amino acid and sugar compositions of the fraction with inhibitory activity against MMPs-2 and 9. Man, mannose; Glc, glucose; Gal, galactose; GlcNH$_2$, glucosamine; GalNH$_2$, galactosamine. Uronic acids were not determined.

ampholyte, referred to "Autofocusing" [27], reversed phase (Bakerbond WP Butyl) and size exclusion (Superdex 75 and 200) HPLCs. Consequently, the active component was recovered in acidic, hydrophilic and high molecular weight fractions. As shown in Fig. 2, the active fraction shows compositional feature of chondroichin sulfate-protein conjugate. However, commercially available chondroichin sulfate C prepared from shark cartilage (Maruha and Sigma) showed no significant inhibitory activity against the MMPs-2 and 9 (data not shown). These facts suggest that peptide moiety of the proteoglycan in the shark cartilage has the inhibitory activity. These findings are inconsistent with the previous idea that TIPM-like proteins are responsible for the inhibitory activity against MMP-2 and 9 [19-21]. The inconsistency may be explained by different extraction procedure and start material. The previous extraction procedure could not fully extract chondroichin sulfate, one of major components in the shark cartilage [19]. On the other hand, we used the shark fin which had been dried and heated in water. Then TIMP-like protein might lose the inhibitory activity against MMP by heat denaturation. Our data demonstrate the presence of a heat-stable MMP inhibitor except for TIMP-like protein in shark cartilage.

3. EFFECT OF ORAL ADMINISTRATION OF THE WATER EXTRACT OF SHARK CARTILAGE ON CANCER PROGRESSION.

Lane and his associates [22] have proposed that the oral administration of the shark cartilage can suppress cancer progression. However, the suggested dose (1g / kg body weight) of the dried shark cartilage powder is too large to continue the shark cartilage therapy. Then extraction and concentration of the active component from the shark cartilage have been demanded. As described above, the water-extract of shark cartilage, which contains the TIMP-like proteins, has been prepared and subjected to animal and human trials to examine the anti-cancer effect by oral administration [21,24-26]. On the other hand, our results indicate that the chondroichin sulfate-protein conjugate has MMP inhibitory activity. Then we prepared the proteoglycan-rich fraction from the dried shark fin cartilage. The inhibitory activity in the water extracted was recovered in the precipitate by adding 3 vol. of ethanol. The precipitate was dried under vacuum. The inhibitory activity in the dried precipitate was stable in room temperature. The compositional feature of the extract is shown in Fig. 3. It consists of protein (mainly collagen), chondroichin sulfate, mineral and so on. The recovery was approximately 2% (w/w). The ethanol precipitate was used as the shark cartilage extract to examine the anti-cancer effect by using the hamster with the chemo-induced pancreatic cancer.

Pancreatic duct cancer was induced by the method of Iki et al. [10], which being evaluated as a good experimental model for human pancreatic cancer. After the carcinogenic treatment, the animals were given the experimental diets containing 0, 0.2 and 0.4% of the shark cartilage extract. This dosage of 0.4% in diet corresponds to 2 or 20 g per day for human on the basis of concentration in diet or intake per body weight, respectively. After receiving the experimental diets for 50 days, the animals were scarified and incidence of the tumor in the pancreas was examined. As shown in Fig. 4, incidence of the advanced carcinoma in the pancreas decreased in dose response manner by the oral administration of the shark cartilage extract. This response is approximately

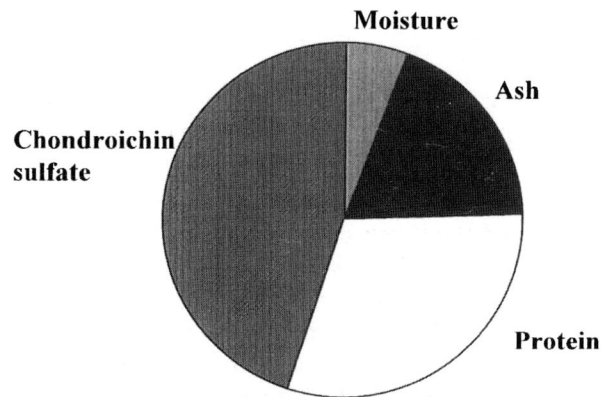

Figure 3. Compositional feature of the shark cartilage extract for the animal experiment.

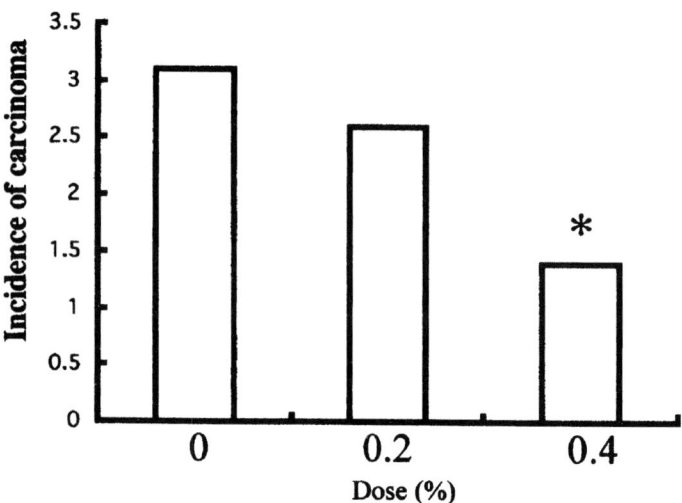

Figure 4. Suppression of chemo-induced pancreatic cancer progression by oral administration of the shark cartilage extract. *, p<0.05.

corresponding to oral administration of a synthetic MMP inhibitor (OPB-3206; 3S-[4-(N-hydroxyamino)-2R-isobutylsuccinyl]amino-1-methoxy-34-dihydrocarbostyril)) in the dose of 0.1% in the same animal model [10]. It is worthy to note that physical performance of the animal was decreased by administration of the OPB-3206 but not by the shark cartilage extract. In another animal model using the mice with the implanted Ehrlich ascites tumor, oral administration of the present preparation also delay growth of the tumor (data not shown). These findings indicate that the oral administration of the water extract prepared from the shark fin cartilage can suppress cancer progression. The anti-cancer component(s) in shark cartilage is stable against heat (50-60°C) and 75% ethanol treatments. Oikawa et al. [18] also reported the presence of the heat-stable angiogenesis inhibitor in shark cartilage.

4. MECHANISM OF THE ANTI-CANCER EFFECT BY ORAL ADMINISTRATION OF THE SHARK CARTILAGE EXTRACT

The hamster with the chemo-induced pancreatic cancer received the basal and shark cartilage diets, which contained the shark cartilage extract at 0 and 0.4%. Serum was collected from the both hamsters and mixed with the MMPs-2, 9 and type V collagen and then incubated at 37°C for 24h. The degradation products of type V collagen by the MMPs were resolved by SDS-PAGE (7.5%) and detected by anti-serum against type V collagen. As shown in Fig. 5, the degradation products by the MMPs-2 and 9 decreased significantly by addition of the serum from the hamsters, which indicates the presence of the MMP inhibitors in serum. The gelatin zymography analysis (Fig. 1) demonstrates that the active form of MMP-2 is expressed in the tumor more abundantly than that of MMP-

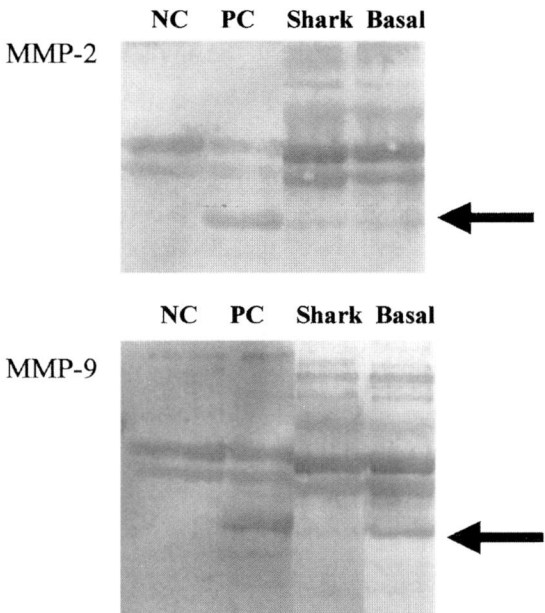

Figure 5. Inhibition of MMP 2 (1.25 mU) and MMP-9 (2.5 mU) by 1 mL of serum from the cancer-bearing hamster receiving the shark cartilage extract (0.4%) and basal diets. NC, negative control; PC, positive control. Degradation products of type V collagen are indicated with arrows.

9. However, the same amount of the serum can inhibits approximately 10 times higher activity of MMP-2 than that of MMP-9, suggesting that the MMP-9 might be involved in progression of pancreatic cancer in spite of smaller expression of the active form than the MMP-2. Very interestingly, the degradation product by MMP-9 almost disappeared by addition of the serum from the hamster receiving the shark cartilage extract, while significant amount of the degradation product was observed by addition of the serum from the hamster receiving the basal diet. Therefore, the oral administration of the water-extract of shark cartilage increased the inhibitory activity against MMP-9 in serum. It is a first direct evidence for the enhancement of the inhibitory activity against the MMPs by oral administration of the shark cartilage and its extract. On the other hand, no significant cyto-toxic effect against cancer cell was observed by the oral administration of the shark cartilage extract. These results might be interpreted by two ways; some component with inhibitory activity against the MMP-9 in the shark cartilage may be adsorbed into blood system or the administration of some components in shark cartilage may enhance the expression of endogenous inhibitor of host animal.

Recently, oral administration of the another type of shark cartilage extract has been demonstrated to suppress the basic fibroblast growth factor (bFGF)-induced angiogenesis by using animal model [28]. In that study, inactivation of the bFGF by some components from shark cartilage is suggested. However, inhibition of the MMPs by the oral administration of the shark cartilage extract also could suppress the bFGF-induced angiogenesis, as the MMP plays critical role in the angiogenesis process.

5. CONCLUSION AND FUTURE PROSPECT

In the present article, we demonstrate the presence of a MMP inhibitor except for TIMP-like protein in the water extract of the dried shark fin, a by-product of fishery industry, and that the MMP inhibitory activity in serum increase significantly by the oral administration of small dosage of the shark cartilage extract. The increased inhibitory activity against MMP in the serum is a good biomaker to evaluate anti-tumor effect produced by foodstuff. Now, further study on identification of the active component by monitoring the MMP inhibitory activity in serum of animal receiving the fractionated extract is in progress. If the MMP inhibitory activity increase in healthy human serum by oral administration of the water-extract of shark cartilage, a clinical trial to demonstrate anti-tumor would be proposed.

REFERENCES

1. Folkman, J. (1989) What is the evidence that tumors are angiogenesis dependent? J. Natl. Cancer Inst., 82: 4-6.
2. Folkman, J. (2001) Angiogenesis-dependent diseases. Semin Oncol., 28: 536-542.
3. O'Reilly, M. S., Holmgren, L., Shing, Y., Chen, C. Rosenthal, R. A., Moses, M., Lane, W. S. Cao, Y., Sage, E. H. and Folkman, J. (1994) Angiostatin: A novel angiogenesis inhibitor that mediates the suppression of metastases by Lewis lung carcinoma. Cell, 79: 315-328.
4. O'Reilly, M. S., Boehm, T., Shing, Y., Fukai, N., Vasios, G., Lane, W. S. Flynn, E. Birkhead, J. R., Olsen, B. R. and Folkman, J. (1997) Endostatin: An endogenous inhibitor of angiogenesis and tumor growth. Cell, 88:277-285.
5. Curran, S. and Murray, G. I. (1999) Matrix metalloproteinases in tumor invasion and metastasis. J. Pathol., 189: 300-3008.
6. Johansson, N., Ahonen, M. and Kahari, V. M. (2000) Matrix metalloproteinases in tumor invasion. Cell Mol. Life Sci., 20: 5-15.
7. McCawley, L. J. and Matrisian, L. M. (2000) Matrix metalloproteinase: multifunctional contributors to tumor progression. Mol. Med. Today, 6: 149-156.
8. Foda, H. D. and Zucker, S. (2001) Matrix metalloproteinases in cancer invasion, metastasis and angiogenesis. Drug Discov. Today, 6: 478-482.
9. van der Rest, M. and Garrone, R. (1991) Collagen family of proteins. FASEB J., 5: 2814-2823.
10. Iki, K., Tsutsumi, M., Kido, A., Sakitani, H., Takahama, M., Yoshimoto, M., Motoyama, M., Tatsumi, K., Tsunoda, T. and Konishi, Y. (1999) Expression of matrix metalloproteinase 2 (MMP-2), membrane-type I MMP and tissue inhibitor of metalloproteinase 2 and activation of proMMP-2 in pancreatic duct adenocarcinomas in hamsters treated with N-nitrosobis(2-oxopropyl)amine.Carcinogenesis,20:1323-1329.

11. Wojtowic-Praga, S., Torri, J., Johnson, M., Steen, V., Marshall, J., Ness, E., Dickson, R., Sale, M., Rasmussen, H. S., Chiodo, T. A. and Hawkins, M. J. (1998) Phase I trial of matristat, a novel matrix metalloproteinase inhibitor, administrated orally to patients with advanced lung cancer. J. Clin. Oncol., 16: 2150-2156.
12. Wojtowic-Praga, S., Low, J., Marshall, J., Ness, E., Dickson, R., Barter, J., Sale, M., McCann, P., Moore, J., Cole, A. and Hawkins, M. J. (1996) Phase I trial of a novel matrix metalloproteinase inhibitor batimastat (BB-94) in patients with advanced cancer. Invest. New Drugs, 14: 193-202.
13. Kahari, V. M. and Saarialho-Kere, U. (1999) Matrix metalloproteinases and their inhibitors in tumor growth and invasion. Ann. Med., 31: 34-45.
14. Heath, E. I., O'Reilly, S., Humphrey, R., Sundaresan, P., Donehower, R. C., Sartorius, S., Kennedy, M. J., Armstrong, D. K., Carducci, M. A., Sorensen, J. M., Kumor, K., Kennedy, S. and Grochow, L. B. (2001) Phase I trial of the matrix metalloproteinase inhibitor BAY12-9566 in patients with advanced solid tumors. Cancer Chemother. Pharmacol., 48: 269-274.
15. Berm, H. and Folkman, J. (1975) Inhibition of tumor angiogenesis mediated by cartilage. J. Exp. Med., 141: 427-439.
16. Langer, R., Brem, H., Falterman, K., Klein, M. and Folkman, J. (1976) Isolation of a cartilage factor that inhibits tumor neovascularization. Science, 193: 70-72.
17. Mose, M. A. and Langer, R. (1991) A metalloproteinase inhibitor as an inhibitor of neovascularization. J. Cellular Biochem., 47: 230-235.
18. Oikawa, T., Ashino-Fuse, H., Shimamura, M., Koide, U. and Iwaguchi, T. (1990) A novel angiogenetic inhibitor derived from Japanese shark cartilage (I). Extraction and estimation of inhibitory activities toward tumor and embryonic angiogenesis. Cancer Letters, 51: 181-186.
19. Lee, A. K., van Beuzekom, M., Glowacki, J. and Langer, R. (1984) Inhibitors enzymes and growth factors from shark cartilage. Comp. Biochem. Physiol., 78B; 609-616.
20. Moses, M. A., Sudhalter, J. and Langer, R. (1990) Identification of an inhibitor of neovasucularization. Science, 248: 1408-1410.
21. Gingras, D., Renaud, A., Mousseau, N., Beaulieu, E., Kachra, Z. and Beliveau, R. (2001) Matrix protease inhibition by AE-941, a multifunctional angiogenetic compound. Anticancer Res., 21: 145-156.
22. Lane, I. W. and Contreras, Jr. E. (1992) High rate of bioactivity (reduction in gross tumor size) observed in advanced cancer patients treated with shark cartilage material. J. Naturopathic Medicine, 3: 8-88.
23. Miller, D. R., Anderson, G. T., Stark, J. J., Granick, J. L. and Richardson, D. (1998) Phase I/II trial of the safety and efficiency of shark cartilage in the treatment of advanced cancer. J. Clin. Oncol., 16: 3649-3655.
24. Falardeau, P., Champagne, P., Poyet, P., Hariton, C. and Dupont, E. (2001) Neovastat, a naturally occurring multifunctional antiangiogenic drug, in phase III clinical trial. Semin. Oncol., 28: 620-625.
25. Dupont, E., Falardeau, P., Mousa, S. A., Dimitriadou, V., Pepin, M. C., Wang, T. and Alaoui-Jamali, M. A. (2002) Antiangiogenic and antimetastatic properties of

Neovastat (AE-941), an orally active extract derived from cartilage tissue. Clin. Exp. Metastasis, 19: 145-153.
26. Weber, M. H., Lee, J. and Orr, F. W. (2002) The effect of Neobastat (AE-941) on an experimental metastatic bone tumor model. Int. J. Oncol., 20: 299-303.
27. Yata, M., Sato, K., Ohtsuki, K. and Kawabata, M. (1996) Fractionation of peptides in protease digests of proteins by preparative isoelectric focusing in the absence of added ampholyte: A biocompatible and low-cost approach referred to as Autofocusing. J. Agric. Food Chem., 44: 76-79.
28. Gonzalez, R. P., Soares, F. dos S. D., Farias, R. F., Pessoa, C., Leyva, A., Viana, G. S. de B. and Moraes, M. O. (2001) Demonstration of inhibitory effect of oral shark cartilage on basic fibroblast growth factor-induced angiogenesis in the rabbit cornea. Biol. Pharm. Bull., 24: 151-154

POLYPHENOLIC COMPOUNDS FROM SEAWEEDS: DISTRIBUTION AND THEIR ANTIOXIDATIVE EFFECT

Joko Santoso[a,b], Yumiko Yoshie[a] and Takeshi Suzuki[a]
[a]Department of Food Science and Technology, Tokyo University of Fisheries, 4-5-7 Konan, Minato, Tokyo 108-8477, Japan
[b]Department of Fisheries Processing Technology, Bogor Agricultural University (IPB), Darmaga, Bogor 16680, Indonesia

Abstract

There is no information on presences of polyphenolic compounds in various seaweeds, and also their mechanism as an antioxidant or pro-oxidant. The distribution and antioxidative activities of polyphenolic compounds of twelve species of Indonesian and Japanese seaweeds were studied, comparing with those of commercial polyphenolic compounds. Fish oil emulsion was used in this study, and incubated at 50°C for 3 hours. Peroxide value (POV), scavenging effect and Fe^{2+} chelating were determined as oxidation markers.

Eisenia bicyclis contained catechin and its isomers, whereas *Caulerpa racemosa*, *Kappaphycus alvarezii*, *Monostroma nitidum*, *Undaria pinnatifida* and *Laminaria religiosa* did not contain catechin and its isomers. Catechol was found in all Japanese seaweed samples, except *E. bicyclis*. The highest concentrations of flavonoids were found in *M. nitidum*, whereas almost all of flavonoids could be found in *U. pinnatifida*. Extracts of *Hizikia fusiformis* had the best antioxidant power in both absence and presence of Fe^{2+} in fish oil emulsion. The highest of scavenging effects was found in morin, followed by rutin and extract of seaweeds (*H. fusiformis* and *U. pinnatifida*). Catechin and gallic acid had the highest effects of Fe^{2+} chelating, and the smallest Fe^{2+} chelating was found in both of extract seaweeds (*H. fusiformis* and *U. pinnatifida*).

1. INTRODUCTION

The consumption of foods from vegetable origin has been associated with reducing risk of a range of chronic diseases [1,2]. Owing to their antioxidative properties, the components such as polyphenols may play a role in the aetiology of chronic disease through oxidative damage to body cells and molecules.

Flavonoids are a group of polyphenolic compounds diversified in chemical structures and characteristics. They occur naturally in fruits, vegetables, nuts, seeds, flowers, and bark and are integral parts of human diets [3].

There are many reports that polyphenolic compounds in tea [4,5], wine [6,7], cacao [8], fruits [9] and vegetables [10] have beneficial effects for our health, especially as an antioxidant. However, there is no information on the presence of polyphenolic compounds in various seaweeds. The first purpose of this research is to study the distribution of catechins and another polyphenolic compounds in some edible Indonesian and Japanese seaweeds.

Many researchers explained the antioxidative effects of polyphenolic compounds by the decrease of several oxidation marker values. However, some antioxidants were also reported to acts as a pro-oxidant. Huang and Frankel [4], Ohshima et al. [11] in the previous research reported that tea catechins and some flavonoids had both antioxidant and pro-oxidant activities, depending on the condition. Their functions to act as antioxidants or pro-oxidants are still unclear, but we often consume polyphenol-rich like tea, wine, fruits, vegetables, etc in our life. Since lipids were absorbed after making micelle by cholic acids in our body, the oil emulsion model was selected for study. Therefore, the second objective of this research is to estimate the function of polyphenolic compounds in the oxidation system by using peroxide value, metal chelating and scavenging effects as markers.

2. MATERIALS AND METHODS

2.1. Materials

Twelve species of Indonesian and Japanese edible seaweeds were used in this experiment. Three green algae (*Caulerpa sertularoides*, *Caulerpa racemosa* and *Monostroma nitidum*), seven brown algae (*Turbinaria conoides*, *Sargassum polycystum*, *Padina australis*, *Undaria pinnatifida*, *Eisenia bicyclis*, *Hizikia fusiformis* and *Laminaria religiosa*) and two red algae (*Kappaphycus alvarezii* and *Porphyra yezoensis*) were collected in each growing districts as shown in Table 1.

After removing sand, the seaweed samples were washed with clean seawater and transported to the laboratory under refrigeration. After washing with tap water and wiping with paper towel, seaweed samples were minced by a food cutter (MK-K75; Matsushita Electric Corp., Osaka, Japan) and stored at -20°C until used.

2.2. Methods

2.2.1 *Analysis of catechin and its isomers*

Seaweeds catechins were extracted according to the method for tea catechins [12] as follows. Each seaweed sample (10 g wet sample) was homogenized with 40 mL of 100% methanol using a mixer (Ultra-Turrax T-25; Janke & Kunkel, IKA-Labortechnik, GmbH Co., Staufen, Germany) at 5,000 - 10,000 rpm for 2 min. The homogenate was centrifuged at 6,000 rpm for 10 min at 4°C and the supernatant was filtered through an Advantec filter paper No. 101 (Toyo Roshi Kaisha, Ltd, Tokyo, Japan). The residue was then homogenized twice with 50 mL of 80% methanol and centrifuged. The combined supernatants were evaporated to remove both methanol and water. The dried residue was redissolved in 100% methanol. After replacing air with nitrogen gas to prevent decomposition of catechins, the extracts were kept at -80°C until analysis. All samples were extracted and analyzed in triplicate.

Catechins were determined by high-performance liquid chromatography (HPLC) according to the modified methods of Terada et al. [13] and Suematsu et al. [14]. Catechins were separated by an ODS column (Inertsil ODS-2, ϕ5 mm, 250×4.6 mm ID, GL Science Inc., Tokyo, Japan) in column oven at 40°C fitted with a guard column (10 ×4.0 mm ID) using acetonitrile/ethyl acetate/0.1% phosphoric acid (85/20/895) as mobile phase, flow rate at 1 mL/min, and analyzed at 280 nm with a spectrophotometer detector SPD-6A UV (Shimadzu Corp., Kyoto, Japan). Authentic gallocatechin (GC), epigallocatechin (EGC), catechin (C), epicatechin (EC), epigallocatechin gallate (EGCg), gallocatechin gallate (GCg), epicatechin gallate (ECg) and catechin gallate (Cg) were purchased from Sigma Chemical Corp., (St Louis, MO, USA). Standard solutions for all catechins were prepared in methanol. Catechin concentrations in seaweeds were calculated using a standard curve at the concentration of 0-50 μg/mL. A peak of each catechin was identified by the retention time and co-injection with the standard solution.

2.2.2. Analysis of flavonoids and related compounds

Total flavonoids were extracted according to the method of Hertog et al.[15] as follows. Each minced fresh seaweed sample (5 g) was homogenized with 40 mL of 75 % methanol including 2 g/L TBHQ (t-butylhydroquinone) by a mixer (Ultra-Turrax T-25; Janke & Kunkel, IKA-Labortechnik, GmbH Corp., Staufen, Germany) at 5,000 - 10,000 rpm for 60 s. Ten milliliters of 6 mol/L hydrochloric acid was added and carefully mixed. The homogenate was refluxed at 90°C for 2 hr. After cooling, the supernatant was filtered through Advantec filter paper No. 101 (Toyo Roshi Kaisha, Ltd, Tokyo, Japan) and transferred to the volumetric flask with methanol. After replacing air with nitrogen gas in order to prevent decomposition of flavonoids, the extracts were kept at -80°C until analysis. All of samples were extracted and analyzed in triplicate.

Flavonoids were determined by HPLC modified from the methods of Hertog et al.[15] and Vinson et al.[16] Flavonoids were separated by a C_{18} column (Nova-Pak C_{18}, 4 mm, 150×3.9 mm ID, Waters Corp., Mildford, MA, USA) fitted with a guard column (20×3.9 mm ID), using 25 % acetonitrile in 0.025 mol/L KH_2PO_4 at pH 2.4 as mobile phase, flow rate at 0.9 mL/min and analyzed by a diode array detector SPD-M10Avp (Shimadzu Corp., Kyoto, Japan). Original standard of flavonoids, rutin (Ru), caffeic acid (CA), catechol (Ct), quercetin (Qr), hesperidin (Hs) and morin (Mr) were purchased from Sigma Chemical Corp., (St Louis, MO, USA). Methanolic standard solutions for all flavonoids were prepared. Flavonoid concentration in the seaweed was calculated using standard curve within concentration 0-200 μg/mL. Separation, identification, and linear curve of standard solution with concentration were ascertained by retention time, spectra and peak area, respectively.

2.2.3. Lipid oxidation model system

Emulsions were prepared with Menhadden fish oil (3%) and water (97%) containing

Tween 20 (0.3%). For 30 g of oil emulsion, 16 mg of pure polyphenolic or methanol extract from 10 g of seaweed was added at the presence of 0.5 mmol/L of ferrous chloride as catalyst. From the preliminary experiment the time incubation was fixed from short (3 hr) to long time (24 hr). In this experiment, the emulsion was incubated at 50°C for 3 hr. Two species of seaweeds extract (*U. pinnatifida*, Wakame and *H. fusiformis*, Hijiki) were used to study the behavior of antioxidant activity in this system. All of experiments were carried out in triplicate. Lipid oxidation was measured by peroxide value (POV) according to the method of Takagi et al. [17]. The scavenging effects of ferrous ion by phenolic compounds was analyzed according to the method of Dinis et al. [18] using 3-(2-pyridyl)-5-6-bis(4-phenyl-sulfonic acid)-1,2,4-triazine (ferrozine) solution from Sigma Chemical Co. and measuring chelate by a spectrophotometer at 510 nm. The chelating effects of Fe^{2+} was analyzed by the method of Fukuzawa and Fujii [19] using 2,2-dipyridil (Sigma Chemical Co.,) in ferrozine solution and measuring chelate by a spectrophotometer at 520 nm.

3. RESULTS

The distribution of catechin, its isomer and flavonoids and related compounds in Indonesian and Japanese edible seaweeds are presented in Table 1. Green alga *C. racemosa* did not contain catechin and its derivatives, whereas another green alga *C. sertularoides* contained gallocatechin, epicatechin and catechin gallate. Little amount of catechin was found in brown algae *S. polycystum* and *P. australis*. In another Indonesian brown alga *T. conoides* had a certain amount of epigallocatechin. In red alga *K. alvarezii* did not contain catechin and its derivatives. All of catechin and its isomers were found in the high concentration of Japanese brown alga *E. bicyclis* (Arame), except catechin gallate. However, in Japanese green algae *M. nitidum* (Hitoegusa) and brown algae *U. pinnatifida* (Wakame) and *L. religiosa* (Hosomekombu) did not contain catechin and its isomers. Brown alga *H. fusiformis* (Hijiki) only contained epigallocatechin, and little amounts of catechin and epigallocatechin gallate were found in red alga *P. yezoensis* (Susabinori). With the difference of catechin and its isomer contents, almost all of flavonoids were found in Japanese seaweed samples with the high concentration. The highest concentrations of flavonoids were found in *P. yezoensis* (Susabinori), followed by *M. nitidum* (Hitoegusa), *U. pinnatifida* (Wakame) and *E. bicyclis* (Arame).

Table 2 shows the POV changes of polyphenolic compounds and seaweed extracts in fish oil emulsion system after incubating at 50°C for 3 hr. In the absence of Fe^{2+} in this emulsion system, only methanol extract from *H. fusiformis* (Hijiki) showed the highest antioxidant activities and significantly different from another polyphenolic compounds, except gallic acid and morin. It decreased POV from 20.8 meq/kg to 16.7 meq/kg, whereas another polyphenolic compound and methanol extract from *U. pinnatifida* (Wakame) acted as pro-oxidant.

The POV increased in accordance with the presence of Fe^{2+} as a catalyst. In this system, all of polyphenolic compounds and methanol extracts from seaweeds became pro-oxidant. Catechin had the highest increment of POV (13.7 meq/kg) and significantly different with another polyphenolic compounds. Although the POV of quercetin in the

Table 1
The distribution of polyphenolic and related compounds in seaweeds (mean ± SD μg/g dry weight)

Scientific Name	Taxonomy	Growing district	Polyphenolic and related compounds													
			GC	EGC	C	EC	EGCg	GCg	ECg	Cg	Ru	CA	Ct	Qr	Hs	Mr
Indonesian seaweeds																
Caulerpa sertularoides	Green	Seribu Island, Jakarta Pref.	98±19	–	–	91±33	–	–	–	17±9	NA	NA	NA	NA	NA	NA
Caulerpa racemosa	Green	Seribu Island, Jakarta Pref.	–	–	–	–	–	–	–	–	NA	NA	NA	NA	NA	NA
Turbinaria conoides	Brown	Seribu Island, Jakarta Pref.	–	170±49	+	+	–	–	–	–	NA	NA	NA	NA	NA	NA
Sargassum polycystum	Brown	Seribu Island, Jakarta Pref.	–	–	18±16	–	–	–	–	–	NA	NA	NA	NA	NA	NA
Padina australis	Brown	Seribu Island, Jakarta Pref.	–	–	52±25	+	–	–	–	–	NA	NA	NA	NA	NA	NA
Kappaphycus alvarezii	Red	Seribu Island, Jakarta Pref.	–	–	–	–	–	–	–	–	NA	NA	NA	NA	NA	NA
Japanese seaweeds																
Monostroma nitidum (Hitoegusa)	Green	Ishigaki, Okinawa Pref.	NA	+	–	–	–	NA	+	–	2700±180	–	10200±810	–	7500±250	1320±38
Undaria pinnatifida (Wakame)	Brown	Chikura, Chiba Pref.	NA	–	–	–	–	NA	–	–	457±6.3	53.6±50	1830±290	202±26	–	1020±110
Eisenia bicyclis (Arame)	Brown	Chikura, Chiba Pref.	NA	4770±580	1240±310	3860±700	280±230	NA	290±120	–	–	–	–	–	6930±430	1860±130
Hizikia fusiformis (Hijiki)	Brown	Chikura, chiba Pref.	NA	3770±1200	–	–	–	NA	–	–	–	–	749±540	–	–	1010±11
Laminaria religiosa (Hosomekombu)	Brown	Kamaishi, Iwate Pref.	NA	–	+	–	–	NA	–	–	–	–	241±33	–	–	1470±120
Porphyra yezoensis (Susabinori)	Red	Futtsu, Chiba Pref.	NA	–	36±4.0	–	32±4.0	NA	–	–	–	46.8±3.8	8000±4100	–	51300±13000	771±51

GC, gallocatechin; EGC, epigallocatechin; C, catechin; EC, epicatechin; EGCg, epigallocatechin gallate; GCg, gallocatechin gallate; ECg, epicatechin gallate; Cg, catechin gallate; Ru, rutin; CA, caffeic acid; Ct, catechol; Qr, quercetin; Hs, hesperidin ; Mr, morin; +, trace amount; –, not detected; NA, not analyzed.

Table 2
Changes in POV of polyphenolic compounds and extract seaweeds in fish oil

Polyphenolic compounds	Without Fe^{2+}	With Fe^{2+}
Control (only fish oil)	20.8 ± 0.5[bc]	29.8 ± 1.9[a]
Catechin	22.2 ± 2.6[c]	35.9 ± 1.9[b]
Rutin	20.2 ± 1.8[bc]	31.4 ± 0.7[a]
Quercetin	27.3 ± 3.3[d]	28.7 ± 6.2[a]
Gallic acid	18.4 ± 0.8[ab]	30.1 ± 1.8[a]
Morin	19.6 ± 1.6[ab]	30.8 ± 3.0[a]
Hizikia fusiformis (Hijiki) extract	16.7 ± 1.5[a]	27.7 ± 4.1[a]
Undaria pinnatifida (Wakame) extract	20.1 ± 2.0[bc]	30.5 ± 4.6[a]

Emulsion system :mean ± SD meq/kg
Values within columns followed by different superscript letters are significantly different ($p<0.05$).

absence of Fe^{2+} was the highest, but in the presence of Fe^{2+} the increment of POV was the lowest (1.4 meq/kg). Compared with control, beside quercetin, only methanol extract from *H. fusiformis* had the POV value lower than fish oil, even though the increment of POV was little higher (11.0 meq/kg) than control (9.0 meq/kg).

Scavenging effects of ferrous ion by polyphenolic compounds are presented in Fig.1. Generally the scavenging effects increased in accordance with increasing levels of polyphenolic compounds, except gallic acid. Morin had the highest scavenging acvtivities, followed by rutin and both of seaweed extracts (Hijiki and Wakame).

Table 3 shows the effects of Fe^{2+} chelating of polyphenolic compounds. Catechin and gallic acid had the highest Fe^{2+} chelating and significantly different with another polyphenolic compounds. The smallest Fe^{2+} chelating were found in two seaweed extracts (*H. fusiformis* and *U. pinatifida*) even though not significantly different from control (fish oil).

4. DISCUSSION

There is not enough information about the distribution of polyphenolic compounds in seaweeds, especially Indonesian seaweeds. Two of the authors (Yoshie and Suzuki) studied the distribution of catechins in Japanese seaweeds [20]. Various concentration of catechins and its isomers were found in Japanese seaweeds. Generally Japanese seaweeds contained the high concentration of catechins and its isomers than Indonesian seaweeds (same species and/or genus). Furthermore for flavonoid contents and their derivatives in seaweeds, there is no information. Jimenez-Escrig et al. [21] only studied the total of polyphenols in some seaweed species grown in Spain and Portugal.

Polyphenolic compounds in fish oils emulsion system acted as both of antioxidant and pro-oxidant. In the absence of Fe^{2+}, gallic acid, morin and extract of *H. fusiformis* acted as antioxidant, whereas another polyphenolic compounds acted as pro-oxidants. Huang and Frankel [4] reported that all tea catechins, gallic acid and propyl gallate acted as pro-oxidant in corn oil water emulsion. Furthermore, they stated that the relative antioxidant

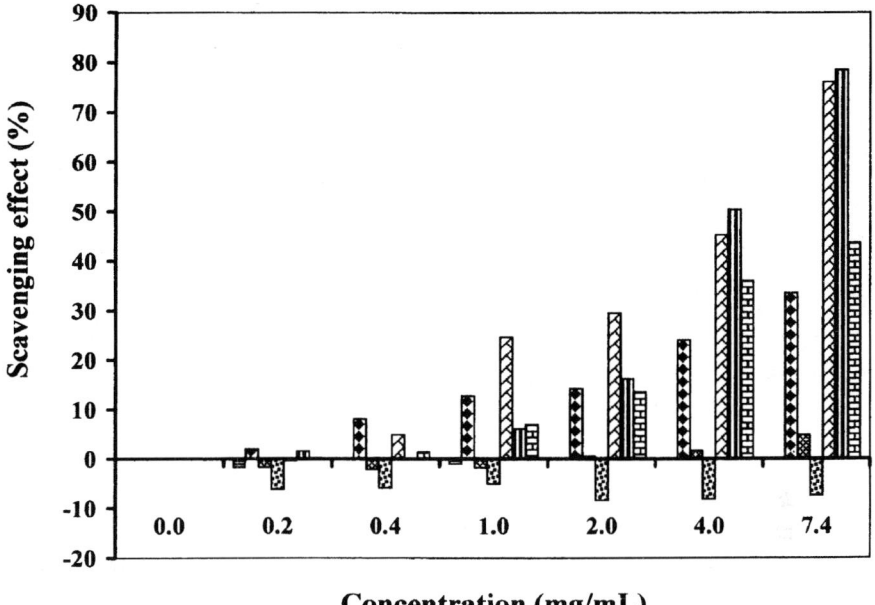

Figure 1. Scavenging effects of polyphenolic compounds and seaweed extracts.
▤, Catechin; ◆, Rutin; ▦, Quercetin; ▩, Gallic acid; ◩, Morin; ▥, Hijiki; ▣, Wakame

Table 3
The effects of chelating Fe^{2+} in fish oil emulsion (mean ± SD mM)

Polyphenolic compounds	Fe^{2+} ion
Control (only fish oil)	0.02 ± 0.01[a]
Catechin	0.47 ± 0.02[d]
Rutin	0.35 ± 0.03[b]
Quercetin	0.41 ± 0.03[c]
Gallic acid	0.47 ± 0.02[d]
Morin	0.35 ± 0.02[b]
Hizikia fusiformis (Hijiki)	0.00 ± 0.01[a]
Undaria pinnatifida (Wakame)	0.00 ± 0.00[a]

Values followed by different superscript letters are significantly different ($p<0.05$).

activities of polyphenolic compounds depending on the lipids system, the presence of metal catalyst, temperature of oxidation, the concentration of antioxidant, the oxidation stage and the methods used to evaluate lipid oxidation. In the presence of Fe^{2+} catalyst, all of polyphenolic compounds acted as pro-oxidant, except extract of *H. fusiformis* and quercetin. The extract of *H. fusiformis* had the highest activity of antioxidant in absence or presence of Fe^{2+} catalyst. This condition may be related to the contents of polyphenolic compounds (epigallocatechin, catechol, morin). Another researcher Yan et al. [22] reported that *H. fusiformis* had the strongest of antioxidant activity, because it contained fucoxanthin as major active compound in acetone extract.

The highest scavenging effects was shown in morin, because it has the total number of OH groups 5 and the presence of a hydroxyl group in position three (3-OH) of the C ring [3]. Since rutin has 4 numbers of OH groups, therefore the scavenging activity was less than morin. Although gallic acid has 3 groups of OH, but percentage of scavenging effects is small, because gallic acid is a monophenol, not as a polyphenol [3]. Since extracts of Hijiki and Wakame contained many kinds of polyphenolic compounds, they had higher scavenging effects. This condition also caused that the effects of Fe^{2+} chelating in fish oil emulsion of both extract seaweeds was the smallest. In contrast all of commercial polyphenolic compounds had the high effects of chelating Fe^{2+} and significantly different with control. Polyphenolic compounds acted as pro-oxidant, usually also have high effects of Fe^{2+} chelating.

REFERENCES

1. Halliwell, B. (1997) Antioxidants and human disease: a general introduction. Nutr. Rev., 55: S44-S52.
2. Temple, N. J. (2000) Antioxidant and disease: more question than answer. Nutr. Res., 2: 449-459.
3. Cook, N. C. and Samman, S. (1996) Flavonoids: Chemistry, metabolism, cardioprotective effects, and dietary sources. Nutr. Biochem., 7: 66-76.
4. Huang, S. W. and Frankel, E. N. (1997) Antioxidant activity of tea catechins in different lipid systems. J. Agric. Food Chem., 45: 3033-3038.
5. Zhang, A, Chan, P. T., Luk, Y. S, Ho, W. K. K. and Chen, Z. Y. (1997) Inhibitory effect of jasmine green tea epicatechin isomers on LDL-oxidation. Nutr. Biochem., 8: 334-340.
6. Carbonneau, M. A., Leger, C. L., Descomps, B., Michel, F. and Monnier, L. (1998) Improvement in the antioxidant status of plasma and low-density lipoprotein in subjects receiving a red wine phenolic mixture. JAOCS, 75: 235-239.
7. Mazza, G., Fukumoto, L., Delaquis, P., Girard, B. and Ewert, B. (1999) Anthocyanins, phenolic and color of cabernet franc, merlot and pinot noir wine from British Colombia. J. Agric. Food Chem., 47: 4009-4017.
8. Osakabe, N., Yamagishi, M., Sanbongi, C., Natsume, M., Takizawa, T. and Osawa, T. (1998) The antioxidative substances in cacao liquor. J. Nutr. Sci. Vitaminol., 44: 313-321.
9. Kahkonen, M. P., Hopia, A. I., Vuorela, H. J., Rauha, J. P., Pihlaja, K., Kujala, T. S. and Heinonen, M. (1999) Antioxidative activity of plant extracts containing phenolic compound. J. Agric. Food Chem., 47: 3954-3962.
10. Chu, Y. H., Chang, C. L. and Hsu, H. F. (2000) Flavonoids content of several vegetables and their antioxidant activity. J. Sci. Food Agric., 80: 561-566.
11. Ohshima, H., Yoshie, Y., Auriol. S. and Gilibert, I. (1998) Antioxidant and pro-oxidant actions of flavonoids: Effects on DNA damage induced by nitric oxide, peroxynitrite and nitroxyl anion. Free Rad. Biol. Med., 25: 1057-1065.

12. Ikegaya, K., Takayanagi, H. and Anan, T. (1990) Quantitative analysis of tea constituents. Chagyo Kenkyu Hokoku, 71: 43-74.
13. Terada, S., Maeda, Y., Masui, T., Suzuki, Y. and Ina, K. (1987) Comparison of caffeine and catechin components in infusion of various tea (green, oolong and black tea) and tea drinks. Nippon Shokuhin Kogyo Gakkaishi, 34: 20-27.
14. Suematsu, S., Hisanobu, Y., Saigo, H., Matsuda, R. and Komatsu, Y. (1995) A new extraction procedure of caffeine and catechins in green tea (Studies on preservation of constituents in canned drinks. part V). Nippon Shokuhin Kagaku Kogaku Kaishi, 42: 419-424.
15. Hertog, M. G. L., Hollman, P. C. H. and Venema, D. P. (1992) Optimalization of a quantitative HPLC determination of potentially anticarcinogenic flavonoids in vegetable and fruits. J. Agric. Food Chem., 40: 1591-1598.
16. Vinson, J. A., Hao, Y., Su, X. and Zubik, L. (1998) Phenol antioxidant quantity and quality in foods: vegetables. J Agric. Food Chem., 46: 3630-3634.
17. Takagi, T., Mitsuno, Y. and Masumura, M. (1978) Determination of peroxide value by the colorimetric iodine method with protection of iodide as cadmium complex. Lipids, 13: 147-151
18. Dinis, T. C. P., Madeira, V. M. C. and Almeida, L. M. (1994) Action of phenolic (acetaminophen, salicylate, and 5-aminosalicylate) as inhibitors of membrane lipid peroxidation and as peroxyl radical scavenger. Arch. Biochem. Biophys., 315: 161-169.
19. Fukuzawa, K. and Fujii, T. (1992) Peroxide dependent and independent lipid peroxidation: site-specific mechanism in initiation by chelated ion inhibition by tocopherols. Lipids, 27: 227-233.
20. Yoshie, Y., Wang, W., Petillo, D. and Suzuki, T. (2000) Distribution of catechins in Japanese seaweeds. Fisheries Sci., 66: 998-1000.
21. Jimenez-Escrig, A., Jimenes-Jimenes, I., Pulido, R. and Saura-Calixto, F. (2001) Antioxidant activity of fresh and processed edible seaweeds. J. Sci. Food Agric., 81: 530-534.
22. Yan, X., Chuda, Y., Suzuki, M. and Nagata, T. (1999) Fucoxanthin as the major antioxidant in *Hizikia fusiformis*, a common edible seaweed. Biosci. Biotechnol. Biochem., 63: 605-607.

POSSIBLE UTILIZATION OF THE PEARL OYSTER PHOSPHOLIPID AND GLYCOGEN AS A COSMETIC MATERIAL

Satoshi Kanoh[a], Kaoru Maeyama[b], Risa Tanaka[a], Tsukasa Takahashi[a], Mayumi Aoyama[a], Mie Watanabe[a], Koichi Iida[b], Seishi Ueda[b], Maki Mae[b], Keiji Takagi[b], Kenji Shimomura[b] and Eiji Niwa[a]

[a] Faculty of Bioresources, Mie University, Tsu, Mie 514-8507, Japan
[b] Research & Development Division, MIKIMOTO PHARMACEUTICAL CO., LTD, Ise, Mie 516-8581, Japan

Abstract

The inner contents of pearl oysters from which pearls of high commercial value are harvested do not find any effective use. We screened the inner contents of pearl oyster *Pinctada fucata* for surfactants or humectants which may be commercially useful as a cosmetic material. Surface tension of the total phospholipid fraction and column-separated subfractions from the pearl oyster were 30-48 mN/m and comparable to those of commercially available surfactants for cosmetic use. Furthermore, the cell proliferation rate and collagen synthetic ability were increased by 11-13% and 29-61%, respectively, when crude glycogen from the pearl oyster was supplied to the culture medium for fibroblasts at concentrations of 0.01-0.2%. When β-particles prepared from the pearl oyster crude glycogen were added to the culture medium for fibroblasts at a concentration of 0.1%, the cell proliferation rate and collagen synthetic ability increased by 15% and 38%, respectively. In addition, the recovery from ultraviolet damage for epidermal keratinocytes was increased by 15-21%, when the crude glycogen from the pearl oyster was added at concentrations of 0.05-0.20% to the culture medium. These results suggest that the phospholipid and crude glycogen from pearl oyster are promising candidate as a cosmetic material.

1. INTRODUCTION

As opposed to the adductor muscle and shell portion, the inner contents of pearl oysters from which pearls of high commercial value are harvested do not find any effective use. Pearls of common sizes are usually hand-picked from the shellfish. Subsequently, while the adductor muscles are preserved in sake lees as "shinjyuzuke" or canned, shells are crushed and used as medicated or food additive calcium. On the other hand, small pearls are isolated from the shucked shellfish by homogenization with slaked lime. The homogenates devoid of pearls are then simply wasted without effective utilization. If useful bio-materials can be isolated from such currently disposed waste, pearl culture and its associated industries will stand to benefit much. Furthermore, a decrease in ocean pollution can also be projected by such a step. Thus, the primary aim of this investigation is to screen the inner contents of Japanese pearl oyster for the presence of constituents which can be of use as cosmetic material [1, 2].

In this study, we found a high surface activity in the total phospholipid fraction [3] and an activation of human skin cell in the glycogen fraction [4] from the pearl oyster.

2. MATERIALS AND METHODS

2.1. Materials

The live specimens of the pearl oyster *Pinctada fucata* cultured at Ago Bay in Mie Prefecture and at Ohura in Ehime Prefecture were transported to the laboratory of Mie University and their inner contents were kept at -20°C immediately after shell removal.

2.2. Preparation of phospholipid

The inner contents from live shellfish were homogenized with the mixture of water and chloroform-methanol (1 : 2, v/v), and passed through a filter paper. To this filtrate was added a solution containing chloroform, water and methanol, and the mixture was subjected to two phase partitioning. The total lipid collected in a chloroform phase layer was subsequently vacuum-dried [5]. The total lipid fraction was applied to the Wakogel C-200 column and successively eluted with chloroform, acetone, and methanol respectively [6]. The total phospholipid fraction was applied to the same column and eluted successively with chloroform- methanol (95:5, v/v), chloroform- methanol (4:1, v/v), chloroform- methanol (3:2, v/v), and chloroform- methanol (1:4, v/v) [7]. Eluates were collected in 40mL fractions and measured for weight of ingredients after vacuum-drying.

2.3. Preparation of glycogen

Glycogen was prepared from the inner contents of live shellfish according to the method of Abder-Akher and Smith [8], and Kohnosu et al.[9] with some modifications. Glycogen was extracted with hot water , then the protein was removed with trichloroacetic acid. To this extract was added ethanol and the precipitate glycogen was collected. This precipitate was washed with ethanol and aqueous glycogen was subsequently vacuum-dried. This crude glycogen was applied to a Toyopearl HW-65 column (1.5 × 85 cm) equilibrated with 1 % sodium chloride. The flow rate was 15 mL/h and 2 mL fraction were collected. An aliquot of each fraction was assayed for total sugar content by the phenol-sulfuric acid method [10] for drawing the elution curve.

2.4. TLC analysis for phospholipid

The total phospholipid and column-separated subfractions were spotted onto TLC plates (silica gel 60, Merck) and subsequently developed by chloroform-methanol-water

(65 : 25 : 4, v/v) [11], chloroform-methanol-acetic acid -formic acid (50 : 30 : 4.5 : 6.5, v/v) [12], and chloroform-methanol- 28% ammonia (65 : 35 : 8, v/v) [13]. Spots indicative of organic components were visualized by heating the plates at 180°C after spraying with 50% sulfuric acid (v/v) [14]. Phospholipid species were localized by spraying with the modified Dittmer-Lester spray reagent as a color reagent [15], which gave blue spots for phospholipids.

2.5. Emulsification test for the total neutral lipid, glycolipid and phospholipid fractions

To three test tubes containing the dry pellets of total neutral lipid, glycolipid and phospholipid fractions prepared from 450 mg of the total lipid, respectively, 2mL each of water and liquid petrolatum (Silkool p-70, Matsumura Oil Research) were added. The tubes were shaken sufficiently, and allowed to stand for 1 week at room temperature for observation of the emulsion state.

2.6. Measurement of surface tension

For each phospholipid fraction, the surface tension was measured by the plate method at 230°C using a CBVP-Z automatic surface tensiometer (Kyowa Interface Science). The total phospholipid and its associated samples fractionated by the column were dissolved or suspended with sonication in deionized water to 0.1% (w/v), and 15 mL-aliquots were measured for surface activity. For reference, commercially available authentic surfactants (sorbitan monostearate and polyoxyethylene sorbitan monostearate) and phospholipid surfactants (lecithin and hydrogenated lecithin prepared from soybean) were examined as well. Deionized water was used as a control.

2.7. Measurement of cell activation

The utilization of the pearl oyster glycogen as a cosmetic material was examined by measuring the cell prolifiration rate and collagen synthetic ability for normal human dermal fibroblasts NHDF (NB)(Kurabo) and epidermal keratinocytes NHEK (B)(Kurabo) as parameters. Fibroblasts was cultured in Eagle's MEM(Nissui) containing 10% fetal bovin serum(FBS). Cell proliferation rate and collagen synthetic ability was measured after 5 days incubation at 37°C in MEM containing 0.5% FBS, and 0.01-0.1% glycogen sample. Epidermal keratinocytes was also cultured in HuMedia-KG2 (Kurabo) at 37°C. Cell proliferation of Epidermal keratinocytes was also measured after 2days incubation at 37°C in HuMedia-KB2(Kurabo) containing 0.025-0.2% glycogen sample. The cell proliferation and collagen synthetic ability in the absence of glycogen were used as a control. Cell proliferation was expressed as percentage, taking the amount of DNA in the absence of glycogen to be 100. DNA in the cell was estimated by the method using ethidium bromide [16,17]. Collagen synthetic ability was also expressed as percentage taking the amount of collagen in the absence of glycogen to be 100. The amount of collagen was estimated using the collagen stain kit (collagen gijyutsu kennsyuukai) [18].

2.8. Measurement of the recovery from ultraviolet damage

Their recovery from ultraviolet damage was examined after the fibroblasts and epidermal keratinocytes were irradiated with the ultraviolet ray. Fibroblast and epidermal keratinocytes irradiated with the ultraviolet ray were cultured in the MEM containing 0.1% BSA with or without glycogen sample (0.05-0.2%). The recovery was expressed as percentage, taking the survival ratio of the cells in the absence of glycogen to be 100. The survival rate was estimated from the amount of LDH of live cells to the total amounts of LDH.

3. RESULTS AND DISCUSSION

3.1. Screening for surfactants in the total lipid fraction

The total lipid fraction of 450 mg prepared from the homogenates of 73 g live specimens was applied to the Wakogel C-200 column and successively eluted with chloroform, acetone, and methanol. The resulting yields of total neutral lipid, glycolipids and phospholipid fractions amounted to 65.5, 96.1 and 59.1 mg, respectively. Consequently, the phospholipid fraction accounted for 27% of the total lipids. The content of the phospholipid fraction from pearl oyster was similar to that from oyster (27%)[19] but lower than that of prawn (50%) [20]. The resulting three fractions were then evacuated to dryness and subjected to the emulsification test. As shown in Fig. 1, all three fractions were well emulsified within 10 min after shaking, but the emulsion with the total neutral lipid fraction disappeared after 24 h. While the emulsion comprising of total glycolipid fraction was slightly decayed after a week, the total phospholipid fraction maintained a considerably stable emulsification state even after the same period.

3.2. Fractionation of the total phospholipid and analysis for composition

The total phospholipid fraction (1.016g) from live specimens of the pearl oyster was applied to the same column and eluted successively with chloroform- methanol (95:5,

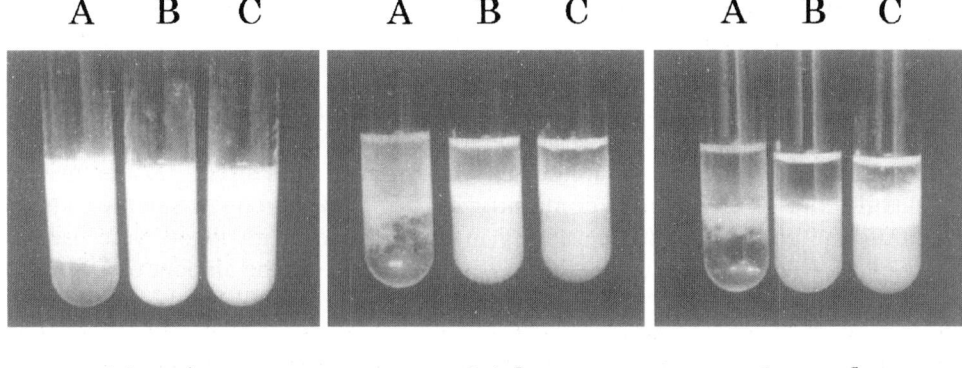

Figure 1. Emulsification test for the total neutral lipid (A), glycolipid (B) and phospholipid (C) fractions from the pearl oyster. Emulsification state was observed at 10 min, 24 h, and 1 week after adding water and liquid petrolatum to each lipid fraction. Reprinted from Kanoh, S. et al. (2000) [3].

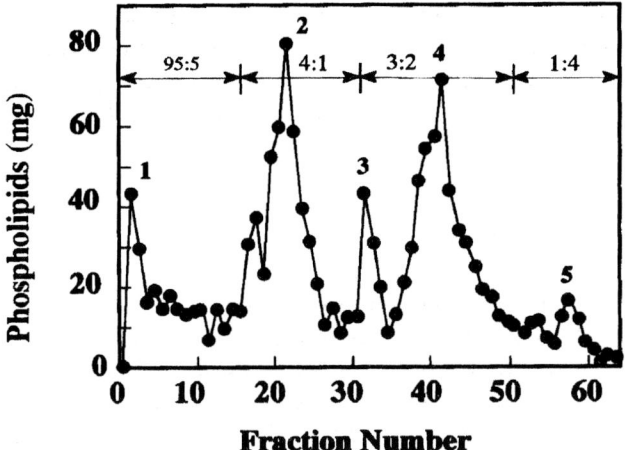

Figure 2. Wakogel C-200 column chromatography of the total phospholipid fraction from the pearl oyster. The total phospholipid fraction was applied to a Wakogel C-200 column (2.0×50 cm) equilibrated with chloroform-methanol (95:5, v/v) and eluted stepwise with 4 solvents with varying ratios of chloroform and methanol (95:5, 4:1, 3:2, and 1:4, v/v). The flow rate was 120 mL/h and each 40 mL fraction collected. Eluates were evaporated to dryness and weighed. Reprinted from Kanoh, S. et al. (2000) [3].

v/v), chloroform- methanol (4:1, v/v), chloroform- methanol (3:2, v/v), and chloroform-methanol (1:4, v/v). Eluates were vacuum-dried and weighed for quantitating the elution profile. As shown in Fig. 2, five major peaks were obtained and their fractions were numbered subfractions 1-5 according to the elution sequence, yielding 53, 285, 80, 314, and 32 mg, respectively. The total amount of these five subfractions eluted from the column was estimated to be 75% of the total phospholipid of 1.016 g applied to column. The elution profile of phospholipid from pearl oyster was similar to that of the phospholipid from rat and beef liver [7], using a silicic acid column. According to Hanahan et al. [7], the peaks eluted with chloroform-methanol (4:1, v/v), chloroform-methanol (3:2, v/v), and chloroform- methanol (1:4, v/v) were presumed to belong to phosphatidylserine/phosphatidylethanolamine, phosphatidylinositol/phosphatidylcholine, and sphingomyelin, respectively.

The subfractions containing respective peaks and the total phospholipid fraction were applied to TLC and developed by the three different methods used for phospholipids from dog lung [11], human platelets [12] and sarcoplasmic reticulum membrane from rabbit muscle [13] as mentioned in Materials and Methods. After developing with three different solvents (Fig. 3, A, B, C), five or six spots were detected for the total phospholipid fraction with sulfuric acid spray on the three TLC plates. These five spots for the total phospholipid fraction were reactive with the modified Dittmer-Lester spray reagent (Fig. 3, D, E, F). A few spots reacting with the same spray reagent were also detected for subfractions 1-5. While subfractions 2 and 3 showed clearly two spots, subfractions 4 and 5 gave nearly one spot. Although a few spots containing organic matter were detected in

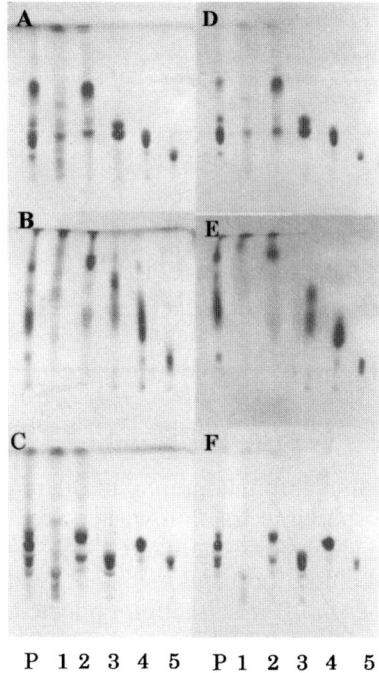

P 1 2 3 4 5   P 1 2 3 4 5

Figure 3. Thin layer chromatograms of the total phospholipid fraction and its associated five subfractions from the Wakogel C-200 column. Lane P represents the total phospholipid fraction, whereas lanes 1-5 correspond to subfractions in Fig.2. The plates consisting of silica gel 60 (Merck) were developed with chloroform-methanol- water (65 : 25 : 4, v/v) (A, D), chloroform-methanol-acetic acid-formic acid (50 : 30 : 4.5 : 6.5, v/v) (B, E), and chloroform-methanol-28% ammonia (65 : 35 : 8, v/v) (C, F). The spots were detected by heating the plates at 180°C after spraying with 50% sulfuric acid (A, B, C) for the total organic matters and with the modified Dittmer-Lester spray reagent (D, E, F) for phospholipids. Reprinted from Kanoh, S. et al. (2000) [3].

subfraction 1, they were hardly reactive with the modified Dittmer-Lester spray reagent. Judging from the elution profile of phospholipids from the silicic acid column [7] and the associated TLC patterns [11-13], the main components of peaks 1 - 5 were presumed to be simple lipids, phosphatidylserine/phosphatidylethanolamine, phosphatidylinositol /phosphatidylserine, phosphatidylcholine and sphingomyelin, respectively, according to the eluting order.

3.3. Surface activity of phospholipids

The surface activity, as indicated by surface tension as a parameter, was measured for the total phospholipid and its associated subfractions 1-5 as well as four commercially available surfactants (Fig. 4). While the surface tension of deionized water (used as a control) was 73 mN/m, commercially available surfactants prepared from soybean

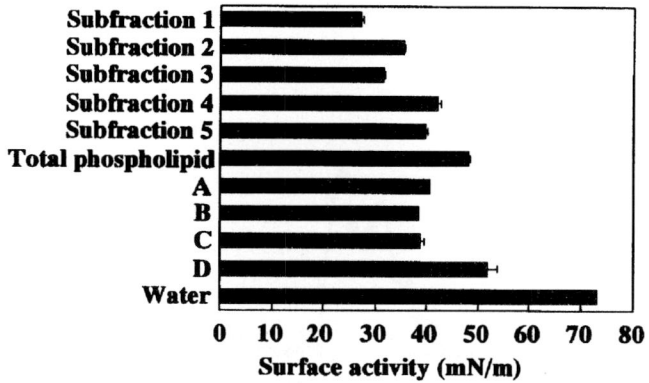

Figure 4. Surface tension of the total phospholipid and its associated subfractions from the pearl oyster. The total phospholipid and subfractions 1-5 in Fig. 2 were measured by the plate method at 23°C together with four commercially available surfactants (A, polyoxyethylene sorbitan monostearate; B, sorbitan monostearate; C, lecithin from soybean; D, hydrogenated lecithin from soybean). Data represent mean±SD for 5 determinations. Reprinted from Kanoh, S. et al. (2000) [3].

lecithin and authentic surfactants showed surface tension values ranging from 38 to 52 mN/m. Similar values for the interfacial tension of modified phospholipids from soybean at water/kerosene interface have been reported by Shimbo et al. [21]. The surface tensions of the total phospholipid and its associated subfractions 4 and 5 (40 - 48 mN/m) were very close to those of commercially available surfactants. The value for subfraction 2 (35.6 mN/m)was slightly lower than those of the total phospholipid and subfractions 4 and 5. However, the surface tension values for subfractions 1 and 3 were about 30 mN/m and much lower than those of commercial surfactants.

3.4. Preparation of pearl oyster glycogen

The crude glycogen was applied to a Toyopearl HW-65 column equilibrated with 1 % sodium chloride. An aliquot of each fraction was assayed for total sugar content by the phenol-sulfuric acid method [10] for drawing the elution curve. As shown in Fig 5, the elution curve from the crude glycogen of pearl oyster had two main peaks, a smaller one with shoulder and a large one. Particle sizes of the components in each peaks were estimated to be 200kDa and 23kDa, respectively [22]. Judging from their sizes, the former was presumed to be α-particle and the latter was β-particle. It has been reported previously that the α-particle appears to be composed of several β-particles linked by α-1,4-glucan chains. It is presumed that the glycogen from pearl oyster has taken similar composition.

As shown in Fig. 6 A, crude glycogen from pearl oyster has the shape of a white granule. Furthermore, α-and β-particle (B, C) is much more white and the flocculent outline is shown. It was estimated that the crude glycogen contained 92% in glucose equivalent.

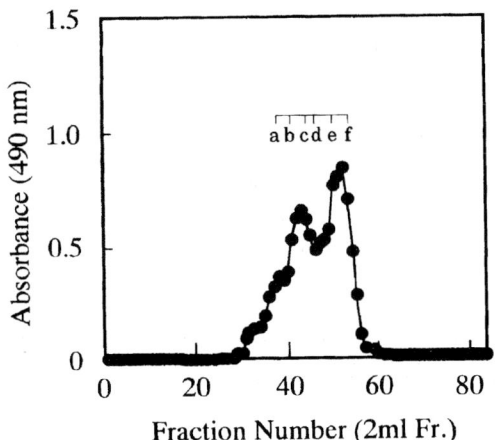

Figure 5. Gel filtration of crude pearl oyster glycogen on a Toyopearl HW-65 column (1.5×85 cm) equilibrated with 1 % sodium chloride. The flow rate was 15 mL/h and 2 mL fraction collected. An aliquot of each fraction was assayed for total sugar content by the phenol-sulfuric acid method for drawing the elution curve. Molecular mass markers (Shodex standard P-82, Showadenko, Tokyo, Japan) for polysaccharides eluted are: a, P-800 (788kDa); b, P-400 (404kDa); c, P-200 (212kDa); d, P-100 (112kDa); e, P-50 (47.0kDa); f, P-20 (22.8kDa). Reprinted from Kanoh, S. et al. (2001) [4].

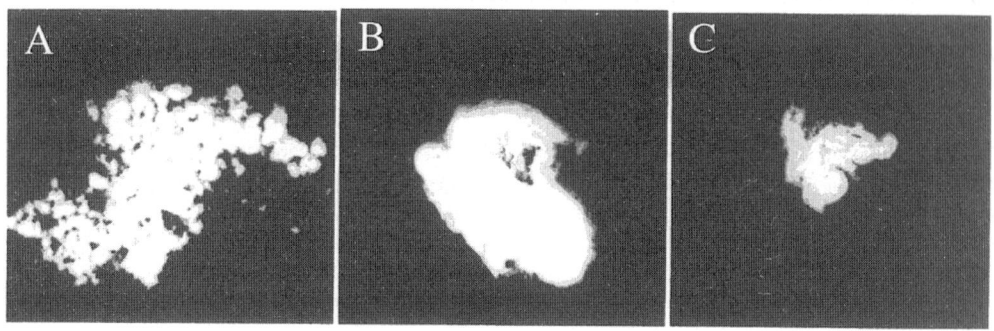

Figure 6. Photographs of pearl oyster glycogen. Crude glycogen (A), purified glycogen α-particles (B), and purified glycogen β-particles (C) from pearl oyster. Reprinted from Kanoh, S. et al. (2001) [4].

3.5. The influence of pearl oyster glycogen on cell proliferation

The utilization of the pearl oyster glycogen as a cosmetic material was examined by measuring the cell proliferation rate of fibroblasts and epidermal keratinocyte. As shown in Table 1, the cell proliferation rate was increased by 11-13% when crude glycogen was supplied to the culture medium for fibroblasts at concentrations of 0.01-0.2%. Furthermore, when glycogen α-particle was supplied to the culture mediuym at concentration of 0.1%, cell proliferation rate was 86%, showing the inhibition of cell

Table 1
Influence of pearl oyster glycogen on cell proliferation rate of normal human dermal fibroblasts NHDF(NB)

Sample	Concentration(%)	Cell proliferation[b]
Control[a]	—	100
Crude glycogen	0.01	113
	0.05	113
	0.10	112
	0.20	111
Glycogen α-particle	0.10	86
Glycogen β-particle	0.10	115

[a]The MEM containing 0.5% FBS was used as a control; [b]Data are given by means with 4 - 6 determinations. Reprined from Kanoh, S. et al. (2001) [4]

proliferation. In contrast, when β-particle was supplied at the same condition the rate was 15% increased. The solubility of α-particle is low for the solvent, and its particle size is larger than that of β-particle. The difference in such quality is considered to be the difference in cell proliferation. Although it was not shown at the table, when the same experiment was conducted using epidermal keratinocytes, the value of 74-98% was observed, showing the inhibition of cell proliferation. By comparison, the cell proliferation rate for fibroblasts was increased by 9-11% when commercially available mussel glycogen was supplied to the culture medium at concentrations of 0.05-0.2%. In contrast, as for epidermal keratinocytes, the rate was decreased by 75-94% when the same glycogen was supplied to the medium at concentration of 0.025-0.1%.

3.6. The influence of pearl oyster glycogen on collagen synthetic ability

The influences of crude glycogen and β-particle on collagen synthetic ability in fibroblasts were examined. As shown in Table 2, the collagen synthetic ability was increased by 29-61% and 8-38% when crude glycogen and β-particle were supplied to the culture medium for fibroblasts at concentration of 0.01-0.1, respectively. Comparing with commercially available mussel glycogen, collagen synthetic ability was increased by 3% when it was supplied at concentration of 0.05-0.2%. This results suggest that the crude glycogen from pearl oyster promises to be a candidate as a humectant with collagen synthetic ability.

3.7. Influence of pearl oyster glycogen on recovery of fibroblasts and epidermal keratinocytes injured by ultraviolet

The recovery from ultraviolet damage for fibroblasts and epidermal keratinocytes were examined. The recovery from ultraviolet damage for epidermal keratinocytes was increased by 15-21%, when the crude glycogen from the pearl oyster was added at concentrations of 0.05-0.20% to the culture medium (Table 3). In contrast, the recovery from ultraviolet damage for fibroblasts was not observed. Furthermore, the recovery from ultraviolet damage for epidermal keratinocytes was not observed, when α- and β-particle were added at the same condition to the culture medium. It is necessary to examine the reason in detail.

Table 2
Influence of pearl oyster glycogen on collagen synthetic ability in normal human dermal fibroblasts NHDF(NB)

Sample	Concentration(%)	Collagen production[b]
Control[a]	–	100
Crude glycogen	0.01	129
	0.05	152
	0.10	161
Glycogen β-particle	0.01	109
	0.05	108
	0.10	138

[a]The MEM containing 0.5% FBS was used as a control; [b]Data are given by means with 3 - 5 determinations. Reprinted from Kanoh et al. (2001) [4]

Table 3
Influence of pearl oyster glycogen on recovery of normal human dermal fibroblasts NHDF(NB) and epidermal keratinocytes NHEK(B) injured by ultraviolet irradiation

Cells	Sample	Concentration(%)	Recovery[b]
Fibroblast	Control[a]	–	100
	Crude glycogen	0.05	97
		0.10	97
		0.20	96
Epidermal Keratinocytes	Control[c]	–	100
	Crude glycogen	0.05	115
		0.10	120
		0.20	121
	Glycogen α-particle	0.10	93
		0.20	89
	Glycogen β-particle	0.10	94
		0.20	93

[a]The MEM containing 0.1% bovine serum albumin was used as a control; [b]Data are given by means with 4 - 7 determinations; [c]Hu Media-KG2 medium was used as a control. Reprinted from Kanoh, S. et al. (2001) [4]

These results of the current study indicate that phospholipids and collagen from the pearl oyster show potential promise as surfactants or humectant with cell activation function for cosmetic ingredients. Additional investigations are now in progress to develop commercial cosmetics using the phospholipid fractions and glycogen fractions in this study.

REFERENCES

1   Sasaki, I. and Suzuki, T. (1992) Research and development of functioned humectants-Mainly phospholipids. Fragrance J., 20: 22-28.

2   Sasaki, I., Arakane, K. and Suzuki, T. (1995) Application of phospholipids as recent humectant formulations. Fragrance J., 23: 56-62.
3   Kanoh, S., Tanaka, R., Takahashi, T., Maeyama, K., Iida, K., Ueda, S., Shimomura, K. and Niwa, E. (2000) Surfactants from the Japanese pearl oyster, *Pinctada fucata*. J. Jpn. Oil Chem. Soc., 49: 495-499.
4   Kanoh, S., Aoyama, M., Watanabe, M., Mae, M., Takagi, K., Shimomura, K. and Niwa, E. (2001) Possible utilization of the pearl oyster *Pinctada fucata* glycogen as a cosmetic material. Nippon Suisan Gakkaishi, 67: 90-95.
5   Bligh, E. G. and Dyer, W. J. (1959) A rapid method of total lipid extraction and purification. Can. J. Biochem. Physiol., 37: 911-917.
6   Rouser, G., Kritchevsky, G. and Yamamoto, A. (1967) Column chromatographic and associated procedures for separation and determination of phosphatides and glycolipids. In: Lipid Chromatographic Analysis. Vol. 1. Marinetti, G.V. (eds), Dekker, New York, pp. 99-162.
7   Hanahan, D. J., Dittmer, J. C. and Warashina, E. (1957) A column chromatographic separation of classes of phospholipides. J. Biol. Chem., 228: 685-700.
8   Abder-Akher, M. and Smith, F. (1951) The repeating unit of glycogen. J. Am. Chem. Soc., 73: 994-996.
9   Konosu, S., Fujimoto, K., Takashima, Y., Matsushita, T. and Hashimoto, Y. (1965) Constituents of the extracts and amino acid composition of the protein of short-necked clam. Bull. Japan. Soc. Sci. Fish., 31: 680-686.
10 Dubois, M., Gilles, K.A., Hamilton, J.K., Rebers, P.A. and Smith, F. (1956) Colorimetric method for determination of sugars and related substances. Anal. Chem., 28: 350-356.
11 Cernansky, G., Liau, D.-F., Hashim, S. A. and Ryan, S. F. (1980) Estimation of phosphatidylglycerol in fluids containing pulmonary surfactant. J. Lipid Res., 21: 1128-1131.
12 Ozawa, A., Jinbo, H., Takahashi, H., Fujita, T., Hirai, A., Terano, T., Tamura, Y. and Yoshida, S. (1985) Determination of higher fatty acids in phospholipid subfractions of human platelets using thin-layer chromatography and gas chromatography. Bunseki Kagaku, 34: 707-711.
13 Herbette, L., Blasie, J. K., Defoor, P., Fleischer, S., Bick, R.J., Van Winkle, W.B., Tate, C. A. and Entman, M. L. (1984) Phospholipid asymmetry in the isolated sarcoplasmic reticulum membrane. Arch. Biochem. Biophys., 234: 235-242.
14 Skipski, V. P. and Barclay, M. (1969) Thin-layer chromatography of lipids. Methods Enzymol., 14: 530-598.
15 Ryu, E. K. and MacCoss, M. (1979) Modification of the dittmer-lester for the detection of phospholipid derivatives on thin-layer chromatograms. J. Lipid Res., 20: 561-563.
16 Prasad, A.S., du Mouchelle, E., Koniuch, D. and Oberlas, D. (1972) A simple fluorometric method for the determination of RNA and DNA in tissues. J. Lab. Clin. Med. 80: 598-602.
17 Goldberg, R.L. and Toole, B.P. (1983) Monensin inhibition of hyaluronate synthesis in rat fibrosarcoma cells. J. Biol. Chem., 258: 7041-7046.
18 Lopez de Leon, A. and Rojkind, M. (1985) A simple micromethod for collagen and

total protein determination in formalin-fixed paraffin-embedded sections. J. Histochem. Cytochem., 33: 737-743.
19 Jeong, Y., Ohshima, T., Koizumi, C. and Kanou, Y. (1990) Lipid deterioration and its inhibition of Japanese oyster *Crassostrea gigas* during frozen storage. Nippon Suisan Gakkaishi, 56: 2083-2091.
20 Muriana, F. J. G., Ruiz-Gutierrez, V. and Bolufer, J. (1993) Phospholipid fatty acid compositions of hepatopancreas and muscle from the prawn, *Penaeus japonicus*. J. Biochem., 114: 404-407.
21 Simbo, K., Gohtani, S., Yamano, Y. and Ina, K. (1993) Emulsifying properties of enzymatically modified phospholipids. Nippon Shokuhin Kogyo Gakkaishi, 40: 755-763.
22 Hata, K., Hata, M., Hata, M. and Matsuda, K. (1984) A proposed model of glycogen particle. J. Jpn. Soc. Starch Sci., 31: 146-155.

UTILIZATION OF WASTES

EXTRACTIVE COMPONENTS OF FISH SAUCES FROM WASTE OF THE FRIGATE MACKEREL SURIMI PROCESSING AND A COMPARISON WITH THOSE OF SEVERAL ASIAN FISH SAUCES

Yasuhiro Funatsu[a], Ken-ichi Kawasaki[a] and Shiro Konagaya[b]
[a]Toyama Prefectural Food Research Institute, 360 Yoshioka, Toyama 939-8153, Japan
[b]Kokugakuin University, Tochigi College, Hirai-cho, Tochigi 328-8588, Japan

Abstract

Extractive components of a fish sauce (WS) prepared from wastes of frigate mackerel, which was processed into surimi, was compared with those of a fish sauce (MS) prepared in the same manner from the minced meat of frigate mackerel, and several Asian fish sauces (nampla, nuoc mam, patis and yeesui). Comparisons of free amino acids and bound amino acids were different with different samples. The major organic acid in WS and MS was lactic acid and that in yeesui was acetic acid. As sugar and sugar alcohol components, erythritol, arabitol, and mannitol were detected in WS and MS and sucrose in nampla and nuoc mam. The results of a taste panel test showed that the elemental taste factors of WS as well as those of MS and nuoc mam were well harmonized and thus tasted mellow. On the other hand, yeesui had a straight salty and peculiar taste.

1. INTRODUCTION

Frigate mackerel caught in a large quantity in Toyama Bay has not always been utilized for human food because of its inherent disadvantage as raw material [1, 2]. Although the processing of surimi from this kind of fish has recently been tried, the production is still very small in scale. Most of the catch is still used for fertilizer [3].

Since the popularity of fish sauces has recently been rising with the growing popularity of ethnic food such as Vietnamese and Thai dishes in Japan [4], the authors attempted to produce fish sauce from frigate mackerel, i.e. from the waste discharged from the initial step of frigate mackerel surimi processing and from the minced meat, using soy sauce koji in fermentation [5]. This paper deals with the extractive components of the fish sauces from the two different fish sauces i.e. the fish sauce from the waste (waste fish-sauce) and that from the minced meat (minced meat fish-sauce), and compared with those of several Asian fish sauces.

2. MATERIALS AND METHODS

2.1. Materials
Fresh frigate mackerel *Auxis rochei* caught in the Toyama Bay in August 1996 were used as raw materials.

2.2. Preparation of fish sauce from fisheries waste
According to the usual process of surimi processing, surimi was processed from frigate mackerel on an experimental scale [6, 7]. The head, bone, skin and intestine discharged at the filleting process were collected as waste. The fillets were minced and the minced meat was submitted to leaching with 3 volumes of cold 0.1% NaCl - 0.3% $NaHCO_3$ solution and ground with a stone mill (MKZA10-10; Masscolloider, Masco Industry Co. Ltd., Saitama). The leached minced meat in the suspension was collected by centrifugation at around 4°C. At this temperature, the lipid component that rose to the surface solidified and removed with a spoon. The rest of the supernatant was put in the waste. The meat particles forming the surface layer of leached minced meat were also put in the waste. Before fermentation, the combined waste was cooked in boiling water. Large quantities of lipid risen to the surface were removed. The remaining materials were cooled to a temperature below 25°C. Moromi (fish sauce mush) was prepared by fermentation of the mixture of the waste, 25% of soy sauce koji on a wet basis and 20% of salt with a purity of 95% at room temperature for one year. The koji used for fermentation was prepared by incubating a 1:1 mixture of steamed defatted-soy bean and roasted wheat with *Aspergillus oryzae* (Ichimurasaki; Bio'c Co., Ltd., Toyohashi). The moromi was centrifuged at 10,000×g for 30 min. The supernatant thus obtained was heated at 90°C for about 30 min and the lipid was removed with a cooking sheet (Leeds Co. Ltd., Tokyo). The defatted sauce thus obtained was filtered with a filter paper (No.5A, Advantec Co. Ltd., Tokyo) and subjected to analyses and sensory evaluation. This sauce is called as waste fish-sauce.

2.3. Preparation of fish sauce from minced meat
The minced meat was fermented to moromi in a similar manner as mentioned above, except that, because there was not much lipid, the lipid was not removed before being subjected to fermentation. The ingredients were otherwise the same as those used above for the waste fermentation. Fish sauce from the minced meat for analyses was prepared similarly as stated above. However, since the filtrate contained a small amount of lipid, the lipid was removed with a cooking sheet at 90°C for about 10 min instead of 30 min. This sauce is called as minced meat fish-sauce.

2.4. Color and chemical components
The values of L*, a* and b* of the specimens were measured by the transmission method. The pH was measured with a glass electrode. The salt content was determined by the Volhart method [8] and Brix was determined according to the soy sauce analysis method [9]. Soluble solids excluding salt was calculated by subtracting salt content from Brix [9]. Total nitrogen (TN) content was measured by the Kjeldahl method [10]. The lipid was determined by the Folch method [11].

2.5. Free amino acid composition and the amino acid composition of peptides including oligopeptides

The specimen was diluted with 200 or 300 volumes of distilled water. The free amino acid composition of the specimen was analyzed as described previously [12]. Substances having molecular weights lower than 5,000 in the filtrate were removed by centrifugal filtration (Ultrafree-MC type; Millipore Co., Ltd., Tokyo). The filtrate thus obtained was hydrolyzed with 6N HCl at 110°C for 24 hours. The hydrolysate, after being filtered with cellulose acetate (0.45μm), was subjected to amino acid analysis using an amino acid analyzer (L-8500A model, Hitachi Co., Ltd., Tokyo). This procedure essentially followed the method of Yoshida et al. [13]. Cys, Met and Trp were analyzed by the method of Moore [14] and Simpson et al. [15]. The peptides including oligopeptides are hereafter called as oligopeptides.

2.6. Organic acid composition

The specimen was diluted with 100 volumes of distilled water. The organic acid composition in this diluted solution was analyzed by HPLC (HPLC Organic Acid Analysis System; Shimadzu Co., Ltd., Tokyo). The analytical conditions of HPLC were as follows: column, Shin-pack SCR-102H (8mm I.D.×300mm L.); mobile phase: 5 mM p-toluenesulfonic acid aqueous solution; flow rate: 0.8 mL/min; temperature: 40°C.

2.7. Free sugars and free sugar alcohols

The specimens were extracted with hot ethanol according to the method proposed by Tajima [16] and the extracts were made up to 50mL. Part of this solution (2mL) was evaporated to dryness under reduced pressure below 60°C. The resulting residue was dissolved in 1mL of TMSI regent (H-type, GL Science Inc., Japan) containing 5μL of 10% β-D-glucoside solution (an internal standard) by stirring for 3 hours at room temperature. The solution was centrifuged at 10,000×g for 10 min at room temperature. The free sugar and free sugar alcohol contents of the supernatant of the specimen were measured using a gas chromatograph (GC) (G188A type, Hewlett Packard Co., Ltd., U.S.) with a slight modification of the method of Yasui et al. [17]. The analytical conditions were as follows: column, HP-5; I.D. 0.25mm×L30m; column temperature: 70-300°C; heating rate: 20°C/min.

2.8. Sensory evaluation of fish sauces

The color and taste panel tests were carried out by a panel consisting of 40 people in all from Toyama Prefectural Food Research Institute and Toyama Prefectural Fisheries Experimental Station, and some specialists from the fish processing industry in Toyama Prefecture. The male-female ratio was 24 to 16. The age distribution was as follows: people in their 20's, 30', 40', 50' and 60' were 10, 8, 12, 8 and 2, respectively. The sensory evaluation of the fish sauces was performed by the SD method [18]. The data were analyzed for mean±(V/n)0.5×t(α, n-1); V: variance; n: number of subjects (n=40); α: significance level (α=0.05) [19].

3. RESULTS AND DISCUSSION

3.1. Color and chemical components of fish sauces

The characteristics of color and chemical components of fish sauces are shown in Table 1 [20]. The colors of waste-fish sauce, minced meat-fish sauce, nampla, nuoc mam, patis and yeesui were deep gray, dark black, dull orange, grayish orange and grayish bright yellow, respectively. The specific gravity of waste-fish sauce was higher than that of patis and yeesui but lower than that of minced meat fish-sauce, nampla and nuoc mam. The pHs and the salt contents of the waste and minced meat fish-sauces were almost the same, while those of nampla, nuoc mam, patis and yeesui were higher than those of the waste fish-sauce. Although the soluble solids excluding salt content of waste-fish sauce, nampla and nuoc mam was about 18-19%, that of the waste fish sauce was lower than that of the minced meat and was higher than that of patis and yeesui. TN content of the waste fish sauce was lower than that of the minced meat fish-sauce, nampla, nuoc mam while that of the waste fish-sauce was higher than that of patis and yeesui. The lipid content of all the fish sauces was below 0.1%.

According to the Japanese Agricultural Standard [21], the TN content and the soluble solids excluding salt of high grade soy sauce (koikuchi) are required to be over 1.5% (w/w) and 16% (w/w), respectively. The waste and minced meat fish-sauces, nampla and nuoc mam satisfied these standards, while those of patis and yeesui considerably met the standard.

3.2. Free amino acid composition and amino acid composition of oligopeptides

The free amino acid compositions of the fish sauces are shown in Table 2 [20]. The total free amino acid content of the minced meat fish-sauce was much higher than that of the waste fish-sauce, and in comparison with the data on nampla, nuoc mam and yeesui, it is considerably high. Glu was the most abundant of the free amino acids in the fish sauces

Table 1
Color and chemical components of fish sauces

	Waste fish-sauce	Minced meat fish-sauce	Nampla	Nuoc mam	Patis	Yeesui
L*	9.76	0.09	48.87	51.73	66.83	60.39
a*	31.58	0.01	23.95	22.00	9.61	13.42
b*	16.70	0.02	74.40	75.67	68.07	69.50
Specific gravity†	1.197	1.244	1.219	1.222	1.159	1.156
pH	4.9	5.0	5.5	5.5	5.4	6.4
Salt (%)	23.5	21.7	24.7	24.5	29.2	28.2
Solble solids excluding salt (%)	18	23	19	19	7	8
Total-N (%)	1.77	3.78	2.08	2.09	1.35	1.25
Lipid (%)	0.01	0.04	0.03	0.03	0.01	0.04

†, values measured at 20°C. Soluble solids excluding salt, Brix [9]-Salt. Reprinted from Funatsu, Y. et al. (2000) [20]

Table 2
Free amino acids composition of fish sauces (mg/100 mL)

	Taste	Waste sauce	Minced meat sauce	Nampla	Nuoc mam	Patis	Yeesui
Tau	–	169.4	206.0	131.8	135.1	137.0	120.2
Asp	U	597.2	1174.8	906.4	903.2	337.7	215.8
Thr	S	309.8	563.2	525.8	562.8	310.8	70.4
Ser	S	319.0	547.4	254.4	224.1	138.8	26.5
Glu	U	855.0	1539.6	1321.2	1381.6	738.8	284.0
Gly	S	264.4	265.4	396.9	397.2	277.7	98.1
Ala	S	434.6	785.6	699.9	811.8	505.8	418.3
Val	B	396.8	685.2	627.4	666.4	419.3	351.6
Cys	–	ND	ND	28.5	32.9	9.6	ND
Met	B	118.8	134.0	235.4	249.0	184.5	98.8
Ile	B	341.6	358.8	387.4	408.3	326.7	331.8
Leu	B	542.6	416.8	479.5	501.7	491.9	555.9
Tyr	B	98.6	159.2	118.2	120.1	116.5	66.2
Phe	B	241.6	374.8	362.8	393.6	232.7	225.1
Trp	–	48.3	23.0	ND	77.1	35.2	13.4
Orn*	B	226.0	12.0	79.4	87.0	147.3	146.9
Lys	B	519.6	1068.6	1069.5	1168.5	735.5	239.3
His	B	326.2	727.4	376.3	390.4	316.8	9.1
Arg	B	50.4	663.6	8.4	25.7	31.8	3.0
Pro	S	321.1	624.4	132.2	55.2	24.5	33.9
Total		6181.0	10329.8	8141.4	8591.7	5518.9	3308.3

ND, not detected; B, bitter; S, sweet; U, umami; *L-Ornithine. Amino acids were classified according to the proposal of Ninomiya et al. [22] with slight modification. Reprinted from Funatsu, Y. et al. (2000) [20]

except yeesui. The Leu, Ala, Ile and Val contents of yeesui were much higher than the Glu contents, and Tau was present in all fish sauces. Though the Arg content of the waste fish-sauce was lower than that of the minced meat fish-sauce, the Orn (L-ornithine) content of the waste fish-sauce was higher.

As for taste, the ratio of amino acids related to umami, bitterness and sweetness to the total amino acid content were almost the same between the waste and minced meat fish-sauces. The amino acids assccociated with umami (Glu+Asp) accounted for about 24-26% of the total amino acids, those associated with bitterness (Arg+Lys+His+Phe+Leu+Ile+Met+Val+Tyr) accounted for about 45-46%, and those associated with sweetness (Pro+Thr+Ala+Gly+Ser) accounted for about 27% [22]. In yeesui, however, umami amino acids did for about 15%, bitter amino acids about 61% and sweet amino acids about 20%. The results are consistent with the sensory evaluation as described below.

The composition of amino acid in oligopeptides in the fish sauces is shown in Table 3 [20]. The contents of Glu, Asp and Gly in all the hydrolysates were higher than those of the other amino acids. On the other hand, the contents of Val, Ile, Leu and Phe which compose bitter peptides were lower in the waste fish-sauce than in the minced meat fish-sauce. The composition of amino acid in oligopeptides in nampla and nuoc mam was

Table 3
Amino acids composition of peptides including oligopeptides in fish sauces (mg/100mL)

	Waste sauce	Minced meat sauce	Nampla	Nuoc mam	Patis	Yeesui
Asp	195.6	617.2	312.0	267.8	167.8	65.4
Thr	16.9	214.6	77.0	68.2	44.8	33.0
Ser	50.7	247.6	58.1	76.1	43.6	21.1
Glu	458.5	1465.6	467.8	550.2	371.2	162.2
Gly	231.1	493.3	254.4	474.4	298.7	107.1
Ala	22.1	182.6	46.7	24.2	26.7	63.7
Val	0.0	100.6	49.2	51.6	11.7	59.7
Cys	3.7	28.5	11.1	0.0	4.3	0.0
Met	16.9	39.6	0.0	0.0	0.0	0.0
Ile	0.0	71.5	14.2	0.0	0.0	25.4
Leu	0.0	80.5	23.3	10.3	0.0	44.1
Tyr	99.6	60.3	0.0	0.0	0.0	0.0
Phe	0.0	53.2	22.5	0.0	0.0	30.6
Trp	0.0	0.0	0.0	0.0	0.0	0.0
Orn*	36.3	23.9	192.2	116.2	40.1	40.1
Lys	58.4	360.4	125.0	72.3	18.0	42.9
His	50.1	270.3	82.2	63.4	39.9	5.2
Arg	32.9	122.1	30.0	25.1	12.3	19.6
Pro	113.2	156.1	130.0	119.3	110.9	65.6
Total	1386.0	4587.9	1895.7	1919.1	1190.0	785.7

*L-Ornithine   Reprinted from Funatsu, Y. et al. (2000) [20]

almost the same except that the latter did not contain Phe, Ile or Cys. With respect to the amino acids in oligopeptides, the total amino acid contents were in the order minced meat fish-sauce>nuoc mam>nampla>waste-fish sauce>patis>yeesui. Fujimaki [23] and Okai [24] pointed out that the saltiness of a certain solution is masked by the taste contributed by peptides, because the peptides enhance sweetness, bitterness and umami taste. These findings suggest that the oligopeptides have a greater effect on the taste in the minced meat fish-sauce than in another fish sauces. Fujimaki [23] also described that umami peptides always contain Glu residues together with a very hydrophilic amino acid residue at the carboxyl terminus. Both of the fish sauces from frigate mackerel might contain umami peptides having Glu in their structure, because Glu was found as the most abundant amino acid in the oligopeptides as mentioned above. However, the amounts of umami peptides might be smaller in the waste fish-sauce than in the minced meat fish-sauce because the Glu content was lower in the waste fish-sauce than in the minced meat fish-sauce. Furthermore, none of nucleotides such as AMP, IMP and GMP, which are known to contribute to umami taste, were detected in any of the fish sauces in this study (data not shown).

3.3 Organic acid compositions

Table 4 shows the organic acid compositions of the fish sauces [20]. Lactic acid in the waste fish-sauce and the minced fish-sauce was found at high concentrations compared with other organic acids. The content of lactic acid of the waste fish-sauce was higher than that of the minced meat fish-sauce, while the contents of acids such as citric,

Table 4
Organic acids composition in fish sauces (mg/100 mL)

	Waste sauce	Minced meat sauce	Nampla	Nuoc mam	Patis	Yeesui
Citric acid	17.3	308.9	118.2	97.1	2.3	ND
Pyruvic acid	ND	ND	ND	2.4	ND	ND
Malic acid	21.9	ND	2.6	22.8	ND	21.2
Succinic acid	39.4	45.5	51.9	61.4	91.2	115.1
Lactic acid	1691.9	1255.7	460.9	305.3	423.6	60.0
Formic acud	26.0	34.2	23.6	18.4	24.4	19.6
Fumalic acid	ND	ND	ND	ND	ND	ND
Acetic acid	70.1	80.5	109.0	154.2	103.6	747.2
Levulic acid	ND	ND	ND	ND	ND	ND
Pyroglutamic acid	129.1	259.5	149.6	207.8	34.0	90.8
Total	1995.7	1984.3	915.8	869.4	679.1	1053.9

ND, not detected   Reprinted from Funatsu, Y. et al. (2000) [20]

pyruvic, malic, succinic, formic, fumaric, acetic and pyroglutamic acids were lower in the minced meat fish-sauce. However, the total organic acid content did not differ between the waste fish-sauce and the minced fish-sauce. The higher amount of lactic acid in the waste fish-sauce was due to the leaching water which contained a large quantity of lactic acid. The total content and composition of organic acid in the waste fish-sauce and the minced fish sauce were almost the same. Only a small amount of pyruvic acid was detected in nuc mam. The citric acid and pyroglutamic acid in patis were found at low concentrations compared with other organic acids. Of all the fish sauces, patis had the lowest total organic acid content. Yeesui had a low content of lactic acid but a high content of acetic acid. This may be due to the difference in fermentation between yeesui and other fish sauces.

3.4. Free sugar and free sugar alcohol compositions of the fish sauces

The free sugar and free sugar alcohol compositions are shown in Table 5 [20]. Glycerol and inositol were detected in all of the fish sauces, and glucose and sucrose were also detected in all of the fish sauces except patis and yeesui. However, erythritol, arabitol and mannitol were found in the waste fish-sauce and minced meat fish sauce. These observations suggested that patis and yeesui were much less sweet than the other fish sauces because of their much lower contents of total free sugars and free sugar alcohols. Based on the total free sugar and free sugar alcohols contents in the samples, one would expect that the sweetness of nuoc mam would be stronger than that of nampla and that the sweetness of the waste fish-sauce would be weaker than that of minced meat fish-sauce. Sugars and sugar alcohols (free and bound forms) such as glucose, ribose, arabinose, galactose, fructose and inositol are generally present in fish meat. Glucose is an especially abundant free sugar. However, considering that the free inositol content of fish meat is only about 2-9 mg/100g [25, 26], and that free sugar and free sugar alcohol compositions are similar to those in soy sauce koji (data not shown), most of the free sugars and free sugar alcohols in the waste and minced meat fish-sauces must have produced from soy sauce koji during fermentation.

Table 5
Free sugars and free sugar alcohols in fish sauces (mg/100mL)

	Waste sauce	Minced meat sauce	Nampla	Nuoc mam	Patis	Yeesui
Glucose	560	1200	32	26	ND	ND
Arabinose	ND	ND	ND	ND	ND	ND
Xyrose	ND	ND	ND	ND	ND	ND
Mannose	ND	ND	ND	ND	ND	ND
Glactose	ND	ND	ND	ND	ND	ND
Sucrose	ND	ND	5900	7300	ND	ND
Glycerol	340	960	130	170	160	18
Erythritol	220	360	ND	ND	ND	ND
Arabitol	110	250	ND	ND	ND	ND
Mannitol	180	500	ND	ND	ND	ND
Inositol	38	53	8	11	8	20
Total	1558	3323	6070	7507	168	38

ND, not detected   Reprinted from Funatsu, Y. et al. (2000) [20]

3.5. Sensory evaluation

Table 6 shows the result of sensory evaluation of the fish sauces conducted by the SD method. The color of the waste fish-sauce was light brown while the minced meat fish-sauce was deep brown. The color of nuoc mam was slightly lighter than that of the waste fish sauce; that of nampla, patis and yeesui was lighter than that of the latter. The taste of the minced meat fish-sauce was slightly stronger than the tastes of the other fish sauces. The taste of yeesui was extremely salty compare with that of the waste and minced meat fish-sauces and nuoc mam.

In this study, it was found that the taste balance of the waste and minced meat fish-sauces prepared with soy sauce koji was better than that of nampla, patis and yeesui. This might be because the organic acid contents of both the fish-sauces were higher than those of several Asian fish sauces. These results indicated that the two kinds of fish sauce prepared with soy sauce koji in this study are suitable for use as condiments.

Table 6
The result of sensory evaluation of the fish sauces

	Waste fish-sauce	Minced meat fish-sauce	Nampla	Nuoc mam	Patis	Yeesui
Depth of color	0.1±0.29	1.7±0.36	-1.5±0.36	-0.9±0.28	-1.6±0.27	-1.4±0.32
Strength of overall taste	0.2±0.29	1.1±0.32	-0.1±0.47	0.2±0.43	0.1±0.44	0.3±0.46
Saltiness	0.8±0.31	0.6±0.37	1.1±0.43	0.4±0.46	1.2±0.41	1.7±0.42
Bitterness	-0.7±0.46	-1.0±0.42	-0.7±0.40	-0.9±0.39	-0.6±0.48	0.0±0.40
Sourness	-0.5±0.42	-0.8±0.42	-0.7±0.45	-0.7±0.41	-0.8±0.45	-0.1±0.41
Taste balance	0.5±0.34	0.4±0.35	-0.3±0.35	0.4±0.28	-0.3±0.35	-1.0±0.39

The data were analyzed for mean$\pm(V/n)$ $0.5 \times t(\alpha, n-1)$. V, variance; n, number of subjects (n =40); α, significance level ($\alpha=0.05$) [19]

4. ACKNOWLEGMENTS

This research was partly supported by a Special Research Development Progress Grants from the Fisheries Agency. The authors are grateful to Mr. S. Kosakai, Himi Fishery Cooperative Association, for providing the information concerning frigate mackerel.

REFERENCES

1. Toyama Prefectural Fisheries Section (1998) Report for Fisheries in Toyama. Toyama Prefecture Government Press, Toyama.
2. Kawasaki, K., Ooizumi, T., Sugano, S. and Motoe, K. (1986) Report for the bases research into available nutrition contents of fisheries resources. Fisheries Agency, Tokyo, 16-34.
3. Funatsu, Y., Nabeshima, Y. and Kawasaki, K. (2001) Report for development of new food materials using unutilized fishes caught along the coast of Japan. Fisheries Agency, Tokyo, pp. 65-80.
4. Ohta, S. (1989) Fish sauce. New Food Industry, 31: 36-48.
5. Nakamura, H., Mohri, Y., Muraoka, I.and Ito, K. (1979) Studies on brewing food containing krill-I. Production of soy sauce with cryo-ground Antarctic krill, *Euphausia superba*. Nippon Suisan Gakkaishi, 45: 1389-1393.
6. Katoh, N., Hashimoto, A., Nakagawa, N. and Arai, K. (1989) A new attempt to improve the quality of frozen surimi from pacific mackerel and sardine by introducing mincing of raw materials. Nippon Suisan Gakkaishi, 55: 507-513.
7. Nonaka, M., Hirata, F., Saeki, H., Sasaki, I. and Matsukawa, M. (1990) An attempt to improve the quality of highly nutritional fish meat for food stuff from sardine by introducing under water mincing of raw material. Nippon Suisan Gakkaishi, 56: 1871-1876.
8. Takasaka, K. (1972) Meat processing technology in practice. Chikyusya, Tokyo.
9. Hirose, Y., Nakajima, T. and Yasaki, H. (1985) Soy Sauce Analysis. In: Soy sauce analysis method. Soy sauce analysis method editing committee (eds), Sanyusya, Tokyo, pp.1-42.
10. Yasui, A. and Tsutsumi, T. (1982) Protein. In: Foods Analytical Methods. Foods analytical method editing committee (eds), Korin, Tokyo, pp.111-113.
11. Folch, J., Lee, M. and Sloane-Staley, G.H. (1957) A simple method for the isolation and purification of total lipids from animal tissues. J. Biol. Chem., 226: 497-509.
12. Kawasaki, K., Funatsu, Y., Ito, Y., Motoe, K. and Nabeshima, H. (1997) Effects of sorbitol and sucrose concentrations in seasoned solution content of taste-active components in seasoned and dried products from walleye pollack. Nippon Shokuhin Kagaku Kougakkaishi, 44: 192-198.
13. Yoshida, M., Hatae, K., Yoshimatsu, F. (1984) Studies on katsuobushi soup stock prepared by extraction with cold water. Kasei Gakkaishi, 35: 529-537.
14. Moore, S. (1963) On the determination of cysteine as cysteric acid. J. Biol. Chem., 238: 235-237.

15. Simpson, R.J., Neuberger, M.R. and Liu, T.Y. (1976) Complete amino acid analysis of protein from a single hydrolysate. J. Biol. Chem., 251: 1936-1940.
16. Tajima, S. (1982) Free amino acid analysis. In: Foods Analytical Methods. Foods analytical method editing committee (eds), Korin, Tokyo, pp. 494-495.
17. Yasui, K., Nikumi, I. and Hase, S. (1979) Quantitative determination of saccharides in foods by gas chromatography. Part 2. Determination of mono- and disaccharides in jams, "yokans", chewing gums, biscuits, crackers and apples juices. Rept. Natl. Food. Res. Inst. 34: 110-115.
18. Nonaka, T. and Yamaguchi, S. (1986) Sensory evaluation of cooking science experiment. In: A cooking science experiment handbook. Fukuba, Y. and Miyakawa, K. (eds), Kenpakusya. Tokyo, pp. 356-394.
19. The Ministry of International and Trade Industry. (1994) Interval estimation of the population mean (Standard deviation unknown). In: JIS handbook. Japan Standards Association (ed), Dainihon Printing Co. Ltd., Tokyo, pp. 629-630.
20. Funatsu, Y., Konagaya, S., Katoh, I., Takeshima, F., Kawasaki, K. and Ino, S. (2000) Extractive components of fish sauce from the waste of frigate mackerel surimi processing and comparison with those of several Asian fish sauces. Nippon Suisan Gakkaishi, 66:1026-1035.
21. The Ministry of Agriculture, Forestry and Fisheries. (1951) In: Department of foods circulation, the ministry of agriculture, forestry and fisheries (ed), Quality standard indicated by Japanese Agricultural Standard, special number of foods, Cyuoh Hoki Syutsupan, Tokyo, pp. 1951-2029.
22. Ninomiya, Y. (1986) The sense of taste phenomenon-various taste reaction induced by umami, sweetness, bitterness, sourness and salt taste. In: Good-taste and sense of taste science. Komata, Y. (ed), Marui Koubunsya, Tokyo, pp. 178-184.
23. Fujimaki, M. (1980) Umami taste in Foods. Jyozou Kyoukaishi, 75: 873.
24. Okai, H. (1988) Development of new tasty peptides. Food and development, 23: 28-33.
25. Konosu, S. (1987) Non-Nitrogen Compounds. In: Fisheries Food Science. Konosu, S. and Suyama, M. (eds), Kouseisha Koseikaku, Tokyo, pp. 59-60.
26. Yamanaka, H. (1988) Sugar and Organic Acid. In: Extractive Components of Fish and Shellfish, Sakaguchi, M. (ed.), Koseisya Koseikaku, Tokyo, pp. 44-55.

DEVELOPMENT OF "KATSUOBUSHI"-LIKE DRIED FISH USING UNDERUTILIZED OVER-MATURED CHUM SALMON

Tsutomu Abe[a], Nobuo Tachibana[b] and Masaaki Tanaka[c]
[a]Hokkaido Food Processing Research Center, Ebetsu, Hokkaido 069-0836, Japan
[b]Edoya Co. Ltd., Obihiro, Hokkaido 080-2469, Japan
[c]Kamada Foods International Ltd., Sakaide, Kagawa 762-0044, Japan

Abstract

New type of salmon product (named Sakebushi) was developed from over-matured chum salmon using a modified process of dried skipjack (Katsuobushi) manufacturing. Mild enzymatic proteolysis and steaming process were introduced to the ordinary process instead of molding and boiling process. Extractive nitrogen in Sakebushi was 2.5 times higher than that of the ordinary process. Glutamic acid in Sakebushi was 12 times higher than that of Katsuobushi. Sensory evaluation indicated that Sakebushi had sufficient palatability with higher umami and lower bitterness.

1. INTRODUCTION

Chum salmon is a major fishery product in Hokkaido, amounting to over 150,000 t per year, accounting for 70% of the total catch of salmon. These products are supplied to commercial areas (Tokyo, Osaka, and other cities) and are domestically consumed.

Recently, the supply of wild chum salmon in Japan has gradually increased by the efficient stocking of its fry. In addition, a large quantity of salmon and trout are constantly imported from North America, Northern Europe, and Chile. With these backgrounds, the price of domestic chum salmon has become lower year after.

The commercial quality grades of chum salmon are judged from the degree of nuptial coloration : S grade, silvery ; A grade, slight nuptial coloration ; B grade, moderate nuptial coloration and C grade, strong nuptial coloration. The ratios are approximately 25%, 30%, 30% and 15%, respectively. Maturation leads to not only nuputional coloration of skin, but also deterioration and discoloration of muscle [1]. The water content increases and the protein and fat content decreases with changes of chemical composition in muscle [2,3]. From the commercial viewpoint of the quality of chum salmon, matured chum salmon is not suitable material for industrial processing such as salting and flaking due to its alternated states. Therefore, C grade chum salmon is priced much lower than S grade and has little commercial value.

The focus of this work is to find out a new usage of matured chum salmon. We tried to make a seasoning for soup applying a process of the dried fish manufacturing of Katsuobushi.

2. PREPARATION OF DRIED CHUM SALMON BY THE TRADITIONAL PROCESS OF KATSUOBUSHI

The defrosted salmon fillet was processed by boiling, smoke-drying, aging, and molding according to the ordinary Katsuobushi manufacturing process [4-6]. Its appearance and flavor were almost the same as those of Katsuobushi. However, it had low concentration of extractive components and umami. Extractive components index of flavor strength was half in concentration of those in Katsuobushi. Inosinic acid in Sakebushi was one-fifth of that in Katsuobushi. Inosinic and glutamic acids increase synergistically umami strength. In the Katsuobushi manufactring, the suitable fat content of skipjack muscle was 1 to 3% and its product was 4 to 7% [7]. The fat content of the matured chum salmon muscle was 1.5% and its product was 5.5%. This evidence suggested that matured chum salmon is a suitable material for processing of dried fish products.

3. MODIFICATION OF THE MANUFACTURING PROCESS

3.1. Modification of process from boiling to steaming
The loss of extractive components occurring in boiling process was greatest among the ordinary Katsuobushi processes. By changing it to a steaming process, the loss of the extractive components from the meat was 10% lower than that of the boiling process.

3.2. Introduction of mild enzymatic proteolysis to muscle
The second modification was a mild enzymatic proteolysis of muscle fiber before the steaming process. The major extractive components consist of amino acids, peptides, and nucleotides. Proteolysis increase amino acids and peptides by hydrolysing muscle fiber.
Effect of protease concentration on total extractive components is shown in Fig.1. Chum salmon muscle during spawning migration has the highest cathepsic activity, at a temperature of 40 to 50°C [2,3]. A non-protease control sample was expected to perform autolysis of the meat by high catepsic activity, but the increase of extractive components was low. In the sample protease added, on the other hand, extractive components gradually increased with incubation time. However, when the total extractive component value exceeded approximately 2.5 mmol/g, it was impossible to manufacture the dried fish because the meat was extremely softened by the action of proteolysis. In the case of using papain, when the protease concentration 0.1% to meat and 5 hours incubation time, salmon meat was possible to handle to manufacture the dried fish.

3.3. Effect of protease on the development of flavor properties
Favorable proteases were selected by sensory evaluation of the prepared extracts in

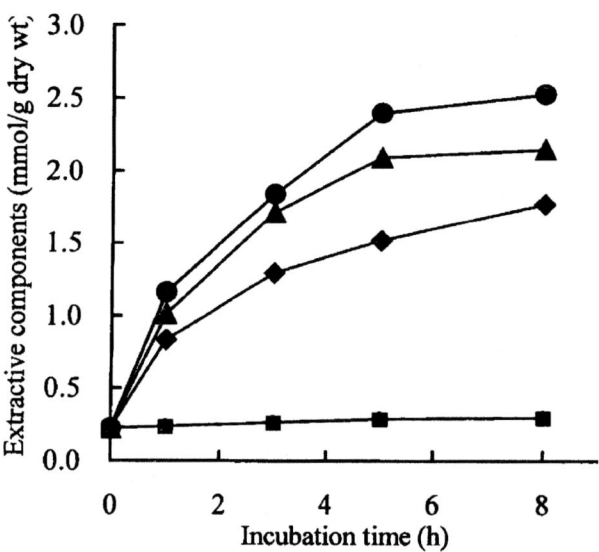

Figure 1. Effect of protease concentration on total extractive components. Proteolysis was conducted at 50°C by adding papain to meat. Concentration (%(w/w)): ■, 0 ; ◆, 0.05 ; ▲, 0.1 ; ●, 0.2.

commercial enzymes. The properties of the selected proteases were those of a mixture of endo-type protease (endopeptidase) and exo-type protease (exopeptidase). Endopeptidase frequently makes bitter poly-peptides. Exopeptidase makes favorable amino acids, but its proteolystic activity is weak, because exopeptidase hydrolyses peptides only from each end of the peptide. Using a mixture of these proteases, endopeptidase produces poly-peptides and exopeptidase hydrolysates the poly-peptides into amino acids. The mixture of these proteases thus produced large amount of amino acids and little peptides, resulting in very low bitter taste.

4. PREPARATION OF SAKEBUSHI BY THE MODIFIED PROCESS

The modified manufacturing process of Sakebushi is shown in Fig.2. The points different from the ordinary process were to introduce the proteolysis process and to change boiling to steaming. Comparison of chemical composition of Sakebushi prepared by the ordinary Katsuobushi manufacturing process with that of the commercially available Katsuobushi and Sababushi (dried mackerel) is shown in Table 1. The result of chemical composition of Katsuobushi and Sababushi were almost the same as that of the previous studies [8,9]. The water and total nitrogen contents of Sakebushi were lower than that of Katsuobushi and Sababushi.

The fat content of Sakebushi was 6.8%, a value that was practicable for processing to the dried fish. In addition, the suitability of various ranked chum salmon was investigated

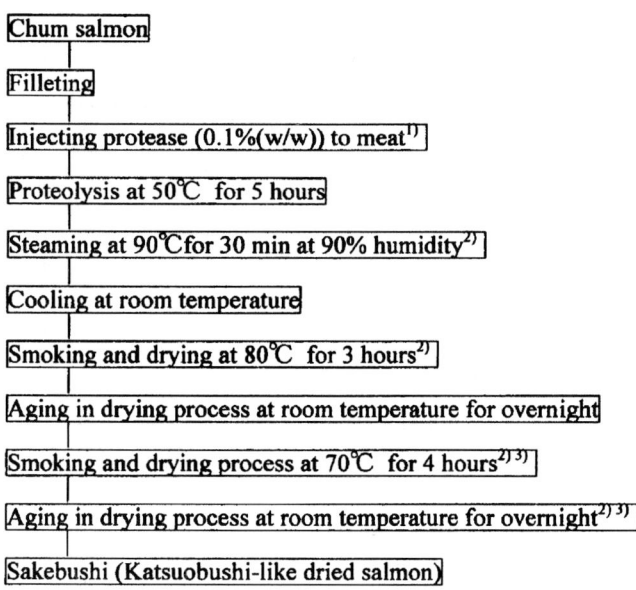

Figure 2. Manufacturing process of Sakebushi. [1] Protease used was flavourzyme (Novo Nordisk Bioindustry Ltd.) dissolved in water. [2] Steaming, smoking, and drying were carried out in an air-conditioned smoke house. [3] These processes were repeated for 4 days.

in the modified process. Sakebushi prepared using high quality chum salmon such as S and A grade had higher fat content which was reflected from the fat content of the raw chum salmon. The fat content is a significantly important concerning to the final products quality and preservation because the higher fat content in dried fish products causes lower stability of oxidation and provides more muddy soup. It suggested that lower ranked chum salmon was better for ingredient.

By the modified process, extractive nitrogen in Sakebushi was 3.2%, a value that was 2.5 times as much as that produced by the ordinary process and 1.3 and 1.6 times, compared with that in Katsuobushi and Sababushi. Glutamic acid in Sakebushi was 394mg/100g, which was 2.6 times as much as that by the ordinary process, in particular, 11.9 times, compared with that in Katsuobushi. Inosinic acid in Sakebushi was 164mg/100g, being one-fourth and one-sixth of that of Katsuobushi and Sababushi, respectively. Sensory evaluation indicated that Sakebushi had low bitterness and sufficient umami strength, compared with Katsuobushi and Sababushi.

When people eat fish or meat in raw or cooking, higher fat products are generally accepted to be favorable and set higher market prices. On the other hand, poor fat fish and meat are lower price, in spite of the nutritive qualities are almost the same. In this study, we could develop a new salmon product using poor fat and uselessly matured chum salmon. Now, some fishery companies are manufacturing Sakebushi with this new process, and supply them to the market place in Japan.

Table 1
Comparison of chemical composition of Sakebushi prepared by the ordinary Katsuobushi manufacturing process with that of commercially available Katsuobushi and Sakebushi

Content		Modified[a]	Sakebushi Ordinary[b]	Katsuobushi[c] Modified	Sababushi[c] Ordinary
Moisture	(%)	12.7	11.7	13.6	13.9
TN[d]	(%)	12.3	13.0	13.1	11.7
Crude fat	(%)	6.8	5.5	3.5	7.8
EN[e]	(%)	3.2	1.3	2.5	2.0
TAP[f]	(μmol/g)	478.9	185.7	301.4	247.5
IMP[g]	(mg/100g)	163.8	153.1	703.7	993.0
Glutamic acid	(mg/100g)	391.4	177.4	33.0	113.7

[a]Manufacturing process is shown in Fig.1. [b]Process of Katsuobushi manufacturing, which employs boiling and smoke-drying; [c]Purchased at a market; [d]Total nitrogen; [e]Extractive nitrogen; [f]Total amino acids and peptides; [g]Inosinic acid; EN, TAP, IMP and Glu values described above are shown on a dry basis.

REFERENCES

1. Yamashita, M. (1994) Studies on the muscle softening phenomenon of chum salmon caught during spawning migration. Nippon Suisan Gakkaishi, 60: 439-442.
2. Konagaya, S. (1982) Enhanced protease activity of chum salmon *Oncorhynchus keta* during spawning migration. Nippon Suisan Gakkaishi, 48: 1503.
3. Yamashita, M. and Konagaya, S. (1991) Proteolysis of muscle proteins in the extensively softened muscle of chum salmon caught during spawning migration. Nippon Suisan Gakkaishi, 57: 2163.
4. Ohta, S. (1983) Katsuobushi 1. New Food Ind., 25: 35-41.
5. Ohta, S. (1983) Katsuobushi 2. New Food Ind., 25: 17-25.
6. Ohta, S. (1984) Kunsei. New Food Ind., 26: 29-35.
7. Takenaga, F., Itoh, S. and Tsuyuki, H. (1991) Comparison of total lipids between nomal-"Katsuobushi" and "Shirata-Katsuobushi". Nippon Shokuhin Kogyo Gakkaishi, 38: 280-287.
8. Fuke, S., Wtanabe, K., Sakai, H. and Konosu, S. (1989) Extractive components of dried skipjack (Katsuobushi). Nippon Shokuhin Kogyo Gakkaishi, 36: 67-70.
9. Kotani, K., Kawamura, K., Wada, T. and Satani, E. (1984). Comparative studies on chemical components of some kinds of "Fushi". Nippon Shokuhin Kogyo Gakkaishi, 31: 624-629.

VOLATILE COMPOUNDS IN THE HEPATIC AND MUSCULAR TISSUES OF COMMON CARP, JAPANESE FLOUNDER, SPANISH MACKEREL AND SKIPJACK

XingAn Song[a], Takashi Hirata[a], Tetsuo Kawai[b], Norifumi Niizeki[a] and Morihiko Sakaguchi[a]
[a]Division of Applied Biosciences, Graduate School of Agriculture, Kyoto University, Kyoto 606-8502, Japan
[b]R&D Department, Shiono Koryo Kaisha, Ltd., Osaka 532-0033, Japan

Abstract
The hepatic and muscular tissues obtained from four fishes, common carp, Japanese flounder, Spanish mackerel and skipjack were analyzed for volatile compounds by the solid-phase microextraction method and subsequent gas chromatography-mass spectrometry. Both tissues of all fishes tested possessed larger amounts of volatiles and more complicated volatile constitutions were detected in the hepatic tissue than in the muscular tissue. In the hepatic tissue of the above-mentioned fishes, 56, 24, 42 and 38 compounds were identified respectively, while 20, 24, 28, and 37 compounds in the corresponding muscular tissue. Some compounds were common in both tissues, e.g. nonanol, 1-octen-3-ol, (E,E)-2,4-heptadienal, (E,Z)-2,6-nonadienal and octadecanal but a considerable number of compounds remained unknown. Some characteristic odor compounds were detected in the hepatic tissue of all fish species tested: octanol, 1-octen-3-ol, 2,4-decadienal and 2,4-heptadienal, probably contributing to the hepatic tissue odor, while hexanal, heptanal and nonanal were found in the muscular tissue. The result of statistical analysis for the distribution of the volatiles suggested that both tissues are different each other in the quantitative and qualitative aspects of odor.

1. INTRODUCTION

Landed fish and shellfish of many species have not been well utilized for human food resources. In fish as an example, internal organs, skin, head, and bone are often thrown away as raw waste. There are some differences in chemical and physical properties between the muscle and other parts [1]. In our previous study [2], some comparisons were made on the proximate composition and nitrogenous extractive components contained in several tissues of four fishes, yellowtail, Japanese flounder, Spanish mackerel, and common carp. There was no notable difference in the compositions of protein and ash except for fat. In visceral tissues, compared with the muscle, some free amino acids including taurine, glutamine, alanine, and glycine were rich in the visceral tissues. The umami strength of visceral tissues calculated on the basis of glutamic acid and IMP contents was also proved not always weaker than that of the white muscle.

In our organoleptic test for the five basic tastes and flavor, the conspicuous difference in flavor was detected between the muscle and visceral tissues. Detailed comparative study on volatile components was thought necessary to clarify the difference. Many studies performed so far on the above point have focused on analysis of volatiles in the muscle tissue [3-6] and thus a great amount of the information has been accumulated on volatile compounds including so-called off-flavor compounds. We identified volatile components in the hepatic and muscular tissues of 4 fish species, using the methods of solid-phase microextraction (SPME) and gas chromatography-mass spectrometry (GC/MS) and compare the contents between their tissues.

2. MATERIALS AND METHODS

2.1. Fish

Four species of fish, common carp (*Cyprinus carpio*), Japanese flounder (*Paralichthys olivaceus*), Spanish mackerel (*Scomberomorus niphonius*), and skipjack (*Euthynnus pelamis*), were purchased from local fish markets. Common carp and Japanese flounder were obtained alive. Spanish mackerel and skipjack were purchased about 24 hr after the catch.

2.2. Sample preparations and extraction procedures

The hepatic tissue of four fish, common carp (hepatopancreas), Japanese flounder (liver), Spanish mackerel (liver), skipjack (liver), and those of the muscular tissue (white muscle) were separated and sliced into small pieces. Two grams of each tissue slice were put in a 4 mL glass vial (Supelco, USA). A SPME fiber (blue fiber, Supelco, USA) was inserted into the vial through a silicone seal (Supelco, USA). Microextraction was carried out at 40°C for 30 min. A blue SPME fiber, (Polydimethylsiloxane / divinylbenzene, 65μL film thickness, Cat. No. 57310-U, manual version) and an SPME fiber syringe holder (Cat. No. 57330-U, manual version) were purchased from Supelco (Bellefonte, PA, USA). Headspace glass vials (4 mL) with soft silicone rubber seals were also from Supelco. The fiber was then applied to GC/MS for analysis. After a 5 min desorbing, the fiber was conditioned under 260°C for 30 min in another GC injector before the next extraction.

2.3. Gas chromatography-mass spectrometry (GC/MS)

Analysis of volatile compounds was performed out on a Shimadzu GC/MS system (GM5050), equipped with a 60m×0.25mm (i.d.) DB-5 fused silica capillary column (J&W Scientific, USA). Splitless injections were carried out. The oven temperature of the GC was programmed from 40 to 240°C at 5°C/min. Helium carrier gas had a column head pressure of 12 psi. The mass spectrometer was operated in the electron impact mode at

70ev, scanning the range m/z from 20 to 300 in a 1s cycle. Identification of volatiles was achieved by comparison of the GC retention times and mass spectra with those, when available, of the pure standard compounds. Most of mass spectra were also compared with those of the data system library (NIH Library) and other spectra data published.

2.4. Statistical analysis

To examine the significant difference of volatile components between the hepatic and muscular tissues, data were analyzed by discriminate analysis with the SPSS software (SPSS LTD., CO., USA), combining the two factors of peak area and the retention time of the hepatic and the muscular tissues in each fish species.

3. RESULTS AND DISCUSSION

3.1. GC patterns of volatiles contained in the hepatic and muscular tissues

Total ion chromatograms of the volatile compounds in hepatic and muscular tissues of four fishes obtained by SPME and GC/MS are shown in Fig. 1 to 8, where Fig. 1, 3, 5 and 7 and Fig. 2, 4, 6 and 8 indicate those of hepatic and muscular tissues, respectively. The difference in GC patterns between the two tissues probably reflect the difference between their odors. These figures show that much more peaks occurred in the hepatic tissue than those in the muscular tissue, corresponding with the preliminary odor observation that the hepatic tissue had a stronger odor than the muscular tissue.

Among the four pairs of GC patterns, common carp showed the largest difference in the patterns between the two tissues, followed by Spanish mackerel, skipjack, and Japanese flounder, as shown in Fig. 1 to 8. It was indicated that the carp hepatic tissue gave the most complicated volatile composition among the hepatic and muscular tissues of fish tested (Fig. 1), although the muscular tissue did not so much (Fig. 2). Almost the same difference in pattern was observed in the mackerel tissues (Fig. 5 and 6). Skipjack showed a different type from other pairs; the muscular tissue contained fewer light-volatile compounds together with heavy-volatile compounds (Fig. 8), whereas the hepatic tissue behaved in manner similar to other hepatic tissues (Fig. 7). Flounder showed the smallest difference among the pairs (Fig. 3 and 4).

The numbers of peaks appeared on GC charts were summarized in Table 1. There

Table 1
Peaks detected in the hepatic and muscular tissues of four fishes

Fish species	Tissue	Total peaks detected	Peaks of compounds identified	Peaks of unknown compounds
Common carp	Hepatopancreas	89	56	33
	White muscle	44	20	24
Japanese flounder	Liver	43	24	19
	White muscle	34	24	10
Spanish mackerel	Liver	82	42	40
	White muscle	41	28	13
Skipjack	Liver	49	38	11
	White muscle	65	37	2

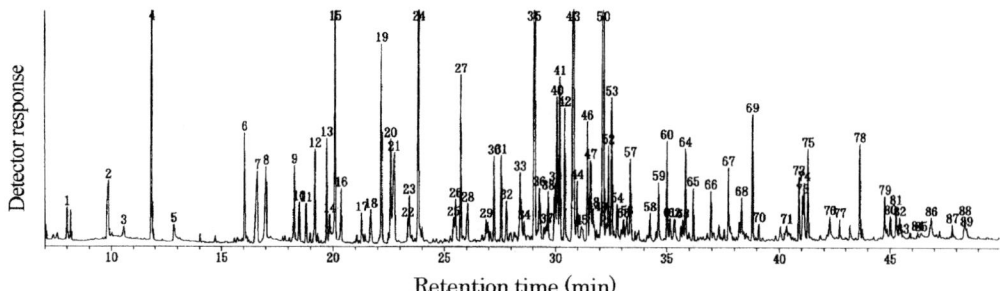

Figure 1. GC chromatogram of volatile compounds in the hepatic tissue of common carp.

Peak number	Compound	Retention time (min)	GC(%)	Peak number	Compound	Retention time (min)	GC(%)
1	pentanal	8.00	0.33	38	butylcyclohexane	29.68	0.73
2	2-ethylfuran	9.88	1.43	39	2-undecanone	30.00	1.11
3	pentanol	10.56	0.38	40	(E,E)-2,4-decadienal	30.08	2.29
4	hexanal	11.86	3.37	42	undecanal	30.43	1.66
6	heptanal	16.06	1.18	45	3,7-dimethyl-2,6-octadienal	31.19	0.38
8	butyrolactone	17.02	1.67	46	3-decen-2-one	31.45	1.38
9	(Z)-2-heptenal	18.28	0.91	47	8-methyl-1-undecene	31.60	1.64
10	benzaldehyde	18.50	0.57	48	3-methyl-1,4-heptadiene	31.70	0.42
11	heptanol	18.80	0.47	51	undecanol	32.32	0.39
12	1-octen-3-ol	19.20	1.25	52	(E)-2-hexenoic acid	32.41	1.33
13	2-pentylfuran	19.73	1.20	53	(1-methylethyl)-cyclopentane	32.56	2.05
16	(E,E)-2,4-heptadienal	20.37	0.85	56	4-(1-methylethenyl)-1-cyclohexene-1-methanol	33.24	0.59
17	(E)-1-butenylcyclo-pentane	21.29	0.35	57	dodecanal	33.40	1.19
18	phenylacetaldehyde	21.70	0.58	58	1-tetradecene	34.27	0.47
19	(E)-2-octenal	22.17	2.56	62	3,4-dimethyl-1,2-cyclopentanedione	35.35	0.66
20	octanol	22.58	1.42				
22	2-nonanone	23.36	0.31	64	4-(2,6,6-trimethyl-1-cyclohexen-1-yl)-3-buten-2-one	35.84	1.23
23	3,5-octadien-2-one	23.42	0.59				
24	nonanal	23.84	4.14	71	tetradecanol	40.34	0.34
26	(E,Z)-2,6-nonadienal	25.52	0.66	72	heptadecane	40.89	0.93
27	2-nonenal	25.74	1.89	73	2,6,10,14-tetramethyl-pentadecane	41.05	0.63
28	nonanol	26.05	0.84				
29	2,4-undecadienal	26.88	0.37	75	octadecanal	41.29	1.18
30	decanal	27.24	0.96	76	tetradecanoic acid	42.28	0.77
31	2,4-nonadienal	27.56	1.06	78	pentadecanal	43.64	1.32
33	1-dodecene	28.42	1.58	80	heptadecanol	45.01	0.62
34	(E)-2-decenal	28.61	0.37	82	9-octadecenal (an isomer)	45.44	0.41
35	(Z)-2-decenol	29.09	4.74	86	hexadecanoic acid	46.86	0.96
36	2-undecanol	29.29	1.09	87	9-octadecenal (an isomer)	47.81	0.32
37	octadecane	29.61	0.33		Total		62.12

remained many volatile compounds unknown (Table 1), due to the lack of comparable mass spectral data and GC retention time data.

Table 1 indicates that the peak numbers detected in the hepatic tissue were greater than those in the hepatic tissue in three of the four fishes tested. Particularly, in common carp

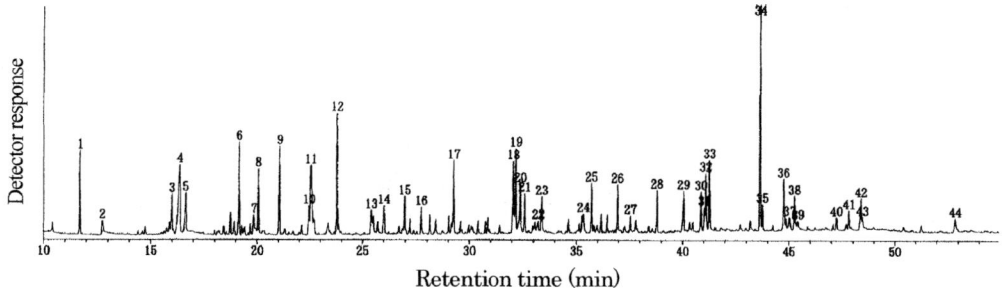

Figure 2. GC chromatogram of volatile compounds in the muscular tissue of common carp.

Peak number	Compound	Retention time (min)	GC(%)	Peak number	Compound	Retention time (min)	GC(%)
1	hexanal	11.71	2.57	19	(E,E)-2,4-nonadiene	32.16	3.98
3	heptanal	15.99	1.53	21	2-methylbutyl propanoate	32.60	1.87
5	butyrolactone	16.64	2.36	23	(E)-3-octen-1-ol	33.38	1.92
6	2-ethylhexanal	19.15	3.43	30	heptadecane	40.87	1.89
8	octanal	20.06	2.47	31	2,6,10-trimethyldodecane	41.04	1.06
9	2-ethylhexanol	21.04	3.21	33	octadecanal	41.28	3.53
10	(E)-2-octenal	22.47	1.16	34	pentadecanal	43.64	8.53
12	nonanal	23.79	4.07	37	1-heptadecanol	45.02	1.54
14	2,5-dimethylbenzaldehyde	25.99	1.63	39	9-octadecenal (an isomer)	45.41	1.09
15	9,12,15-octadecatrienoic acid	26.96	1.59	41	9-octadecenal (an isomer)	47.81	1.09
						Total	46.54

and Spanish mackerel, the peak numbers in the hepatic tissue were approximately twice greater than those in the muscular tissue. The peak numbers and GC patterns suggest that the muscular tissue is almost odorless and in contrast, the hepatic tissue of fish is disliked in general due to its odor.

In carp, there were 89 peaks as the hepatic tissue volatile compounds detected, indicating almost twice as much as 44 peaks from in muscular tissue volatiles (Table 1). The number was the greatest among those of the tissues tested, supporting an idea that the hepatic tissue of the carp emits a strong muddy smell. In the hepatic and muscular tissues of flounder, on the other hand, 43 and 34 peaks occurred, respectively (Table 1). The smaller difference in numbers suggested that the hepatic tissue of this fish species should be the most suitable for foods for its weak odor.

3.2. Volatile compounds in the hepatic tissues

The identified volatile compounds in both hepatic and muscular tissues of the four fishes are listed in the legends of Fig. 1 to 8. Each legend gives the volatile compounds with their corresponding GC peak numbers in the figure, retention time data and GC peak areas (%).

Many compounds was detected in the hepatic tissues of the fishes tested: heptanol, octanol, pentanol, 2-pentylfuran, (E,E)-2,4-decadienal, phenylacetaldehyde, and

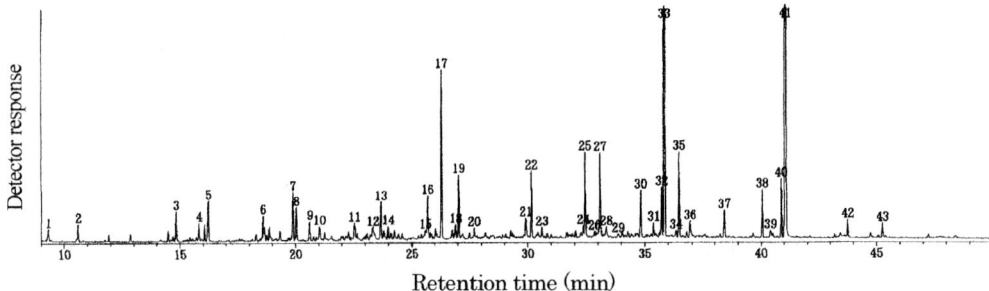

Figure 3. GC chromatogram of volatile compounds in the hepatic tissue of Japanese flounder.

Peak number	Compound	Retention time (min)	GC(%)	Peak number	Compound	Retention time (min)	GC(%)
1	3-methylbutanol	9.32	0.56	21	1-undecene	29.93	1.07
2	pentanol	10.61	0.54	25	2,7,10-trimethylnonadecane	32.46	3.14
3	1,2-dimethylbenzene	14.84	1.13	26	2-methylcyclohexanol	32.87	0.62
4	1,3-dimethylbenzene	15.83	0.66	27	hexadecanal	33.10	3.84
6	1-ethyl-2-methylbenzene	18.62	0.87	31	1-octadecene	35.39	0.68
7	1-ethyl-3-methylbenzene	19.89	1.74	33	2-methyl-octadecane	35.85	17.31
8	decane	20.03	1.15	35	BHT (2,6-di-*tert* -butyl-4-methylphenol)	36.47	3.10
9	1,4-dichlorobenzene	20.60	0.74				
10	1,2,4-trimethylbenzene	21.02	0.84	37	pentadecane	38.43	1.32
12	2-nonanone	23.33	0.66	39	1-pentadecene	40.41	0.58
13	nonanal	23.68	1.50	40	heptadecane	40.88	2.17
15	1-decene	25.59	0.59	41	2,6,10,14-tetramethyl -pentadecane	41.07	27.29
18	naphthalene	26.91	0.77			Total	72.87

hexadecanoic acid. Also, some compounds were identified with higher frequency in this tissues; nonanol, 1-octen-3-ol, (E,E)-2,4-heptadienal, (E,Z)-2,6-nonadienal, 2-nonenal, 2-nonanone, 2-undecanone, 3,5-octadien-2-one, benzaldehyde and decanal. The following compounds were found in the hepatic tissue with high GC areas: 1-octen-3-ol, 2,4-decadienal, 2,4-heptadienal, 2,6-nonadienal, 2-undecanone, 3-decen-2-one, benzaldehyde, decanal, heptanal, hexadecanal, octadecanal, and undecanal. It was noteworthy that nonanal was characterized in the hepatic tissue of all fish species tested.

Among the 89 peaks in the carp hepatic tissue, 56 of them were identified (Fig. 1), including several aldehydes (decanal, phenylacetaldehyde, hexanal, octadecanal, nonanal, undecanal, (E)-2-octenal, (E,E)-2,4-heptadienal, and (E,E)-2,4-decadienal), alcohols (dodecanol, heptanol, 1-octen-3-ol, nonanol, octanol, pentanol, and undecanol), furans (2-ethylfuran and 2-pentylfuran), and acids (hexadecanoic and tetradecanoic acids). They had relatively higher GC proportions. Such volatiles were detected in the hepatic tissue of carp but not in the muscular tissue as 2,4-nonadienal, 2-decenal, 2-ethylfuran, octadecane, pentanal and tetradecanoic acid. 1-Tetradecene and 2,4-undecadienal were found only in the carp hepatic tissue among the four hepatic tissues.

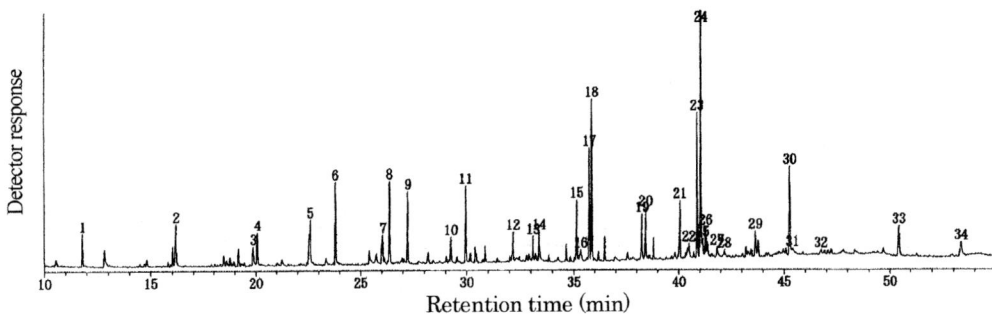

Figure 4. GC chromatogram of volatile compounds in the muscular tissue of Japanese flounder.

Peak number	Compound	Retention time (min)	GC(%)	Peak number	Compound	Retention time (min)	GC(%)
1	hexanal	11.81	1.41	16	2-butenylbenzene	35.34	1.29
3	1,3,5-trimethylbenzene	19.88	1.40	18	2-methyloctadecane	35.86	6.09
4	octanal	20.09	1.46	19	ethyl dodecanoate	38.27	2.56
6	nonanal	23.81	2.99	20	pentadecane	38.45	2.51
7	nonanol	26.06	2.09	22	1-pentadecene	40.51	1.25
8	5-methyl-2-(1-methylethyl)-cyclohexanol	26.39	3.25	23	heptadecane	40.90	6.02
				24	2,6,10,14-tetramethyl-pentadecane	41.07	19.02
9	decanal	27.24	3.03				
10	octadecadienal	29.27	1.44	25	tridecanol	41.16	1.32
11	1-methoxy-4-(2-propenyl)benzene	29.97	4.10	26	octadecanal	41.30	1.63
				27	10-undecenyl hexanoate	41.39	1.38
13	tetradecane	33.11	1.35	28	methyl 3-hydroxydodecanoate	41.85	1.51
14	dodecanal	33.40	1.79	29	pentadecanal	43.65	1.78
15	dodecanol	35.16	2.74			Total	73.41

In the flounder hepatic tissue, 24 of the 43 peaks were identified (Fig. 3); 1-pentadecene, pentanol, 2-nonanone, nonanal and pentadecane occurred as common compounds. 1-Octadecene, 3-methylbutanol, and naphthalene were detected only in the hepatic tissue. There were found a few odor compounds with small GC areas.

In the mackerel hepatic tissue, 42 of the 82 peaks were assigned (Fig. 5), containing, as dominant compounds, some alcohols (heptanol, octanol, nonanol and undecanol) and some aldehydes (hexanal, heptanal, nonanal, undecanal, and dodecanal). 4-Heptenal and limonene arose from the hepatic tissue but not from the muscular tissue. (E,Z)-2,4-Decadienal and pentadecanoic acid were found only in the mackerel among the 4 fishes.

Identified peak numbers were 38 of the 49 in the hepatic tissue of skipjack (Fig. 7), including heptanol, nonanol, phenylacealdehyde, 1-octen-3-ol, (E,E)-2,4-heptadienal, (E,Z)-2,6-nonadienal, 2-nonanone, 2-undecanone, (E)-2-octenal, benzaldehyde, decanal, heptanal, hexanal, nonanal, pendadecane and undecanal. 1-Penten-3-ol, octanal, tetradecane and trimethylamine were detected in skipjack alone.

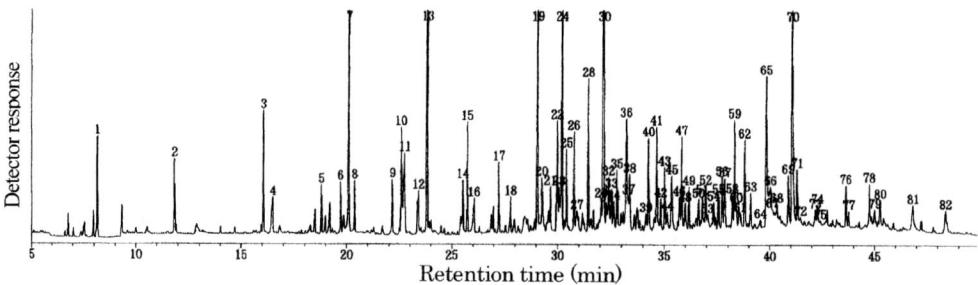

Figure 5. GC chromatogram of volatile compounds in the hepatic tissue of Spanish mackerel.

Peak number	Compound	Retention time (min)	GC(%)	Peak number	Compound	Retention time (min)	GC(%)
1	furan	8.16	0.90	25	(E,Z)-2,4-decadienal	30.42	1.10
2	hexanal	11.82	0.84	26	4-(1-methylpropyl)phenol	30.78	1.46
3	heptanal	16.06	1.38	28	8-methyl-(Z)-2-undecene	31.45	1.95
4	4-heptenal	16.49	0.87	31	undecanol	32.30	0.56
5	heptanol	18.80	0.63	34	3,3,6-trimethyl-2,5-heptanedione	32.64	0.63
6	2-pentylfuran	19.72	0.73	35	1,1-dimethoxyhexane	32.80	1.15
7	limonene	20.11	4.14	38	dodecanal	33.39	0.93
8	(E,E)-2,4-heptadienal	20.38	0.84	39	2-hexenyl hexanoate	34.15	0.55
9	(E)-2-octenal	22.15	0.77	47	4-(2,6,6-trimethyl-1-cyclohexen-1-yl)-3-buten-2-one	35.84	1.39
10	octanol	22.59	1.61				
12	3,5-octadien-2-one	23.42	0.58	50	2-methyl-2-nonen-4-one	36.63	0.58
13	nonanal	23.84	4.09	51	tridecadiene	36.81	0.81
14	(E,Z)-2,6-nonadienal	25.52	0.81	56	2-undecenal	37.74	0.81
15	2-nonenal	25.74	1.34	65	hexyl decanoate	39.85	3.45
16	nonanol	26.05	1.01	69	9-methyldecanol	40.88	1.25
17	decanal	27.23	0.88	70	5-propyltridecane	41.07	5.43
18	(Z)-2-decenal	27.79	0.68	71	octadecanal	41.29	1.23
19	(Z)-2-decenol	29.08	3.95	74	pentadecanoic acid	42.26	0.83
20	butylcyclohexane	29.27	1.13	76	octadecanal	43.64	1.00
21	2-undecanone	29.67	0.62	79	heptadecanol	45.00	0.58
23	(E,E)-2,4-decadienal	30.08	0.98	81	hexadecanoic acid	46.83	1.26
24	undecanal	30.21	3.11		Total		58.84

3.3. Volatile compounds in the muscular tissues

Nonanal was characterized as the sole common volatile compound in both tissues of all species tested. 2-Undecenal, 2,6,10-trimethyldodecane and heptadecane were common compounds in the muscular tissues and octanal was detected with higher frequency.

In the carp muscular tissue, 20 of the 44 peaks were identified (Fig. 2); many

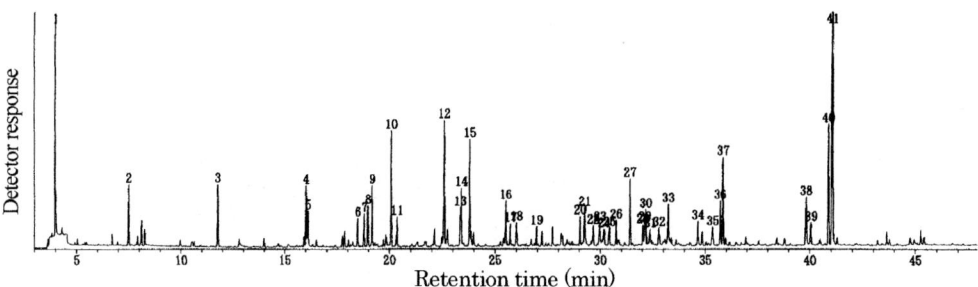

Figure 6. GC chromatogram of volatile compounds in the muscular tissue of Spanish mackerel.

Peak number	Compound	Retention time (min)	GC(%)	Peak number	Compound	Retention time (min)	GC(%)
1	trimethylamine	3.99	7.75	18	2,5-dimethylbenzaldehyde	26.01	1.68
2	1-penten-3-ol	7.50	1.80	20	(E)-2-decenol	29.04	1.28
3	hexanal	11.78	2.12	23	2-undecanone	29.98	1.40
4	heptanal	16.03	2.26	25	undecanal	30.45	1.21
6	benzaldehyde	18.48	1.14	26	(E,Z)-2,4-decadienal	30.77	1.19
7	heptanol	18.78	1.30	27	8-methyl-1-undecene	31.44	2.70
8	3,5,5-trimethyl-2-hexene	18.97	1.49	30	1,3,5,8-undecatetraene	32.17	1.53
9	1-octen-3-ol	19.17	2.46	32	3-ethyl-3-methylpentane	32.80	1.15
10	octanal	20.08	4.47	35	2,4,6-trimethyl-1,3,6-heptatriene	35.36	1.31
11	(E,E)-2,4-heptadienal	20.36	1.16	37	4-(2,6,6-trimethyl-1-cyclohexen-1-yl)-3-buten-2-one	35.85	3.54
13	2-nonanone	23.36	1.57				
14	3,5-octadien-2-one	23.41	2.17	38	hexyl decanoate	39.82	2.65
15	nonanal	23.81	4.22	40	heptadecane	40.88	5.18
16	(E,Z)-2,6-nonadienal	25.51	1.92	41	2,6,10-trimethyldodecane	41.06	15.50
17	2-nonenal	25.73	0.95			Total	77.10

aldehydes (hexanal, 2-ethylhexanal, octanal, nonanal, octadecanal, pentadecanal and (E)-2-octenal), some alcohols (2-ethylhexanol, 3-octen-1-ol, and 1-heptadecanol) and an acid (9,12,15-octadecatrienoic acid) were detected in higher GC peak area proportions. The acid was found only in the muscular tissue in carp. (E)-2-Octenal occurred in carp alone.

Twenty four of the 34 peaks in the flounder muscular tissue were assigned (Fig. 4), including nonanol, 1-pentadecene, decanal, dodecanal, heptadecane, hexanal, nonanal, octadecanal and octanal. Nonanol, 1-pentadecene and decanal were detected only in the flounder muscle.

In the mackerel muscular tissue, 28 of the 41 peaks were identified (Fig. 6): 1-octen-3-ol, (E,Z)-2,4-decadienal, (E,E)-2,4-heptadienal, (E,Z)-2,6-nonadienal, 2-nonanone, 3,5-octadien-2-one, benzaldehyde, 2-undecenal, heptanol, trimethylamine, etc.

In the skipjack muscular tissue, 37 of the 65 peaks were identified (Fig. 8); aldehydes (heptanal, nonanal, dodecanal and undecanal) and hydrocarbons (heptadecane, pentadecane, 1-tetradecene and tetradecane) were detected in higher GC peak area proportions. 1-Tetradecene, 1-undecanol and pentadecanoic acid were detected only in the muscular tissue of the fish species.

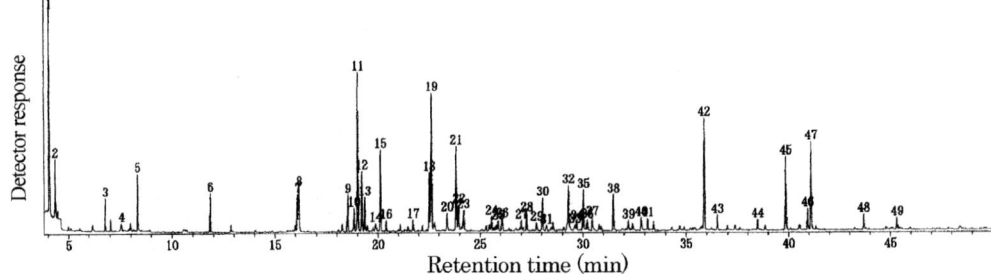

Figure 7. GC chromatogram of volatile compounds in the hepatic tissue of skipjack.

Peak number	Compound	Retention time (min)	GC(%)	Peak number	Compound	Retention time (min)	GC(%)
1	trimethylamine	4.06	6.47	26	nonanol	26.09	0.83
2	acetone	4.36	2.77	28	decanal	27.28	0.90
3	3-methylbutanal	6.79	1.21	30	2,6,6-trimethyl-1-cyclohexene-1-acetaldehyde	28.06	1.87
4	1-penten-3-ol	7.56	0.50				
6	hexanal	11.85	1.74	31	benzothiazole	28.25	0.48
7	heptanal	16.09	1.93	35	2-undecanone	30.03	2.25
9	benzaldehyde	18.53	1.98	36	tridecane	30.23	0.80
10	heptanol	18.84	1.18	37	undecanal	30.46	0.80
11	4-methyl-1-hexene	19.02	8.31	38	3-decen-2-one	31.49	1.78
12	1-octen-3-ol	19.22	3.00	39	2,6-dimethyl-1,7-octanediol	32.22	0.64
13	2,4-dimethyl-3-pentanone	19.36	1.61				
15	octanal	20.13	3.95	40	1,1-dimethoxyhexane	32.84	0.81
16	(E,E)-2,4-heptadienal	20.41	0.57	41	tetradecane	33.15	0.62
17	phenylacetaldehyde	21.73	0.67	43	BHT (2,6-di-*tert*-butyl-4-methylphenol)	36.53	0.81
18	(E)-2-octenal	22.54	2.97				
20	2-nonanone	23.40	1.06	44	pentadecane	38.49	0.51
21	nonanal	23.85	4.19	45	isooctyl 2-propenoate	39.87	4.34
22	1,2-dimethyl-3-(1-methyl-ethenyl)cyclopentanol	23.97	1.56	46	nonadecane	40.93	1.07
				47	2,6,10,14-tetramethyl-pentadecane	41.10	4.70
23	3,5-dimethylcyclohexanol	24.21	1.34				
24	(E,Z)-2,6-nonadienal	25.56	0.74	48	pentadecanal	43.69	0.86
25	1,4-dimethylcyclohexane	25.88	0.54			Total	72.36

3.4. Differences in volatile compounds between the hepatic and muscular tissues

In order to clarify more comprehensively the differences in volatile compounds between the hepatic and muscular tissues, the odor profile of the compound was inserted into the square brackets, followed by the compound name.

In carp, some marked differences in the volatile compounds between the hepatic and the muscular tissues were found; the following odor compounds occurred only in the hepatic tissue but not in the muscular tissue; heptanol, nonanol [green grass], octanol [woody], 1-octen-3-ol [mushroom-like], pentanol, (E,E)-2,4-decadienal [fatty], (E,E)-2,4-heptadienal [fishy], 2,4-nonadienal [fatty], (E,Z)-2,6-nonadienal [cucumber-like], benzaldehyde [cherry-like], phenylacetaldehyde [green grass], decanal [green grass], 2-ethylfuran, 2-pentylfuran and pentanal [fruit], although the following other odor compounds arose from both tissues: (E)-2-octenal [fishy], heptanal [burnt fat], hexanal

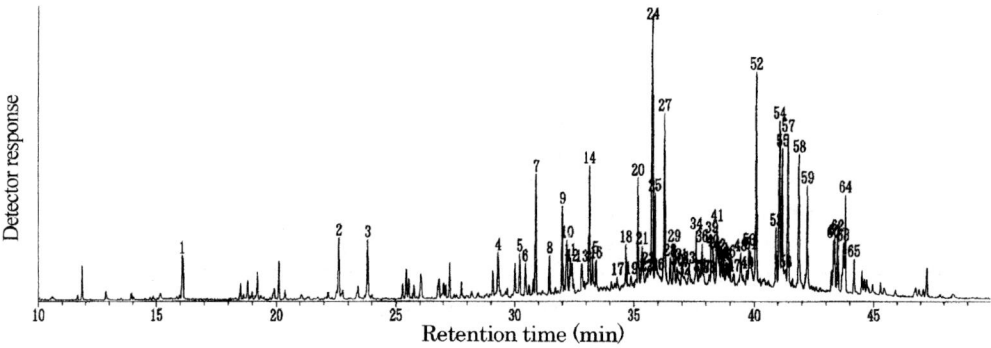

Figure 8. GC chromatogram of volatile compounds in the muscular tissue of skipjack.

Peak number	Compound	Retention time (min)	GC(%)	Peak number	Compound	Retention time (min)	GC(%)
1	heptanal	16.08	1.04	39	5-tetradecanol	38.24	1.54
2	3-nonyl-1-cyclohexene	22.61	1.30	41	pentadecane	38.47	2.19
3	nonanal	23.83	1.22	42	2-tetradecanol	38.56	0.70
4	(E)-2-decenol	29.30	0.96	43	2-methyl-5-undecanol	38.65	1.14
6	undecanal	30.45	0.74	46	4-dodecanol	38.98	0.96
8	3-decen-2-one	31.47	0.71	48	2-dodecanol	39.44	1.59
10	2,6-dimethyl-1,7-octanediol	32.20	1.58	49	2,6,10-trimethyl-dodecane	39.70	0.81
11	undecanol	32.33	0.88	53	heptadecane	40.92	1.85
13	2-butyloctanol	32.82	0.86	54	6-tridecanol	41.06	4.01
14	tetradecane	33.13	2.35	55	2-tridecanol	41.18	3.00
16	dodecanal	33.42	0.92	56	1,11-undecanediol	41.32	0.78
17	1-tetradecene	34.29	0.75	57	1-tridecanol	41.41	3.38
19	2,9-dimethylundecane	34.88	0.70	58	3-heptadecanol	41.87	3.16
20	1-dodecanol	35.18	2.15	59	pentadecanoic acid	42.22	2.85
22	3,7-dimethyl-6-octenal	35.53	0.82	60	7-tetradecanol	43.30	0.90
27	ethyl 2,3-dimethylbutanoate	36.28	4.52	61	6-tetradecanol	43.36	1.06
28	BHT (2,6-di-*tert*-butyl-4-methylphenol)	36.51	1.25	62	7-ethyl-2-methyl-4-undecanol	43.49	1.39
30	10-methyl-1-undecanol	36.80	0.95	65	1-hexadecanol	44.18	0.78
35	2,6,10,14-tetramethyl-hexadecane	37.75	0.72		Total		56.51

[green grass] and nonanal [green grass]. Therefore, the hepatic tissue can be considered to have stronger odor than that the muscular tissue. A similar phenomenon was observed in the mackerel.

Little differences were observed between both tissues of the flounder in the number and GC area of peaks, as mentioned above. The following compounds which always occurred as typical volatile compounds in tissues of other fishes were neither detected in the hepatic nor the muscular tissue of the flounder; octanol, 1-octen-3-ol, (E,E)-2,4-heptadienal, (E,Z)-2,6-nonadienal, (E)-2-octenal, trimethylamine [ammonia] and undecanal. This evidence strongly suggests that both tissues of the flounder have less smell than those of other fishes.

Some aldehydes (hepatanal, nonanal, pentadecanal and undecanal) were detected in both tissues of skipjack. Many typical volatile compounds were identified only in the hepatic tissues but not in the muscular tissue: heptanol, nonanol, 1-octen-3-ol, 1-penten-3-ol [green grass], (E,E)-2,4-heptadienal, (E,Z)-2,6-nonadienal, (E)-2-octenal, benzaldehyde, phenylacetaldehyde, decanal, hexanal, octanal [solvent-like] and trimethylamine.

From these lines of evidence, the hepatic tissue was considered to give stronger odor than the muscular tissue. Furthermore, because many peaks in the muscular tissue, arising from a higher temperature GC range, belonged to the peaks of compounds having high boiling points and little relationship with odors, as mentioned above, the odor of the muscular tissue should possibly be weak, and the odor of the hepatic tissue would not be so stronger than that of carp and the mackerel.

Many compounds, usually detected in tissues of the three seawater fishes tested, were not found in both tissues of the carp, such as 1-penten-3-ol, 1-octadecene, 1-pentadecene, 2-undecenal, (E,Z)-2,4-decadienal, 3-methylbutanol, 3-methylbutanal, 4-heptenal, heptadecane, heptanol, limonene, naphthalene, pentadecane, pentadecanoic acid, tetradecane, tridecane, and trimethylamine.

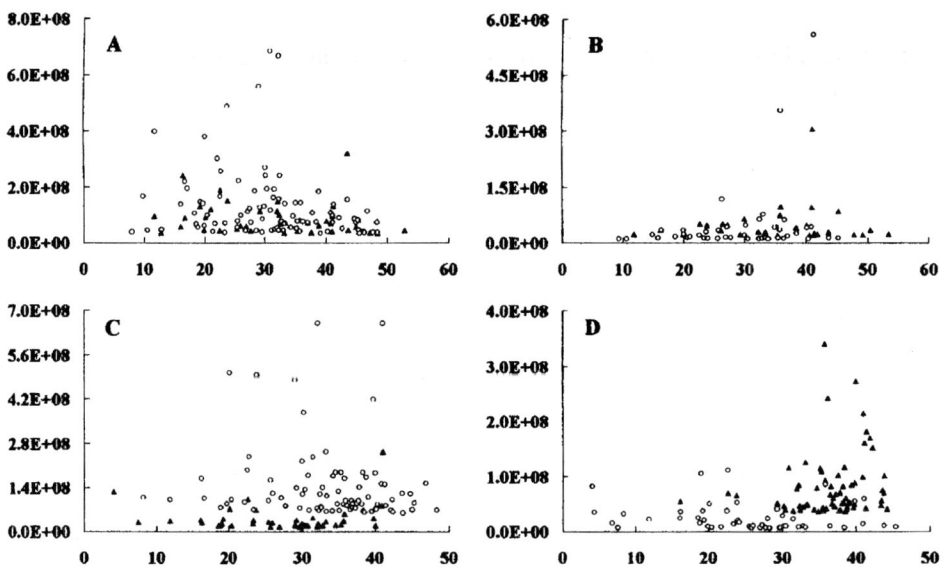

Figure 9. Variances of volatile compounds of hepatic and muscular tissues of four fishes. A, common carp; B, Japanese flounder; C, Spanish mackerel; D, skipjack. The vertical and horizontal axes indicate each GC peak area (%) and retention time (min) of compounds appearing as peaks on the GC charts, respectively. ▲, Muscular tissue; ○, Hepatic tissue. $p<0.05$ for A-D.

On the other hand, some volatiles detected only in the hepatic and / or muscular tissues of carp in comparatively higher portions but not in those of seawater fishes were as follows: 2,4-nonadienal, 2-decanal, 9,12,15-octadecatrienoic acid, 2-ethylfuran, octadecane, pentanal and tetradecanoic acid. These compounds listed in above Tables are considered to have odor specific to the freshwater fish.

As the results of statistical analysis (Fig. 9), significant differences were observed in the distribution pattern of volatile compounds between the hepatic and muscular tissues in all of fishes tested ($p<0.05$): the peak-area contributed to the difference greater than the retention time in the carp and the mackerel, while in the flounder and skipjack, the retention time did greater. The differences, therefore, were considered to give one of the reasons why the muscular tissue more paratable than the visceral tissues in fish.

Many typical volatile compounds reported so far to occur in muscular or hepatic tissues of fish and shellfish [3-11] were also detected in this study, such as 1-octen-3-ol, 2,4-decadienal, 2,4-heptadienal, 2,6-nonadienal, 2-nonenal, 2-nonanone, benzaldehyde, decanal, heptanal, hexanal, nonanal, octanal, trimethylamine, undecanal, etc. Only nonanal was detected in both tissues of all fishes tested. In both of the hepatic and muscular tissues of skipjack and in the hepatic tissues of Japanese flounder, BHT (2,6-di-*tert*-butyl-4-methylphenol) was detected. This compound has not been recognized to consist of volatiles in fish tissues but was detected in the present study, possibly originating from food additives through the food chain. Further detailed investigation is needed to know whether the volatile compound relates to the odor development of fish tissues.

It was concluded that the hepatic tissue contained greater numbers of volatiles than the muscular tissue in all of the four fishes. The hepatic tissue, containing various volatiles, has an odor different from that of the muscular tissue in both quality and quantity.

REFERENCES

1. Konosu, S. and Hashimoto, K. (1992) The Uses of Marine Products, New Series of Fisheries Science, No.24, Koseisha Koseikaku, Tokyo, pp. 187-243.
2. Song, X.A., Hirata, T. and Sakaguchi, M. (2000) Proximate composition and nitrogenous extractive components in fish muscle and internal organs. Nippon Suisan Gakkaishi, 66: 282-290.
3. Shahidi, F. and Cadwallader, K.R. (eds.) (1997) Flavor and Lipid Chemistry of Seafoods, ACS Symposium Series No. 674, American Chemical Society, Washington, DC.
4. Kawai, T. (1996) Fish flavor. Critical Reviews in Food Science and Nutrition, 36: 257-298.
5. Milo, C. and Grosch, W. (1996) Changes in the odorants of boiled salmon and cod as affected by the storage of the raw material. J. Agric. Food Chem., 44: 2366-2371.
6. Nijssen, L.M., Visscher, C.A., Maarse, H., Willemsens, L.C. and Boelens, M.H. (eds) (1996) Volatile Compounds in Food-Qualitative and Quantitative Data, 7th Edition. TNO Nutrition and Food Research Institute, Zeist, The Netherlands.

7. Schultz, T.H., Flath, R.A., Mon, T.R., Eggling, S.B. and Teranishi, R. (1977) Isolation of volatile components from a model system. J. Agric. Food Chem., 25: 446-449.
8. German, J.B., Berger, R.G. and Drawert, F. (1991) Generation of fresh fish flavor: rainbow trout (*Salmo gairdenri*) gill homogenate as a model system. Chem. Mikrobiol. Technol. Lebensm., 13: 19-24.
9. Cha, Y.J., Baek, H.H. and Hsieh, T.C.Y. (1992) Volatile components in flavour concentrates from crayfish processing waste. Sci. Food Agric., 58: 239-348.
10. Kim, H.R., Baek, H.H., Meyers, S.P., Cadwallader, K.R. and Godber, J.S. (1994) Crayfish hepatopancreatic extract improves flavor extractability from a crab processing by-product. J. Food Sci., 9: 91-96.
11. Chung, H.Y. and Cadwallader, K.R. (1994) Aroma extract dilution analysis of blue crab claw meat volatiles. J. Agric. Food Chem., 42: 2867-2870.

PURIFICATION AND SOME PROPERTIES OF CUTTLEFISH INK POLYPHENOL OXIDASE

Peigen Zhou, Xiaoyu Qi and Xiaoxian Zheng
College of Food Science and Technology, Shanghai Fisheries University Shanghai, 200090, China

Abstract

Polyphenol oxidase (PPO) from cuttlefish (*Sepia esculeuta* Hoyle) ink was isolated and purified by DEAE-Sepharose CL-6B and Sephadex G-150 column chromatography. The specific activity of the enzyme was increased to nearly 11-fold. The optimum temperature and pH for PPO-DOPA reaction were 55°C and 7.5, respectively. The purified enzyme was quite stable at the alkaline pHs, very unstable at the acid pHs and the most stable at pH 7.0. The enzyme was heat stable up to 60°C, but was rapidly inactivated by 30 min incubation at temperature higher than 60°C. The apparent K_m values for DL-DOPA and for pyrogallol as a substrate were 2.46 mM and 11.37 mM, respectively. The results indicated that the PPO had a higher affinity for DL-DOPA than for pyrogallol. 2-Mercaptoethanol and sodium hydrogen sulfite were found to be the most potent inhibitors of the enzyme.

1. INTRODUCTION

Polyphenol oxidase (PPO, E.C.1.14.18.1) is widely distributed in nature. It is probably present in all plants [1] found in microorganisms, especially fungi, and in some animal organs [2]. Enzymatic browning of fruits, vegetables, and crustaceans due to PPO activity has been extensively studied [3].

Some bioactive functions of squid and cuttlefish ink have been reported in addition to as food materials. Yamanaka et al. have reported on the antibacterial properties of squid ink and a research group led by Hajime Matsue at the Aomori Prefecture Industrial Technology Development Center first announced in 1990 that squid ink contains substances with tumour inhibiting activity (Okuzumi and Fujii,T. [4]).

Although it has been reported that the black color of squid and cuttlefish ink is due to its large content of dark color pigment called melanin which is synthesized from the amino acid tyrosine, the mechanism of its formation is still unclear. In the present study the PPO was extracted from cuttlefish ink and characterized with respect to its kinetic properties and responses to pH, temperature, and some inhibitors.

2. MATERIALS AND METHODS

2.1. Materials

Fresh cuttlefishes (*Sepia esculeuta* Hoyle) were obtained from a local market. The ink was squeezed out its ink sac immediately and stored at 4°C.

2.2. Extraction and partial purification of PPO

PPO was extracted according to the procedure of Simpson et al. [5] with a slight modification. Thirty gram of cuttlefish ink powder was extracted with 150 mL of 0.05 M sodium phosphate buffer, pH 7.2, containing 1.0 M NaCl and 0.2% Brij 35 (W/V, Nacalai Chemicals, LTD. Japan) for 1h while stirring gently with a magnetic stirrer. The extract was centrifuged at 20,000×g for 30 min, and the supernatant was fractionated with solid ammonium sulfate. The fraction precipitating between 30% and 80% saturation was collected by centrifugation at 12,500×g for 30 min, redissolved in minimum volume of the extraction buffer, and then dialyzed overnight against 0.01 M sodium phosphate buffer, pH 7.2.

The dialyzed enzyme was chromatographed on a DEAE-Sepharose CL-6B column (1.6 × 60 cm) equilibrated with 0.05 M sodium phosphate buffer, pH 7.2. After sample loading, the column was eluted with a linear gradient of 0 -1.0 M NaCl in 0.05 M sodium phosphate buffer, pH 7.2, at a flow rate of 0.3 mL/min. Fractions (6.8 mL) were collected and assayed for PPO activity. The most active fractions were combined and dialyzed against 3L of 0.05 M sodium phosphate buffer, pH 7.2.

After dialysis, the PPO-containing fraction from DEAE-Sepharose CL-6B column chromatography was applied onto a Sephadex G-150 column (1.6 × 60 cm) and eluted with 0.05 M sodium phosphate buffer, pH7.2, at a flow rate of 0.18 mL /min. Fractions (5 mL) were collected and assayed for PPO activity. The most active fractions were combined and used for characterization of the enzyme.

All procedures above were conducted at 4°C unless otherwise noted.

2.3. PPO Activity and protein determination

The PPO activity was assayed with DL-DOPA (DL-β-3,4-dihydroxyphenylalanine) as a substrate according to the procedure of Zhou et al. [6] with a slight modification. The assay was performed using 0.02 mL of enzyme solution and 2.98 mL of 5.0 mM DL-DOPA in 0.05 M sodium phosphate buffer, pH6.5. The increase of absorbance at 475 nm at 30°C was measured against a reference of the same buffer without the enzyme on a spectrophotometer. One unit of PPO activity is defined as the amount of the enzyme which caused a change in absorbance of 0.001 /min.

Protein concentration was determined according to the method of Lowry at al. [7] with bovine serum albumin as a standard.

2.4. Effect of pH on PPO activity and stability

Three kinds of buffer solutions used for this study were as follows: 0.1 M citric acid -0.2 M Na_2HPO_4 for pH 3.0-5.0, 0.1 M NaH_2PO_4 -0.1 M Na_2HPO_4 for pH 6.0-8.0 and 0.1 M glycine-0.1 M NaOH for pH 8.5-10.0. To determine the effect of pH on PPO activity, the substrate, DL-DOPA (5.0 mM), was prepared in various buffers. PPO activity was measured at 475 nm and 30°C using the spectrophotometer procedure as described above.

To determine the effect of pH on PPO stability, 0.1 mL of the enzyme solution was preincubated in 0.4 mL of various buffer solutions ranging from pH 3.0 to pH 10.0 for 30 min at 30°C. The residual PPO activity was assayed by mixing 2.9 mL of 5 mM DL-DOPA in 0.05 M sodium phosphate buffer solution, pH 6.5, with 0.1 mL of the preincubated PPO solution and by measuring the increase of absorbance at 475 nm and 30°C.

2.5. Effect of temperature on PPO activity and stability

The effect of temperature on PPO activity was determined by preequibrating the substrate solution, 5.0 mM DL-DOPA in 0.05 M sodium phosphate buffer (pH 6.5), at various temperatures (20-80°C) for 5 min prior to the addition of the enzyme. The PPO activity was assayed as described previously.

In the thermal stability studies, the enzyme solution was incubated at various temperatures (20-80°C) for 30 min and rapidly cooled in an ice bath for 5 min. 0.02 mL of the heat-treated enzyme solution was added to 2.98 mL of 5 mM DL-DOPA solution, pH 6.5, and the residual PPO activity was measured.

2.6. Substrate specificity

In the studies of the substrate specificity, the reaction system consisted of 0.02 mL of the enzyme solution and 2.98 mL of various substrate solutions prepared in 0.05 M sodium phosphate buffer, pH 6.5. The increase in absorbance at the optimum wavelength for each substrate was measured [6].

For the determination of Km, DL-DOPA solutions at concentration from 0.3 to 5.0 mM in 0.05 M sodium phosphate buffer, pH 6.5, and pyrogallol solutions from 2.0 to 15 mM in the same buffer were used as substrates. The reaction system was the same as the PPO assay. Initial rates were measured as the increase in absorbance at 475 nm for DOPA-PPO reaction and at 420 nm for pyrogallol-PPO reaction, respectively. The apparent Km value was calculated from the Lineweaver Burk plots [8].

2.7. Effect of inhibitors on PPO activity

The effect of various inhibitors (ascorbic acid, sodium hydrogen sulfite, mercaptoethanol, benzoic acid and EDTA) on activity of Cuttlefish Ink PPO was investigated as follows: 0.04 mL of inhibitor solutions with different concentrations (0.15, 1.5, and 15 mM) in 0.05 M sodium phosphate buffer (pH 6.5) was added to 0.02 mL of the enzyme solution, and then incubated at 30°C for 30 min. The residual PPO activity was assayed by mixing 2.94 mL of 0.05 M DL-DOPA solution (pH 6.5) as a substrate with the incubated PPO-inhibitor solution and measuring the change in absorbance at 475 nm and 30°C [6].

Table 1
Purification scheme for cuttlefish ink polyphenol oxidase

Step	Total Vol. (mL)	Total act. (units)	Total Protein (mg)	Specific act. (units/mg)	Yield (%)	Purification (fold)
Crude extract	97.0	305229.9	121.3	2516.3	100	
30-80% $(NH_4)_2SO_4$	14.0	386400.0	59.8	6461.5	126.6	2.6
DEAE-Sepharose CL-6B	21.0	263436.6	16.0	16464.8	86.3	6.5
Sephadex G-150	26.0	183630.2	6.8	27004.4	60.2	10.7

3. RESULTS AND DISCUSSION

3.1. Partial purification of PPO

A summary of the partial purification scheme for cuttlefish ink PPO is shown in Table 1. The specific activity of the precipitate resulting from 30-80% ammonium sulfate fractionation was increased by 2.57-fold by comparison with the crude extract. The purification of PPO was completed by DEAE-Sepharose CL-6B column chromatography. The elution profile of PPO from the chromatographic column was presented in Fig. 1. A peak of PPO activity was eluted at about 0.6 M NaCl. The fractions corresponding to the peak of PPO activity were pooled and then dialyzed. A 6.55-fold purification for the PPO enzyme was achieved. After that, a gel chromatography was used for the further purification of PPO preparation. The elution curve of PPO on the Sephadex G-150 column was shown in Fig. 2. A single protein peak with PPO activity was obtained in this step. The specific activity of the purified PPO was about 11-fold greater than that of the crude extract.

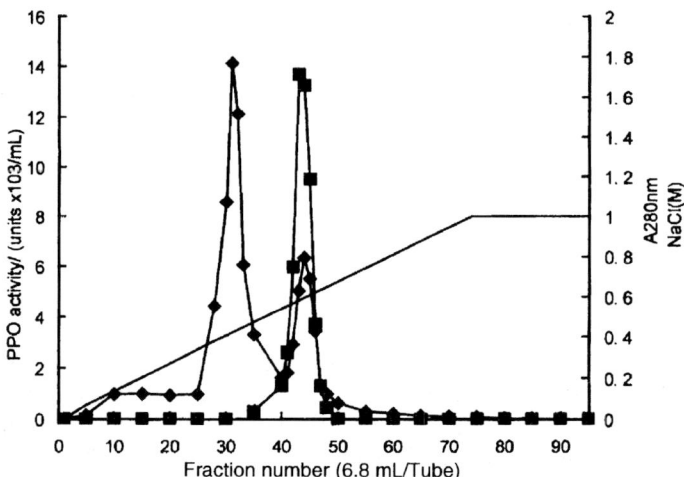

Figure 1. Elution profile of cuttlefish ink PPO on DEAE-Sepharose CL-6B column. ─■─, PPO; ───, NaCl; ─◆─, A_{280nm}.

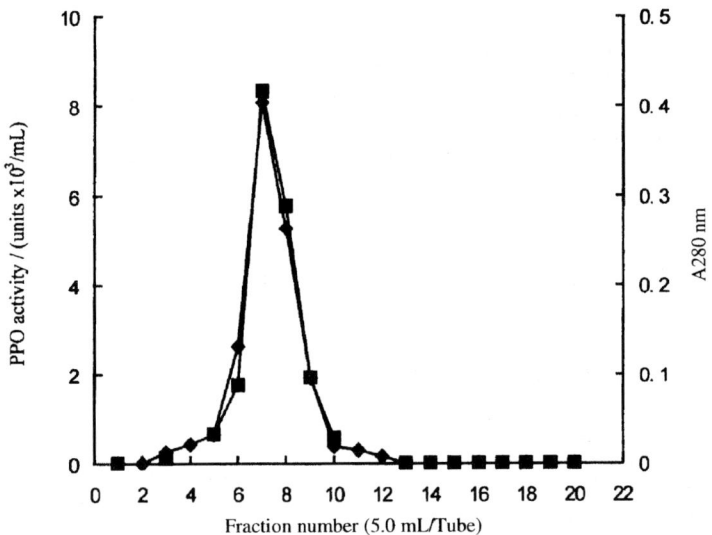

Figure 2. Elution profile of PPO from Sephadex G–150 column. ◆ , PPO; ■ , A$_{280nm}$

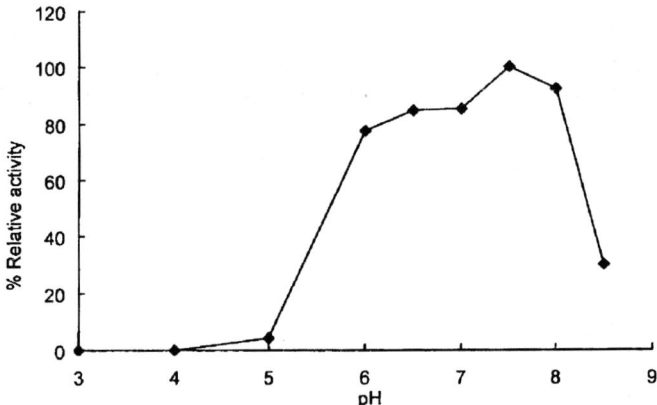

Figure 3. Effect of pH on activity of cuttlefish ink PPO.

3.2. Optimum pH and stability

The pH-activity profile for the oxidation of DL- DOPA by cuttlefish ink PPO was shown in Fig. 3. The PPO was active between pH 6.0 and 8.0 and most active at pH 7.5. It was similar to the results obtained from shrimp PPO (pink shrimp, white shrimp, brown shrimp, Taiwanese black tiger shrimp and Japanese prawn). PPO enzymes from all the species exhibited the optimal activity within the pH values ranging from neutral to alkaline (pH 6-8) [6, 9].

The pH – stability curve for cuttlefish ink PPO was shown in Fig. 4. The PPO showed

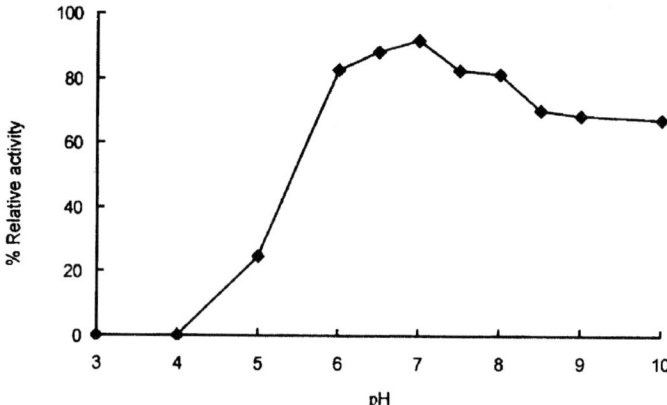

Figure 4. Effect of pH on cuttlefish PPO stability.

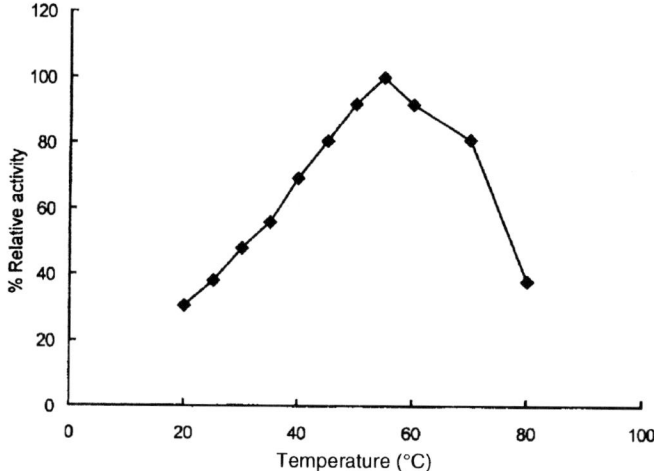

Figure 5. Effect of temperature on cuttlefish ink PPO activity.

maximal stability at pH 7 and appeared to be less susceptible to activation at alkaline pHs (>8) than at acidic pHs (<6). For example, the PPO retained about 70% of its original activity after 30 min preincubation at pH 9, while 30% at pH 5. In this case, cuttlefish ink PPO is considerably similar to shrimp PPO [10].

3.3. Optimum temperature and stability

Figure 5 showed the effect of temperature on cuttlefish ink PPO activity at pH 6.5. The activity for PPO-DOPA reaction was increased almost linearly with the increase of temperature under 55°C and decreased with the increase of temperature over 55°C. The temperature optimum (55°C) was higher than that observed for shrimp. For example, the

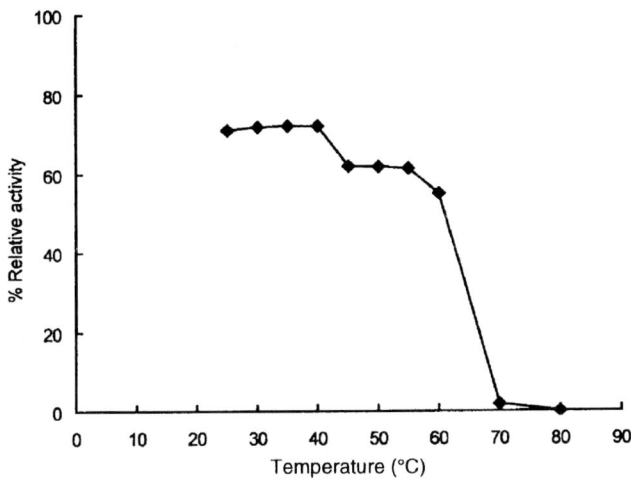

Figure 6. Effect of temperature on cuttlefish ink PPO stability.

temperature optimum for pink shrimp PPO was 40°C (Simpson et al.1988) [5], 45°C for white shrimp PPO [10] and 40°C for prawn PPO [11].

The heat inactivation of the PPO was shown in Fig. 6. This enzyme exhibited a high thermostability at lower temperatures up to 40°C and a certain thermostability between 40 and 60°C, but was unstable at higher temperatures. For instance, the PPO activity was almost same as its original activity after 30 min at 40°C and about 55% of the original activity of the PPO was retained after 30 min at 60°C. However, the enzyme was inactivated rapidly over 60°C and no activity was detected at 80°C. The range of maximal temperature stability for cuttlefish ink PPO was comparable with that for shrimp PPO (20-30°C for pink shrimp PPO, 25-50°C for white shrimp PPO and 30-35°C for Taiwanese black tiger shrimp PPO) [9,10,12].

3.4. Substrate specificity

The PPO activities using different substrates were shown in the Table 2. Relative activities of the enzyme measured at the optimum wavelength of each substrate were calculated using DL-DOPA as the basis of comparison. Maximum activity was detected

Table 2
Substrate specificity of cuttlefish ink polyphenol oxidase

Substrate	Conc. (mM)	Wavelength (nm)	Activity (units/mL)	Relative activity (%)
DL-DOPA	5.0	475	11200.0	100
Pyrogallol	10.0	334	7883.5	70.4
L-Tyrosine	2.5	472	0	0

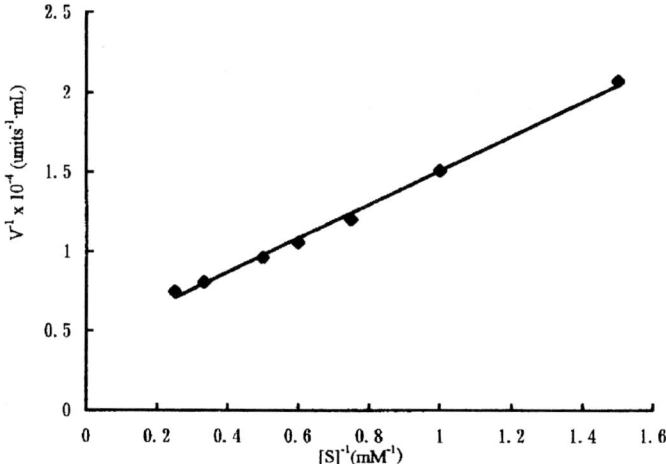

Figure 7. Double reciprocal plot of PPO – DL- DOPA reaction.

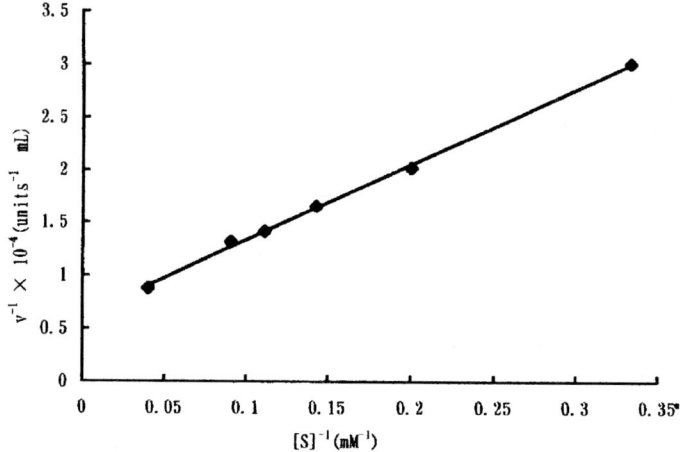

Figure 8. Double reciprocal plot of PPO–pyrogallol pyrogallicacid reaction.

toward catechol, followed by DL-DOPA and pyrogallol, and no tyrosinase activity was detected in the PPO preparation.

The Km values as calculated by the Lineweaver-Burk plot were 2.46 mM for DL-DOPA (Fig. 7) and 11.37 mM for pyrogallol (Fig. 8). This result indicated that the PPO had a higher affinity for DL-DOPA than for pyrogallol.

3.5. Inhibition of PPO

The effect of different inhibitors on the cuttlefish ink PPO was shown in Table 3. Of all inhibitors used in the study, 2-mercaptoethanol was most effective for inhibition of the

Table 3.
Effect of inhibitors on activity of cuttlefish ink polyphenol oxidase

Inhibitor	Conc. (mM)	Inhibition (%)
Ascorbic acid	15	39.6
	1.5	25.6
	0.15	23.8
EDTA (Na)	15	36.9
	1.5	33.5
	0.15	15.8
Sodium hydrogen sulfite	15	100
	1.5	93.9
	0.15	19.0
2-Mercaptoethanol	15	100
	1.5	100
	0.15	85.5
Benzoic acid	15	
	1.5	8.2
	0.15	8.2

cuttlefish ink PPO, followed by sodium hydrogen sulfite, ascorbic acid, EDTA and benzoic acid.

REFERENCES

1. Whitaker, J.R. (1972) Principles of Enzymology for the Food Scinence. Marcel Dekker, New York.
2. Brown, B.R. (1967) Biochemical aspects of oxidative coupling of phenols. In:Oxidative Coupling Phenols. Taylor, W.I. and Battersby, A.R. (eds.), Marcel Dekker, New York, p.167.
3. Chen, J.S., Preston, J.F., Wei, C.I., Hooshar, P., Gleeson, R.A. and Marshall, M.R. (1992) Structural comparison of crustacean, potato, and mushroom. J. Agric. Food Chem., 40: 1326-1330.
4. Okuzumi, M. and Fujii, T. (eds.) (2000) Nutritional and Functional Properties of Squid and Cuttlefish. National Cooperative Association of Squid Processors, Tokyo.
5. Simpson, B.K., Marshall, M.R. and Otwell, W.S. (1988) Phenoloxidase from pink and white shrimp: kinetic and other properties. J. Food Biochem., 12: 205-217.
6. Zhou, P.G., Smith, N.L. and Lee, C.Y. (1993) Potential purification and some properties of Monroe apple peel polyphenol oxidase. J. Agric. Food Chem., 41: 532-536.
7. Lowrry, O.H., Rosebrough, N.J., Farr, A.L. and Randall, R.J. (1951) Protein measurement with Folin phenol reagent. J. Biol.Chem., 193: 265-275.
8. Lineweaver, H. and Burk, B. (1934) The determination of enzyme dissociation constants. J. Am. Chem. Soc., 56: 658-666.
9. Rolle, R.S., Guizani, N., Chen J.S., Marshall, M.R., Yang, J.S. and Wei, C.I. (1991)

Purification and characterization of phenoloxidase isoforms from Taiwan black tiger shrimp (*Penaeus monodon*). J. Food Biochem., 15: 17-32.
10. Simpson, B.K., Marshall, M.R. and Otwell, W.S. (1987) Phenoloxidase from shrimp (*Penaeus setiferus*): purification and some properties. J. Agric. Food Chem., 35:918-921.
11. Zhou, P.G., Zhao, J., Wang, J.X., Qi, X.Y. and You, Y.M. (1997) Characterization of polyphenol oxidase from Japanese prawn (*Penaeus japonicus*). In: Proceedings of the Third International Symposium on the Efficient Application and Preservation of Marine Biological Resources, Nishida M.(ed), Fukui Prefectural University, Obama, Fukui, Japan., pp.33-43.
12. Madero, C.F. and Finne, G. (1982) Properties of phenoloxidase isolated from Gulf shrimp. Proc. Seventh Annual Trop. Subtrop. Fish. Technol. Conf. Americas., 7: 328.

BIOLOGICAL ACTIVITIES OF THE COMPONENTS FROM SCALLOP SHELLS

Y. C. Liu, K. Uchiyama, N. Natsui and Y. Hasegawa
Muroran Institute of Technology, Department of Applied Chemistry, Muroran, Hokkaido 050-8585, Japan

Abstract

Biological activities of the components which were extracted from the scallop shells were investigated for the effective utilization of the scallop shells. Activation was observed for α-chymotrypsin activity, while the scallop shell extract showed strong inhibitory activities for elastase and trypsin. The scallop shell extract inhibited generation of superoxide anion generated by xanthine and xanthine oxidase. When the scallop shell extract was supplied to culture medium for skin fibroblast cells, the cell growth rate was increased. These results suggest that scallop shells contain some bioactive substances.

1. INTRODUCTION

Scallop is one of the major marine products industry in Hokkaido, Japan. The scallop shells of about 300 thousands ton per a year are generated as industrial wastes.

The molluscan shells are mineralized by composites of $CaCO_3$ and organic proteins, which exhibit nanoscale regularity and strength. Although proteins as an organic matter constitute only 1-5% of weight of biomineralized composite material, they are responsible for its organization and the resulting enhancement of strength of $CaCO_3$ crystal. In recent years, there have been several reports about the cloning of the shell organic matrix proteins [1-3]. However, the components which have not been identified appear to occur in the shells.

The shells are composed of two kinds of $CaCO_3$ polymorphs of primatic layer and nacreous layer. The nacreous layer has been used as chinese medicine for keeping skin moist, counteracting the poison, and making the spirit stable. The nacreous layer is also used as a cosmetic in Japan. However, few studies about the bioactive substances in the shell have not been reported [4]. In this study, we investigated biological activities of components which was extracted from scallop shells, and the possibility of the utilization of scallop shells is discussed.

2. MATERIALS AND METHODS

2.1. Extraction of scallop shell components

Scallop shells were isolated from *Patinopecten yessoensis* and the shells were cleanly brushed to remove an adhered materials, and then crushed to a powder. The powdered shells were dialyzed against 1000mL of 5% acetic acid to decalcify absolutely. This was followed by exhaustive dialysis against 1000mL of deionized water to remove the acetic acid. After dialysis, the water-soluble and -insoluble fractions were obtained by centrifugation at 16,000×g for 20min at room temperature. The water-soluble fraction was concentrated and loaded onto Sephacryl S-200 gel filtration column (0.5×35cm) previously equibrilated with deionized water, and the fractions of 1mL were collected. Each fraction was employed as the scallop shell extract for protease inhibitory assay and free radical scavenging assay. The water-insoluble fraction was extracted with 70% methanol again. The methanol-soluble fraction was concentrated and dissoved at a concentration of 153 mg/mL (w/v) in 10% dimethyl sulfoxide. The methanol-soluble fraction was employed for growth assay of skin fibroblast cells.

2.2. Measurement of protease activities

Elastase activity was measure by the method of Ito et al. [5]. The reaction mixture was 50mM Hepes-NaOH (pH 7.5) and 0.3mM butoxycarbonyl-alanine-4-nitrophenol as substrate in the presence and the absence of the scallop shell extract. After 1μg of pancreas elastase was added to the reaction mixture, the developmented color was measured at 405 nm. Trypsin activity was measured in a solution containing 12.5mM Tris-HCl (pH8.0) and 0.0625mg/mL α-N-benzoyl-DL-arginine-p-nitroanilide in the absence and the presence of the scallop shell extract [6]. The reaction was started by adding trypsin to be 0.5mg/mL, and the optical density at 405nm was measured. In the α-chymotrypsin inhibitory assay, 0.2 mM N-succinyl-Ala-Ala-Pro-Phe-p-nitroanilide was dissolved in 10mM Hepes-NaOH (pH 7.5) in the absence and the presence of scallop shell extract. To start the reaction, α-chymotrypsin was added to be 0.0013mg/mL and the optical density at 595nm was measured.

2.3. Free radical scavenging activity

Superoxide anion was generated by xanthine and xanthine oxidase. Xanthine oxidase activity was measured according to the method of Wede et al.[7]. The reaction mixure was 340μM xanthine and 20mM potassium phosphate buffer (pH6.5) in the presence and the absence of the scallop shell extract. After addition of 0.05U/mL xanthine oxidase, absorbance at 290nm was measured. Production of superoxide radical was measured by reduction of nitroblue tetrazolium [7]. The reaction mixture contained 170μM xanthine, 320μM nitroblue tetrazolium, and 20mM potassium phosphate (pH 6.5). Various concentrations of the scallop shell extracts were added to the reaction mixture. The reaction was started by adding 0.05 U/mL xanthine oxidase and the reduction of nitroblue tetrazolium was detected by measuring absorbance at 560nm.

2.4. Protein concentrations

Protein concentrations were measured according to the method of Lowry et al. [8].

2.5. Cell culture

Skin fibroblast cells were purchased from JCRB cell bank. Cells were maintained in Dulbecco's modified Eagle's medium supplemented with 5% fetal calf serum (FCS) under a gas mixture of 95% air/5% CO_2.

2.6. Growth assay of skin fibroblast cells

Skin fibroblast cells were seeded at a density of 4×10^4 cells in a volume of 50 µL per well. After 24h, the scallop shell extract (methanol-soluble fraction) was added to culture medium with various kinds of concentrations. The cells were treated for 24h with the scallop shell extract, and then the cell numbers was quantified by the 3-(4,5-dimethyl) ethiazole (MTT) assay [9].

3. RESULTS

After decalcifying the scallop shells using 5% acetic acid, water-soluble fraction was pooled as described in Materials and Methods. The water-soluble fraction was separated by a Sephacryl S-200 gel filtration column (Fig. 1), and each fraction was employed for the following assay.

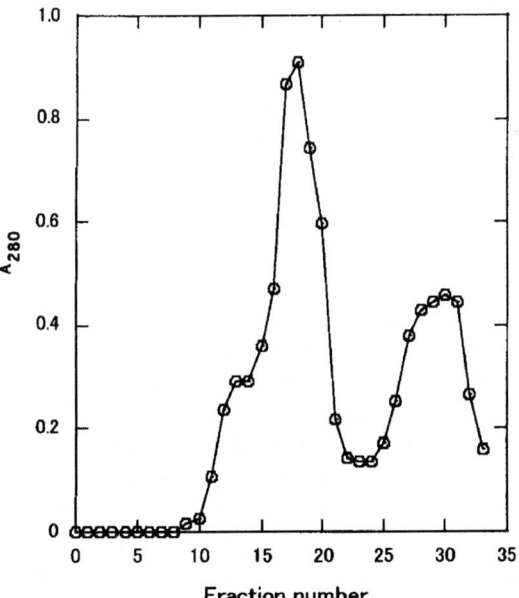

Figure 1. Sephacryl S-200 gel filtration column chromatography of scallop shell extract. The water-soluble fraction from the scallop shells was separated by Sephacryl S-200 gel filtration column equilibrated with deionized water at a flow rate of 5 mL/h. After absorbance at 280nm of each fraction was measured, each fraction was employed for protease inhibitory assay and free radical scavenging assay.

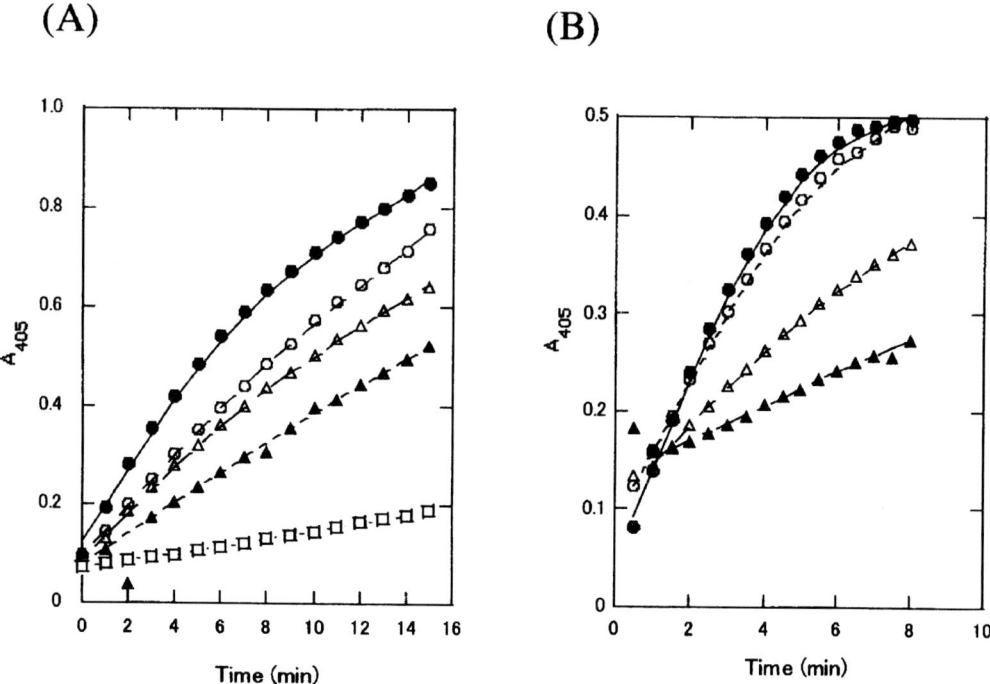

Figure 2. Inhibition of elastase and trypsin activities. (A) Elastase activity was measured as described in Materials and Methods. The reactions were performed in the absence (●) and the presence of 0.06mg/mL (□), 0.03mg/mL (▲), 0.015mg/mL (△) and 0.006mg/mL (○) of the scallop shell extract (fraction number 18). (B) Trypsin activity was measured in a solution containing 12.5mM Tris-HCl (pH8.0) and 0.0625mg/mL α-N-benzoyl-DL-arginine-p-nitroanilide in the absence (●) and the presence of 0.087mg/ml (▲), 0.05mg/mL (△), and 0.025mg/mL (○) of the scallop shell extract (fraction number 18).

3.1. Protease inhibitory and activation assays

The scallop shell extract (fraction number 18) after gel filtration inhibited elastase activity in a does-dependent manner (Fig. 2A), and the activity was inhibited to about 80% at the protein concentration of 0.06mg/mL. When the elastase activities were measured in solutions containing various concentrations of substrate in the presence and the absence of the scallop shell extract, double reciplocal plots of elastase activities indicated non-competitive inhibition (data not shown). In trypsin inhibitory assay, the activity was detected in several fractions (fraction number 15-20). The inhibition was about 70% at the protein concentration of 0.087mg/mL (fraction number 18) (Fig. 2B). However, any fractions did not exhibit the inhibitory activity for α-chymotrypsin (data not shown). On the other hand, several fractions (fraction number 25-30) enhanced activity for α-chymotrypsin (Fig. 3), but not for elastase and trypsin. These results

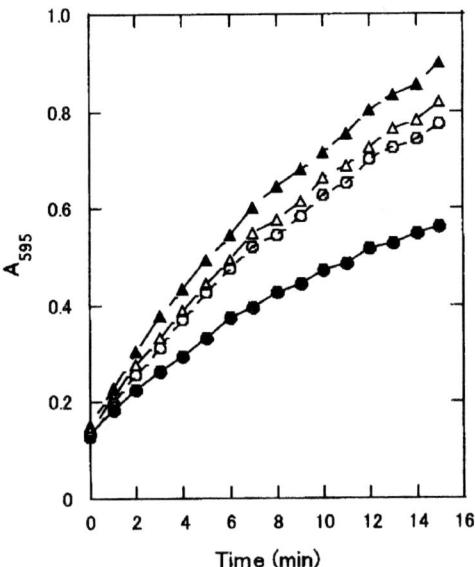

Figure 3. Activation of α-chymotrypsin activity. α-chymotrypsin activity was measured in a solution containing 10mM Hepes-NaOH (pH 7.5) and 0.2mM N-succinyl-Ala-Pro-Phe-p-nitroanilide in the absence (●) and the presence of 0.017mg/mL (▲), 0.0085mg/mL (△) and 0.007mg/mL (○) of scallop shell extract (fraction number 27).

suggest that the scallop shells contain the bioactive substances which inhibit or activate serine protease activities.

3.2. Free radical scavenging assay

The ability to scavenge superoxide anion generated by xanthine and xanthine oxidase was investigated. The activity of xanthine oxidase was measured by production of uric acid from xanthine, and generation of superoxide anion was measured by formazan production from nitroblue tetrazolium. The scallop shell extract (fraction number 13) inhibited generation of superoxide anion in a does-dependent manner (Fig. 4A). The inhibition was about 67% at the protein concentration of 0.06mg/mL. On the other hand, the scallop shell extract did not inhibit the xanthine oxidase activity even in the concentration (0.06mg/mL) which inhibit the xanthine oxidase activity (Fig. 4B), suggesting that the scallop shells contain free radical scavenging substances.

3.3. Cell growth assay

We investigated the effect of the scallop shell extract on the growth rate of skin fibroblast cells as described in MATERIAL AND METHODS (Fig. 5). The scallop shell extract enhanced significantly the growth rate of skin fibroblast cells. The number of cells compared to the control culture was raised to about 137% at a does of 1.39mg/mL (w/v). When the cell growth assay was performed in serum-free medium to exclude an influence

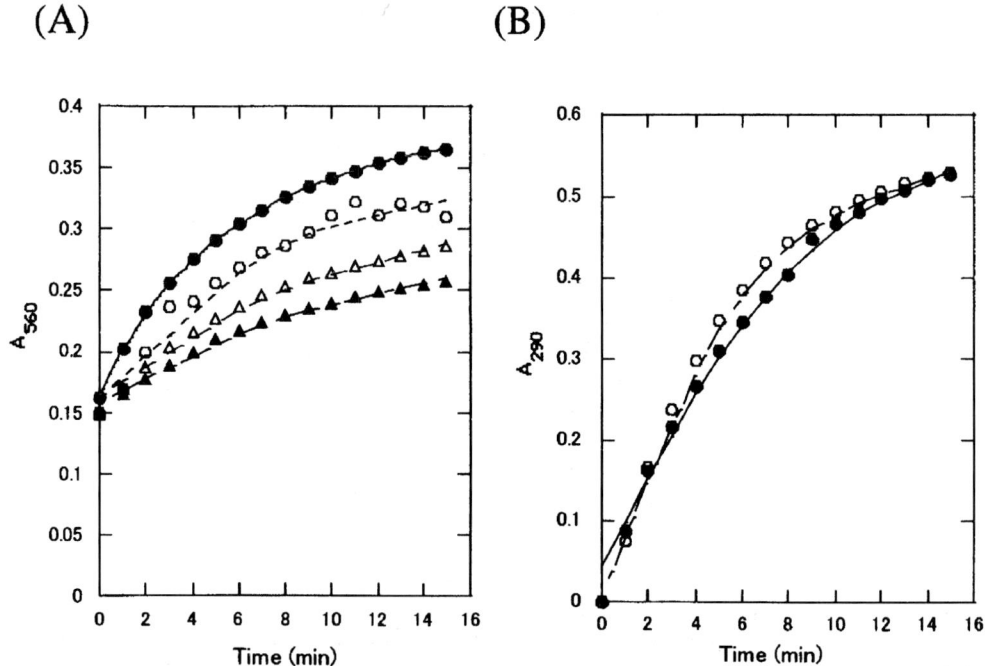

Figure 4. Inhibition of superoxide anion generation. (A) Superoxide anion generation was measured in a solution containing 170μM xanthine, 320μM nitroblue tetrazolium, and 20mM potassium phosphate (pH 6.5) in the absence (●) and the presence of 0.06mg/mL (▲), 0.023mg/mL (△), and 0.0025mg/mL (○) of scallop shell extract (fraction number 17). The reaction was started by adding 0.05U/mL xanthine oxidase. (B) Xanthine oxidase activity was measured in a solution containing 340μM xanthine and 20mM potassium phosphate buffer (pH6.5) in the absence (●) and the presence (○) of 0.06mg/mL of the scallop shell extract.

of FCS, similar result was found (data not shown). These results suggest that the substance in the scallop shell extract may act as growth factor for skin fibroblast cells. In this assay, the methanol-soluble fraction was employed as scallop shell extract as described in Materials and Methods, because the water-soluble fraction did not show growth-promoting activity.

4. DISCUSSION

Skin, which has a highly differentiated and complex structure, is vulnerable to free radical damage because of its contact with oxygen. The free radical damage of skin causes alteration in extracellular matrix of connective tissue such as collagen and elastin, resulting in loss of skin tone and wrinkles [10-12]. It is generally believed to reverse skin damage like photoaging by increasing collagen synthesis in skin. The increase of collagen

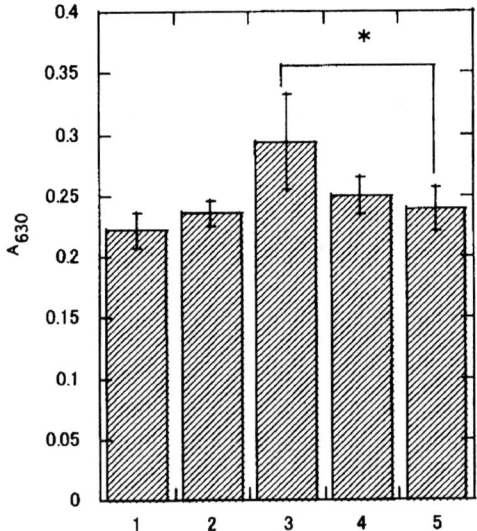

Figure 5. Growth-promoting activity of skin fibroblast cells. Cells were treated in the absence (5) or the presence of the scallop shell extract at concentrations of 0.01mg/mL (1), 0.14mg/mL (2), 1.39mg/mL (3) and 13.91mg/mL (4). Number of cells was measured by MTT assay after 24h treatment. Data were combined from 9 wells of a 96-well plate and the bars show the standard errors of mean (S. E. M.). Statistical significance was determined by Student's T-test. *, $p < 0.05$ relative to control.

synthesis may be achieved by functional activation or proliferation of skin fibroblast cells [13]. Free radical scavenging activity and growth-promoting activity for skin fibroblast cells of the scallop shell extract may be effective for protecting skin. In addition, the inhibitory effect of the scallop shell extract for elastase activity may be also effective for the protection of skin. The increase of elastase activity in skin destructs elastic fibres, resulting in reduced skin elasticity [14]. Therefore, the elastase inhibitor has been mixed in some cosmetics. These results may raise possibility of the utilization of the scallop shells as a cosmetic for protecting skin.

Proteases are involved in blood coagulation and pathogenicity of a variety of human infectious agents. Elastase is responsible for the abnormal turnover of the elastin associated with the development of pulmonary emphysema and rheumatoid arthritis [15]. The free radical is also involved in various kinds of degenerative diseases such as atherosclerosis and neurodegeneration [16]. The components of scallop shells may be of value in diseases.

The shell organic matrix constitutes of proteins, polysaccharides, lipids, and so on. Analysis of the water-soluble fraction by SDS-PAGE revealed at least three kinds of proteins with molecular masses of 90kDa, 20kDa and 17kDa (data not shown). The existence of saccharide in the water-soluble fraction was also found by orcinol-H_2SO_4 method. However, it is unclear whether bioactive substances are proteins or saccharides or the other components. It will be necessary to isolate and identify these substances in future.

REFERENCES

1. Sudo, S., Kujikawa, T., Nagakura, T., Okubo, T., Sakaguchi, K., Tanaka, M., Nakashima, K. and Takahashi, T. (1997) Structure of mollusc shell framework proteins. Nature, 387: 563-564.
2. Miyashita, T., Takagi, R., Okushima, M., Nakano, S., Miyamoto, H., Nishikawa, E. and Matsushiro, A. (2000) Complementary DNA cloning and characterization of pearlin, a new class of matrix protein in the nacreous layer of oyster pearls. Mar. Biotechnol., 2: 409-418.
3. Samata, T., Hayashi, N., Kono, M., Hasegawa, K., Horita, C. and Akera, S. (1999) A new matrix protein family related to the nacreous layer formation of *Pinctada fucata*. FEBS Lett., 462: 225-229.
4. Huibi, X., Kaixun, H., Qiuhua, G., Zhonghong, G. and Xiuxian, H. (2001) A study on the prevention and treatment of myopia with nacre on chicks. Pharm. Res., 44: 1-6.
5. Ito, N., Iwamori, Y., Hanaoka, K. and Iwamori, M. (1998) Inhibition of pancreatic elastase by sulfated lipids in the intestineal mucosa. J. Biochem., 123: 107-114.
6. Vanderjagt, D.J., Freiberger, C., Vu, H.T.N., Mounkaila, G., Glew, R.S. and Glew, R.H. (2000) The trypsin inhibitor content of 61 wild edible plant foods of Niger. Plant Foods for Human Nutriton, 55: 335-346.
7. Wede, I., Altindag, Z.Z., Widner, B., Wachter, H. and Fuchs, D. (1998) Inhibition of xanthine oxidase by pterins. Free Rad. Res., 29: 331-338
8. Lowry, O.H., Rosebrough, N.J., Farr, A.L. and Randall, R.J. (1951) Protein measurement with the Folin phenol reagent. J. Biol. Chem., 193: 265-275.
9. Manthorpe, M., Fagnani, R., Skaper, S.D. and Varon, S. (1986) An automated colorimetric microassay for neuronotrophic factors. Brain Res., 390: 191-198.
10. Calabrese, V., Scapagnini, G., Randazzo, S.D., Randazzo, G., Catalano, C., Geraci, G. and Morganti, P. (1999) Oxidative stress and antioxidants at skin biosurface: a novel antioxidant from lemon oil capable of inhibiting oxidative damage to the skin., Drugs Exp. Clin. Res., 25: 281-287.
11. Kohen, R. (1999) Skin antioxidants: their role in aging and in oxidative stress-new approaches for their evaluation. Biomed. Pharmacother, 53: 181-192.
12. Scharffetter-Kochanek, K., Brenneisen, P., Wenk, J., Herrmann, G. and Ma, W. (2000) Photoaging of the skin from phenotype to mechanisms. Exp. Gerontol., 35: 307-316
13. Kim, S.J., Park, J.H., Kim, D.H., Won, Y.H. and Maibach, H.I. (1998) Increased in vivo collagen synthesis and in vitro cell proliferative effect of glycolic acid. Dermatol. Surg., 24: 1054-1058.
14. Bouloc, A., Godeau, G., Zeller, J., Wechsler, J., Revuz J. and Cosnes, A. (1999) Increased fibroblast elastase activity in acquired cutis laxa. Dermatology, 198: 346-350.
15. Power, J.C., Plaskon, R.R. and Kam, C.M. (1996) Low-molecular-weight inhibitors

of neutrophil elastase. Lung Biol. Health Dis., 88: 341-370.
16. Young, I.S. and Woodside, J.V. (2001) Antioxidants in health and disease. J. Clin Pathol., 54: 176-186.

USEFULNESS OF WASTE ALGAE AS A FEED ADDITIVE FOR FISH CULTURE

Heisuke Nakagawa

Graduate School of Biosphere Science, Hiroshima University
1-4-4, Kagamiyama, Higashi-hiroshima, Hiroshima 739-8528, Japan

Abstract

To assess the usefulness of *Porphyra* as a feed additive in fish feed, waste *Porphyra* was supplemented to an artificial diet at levels of 5 % and 3 %, and fed to 0-year and 1-year red sea bream *Pagrus major*, respectively. The effects of *Porphyra*-meal supplementation were evaluated by growth, feed efficiency, lipid accumulation, serum parameters, and lipid mobilization during starvation. The *Porphyra*-meal supplemented diet improved growth and feed efficiency. In addition, the muscle RNA/DNA ratio as a parameter of protein synthesis was higher, and acid proteinase activity as protein degradation was less active than in the control group. The *Porphyra*-meal accelerated triglyceride accumulation. The phenomenon that starvation for 10 days after the feeding experiment resulted in low body weight loss was caused by preferential lipid mobilization to energy and suppression of muscle protein consumption. Liver function assessed by the recovery time from anesthetic conditions with 0.1 % 2-phenoxyethanol was improved by *Porphyra*-meal supplementation. Resistance to air-dipping and low-oxygen tolerance was also improved. The results support the practical use of waste *Porphyra*-meal as a feed additive in fish diet.

1. INTRODUCTION

"Nori" is the Japanese name for purple laver belonging to the genus *Porphyra*. Large-scale culture of purple laver *Porphyra yezoensis* for daily food intake has been carried out in Japan. The recent total output is more than 7 billion sheets per year. The *Porphyra* is usually sold as a dried rectangular sheet. Low quality *Porphyra* (ca. 800 kg) sheets and waste piece of sheets disused from food companies (ca. 110kg) are annually disposed of as industrial waste and burned in Japan. Recycling of industrial waste materials should be considered from the point of effective measures against environmental pollution and the efficient use of resources. Therefore, the usefulness of the *Porphyra* waste was examined for its application to a feed additive for fish diet.

The application of algae for fish feed has been determined on protein sources and feed additives. High protein content and high production of algae in tropical countries were considered as protein sources for fish feed. However, attention should be given to use algae as the sole protein source in fish feed due to the short supply of algae. Furthermore, the feeding with large amount of algae to fish might result in malformation and impaired growth [1]. Mustafa and Nakagawa [2] investigated the nutritional merits of algae as a feed additive for fish culture. The use of algae as a feed additive resulted in positive effects on growth performance, feed utilization, carcass quality, physiological conditions, stress response, disease resistance, lipid metabolism, protein metabolism, etc.

Accordingly, the studies examined the possibility of using *Porphyra* disposed of as industrial waste as a feed additive. The effects were assessed by growth performance, stress response, and lipolysis activity.

2. MATERIALS AND METHODS

2.1. Fish and rearing condition

Zero-year red sea bream *Pagrus major* were used after a two-week acclimatization period in the Fisheries Laboratory of Hiroshima University. Fingerlings of 2.1 g average body weight were divided into two duplicate groups. The fish were reared for 41 days in one-ton plastic tank that contained 200 fish. The water temperature and salinity ranged between 22.2-26.5°C and 36.0-37.5 ppt, respectively. The fish were satiated four times a day. One-year fish were obtained from a fish farmer. The fish averaged 59.5 g were divided into two duplicate groups and reared for 103 days after a two-week acclimatization in the Fisheries Laboratory. A one-ton tank contained 25 fish. The water temperature and salinity ranged between 18-26°C and 26-34 ppt, respectively. The feeding was carried out twice daily. Following the feeding experiment, the fish were kept in the tanks and starved for 10 days. Body weight loss and reduction of body constituents were monitored.

2.2 Diet preparation

Waste dry *Porphyra yezoensis* obtained from a laver processing company was pulverized. The algae meal was supplemented to an Oregon moist-type diet at a level of 5 % for the zero-year fish and 3 % for the one-year fish, respectively. Diet composition and proximate composition are shown in Tables 1 and 2. The crude protein and crude lipid content of the *Porphyra*-meal were 19.4 % and 1.8 %, respectively. For comparison of liver function and stress response, an *Ascophyllum nodosum* meal and an *Ulva* sp. meal were used as a feed additive in addition to *Porphyra* meal.

2.3. Growth and biometric parameters

Feed utilization and morphological measurements were carried out by the following equations.

Feed efficiency (%) = weight gain / diet given×100
Protein efficiency ratio = weight gain / protein given
Muscle protein deposition (%) = muscle protein gain / total protein given×100
Muscle ratio (%) = muscle weight / body weight×100

Tabe 1
Formulation of diet for red sea bream

Ingredients (g/kg diet)	Zero-year fish		One-year fish	
	Control	Experimental	Control	Experimental
Composed diet[a]	50.0	47.5	47.5	47.5
Sand lance	33.0	31.3	23.7	23.7
Krill	17.0	16.2	23.8	23.8
Porphyra-meal	0	5.0	0	3.0
Cellulose			3.0	0
Vitamin mixture[b]			1.0	1.0
Mineral mixture[c]			1.0	1.0
Total	100	100	100	100

[a]Moist mash diet (Kaneko Sangyo Co. Ltd, Japan) composed of 50% fish meal, 19% corn gluten meal, 16% wheat meal, 10% rice bran and 5% vitamin-mineral mixture
[b]Halver's vitamin mixture; [c]Salt mixture No. 2 (ICN Nutritional Biochemical)

Table 2
Proximate composition of diet (%)

	Zero-year fish		One-year fish	
	Control	Experimental	Control	Experimental
Moisture	42.7	40.2	46.1	46.2
Crude protein	39.3	38.5	30.4	31.9
Lipid	4.0	4.4	4.0	4.1
Ash	6.4	6.6	6.8	7.0

Hepatosomatic index (HIS, %) = liver weight / body weight×100
Intraperitoneal fat body ratio (IPF ratio, %) = IPF weight / body weight×100

2.4. Blood analyses

The blood collected from the caudal artery was analyzed. Hematocrit was measured by centrifugation for 10 min at 11,000 rpm. The serum was obtained by centrifugation at 3000 rpm for 10 min. Serum total protein and total lipid content were measured by Biuret and sulfo-phospho-vanillin methods [3], respectively. Non-esterified fatty acids (NEFA) were measured by enzymatic assay with a NEFA C test kit (Wako Pure Chemical Co. Ltd.) The albumin / globulin (A/G) ratio was estimated by the ammonium sulfate salting-out method (Al-Glo ratio test kit, Kyokuto Seiyaku Kogyo Co. Ltd.). Amino nitrogen was measured according to the method of Goodwin [4].

2.5. Biochemical analyses

Crude protein content was measured by the Kjeldahl method. Lipid was extracted with methanol-chloroform. Lipid class composition was measured by an Iatroscan TH10 (Iatron Co. Ltd.). For determination of cellular growth parameters, dorsal white muscle was frozen in liquid nitrogen. As a parameter of protein synthesis, RNA and DNA were measured according to Munro and Fleck [5]. Protein was determined by the Pholin method. Muscle acid proteinase activity was measured by the method of Makinodan et al.

[6]. The activity was expressed as nanomoles of tyrosine released from the substrate per hour.

2.6. Liver function and vitality tests

Zero-year fish were subjected to the following tests. For comparison of liver function and vitality tests, *Ulva* sp. and *Ascophyllum nodosum* meal-supplemented diets were used. Liver function was examined by the recovery time from anesthetization with 2-phenoxyethanol [7]. Twenty fish were immersed for 30 s in seawater containing 1% of 2-henoxyethanol. The recovery was recorded and was the time required for the fish to start swimming again [8].

Tolerance to hypoxia was assessed by the method of Nakagawa et al. [9]. Ten fish were placed in a 5 L closed container filled with oxygen-saturated water and were kept there for 35 min. The oxygen tension and the number of fish that could not maintain a normal upright orientation were monitored. In the air-dipping tolerance test, 10 fish were exposed to air for 5 min and replaced in oxygen-saturated water. The fish settled on the bottom. The response to air-dipping was defined as the state when the fish swam laterally along the bottom.

2.7. Statistical analyses

The data were analyzed for significant difference using the Duncan's multiple range test. Probabilities of <0.05 were considered significant.

3. RESULTS

Table 3 shows the effects of the *Porphyra*-meal as a feed additive on growth and feed utilization. The total diet given was significantly higher and corresponding to high growth in the *Porphyra*-fed group in 0-year fish. However, although the total diet given was not significantly differentiated in both groups in the 1-year fish, body weight gain and biomass gain were higher in the experimental group. Significantly high feed efficiency and protein efficient ratio were found only in the 1-year fish.

Table 3
Effect of a feed additive on growth and feed utilization in red sea bream

	Zero- year fish		One-year fish	
	Control	Experimental	Control	Experimental
Total diet given (kg)	3.56	3.85*	14.3	14.4
Survival (%)	77.8	87.8	100	98.0
Biomass increase (g)	1595	2281*	3832	4269*
Weight gain (g/fish)	10.9	13.3*	153	175*
Feed efficiency (%)	51.5	62.3	26.9	29.8*
Protein efficiency ratio	1.31	1.60	0.88	0.95*
Muscle protein deposition (%)	9.3	12.5	11.7	13.5
Protein sparing (%)	–	14.9	–	9.6

Mean of duplicate tanks ; *$p<0.05$ to control

Table 4
Effect of a feed additive on biometric parameters in red sea bream

	Zero-year fish		One-year fish	
	Control	Experimental	Control	Experimental
Body weight (g)	13.2±1.0	17.2±5.2*	214±31	233±29*
Body length (cm)	73.0±6.8	78.2±7.1*	18.7±1.1	19.4±1.1
Muscle ratio (%)	35.2±2.8	37.4±4.0*	46.6±1.4	48.8±1.6*
Hepatosomatic index (%)	1.12±0.41	1.51±0.37*	2.10±0.28	2.02±0.32
IPF ratio (%)	0.23±0.26	0.51±0.51*	2.17±0.69	2.11±0.58

*$p<0.05$ to control ; IPF, intrapertoneal fat body

Table 5
Effect of a feed additive on muscle constituents in red sea bream

	Zero year fish		One year fish	
	Control	Experimental	Control	Experimental
Protein (mg/g)	175±6.5	178±7	168±16	174±13
RNA/DNA ratio	3.14±0.38	4.17±0.54*	2.09±0.18	2.12±0.45
Protein/DNA	4.47±0.34	5.03±0.41*	76.5±3.2	81.1±8.6
Acid proteinase#	nd	nd	15.5±4.7	7.3±2.0*

*$p<0.05$ to control (n=6) ; nd, not determined ; #, Activity was expressed as nanomoles of tyrosine released per mg protein per hour.

Table 6
Effect of a feed additive on serum properties in the one year-red sea bream

	Control	Experimental
Hematocrit (n=25; %)	39.2±4.8	39.5±4.9
Total protein (n=10; g/100ml)	3.78±0.80	3.92±0.99
Albumin (n=10; g/ml)	1.66±0.36	1.85±0.85
A/G ratio (n=10)	0.80±0.15	0.91±0.09
Glucose (n=10; mg/100ml)	81.1±27.3	72.9±18.2
Total lipid (n=10; mg/100ml)	668±141	562±231
NEFA (n=10; mEq/l)	0.19±0.05	0.12±0.03*
Amino-N (n=10; mg/100ml)	10.4±4.2	12.0±2.9

*$p<0.05$ to control ; NEFA, Nonesterified fatty acids; Amino-N, Amino-nitrogen

The average body weight and muscle ratio were significantly higher in the experimental groups of both 0-year and 1 year fish, as shown in Table 4. Both HSI and IPF ratio were significantly higher in the experimental group of the 0-year fish, but there were no differences in the 1-year fish. Table 5 shows muscle constituents concerning to somatic growth. While the RNA/DNA ratio as a parameter of protein synthesis was significantly higher in the *Porphyra*-fed group in the 0-year fish, no effect was observed in the 1-year fish. Acid proteinase activity as a parameter of protein degradation was significantly reduced by the *Porphyra*-meal supplementation in the 1-year fish.

Table 6 shows the serum properties. Hematocrit, serum albumin, glucose, total lipid, and amino nitrogen were the same between the two groups. Total serum lipid tended to

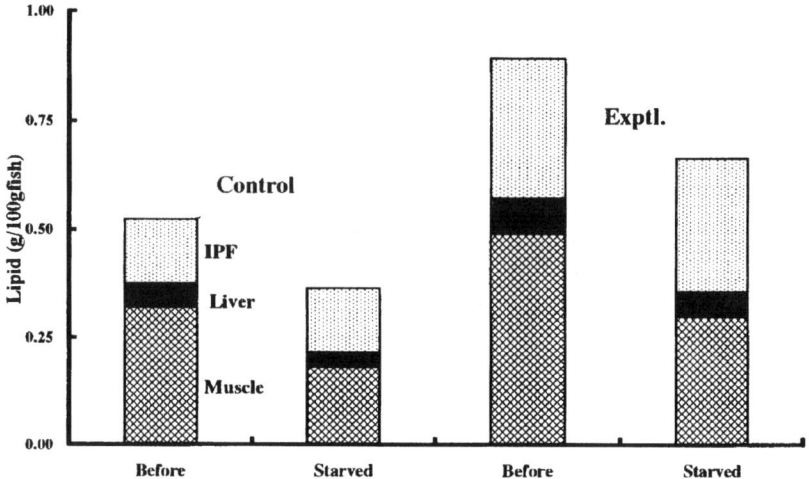

Figure 1. Effects of *Porphyra* supplementation to the diet on lipid reserves in the muscle, liver, and intraperitoneal fat body (IPF) of the 0-year red sea bream before and after starvation.

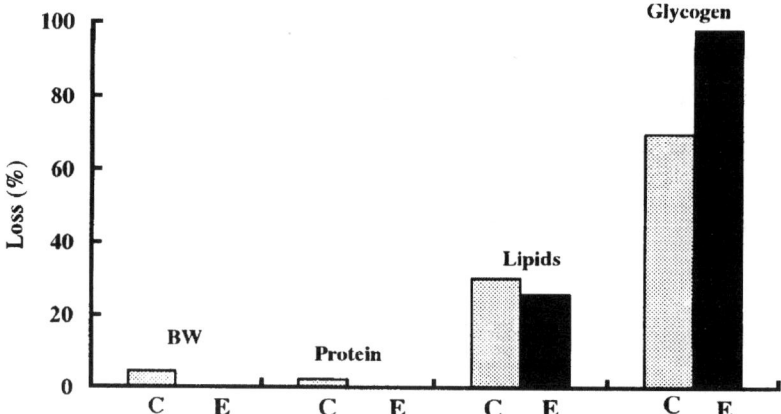

Figure 2. Effects of *Porphyra* supplementation to the diet on loss in body weight (BW), muscle protein (Protein), muscle lipids (Lipids), and liver glycogen (Glycogen) of the 0-year red sea bream by starvation. C, control group; E, experimental group.

decrease by the *Porphyra*-meal supplementation. Nevertheless, feeding with the *Porphyra*-meal reduced NEFA.

Figure 1 shows the total reserved lipids calculated from the sum of the muscle, liver, and IPF lipid in the 0-year fish. The lipid accumulation was higher in the *Porphyra*-supplemented group. Although the phospholipid level was irrespective of the *Porphyra*-meal, triglycerides were markedly increased in the muscle and IPF. Following the feeding experiment, the 0-year fish were starved for 10 days to examine lipolysis activity.

Figure 2 shows the percentage loss of body weight, muscle protein, muscle lipid, and liver glycogen of the 0-year fish. Body weight and muscle protein were lost only in the control group. The starvation depressed the lipid reserves, mainly in the muscle. While consumption of the muscle lipid during starvation was slightly suppressed by the *Porphyra*-meal supplementation, liver glycogen was almost depleted. A tendency toward the preferential use of liver glycogen (98 % decline) was observed in the *Porphyra*-fed group compared to the control group (70% decline).

Improvement of liver function, defined as reduced recovery time from anesthesia, showed a considerable difference between the control and the *Porphyra*-fed groups (Fig. 3). The *Porphyra*- and *Ulva*-fed groups were quicker in recovery time from the anesthesia

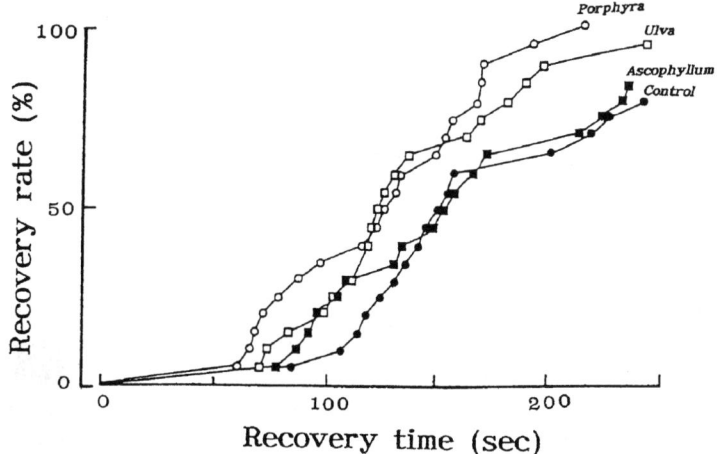

Figure 3. Effects of feed additives on recovery from anesthesia by 2-phenoxyethanol in the 0-year red sea bream.

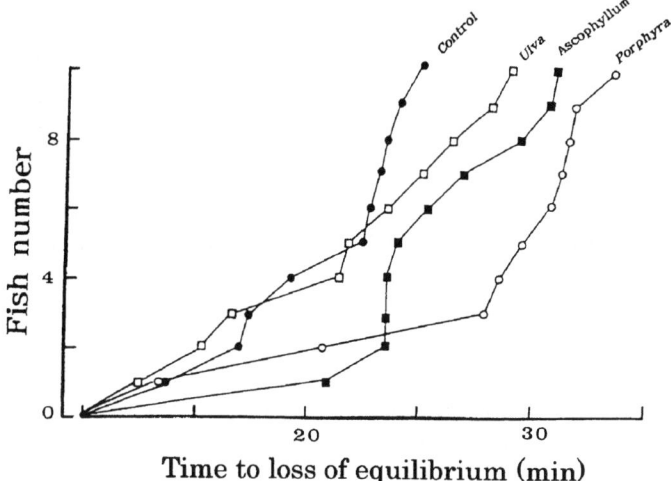

Figure 4. Effects of feed additives on hypoxia tolerance in the 0-year red sea bream.

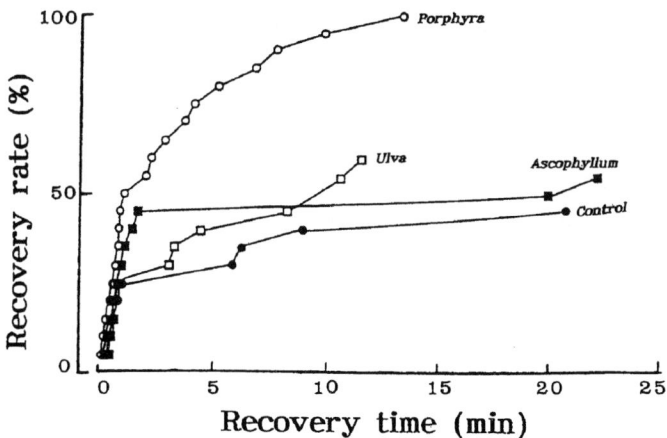

Figure 5. Effects of feed additives on recovery from air-dipping in the 0-year red sea bream.

than the *Ascophyllum*-fed and control groups. The *Porphyra*-fed group completely recovered from the anesthesia, but some fish died without recovery in the other groups. Liver function was markedly improved by the *Porphyra*- and *Ulva*-meal supplementation. When the fish were kept in a water-filled container without a fresh oxygen supply, the fish gradually lost balance with decreasing oxygen tension (Fig. 4). The fish of the control group lost equilibrium earlier than the algae-fed groups. The mortality of the control group was higher than the *Porphyra*-fed group. Figure 5 shows recovery of fish from air-dipping. After air-dipping for 5 min, the fish that were transferred to oxygen-saturated water could not swim soon. The *Porphyra*-fed fish recovered rapidly from succumbed condition. The survival rate after the experiment was 100, 60, 55, and 45 % in the *Porphylra, Ulva, Ascophyllum,* and control groups, respectively. Based on the recovery time from the stressful conditions, the *Porphyra* feeding appeared to reduce the sensitivity.

4. DISCUSSION

The usefulness of algae as a feed additive has been investigated in fish in terms of improvement of lipid metabolism, disease resistance, stress response, protein deposition, etc. [2]. The present study proved the utility of the *Porphyra*-meal as an ingredient of composed feed for fish culture.

Feeding algae as a feed additive improved the growth performance and feed utilization [10,11]. According to the RNA/DNA ratio, protein/DNA ratio, and acid proteinase activity, algae supplementation influenced protein metabolism, cellular growth, and protein deposition. The growth performance in the *Porphyra*-fed fish might reflect these parameters. The phenomenon that dietary algae contribute to the absorption of dietary carbohydrate and protein assimilation of nutrients into body constituents [12] might tightly correlate to the acceleration of growth by the *Porphyra*-feeding. Triglycerides increased in the muscle by the algae supplementation, although phospholipids were not influenced. The *Porphyra* exerted a higher protein deposition in the muscle and accumulated surplus energy as glycogen and triglycerides. Accumulation of lipids by algae-supplementation has been reported in red sea bream and black sea bream [10,13,14]. The *Porphyra*-fed fish consumed efficiently the reserved lipids and suppressed muscle protein consumption during starvation. As the result, body weight loss could be minimized. As enhanced lipolysis was accompanied by a decrease in the NEFA level [15-17], the decrease in serum NEFA, which was lower in the *Porphyra*-fed group, implied active lipolysis.

The short recovery time from 2-phenoxyethanol-anesthesia in the fish fed the *Porphyra* implied an improvement of liver function, as the alcohol is metabolized in the liver [7]. The dietary *Porphyra* also improved the resistance to air-dipping and low oxygen. The resistance against low oxygen was elevated by feeding with *Ulva*-meal in black sea bream *Acanthopagrus schlegeli* [9]. The dietary *Porphyra*-meal as a feed additive was advantageous in fish exposed to critical environmental stressors. The data from the present experiment suggest that *Porphyra* could improve the stress responses and starvation tolerance in red sea bream. The *Porphyra* showed a pronounced effect on the improvement of vital activities similar to *Ascophyllum* and *Ulva*.

As feed additives in fish diets, *Ascophyllum nodosum*, *Ecklonia maxima*, *Laminaria digitata*, and *L. pallida* are widely used. The optimum supplementation level of algae to the diet is 2-5 % [16]. The results support the practical use of waste *Porphyra*-meal as a feed additive for fish diets. Nevertheless, supply and cost problems restrict its use for fish diets. Considering the expense of collection from purple laver companies, the overall price of waste *Porphyra* (ca.4US$) is estimated at about double the cost of actually importing the algae. Therefore, the collection cost of algal waste from output should be discussed. If the *Porphyra* is used for the diet at the breeding stage, the output of waste *Porphyra* is not sufficient to meet the demands of the diet companies. Supplementation of the *Porphyra*-meal to a larval diet could reduce the demand and to an amount equivalent of the discharged waste, and consequently reduce the collection cost. Accordingly, considering the quantity of supply and cost, the waste *Porphyra*-meal would be applicable for a larval diet prepared by a company.

5. ACKNOWLEDGEMENTS

We wish to thank Mr. S. Izumi of the Fisheries Laboratory of Hiroshima University for his cooperation during the experiment.

REFERENCES

1. Meske, C. and Pfeffer, E. (1978) Growth experiments with carp and grass carp. Arch. Hydrobiol. Beih., 11: 98-107.

2. Mustafa, M. G. and Nakagawa, H. (1995) A review: Dietary benefits of algae as an additive in fish feed. Israeli J. Aquaculture - Bamindgeh, 47: 155-162.
3. Frings, C. S. and Dunn, R. T. (1970) A colorimetric method for determination of total serum lipids based on the sulfo-phospho-vanillin reaction. Am. J. Clin. Path., 53: 89-91.
4. Goodwin, J. F. (1968) The colorimetric estimation of plasma amino acid nitrogen with DNPB. Clin. Chem., 14: 1080-1090.
5. Munro, H. N. and Fleck, A. (1966) The determination of nucleic acids. Meth. Biochem. Analysis, 14: 113-178.
6. Makinodan, Y., Akasaka, T., Toyohama, H. and Ikeda, S. (1982) Purification and properties of carp cathepsin D. J. Food Sci., 47: 647-652.
7. Hilton, J. W. and Dixon, D. G. (1982) Effect of increased liver glycogen and liver weight on liver function in rainbow trout, *Salmo gairdneri*, Richardson: recovery from anesthesia and plasma ^{35}S-sulphobromophthalein clearance. J. Fish Dis., 5: 185-195.
8. Nakagawa, H., Umino, T. and Tasaka, Y. (1997) Usefulness of *Ascophyllum* meal as feed additive for red sea bream, *Pagrus major*. Aquaculture, 151: 275-281.
9. Nakagawa, H., Kasahara, S., Sugiyama, T. and Wada, I. (1984) Usefulness of *Ulva*-meal as feed supplementary in cultured black sea bream. Suisanzoshoku., 32: 20-27.
10. Mustafa M. G., Wakamatsu S., Takeda T., Umino T. and Nakagawa H. (1995) Effects of algae meal as feed additive on growth, feed efficiency, and body composition in red sea bream. Fish. Sci., 61: 25-28.
11. Mustafa, M. G, Wakamatsu, S., Takeda, T., Umino, T. and Nakagawa, H. (1995) Effects of algae as a feed additive on growth performance in red sea bream, *Pagrus major*. Proc. Symp. Trace Nutri., 12: 67-72.
12. Yone, Y., Furuichi, M. and Urano, K. (1986) Effects of wakame *Undaria pinnatifida* and *Ascophyllum nodosum* on absorption of dietary nutrients, and blood sugar and plasma free amino-N levels of red sea bream. Bull. Jpn. Soc. Sci. Fish., 52: 1817-1919.
13. Yone, Y., Furuichi, M. and Urano, K. (1986) Effects of dietary wakame *Undaria pinnatifida* and *Ascophyllum nodosum* supplements on growth, feed efficiency, and proximate compositions of liver and muscle of red sea bream. Bull. Jpn. Soc. Sci. Fish., 52: 1465-1468.
14. Nakagawa, H., Kasahara S. and Sugiyama, T. (1987) Effect of *Ulva* meal supplementation on lipid metabolism of black sea bream, *Acanthopagrus schlegeli* (Bleeker). Aquaculture, 62: 109-121.
15. Nematipour, G. R., Nakagawa, H., and Ohya, S. (1990) Effect of *Chlorella*-extract supplementation to diet on *in vitro* lipolysis in ayu. Nippon Suisan Gakkaishi, 56: 777-782.
16. Nakagawa, H., Nematipour, G. R., Yamamoto, M., Sugiyama, T. and Kusaka, K. (1986) Optimum level of *Ulva* meal diet supplement to minimize weight loss during wintering in black sea bream, *Acanthopagrus schlegeli* (Bleeker). Asian Fish. Sci., 6: 139-148.
17. Nakagawa, H. and Kasahara, S. (1986) Effect of *Ulva* meal supplementation on the lipid metabolism of red sea bream. Bull. Jpn. Soc. Sci. Fish., 52: 1887-1893.

DISPOSAL AND RECYCLING OF FISHERIES PLASTIC WASTES:FISHING NET AND EXPANDED POLYSTYRENE

Haruyuki Kanehiro
Department of Marine Science and Technology, Tokyo University of Fisheries, Minato, Tokyo 108-0075, Japan

Abstract

Fishing net wastes generating from fisheries amounts to about 2,000 tons/year. Most fishing net wastes are made with only a single type of plastic and they are suited to recycling. As for disposal of fishing net wastes, "material recycling" and "chemical recycling" are now positioned as a rational and efficient method. Of the wasted netting materials, polyethylene and nylon are mostly conductive to material and chemical recycling. At present, the recycling of fishing net wastes is still limited in quantity, but has been increasing inch by inch in these years. Expanded polystyrene is largely utilized for containers and packaging materials in fishery, agriculture and electrical appliances, and its output amounts to 209,000 tons in 2000. Of this, 183,000 tons are discharged after use and disposed. About 45% of total wastes were disposed by incineration and landfill. The remaining 55% (100,000 tons) are being used to recover resources from wastes. Of the remaining portion, 33% (60,400 tons) is used in material recycling and 23% (39,600 tons) in thermal recycling. Recycled products have frequently been used again as packaging material for household and electrical appliances, and gardening pots, etc. A part of the recycled products is exported to Asian countries as raw materials to manufacturing other products.

1. INTRODUCTION

The economic growth in recent years brought drastic changes of wastes in quantitative and qualitative terms. Particularly, plastics came to hold an increasing share in wastes and its The disposal was highlighted as a big social issue. In this review, present situations of fisheries plastic wastes (such as fishing nets and foamed polystyrene) disposal and recycling in Japan are described.

2. FISHING NETS PRODUCTION AND CONSUMPTION

Figure 1 shows Japan's fish catches and fishing nets output. The output values of nets in Fig.1 were for all of the industrial nets including fishery. It reveals that fish catches had grown along with the development of fishing industry after World War II. The production of fishing nets has been rising year by year in proportion to the fish catches increasing. But, Japan's fishing nets output has been decreasing year by year since 1985.

Table 1 presents the estimated consumption of fishing nets by different fishing methods. Fishing nets consumption in 1998 is estimated to be 7,700 tons. (This is 40% of the largest in 1985). Of this, 2,700 tons are used for stationary nets, 1,780 tons for trawl nets, 1,400 tons for drift/gill nets, and 770 tons for purse seine nets.

Among resins, nylon holds the largest 34%, followed by polyethylene 28% and polyester 18% (Fig.2). These three resins amounted to more than 80% of the total output.

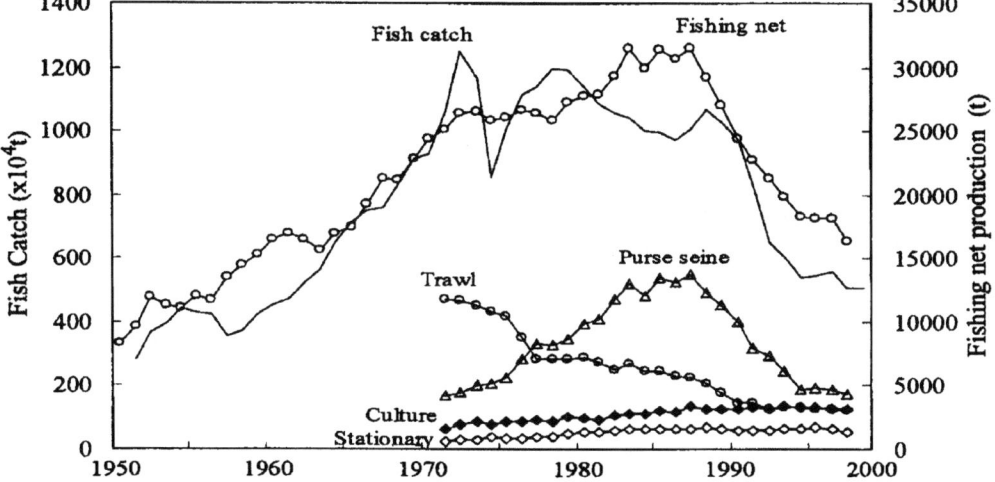

Figure 1. Changes in fish catch and fishing net production.

Table 1
Estimated fishing net generation (ton)

	1994		1996		1997		1998	
Stationary net	2,860	33.70(%)	3,110	38.20(%)	2,780	36.20(%)	2,690	35.20(%)
Trawl net	2,040	24.10	1,850	22.70	1,790	23.30	1,750	22.90
Drift, Gill net	1,850	21.80	1,640	20.10	1,550	20.20	1,400	18.30
Purse seine net	940	11.10	800	9.80	770	10.00	760	9.90
Others	130	1.50	140	1.70	140	1.80	150	2.00
Culture net	660	7.80	600	7.40	640	8.30	900	11.80
Total	8,480	100	8,140	100	7,670	100	7,650	100

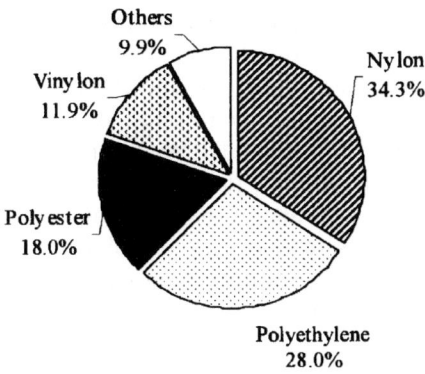

Figure 2. Fishing net production (1998).

3. Wastes, disposal and recycling of used fishing nets

Of the total output, fishing net wastes generating from fisheries is estimated to be about 2,000 tons a year. Present situations of fishing net wastes disposal are shown in Fig.3. Most fishing nets are made from only a single type of resin and they are suited to recycling. Formerly, however, most of fishing net wastes had been disposed by landfill and incineration. Recently, fishing net wastes disposal had been changed to the way of resource recovery due to a shortage of landfill sites and environmental problems. As for resource recovery of fishing net wastes, "material recycling" and "chemical recycling" are now positioned as a rational and efficient method.

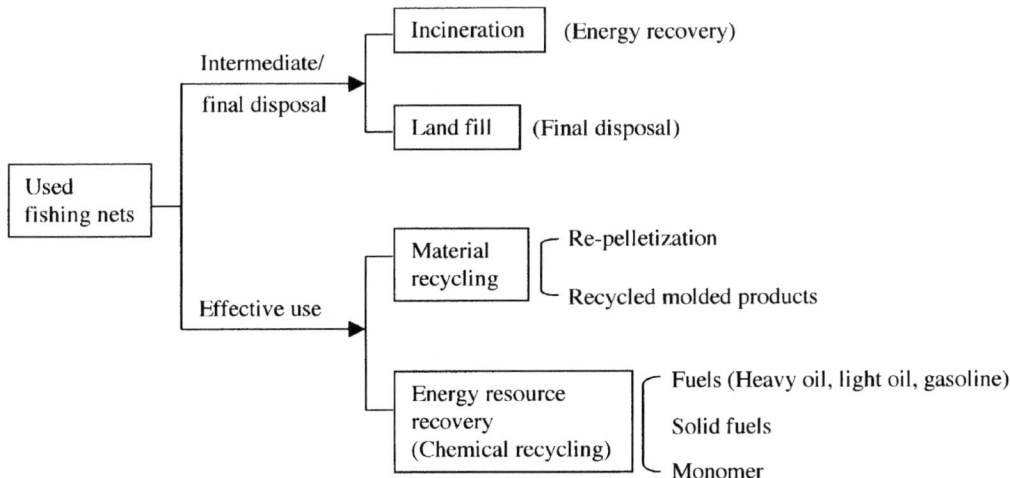

Figure 3. Flow of fishing net wastes disposal and recycling.

Polyethylene and nylon are mostly conductive to material and chemical recycling of the wasted netting materials. Used fishing nets were collected and sold to fiber manufacturer or recycling business operators, and then recycled. Recycling methods of fishing net wastes are divided into re-pelletization and production of recycled molded products. Lately, recycling into fuels also have been gaining attention.

Re-pelletization is the basic method of fishing net wastes recycling. In re-pelletization, good quality waste nets alone must be used. When wasted fishing nets are not homogeneous, consist of various resins, vary in color, contain foreign matters, or are contaminated, they can not be recycled into pellets. Wasted fishing nets of such grade are reused in the form of recycled molded products. Recycled molded products should be poor in quality and performance, thus losing their commercial value.

Wasted fishing nets subject to re-pelletization largely come from the fishing net manufacturing. And some from used fishing nets are also included to some extent. The end result of material recycling is a variety of new plastic products. To give some examples of material recycling, nylon and polyethylene fishing nets give daily necessities, electrical appliances, and materials for the agricultural and fishing industries.

There are two ways in chemical recycling of wasted fishing nets. One is the material recovery providing raw chemicals (monomer) to manufacturing new fiber materials (fishing net), again. For example, thermal decomposition of nylon fishing nets to produce effectively the raw chemicals, that is nylon-6 monomer (caprolactam).

The other is the energy resource recovery to recover energy from wastes in such forms as heat, electricity or oil. Some example is thermal decomposition, that is "liquefaction", of polyethylene fishing nets producing fuels or oils (heavy oil, light oil and gasoline).

4. Energy consumption and recycling cost of fishing net wastes

Energy consumption and cost performance of wasted nets by recycling methods were evaluated by LCA (Life Cycle Assessment) analysis. An example of manufacturing and recycling flow of synthetic fiber products (clothes and fishing nets) is shown in Fig.4. In the figure, values of energy consumption and cost in each process estimated by LCA analysis are indicated [1]. Comparison of energy consumption between virgin polymer

Table 2
Comparison of energy consumption by recycling methods

Products	Material	Recycling method	Energy to manufacturing virgin ptoduct	Recycling energy (10^6Kcal/t)	Recycling merit
Clothes[a]	NY-6	Chemical[b]	14.0	9.2	○
		Material[c]	16.0	2.2	◎
	PET	Chemical[b]	4.3	4.4	×
		Material[c]	6.3	2.2	△
Fishing net	NY-6	Chemical[b]	14.0	9.5	○
		Material*[c]	16.0	2.2	◎

[a]Easy-to-recycle products; [b]Recycling to monomer; [c]Re-pelletization JCFA (1997) [1]

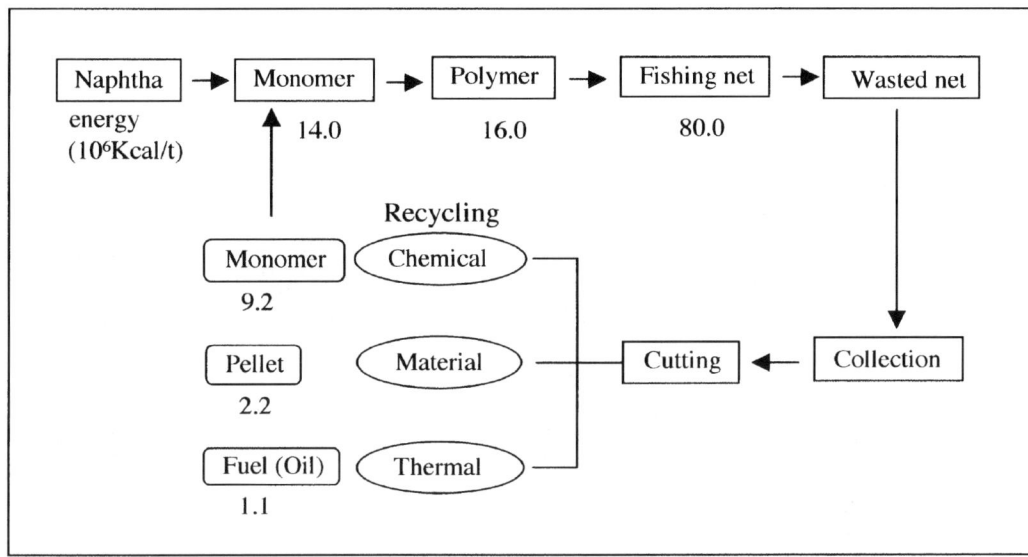

Figure 4. Recycle energy and cost in recycling processes of wasted nylon net.
JCFA (1997) [1].

and recycled polymer by different recycling methods (re-pelletization as "material recycling" and recycling to monomer as "chemical recycling") is shown in Table 2. Energy of manufacturing monomer and virgin polymer (pellet) from raw material naphtha for nylon-6 and polyester products are estimated to 14.0 and 16.0×10^{6}kcal/ton, and 4.3 and 6.3×10^{6}kcal/ton, respectively. Recycling energy from wasted polymer, on the other

Table 3
Value of recycled polymer

Material	Commercial value*
PET	(Virgin polymer cost)×0.3
Nylon-6	(Virgin polymer cost)×0.4

*Recycled polymer (pellet) is poor in quality and performance. JCFA (1997) [1].

Table 4
Recycling cost balance by recycling methods

Recycling method	Recycling cost balance[a] (yen/kg)	
	Ny-6	PET
Chemical[b]	30	-65
Material[c]	-25	-95

[a](Recycling cost balance)=(value of recycled polymer)− (recycling cost) [b]Recycling to monomer ; [c]Re-pelletization JCFA (1997) [1]

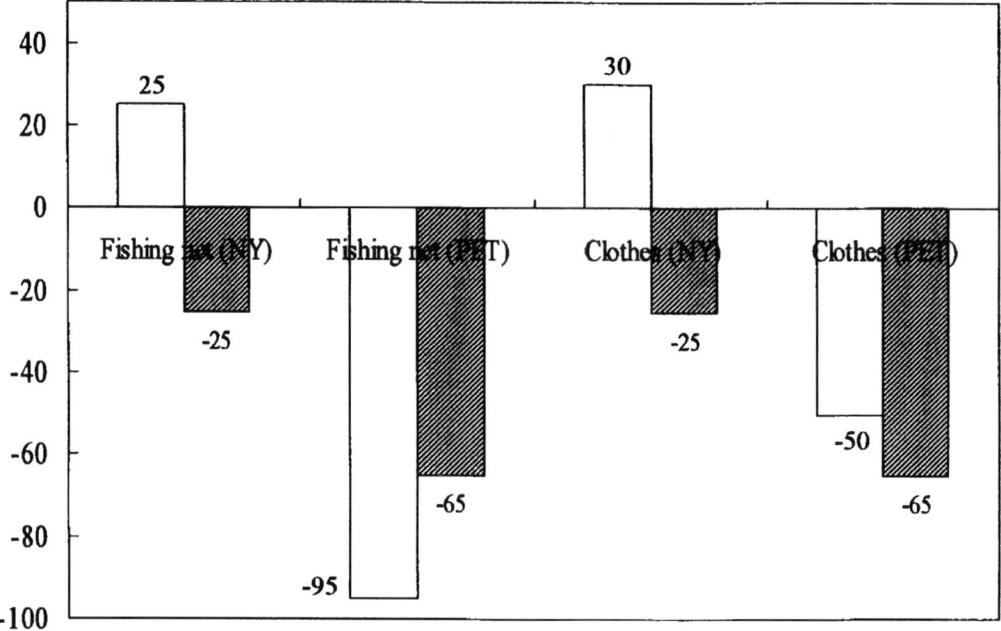

Figure 5. Recycling cost performance by recycling methods. JCFA (1997) [1]. □, Material; ▨, Chemical; NY, Nylon; PET, Polyester.

hand, are estimated to 9.2-9.5 and 2.2×10^6 kcal/ton, respectively for monomer recycling and re-pelletization. For nylon-6 materials, energy consumption of recycling from wasted polymer is lower than that of virgin polymer, thus showing advantage of recycling. However, for polyester, it reveals that recycling is not always advantageous.

With comparison between recycling methods (material and chemical), the former is superior to the latter. However, it must pay attention that recycled polymer (pellet) should be poor in quality, thus losing their commercial value. In Table 3, the commercial values of recycled polymer (pellet) are indicated. It reveals that commercial value of recycled polymers become 60 to 70 % lower than that of virgin polymer.

Table 4 shows total cost balance estimated from value of recycled polymer and recycling cost by recycling methods. Where, total recycling cost is represented by the following equation.

(Total recycling cost balance) = (Value of recycled polymer) - (recycling cost)

Figure 5 shows comparison of total cost performance in chemical and material recycling. As for polymer material, total cost performance is estimated as "+" for nylon-6 in both netting products of clothes and fishing nets, but "-" for polyester. However, in comparison between recycling methods, only chemical recycling is "+", thus showing higher cost performance as recycling method.

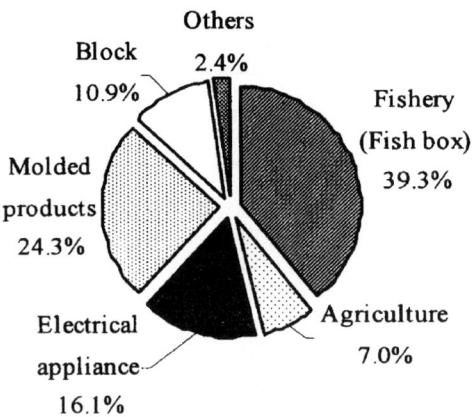

Figure 6. Production of foamed polystyrene. 209,000 tons (2000).

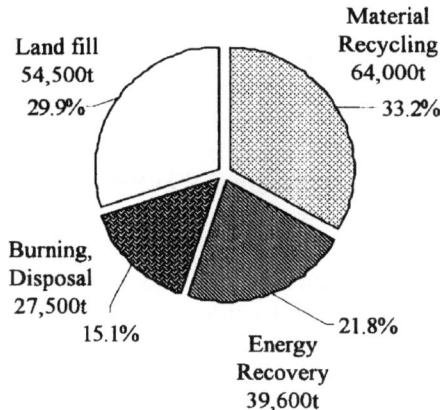

Figure 7. Recycling and treatment of used EPS (Expanded Polystyrene). 182,000 tons (1999).

5. Disposal and recycling of foamed polystyrene

Expanded polystyrene (EPS) is largely utilized for containers and packaging materials in fishery, agriculture and electrical appliances, and its output amounts to 209,000 tons in 2000 (Fig.6). The share of fishery held the largest portion, amounting to 82,000 tons (39.2%). Of total output of EPS, 183,00 tons is discharged after use and disposed.

Waste disposal methods of EPS popularly in practice are incineration, landfill and resource recovery (material and energy). Figure 7 shows the disposal/treatment of used EPS. About 30% and 15% of total wastes were disposed by incineration and landfill,

Table 5
Recycling rate of polystyrene form

	Production (ton)	Amounts circulated domestically (ton)	Recycled (ton)	Recycling rate (%)
1991	237,000	171,000	21,500	12,6
1992	232,000	170,000	29,600	17.4
1993	225,000	165,000	33,200	20.1
1994	223,000	158,000	38,200	24.2
1995	224,000	179,000	48,900	27.3
1996	225,000	180,000	51,700	28.7
1997	226,000	182,000	54,900	30.2
1998	213,000	182,000	56,800	31.2
1999	212,000	182,000	60,400	33.2

respectively. The remaining 55% (100,000 tons) are being used to recover resources from wastes.

There are two ways to recover resources from wasted EPS. One is the material recovery, that is the recycling of EPS into other products (plastics into plastics). The other is the energy recovery from wastes in such forms as heat, electricity or oil. Of the resource recovery 55%, 33% (60,400 tons) is used in material recycling and remaining 23% (39,600 tons) in energy recovery (thermal recycling).

Table 5 shows the changes of waste amounts and material recycling rate of EPS. It reveals that recycling rate has been rising year by year (12.6% in 1991 to 33.2% in 1999).

Recycled products have frequently been used again as packaging material for household and electrical appliances, and gardening pots, etc. A part of the recycled EPS (in such forms as pellet and ingot) is exported to Asian countries as raw material to manufacturing other products.

6. Problems and key points in waste disposal

At present, recycling of fisheries plastic products, especially in fishing nets, is still limited in quantity. In the near future, more progressive promotion in waste reduction and recycling is expected.

In order to solve the waste problems, broad cooperation among industry, consumers and administration is required. There are many important factors to solve waste disposal problems and promoting recycling. These are summarized as follows:
① Waste reduction
- To develop plastic products with longer service life by manufacturers and an effort to decrease wastes

② Promotion of resource recovery (material and chemical recycling)
- Change of materials to usable resins for easier recycling

In case of fishing net materials, nylon and polyethylene are suited for recycling than other materials.
- More efforts to promote the use of recycled resources as raw materials
- To secure more efficient collection and transport system of wastes
  Collection of high quality waste
  Stable ensuring of wastes quantity from the whole country
- Development of more efficient resource recovery methods
  Effective resource recovery can cause changing wastes in quality, thus producing influences on disposal methods
- Efforts to the cost-down of recycling
  Technological development to improve the recycling efficiency
- Support by administration for the establishment of recycling business

③ Others
- Development and use of new plastic materials
  "Biodegradable plastics", which are decomposed by bacteria present in natural environment (soils and sea water)

REFERENCE

1. Japan Chemical Fibers Association (JCFA), Tokyo, Japan. (1997) A report of the examination on recycling of synthetic fiber.

SHELL NURSERIES: ARTIFICIAL REEFS USING WASTE SHELLS

Keiichi Katayama[a], Minoru Tahara[a], Ken Tsumura[b] and Hiroshi Kakimoto[c]
[a]Kaiyo Kensetsu Co., Kojima Ekimae, Kurashiki, Okayama 711-0921, Japan
[b]Hokkaido Tokai University, Minamisawa, Sapporo 005-8601, Japan
[c]National Coastal Fisheries Development Association, Kamakuragashi Bld., Uchikanda, Tokyo 101-0047, Japan

Abstract

In response to the worldwide concern of sustaining and increasing marine life, a novel structure, called a shell nursery, was made of waste shells to develop artificial reefs and was tested for its effect on the growth of fish, polychaete, amphipod, decapod and seaweed.

The result of this test showed that the shell nurseries had more seaweed, creeping polychaete, amphipod and decapod, all of which are good diets for small fish, than the concrete test pieces used for comparison. There was also observed an increase in the number of fish gathering around the reef area. These findings suggest that the shell nurseries produced from waste shells are effective to meet the goal of creating places where a wide variety of marine organisms can multiply.

1. INTRODUCTION

Maintenance and increase of marine bioresources are crucial for securing food for mankind in the future. The results of various experiments in aquatic tanks as well as investigations of artificial reefs in coastal waters show that artificial reefs are used as "feeding ground," "hiding ground," and for some organisms "spawning ground" [1-3].

Another important consideration is that shells which are the byproduct of various shellfish industries are a troublesome waste because there is no known method of disposal.

We have, therefore, developed artificial fish reefs with the maximum possible functions, using these waste shells [4]. These artificial reefs provide food, hiding ground and spawning ground for fish of different stages, from juvenile fish, which are badly depleted, to mature fish. We call these structures "shell nurseries."

Here, we will describe how the structures using shell nursery tubes attract fish and other marine organisms and are utilized as a spawning ground. This will prove that the shell nurseries are effective in attracting and increasing a wide variety of marine life.

2. MATERIALS AND METHODS

2.1. Materials

The shell nursery structures which we have developed are made of 2.0 × 2.5cm mesh polyethylene net tubes (15cm in diameter, 100 cm in length) which are filled with shells (Fig. 1) and assembled into panels. The panels are attached to a steel frame or a concrete frame to form a cube-like structure.

Smaller test pieces were used for the test of the effectiveness of shell nurseries for cultivating food sources. These test pieces were 2.0 × 2.5cm mesh tubes, 15cm in diameter and 30cm in length, and they were packed with one of the following: oyster shells, crushed stones, pearl oyster shells, or scallop shells. Concrete test pieces of the same dimensions were also used for comparison.

A complete artificial reef (Shell Nurse 6.0 type) was also tested. It had a 6.5m high steel frame built on a concrete base measuring 7.8 × 7.8m with shell nursery panels attached to the frame (Fig. 2).

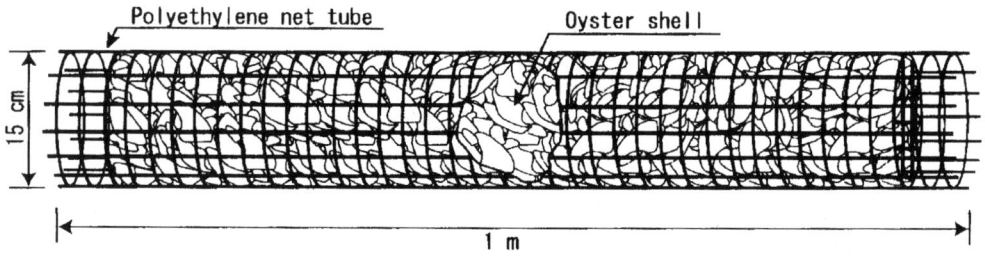

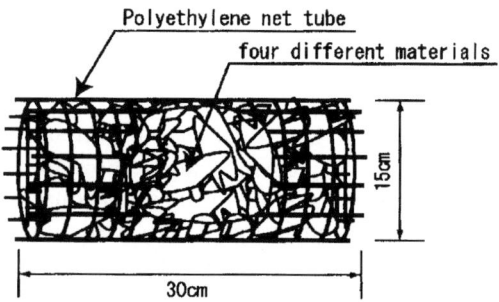

Figure 1. Structure of a shell nursery tube (above). Polyethylene net tube (2.0 × 2.5cm mesh, 15cm in diameter, 1m in length) and a test piece (below) same dimension except 30cm in length.

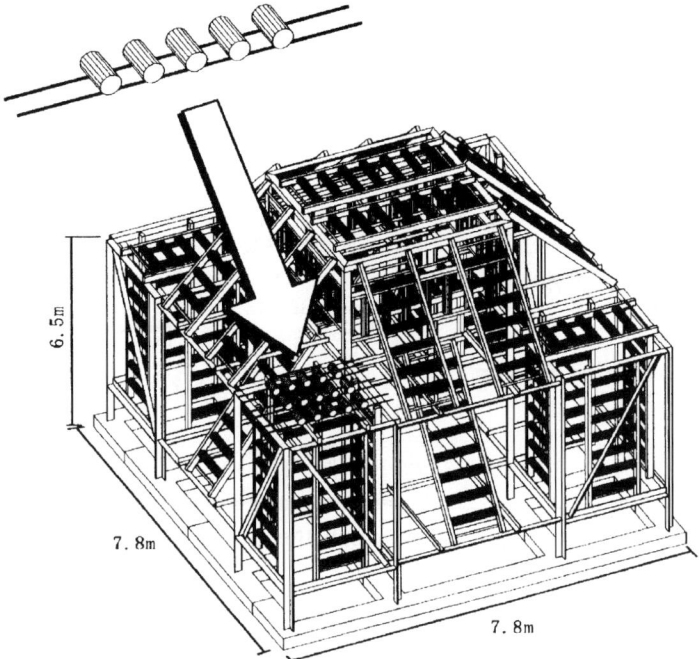

Figure 2. Shell Nurse 6.0 type and test pieces.

2.2. Methods

2.2.1. Location of artificial reefs and recovery of test pieces

The test pieces were attached to artificial reefs placed at three different locations under water at depths of 19m or less in the Seto Inland Sea in Kurashiki City, Okayama Prefecture, in February, 1996. They were recovered regularly over two years by scuba divers. Upon recovery, all the attached organisms were removed, and their wet weights were measured after species identification was carried out (Fig. 3).

2.2.2. Calculation of percentage of empty space

Much of the artificial reef's ability to attract a variety of species and a large amount of small organisms which easily become food for larger organisms depends on the volume of empty space formed in the reef material. The percentage of empty space, which we refer to as vacuity, was calculated based on the bulk volume of the shell nursery test pieces and the volume of water displaced when submerged in an aquatic tank.

2.2.3. Survey of the artificial reefs

An entire test reef (Shell Nurse 6.0 type) was also submerged in the same area at a depth of 19m in February, 1996. For five years after that, the species, the length and the number of fish around the artificial reef were checked visually by divers. From the

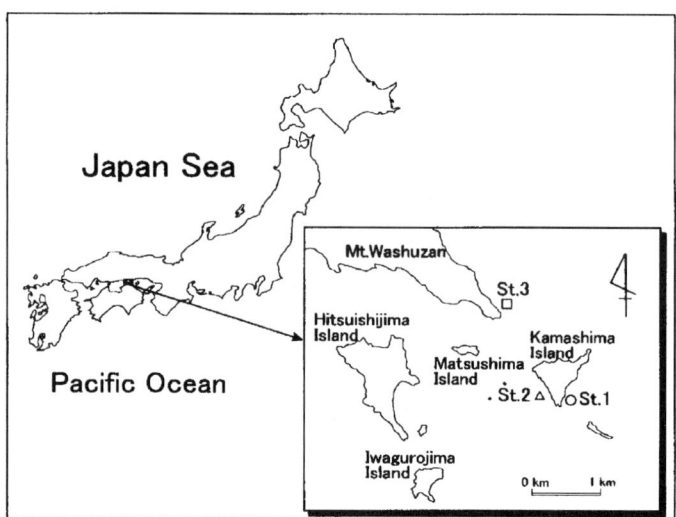

Figure 3. A map of the general area in which the experimental reefs were located. ○, Station 1; △, Station 2; □, Station 3.

measured lengths, biomass was obtained through the formula for relation between body length, total length and body weight [5]. Other diving checks were also performed for the fish around other types of shell nurseries submerged off the coasts of Aichi Prefecture and Osaka Prefecture.

3. RESULTS AND DISCUSSION

3.1. Increase in the number and weight of species attached to the test pieces

The number of various life forms that became attached to the test pieces or sheltered in the small openings of the test pieces increased over an 8-month period and then stabilized. In most cases, a greater number of life forms gathered around the test pieces made of oyster shells (Fig. 4). Polychaete, amphipod, and decapod were especially prevalent on the test pieces made of oyster shells, and much less prevalent on the test pieces made of concrete. Depending on the season, the formation of sargassum beds on the test pieces could be observed. In some seasons, many sargassum beds of different kinds were formed, primarily made up of sea trumpet (*Ecklonia cava*). These seaweeds were seen more on the shell nursery pieces than on the concrete cylinders. This is probably due to the entanglement of the seaweed roots in the polyethylene resin net, making the roots difficult to dislodge.

A comparison of the numbers of life forms attached to the different test pieces or moving within their small openings showed that, in most cases, the test pieces made of oyster shells had the largest number and variety, while the concrete pieces had the smallest. The level of the test pieces made of crushed stones tended to fall between the

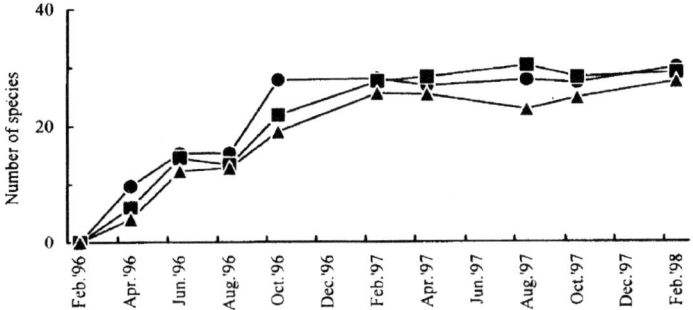

Figure 4. Average monthly change in the number of species attached to or dwelling in three different test pieces for all three test stations. ●, Polyethylene net tube containing oyster shells; ■, Polyethylene net tube containing crushed stones; ▲, Concrete cylinder.

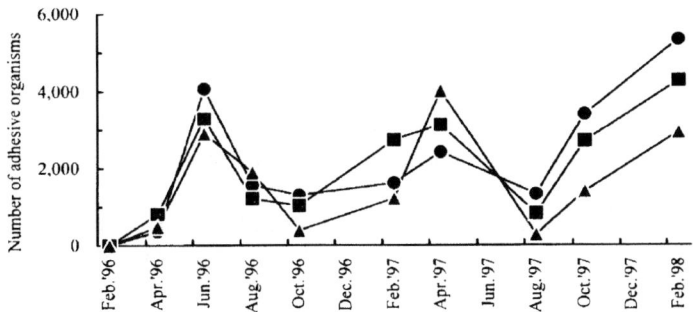

Figure 5. Average monthly change in the number of organisms attached to or dwelling in three different test pieces for all three stations. ●, Polyethylene net tube containing oyster shells; ■, Polyethylene net tube containing crushed stones; ▲, Concrete cylinder.

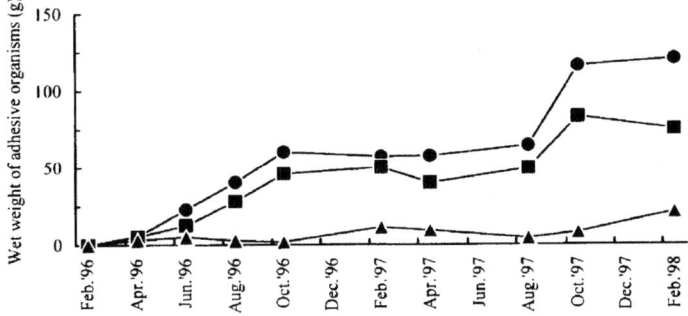

Figure 6. Average monthly change in wet weight of organisms attached to or dwelling in three different test pieces for all three stations. ●, Polyethylene net tube containing oyster shells; ■, Polyethylene net tube containing crushed stones; ▲, Concrete cylinder.

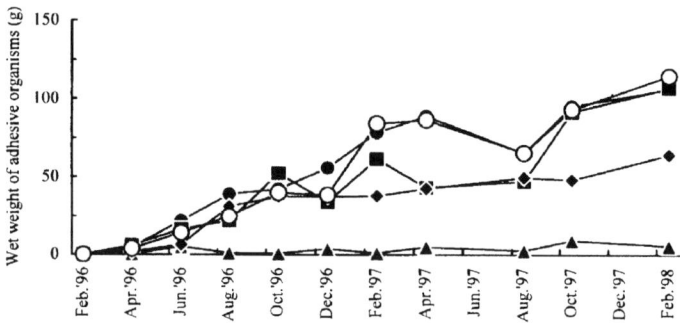

Figure 7. Monthly change in wet weight of adhesive organisms attached to or dwelling in five different test pieces at a single test station. ●, Polyethylene net tube containing oyster shells; ○, Polyethylene net tube containing pearl oyster shells; ◆, Polyethylene net tube containing scallop shells; ■, Polyethylene net tube containing crushed stones; ▲, Concrete cylinder.

oyster shells and the concrete pieces. This difference was especially large after two years (Fig. 5).

A comparison of the total wet weight of all organisms attached to and occupying the small open spaces of the test pieces clearly showed that weights were greatest for the test pieces made of oyster shells, and the least for the concrete test pieces. The test pieces made of crushed stones had a middle value. These differences were especially large after two years (Fig. 6).

At one of the three test stations, test pieces made of pearl oyster shells and scallop shells were tested. The results showed that pearl oyster shells, which are similar to oyster shells in shape and size, displayed about the same level of total wet weight as the oyster shell test pieces. Concrete test pieces had consistently low wet weight levels, while test pieces made of scallop shells showed, in most cases, medium levels. Occasionally, high levels of total wet weight were observed on crushed stones, though mostly medium levels were observed (Fig. 7).

3.2. Vacuity rates of test pieces and their relation to wet weight of food organisms

The vacuity rates (1-(displaced water volume/bulk volume) × 100) calculated for the test pieces were: 88% for scallop shells, 85% for pearl oyster shells, 82% for oyster shells, 48% for crushed stones and 0% for concrete cylinders of the same size (Fig. 8).

Creeping polychaete, amphipod, and decapod easily become food for fish and are thus called preferable food organisms. They tend to live in narrow spaces [4], so one goal of the test was to show a relationship between the vacuity of test pieces and average wet weight of these organisms collected regularly over a two-year period (Fig. 9). The result shows that the test pieces of oyster shells and pearl oyster shells with 82-85% vacuity were the best, which supports the argument that these percentages are the most effective for the habitat of such food organisms.

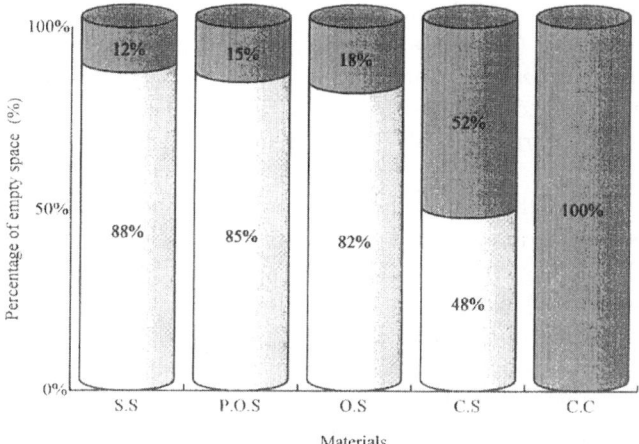

Figure 8. Vacuity rates in five different test pieces. White column: percentage of empty space. Dark column: percentage of solid matter. S.S, Scallop shells; P.O.S, Pearl oyster shells; O.S, Oyster shells; C.S, Crushed stones; C.C, Concrete cylinder.

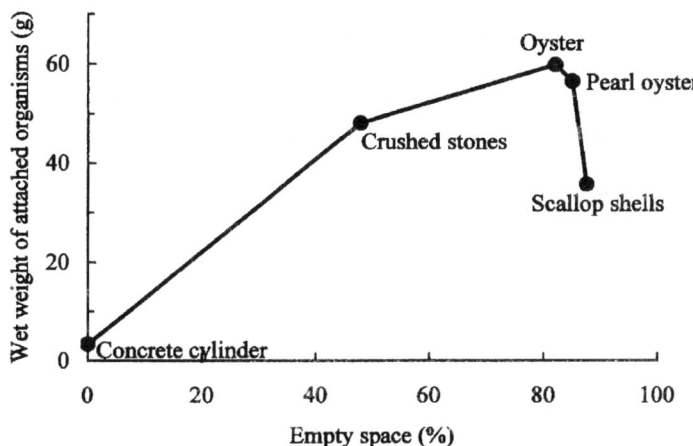

Figure 9. Relation between the percentage of empty space and wet weight of organisms (polychaeta, amphipod, decapod) attached to or dwelling in five different test pieces.

3.3. Utilization of complete reef by various species

It was confirmed that juvenile fish such as hinge-beak shrimp (*Rhynchocimetes uritai*), juvenile octopus (*Octopus vulgaris*) and juvenile sea cucumber (*Apostichopus japonicus*), young scorpion fish (*Sebastiscus marmoratus*), gobies (*Gobiidae*), and red-spotted grouper (*Epinephelus akaara*) sought shelter in small open spaces made of oyster shells at artificial reefs. Bottom perch (*Apogon semilineatus*), spotted cardinal fish (*Apogon notatus*), and mature scorpion fish and rockfish (*Sebastes inermis*) lived in great numbers around the complete reef (Shell Nurse 6.0 type) and its outlying areas. Bamboo leaf wrasse (*Pseudolabrus japonicus wrasse*), filefishes (*Monacanthidae*) and such came to feed on organisms that had attached to the reef. Thus the shell nursery reef created a habitat allowing to the formation of a food chain like that reported by Kakimoto and Ohkubo [1].

The result of counting the fish at the complete reef (Shell Nurse 6.0 type) by diving observation shows that the amount of edible fish gathering around the reef increased over 5 years (Fig. 10). During the observations, wrasses (*Labridae*) and black scraper (*Thamnaconus modestus*) were seen eating the attached organisms, while rockfish (*Sebastes inermis*) were constantly catching plankton. This shows that due to the proliferation of attached food organisms, the number of fish increased and the duration of fish staying in the area was extended.

It was also confirmed that at the reefs in the coastal area of Niigata Prefecture Spear squid (*Loligo bleekeri*) took advantage of artificial reefs to spawn their eggs. Black rockfish (*Sebastes schelegeli*) were observed gathering at an artificial reef to spawn their young [1]. Greenlings (*Hexagrammos otakii*) and broad-mantle squid (*Sepioteuthis lessoniana*) were observed using the reef as a spawning ground at the shell nurseries submerged at 12m under water in Aichi Prefecture and at 10m in the coastal area of Osaka Prefecture (Fig. 11).

The most obvious explanation why shell nurseries showed such effective results as nurseries was that many adequately narrow and complex spaces were formed among the sea shells and provided feeding grounds and hiding grounds for small organisms which need protection, as well as for adhesive life forms (especially small crustaceans) eaten by small organisms.

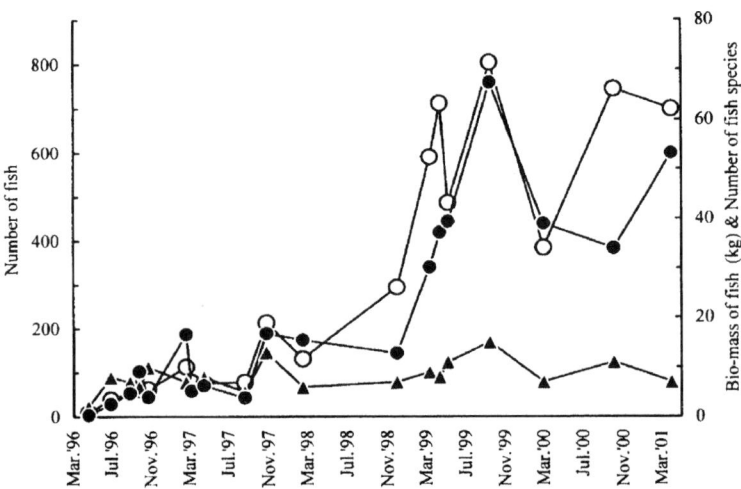

Figure 10. Monthly change in the number of fish (open circle), biomass of fish (closed circle) and number of fish species (triangle).

Figure 11. Eggs of greenling (*Hexagrammos otakii*) on the shell nursery.

Based on the findings stated above, we can confirm the following:
1. Shell nurseries form sargassum beds (or seaweed beds) which are grounds of reproduction and areas of living for juvenile fish.
2. Shell nurseries themselves are effective habitats for small organisms and juvenile fish.
3. Shell nurseries become spawning grounds for organisms laying eggs on reefs.
4. Shell nurseries are very effective for attracting mature fish for the fishing industry.

In addition, shell nurseries can make a great contribution to human society by making good use of waste seashells for artificial reefs.

REFERENCES

1. Kakimoto, H. and Ohkubo, H. (1985) General study on the artificial reefs of Niigata prefecture and realizations. Bull. Niigata Pref. Fish. Exp. Sta., 1-327.
2. Kakimoto, H., Ohgai, M., Tsumura, K. and Noda, M. (1995). The effect of artificial reef and seaweed bed on survival of prey fish. '95 International Conference on Ecological System Enhancement Technology for Aquatic Environments, Tokyo, Proceedings 1, 227-232.
3. Kakimoto, H. (1998) Studies on the biological function of artificial reefs. Fisheries Engineering, 35: 1-7.
4. Kaiyokensetsu Co. (2000) The Studies on the Shell Nursery.
5. Fukuda, T. (1995) Allometory of some fishes –Relation between body length, total length, and body weight –Bull. Okayama Pref. Fish. Exp. Sta., 2: 167-170.

SEAFOOD QUALITY AND THE RELATED ASPECTS

QUALITY ISSUES AND MORE EFFICIENT UTILIZATION OF THE FISH CATCH

H. Allan Bremner

Allan Bremner & Associates, 21 Carrock Court, Mount Coolum, Queensland 4573, Australia

Abstract

The desire for quality is a motivating force for achieving more efficient utilization of the available harvest. The adoption of methods and techniques that improve the quality status of a higher proportion of the catch ensures better returns. This leads, in turn, to greater recognition of the advantages that can be obtained by using a quality approach and by preventing deterioration and consequent down-grading of the product.

To ensure efficient utilization it is necessary to consider the chain of events from raw material to finished product as a whole and to ensure that all the steps benefit from any improvements that are made. A chain can be considered as a 'virtual' company even though it is comprised of independent business units. To achieve efficiency may require re-engineering of the chain. It definitely requires seamless flow of relevant information on traceability, and on product and process that can be incorporated into the business operations of each unit. Parts of this information can also be used for customer and consumer relations and for market enhancement and advertising.

Techniques for evaluating quality attributes must be used that are appropriate to the task and that can be performed quickly. The systems should provide results on current product status and also predict remaining shelf life under standard conditions. The nature and properties of the materials must be taken into account and all the capture, handling, transporting, processing, packaging and distribution systems must be designed to suit the product. This is particularly critical if the product is live, as is the case with some crustacea and high-priced fish.

The adoption of proper conditions means that off-cuts and other materials such as organs are in good condition for manufacture of a whole range of by- products.

1. INTRODUCTION

Issues of quality cannot, and must not, be divorced from the endeavours of making more efficient utilization of the available catch. The desire to produce quality is a potent motivating force for achieving greater efficiencies and thus for obtaining the maximum price possible for the products in a competitive market. To achieve this end it is well recognised that a company, a group of companies or an industry sector must adopt the appropriate approach at the managerial level and provide the leadership and stated commitment to quality issues. This paper will not be limited to the product alone but will include discussion of management and the need for a greater consciousness of more holistic approaches using a 'whole of chain' strategy. Management in this context does not mean management of the fish stocks, although that is a factor in the overall picture, but means the philosophy, tools, practices and techniques used to run the industry that converts the raw material into food for human consumption and for valuable by-products.

2. BUSINESS TOOLS AND QUALITY MANAGEMENT

Total Quality Management (TQM) is one of the leading business tools and it has gradually evolved and developed over the years [1]. It embodies the principle of continuous improvement and so is in complete accord with the idea of obtaining higher quality products and higher yields of these high quality products from available raw materials. TQM is consistent with other systems and requirements such as Hazard Analysis Critical Control Point (HACCP) and with other registered and audited Own Check Systems (EU regulation 91/493). A related tool is that of Business Process Re-engineering (BPR) in which business processes are examined and analysed, then re-engineered to be more effective and efficient. The focus of BPR on improving processes means that it is not a substitute for TQM but that the integration of these two approaches complement one another [2]. It is also recognised that TQM has a somewhat inward focus rather than on external customers and that a major influence on business excellence will be the power in processing provided by Information Technology (IT) [2].

A management strategy known as Six Sigma provides a new way of evaluating and improving businesses to ensure their future and it has achieved considerable success with USA based multinational corporations [3]. By evaluating the true costs of poor quality in products and processes this approach refutes the old belief that continued efforts to improve quality are controlled by the 'Law of diminishing returns'. On the contrary the approach shows that as resources are employed to upgrade quality, then the costs actually diminish as sales increase and efficiency improves. Partly this is achieved by identifying the 'hidden' factory within the factory that deals with re-work of product and process and with true evaluation of yield and cumulative yields at each process step rather than basing calculations on the ratio of 'raw material in' to 'finished goods out'. With manufactured goods it has been established that faults found online are strongly indicative of failure of the product in use and with consequent return of product and loss of brand image[3]. More than good product technology is required though, good processes are needed to make that product. All the operations and systems that go into the final product including concept design, material sourcing, marketing, inventory, and invoicing must be considered and planned, not just the operations of the machines used in manufacture. Producibility, that is the ability to make the product efficiently without failure, is a cornerstone for success since competing corporations may readily have the ability to reproduce an innovation more cheaply. The customer perceptions of quality must be established and expressed in a manner that allows identification of those features that are 'critical to quality'. This must be done in such a way that factors that are critical to quality can be listed in priority so that the appropriate steps in service or manufacturing can be taken to improve them in a manner that is measurable. The quality aspects must be incorporated into the product at the design stages so that materials, methods and equipment chosen are appropriate for the task and provide error free operation, thus eliminating re-work and downtime and resulting in a product with the desired characteristics. At some stage this will require total re-design of the operation, but that will be necessary to stay in business.

3. BUSINESS TOOLS AND THE INDUSTRY

The situation is not so clear in the fish products industry as to what business tools, if any, are being used to ensure quality. It can be assumed many of the larger manufacturing companies, that make fish and other food products for sale on the world market, have a quality policy and employ forms of business tools to improve their operations. These matters are rarely reported in the technical literature. In fisheries magazines, the news items deal more with ISO registrations or adoption of HACCP schemes by various companies but these are merely the first steps in developing a quality based operation. A large proportion of the catching, handling, auction, processing and distribution stages in most countries are in the small to medium enterprise (SME) level. Many now use computers and the Internet and there are a range of marketing sites that have sprung up in the last 2-3 years. Quality assurance is well known and there are specific texts for the seafood industry [4].

The Norwegian industry undertook a Quality Wave initiative to familiarize the industry with TQM and to encourage them to adopt it. As a consequence it was taken up at many levels and even vessels developed their TQM systems for operation [5]. This resulted in improvements in handling procedures and in product yields [6]. A survey of the Danish industry indicated that only about 22% of the processors had registered HACCP schemes, but that at least 95% had Own Check systems that were registered with the local health authorities [7].

The fishing industry and export of seafood products earns Iceland 75% of its total export revenues and 55% of all its foreign currency and employs about 13,000 of its 275,000 population. Increases in efficiency and development of new methods of handling and processing have been responsible for continuing increases in productivity as shown by the decrease in the workforce, handling the same tonnages, in the processing industry from 9000 in 1995 down to 6100 in 1999 (Einarsson H. personal comm. 2001). Higher yields of higher priced products have been obtained. Similarly, in New Zealand more careful catch handling, attention to temperature control and better logistics have meant that a greater proportion of the hoki catch can be processed into prime products for the export market, thus bringing a much greater return. Since the handling procedures are improved the standard of the remaining fillet flesh after removal of the loin is also high and excellent products are made from it. In addition, the fishery has been approved by the Marine Stewardship Council thus furnishing external evidence that sustainable fishing methods are in use and providing a marketing edge [8].

4. WHOLE OF CHAIN

It is becoming well recognised that approaches that look at the chain as a whole are necessary to improve efficient utilization. Gains that can be made at one part of the chain are often lost if improvements are not made in the other steps. Although this is common sense, it has not always been easy to put into practice as the separate steps within a chain are often completely independent of each other and have different aims, goals and operating philosophies. In many instances this situation is likely to continue, as chains form and reform according to supply and demand, and to logistics and market prices. However, many now understand that to ensure continuity of supply and stability of pricing, broader agreements on cooperation may be necessary. A driving force for this is that in many countries consumers make their seafood purchases in supermarkets and supermarkets must have regular supplies of the required quality at known prices in order to provide uniform supply to the consumers.

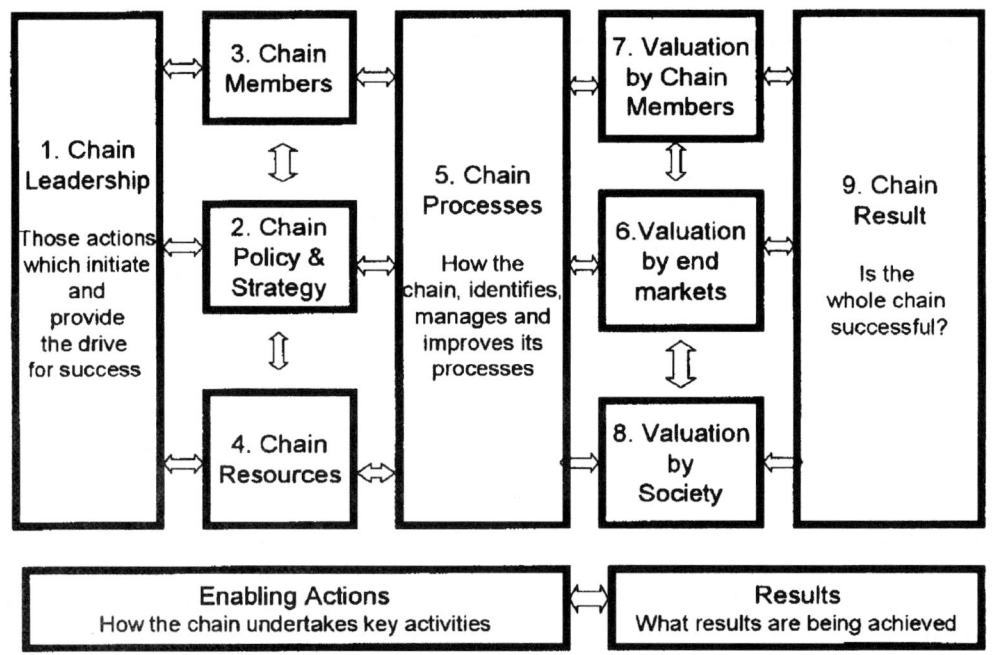

Adapted from suggestion by Folkert et al 1997 Nehem Consulting Group

Figure 1. A model for chain management.

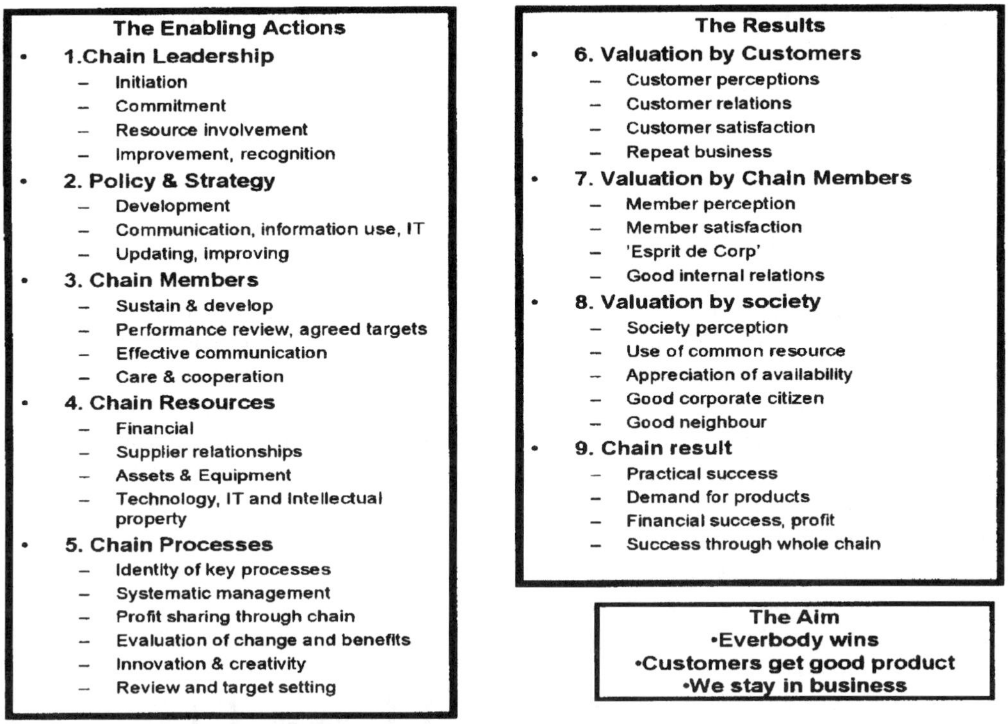

The Enabling Actions	The Results

• 1. Chain Leadership
  − Initiation
  − Commitment
  − Resource involvement
  − Improvement, recognition
• 2. Policy & Strategy
  − Development
  − Communication, information use, IT
  − Updating, improving
• 3. Chain Members
  − Sustain & develop
  − Performance review, agreed targets
  − Effective communication
  − Care & cooperation
• 4. Chain Resources
  − Financial
  − Supplier relationships
  − Assets & Equipment
  − Technology, IT and Intellectual property
• 5. Chain Processes
  − Identity of key processes
  − Systematic management
  − Profit sharing through chain
  − Evaluation of change and benefits
  − Innovation & creativity
  − Review and target setting
	• 6. Valuation by Customers
	  − Customer perceptions
	  − Customer relations
	  − Customer satisfaction
	  − Repeat business
	• 7. Valuation by Chain Members
	  − Member perception
	  − Member satisfaction
	  − 'Esprit de Corp'
	  − Good internal relations
	• 8. Valuation by society
	  − Society perception
	  − Use of common resource
	  − Appreciation of availability
	  − Good corporate citizen
	  − Good neighbour
	• 9. Chain result
	  − Practical success
	  − Demand for products
	  − Financial success, profit
	  − Success through whole chain

The Aim
•Everbody wins •Customers get good product •We stay in business

Figure 2. The elements that determine chain performance.

It is thus necessary to bring all the elements of the chain together to work out means of cooperating so that each step may benefit. This demands a new way of thinking for many operations so that trust is generated between the partners, and this requires that some formal approach must be mutually agreed on.

One such approach can be based on the model developed for evaluation of quality awards by the European Foundation for Quality Management (www.efqm.org). This approach has proved itself to be a most robust system capable of being applied to companies in an enormous variety of industries ranging from small family businesses to large multinational corporations. Whilst it was developed in order to present quality awards, it contains all the elements necessary for evaluating management of a chain.

A modified version, developed from an idea presented to the World Congress on Food Hygiene, The Hague 1997 by Folkerts of Nehem Consulting, has been adapted here to suit the fish industry. It is constructed of 9 basic elements arranged in two parts. The first five are the enabling functions that must be set up within the chain for it to operate satisfactorily, while the remaining four provide the means of evaluating performance (Fig. 1). Each segment contains within it the matters that must be set up and which can be rated (Fig. 2). The total score can be expressed as a value with a maximum total of 1000 points, 500 for the enabling functions and 500 for the performance functions. The balance between enabling functions and performance can clearly be seen and any appropriate remedial steps can be quickly identified, although solutions may not be simple. Consideration needs to be given to the relative values allotted to each of the segments and this should be set by negotiation within the chain at the outset, with the proviso that it can be revisited and altered in the light of experience.

5. VIRTUAL CHAINS, E-AUCTIONS AND E-COMMERCE

The outcome of this is that once a chain is set up and operating it can be thought of as a virtual operation and given consideration as though it were a company in its own right with different operating arms. Indeed this is an apt analogy. When this degree of trust is established then it becomes possible for the companies involved to share information and have access to one another's operating systems to smooth logistics and cater to fluctuating supply and demand. The result is a more efficient order and distribution system with a minimum of delay. Since fundamentally the quality of seafood products depends on time-temperature considerations, it means that products that are commonly referred to as being 'fresher' or of better 'quality' can be placed on the market more rapidly and with greater surety than previously.

Links in some of the parts of these chains are already becoming established with the use of online auctions. The largest of these, PEFA.COM, now auctions online product from 11 auction halls in several different European countries [9]. Thus buyers can source an enormous variety of products online nearly simultaneously to meet their requirements. As yet this represents only a couple of steps in the chain, as the catching and processing steps are not yet involved. But it demonstrates the power and speed of IT systems and provides information on user requirements and specifications from which more extensive operations can be based.

At least 25 other e-commerce websites dealing with seafood (and other food products) can be found which offer products for sale [10]. No doubt there will be further advances in B2B (business to business), B2C (business to consumer), C2B (consumer to business), P2P (peer to peer) and even C2C (consumer to consumer) systems in the near future.

6. RE-ENGINEERING THE SUPPLY CHAIN

Along with this comes the need to completely examine the supply chain and all unit operations within it to determine which may no longer be necessary. The whole chain can be re-engineered since its focus and goals as a chain may now be different to those when it operated as independent organisations. This may involve the deletion of unnecessary steps in handling or in the avoidance of senseless repetition of procedures such as grading and weighing. In turn this requires that proper marking and labelling procedures and the development of fast and reliable traceability systems to ensure integrity of product identity, origin, safety and suchlike.

7. TRACEABILITY

'The ability to trace the history, application or location of an entity, by means of recorded identifications' is the definition that is used by the International Standards Organisation (ISO:1994). Traceability can thus be considered to be a subsystem of a quality management system which, of course, includes safety. A traceability system is more than just a means of protecting product identity, or of proving a product is authentic or 'true to label', and it is not just a brand mark or paper trail or record keeping. It is an active system that can be used both while the product is in existence and after it is consumed or used as an ingredient. So it has a very broad scope, is quite complex and has, as yet, an improperly developed theoretical base.

Some primary reasons for traceability are to provide consumer protection, with everything that involves, and to give brand protection to promote confidence and business, also with all that involves (Table 1).

The first steps in making a traceability system are in deciding what to trace, what is an acceptable degree of traceability, what is the scope of traceability and what is a traceable unit. These are very important business decisions. Generally they can only be taken if there is some clear knowledge about the whole chain and how the products will be used.

Clearly it is ridiculous to attempt to trace a single fish like a sardine, but tracing a high priced tuna weighing 150kg from catch to market is very worthwhile [11]. But should a traceable unit be a single box of fish, a single haul, a day's catch or a trip? And how do the variables of species, size and quality fit into the picture? All these factors enter into consideration and there are different answers according to different circumstances of species, location, market requirements and chain requirements. If the advantages of traceability are not passed along the chain to improve its efficiency, effectiveness and profitability, then the advantages for operators are restricted to improvement in their own business. Effectiveness relates to results, while efficiency focuses on activity. These improvements can be a powerful argument for adopting traceability but the full potential

Table 1
Traceability, consumer safety and brand protection

Consumer safety aspects include	Brand protection includes
• True to label	• Meeting real & perceived customer needs
• Approved ingredients and materials	• Regulatory compliance
• Approved processing	• Marketing advantages
• Packaging	• Product confidence
• Labelling	• Incident management
• Environment	• Crisis management
• Ecology & sustainability	
• Regulatory & statutory requirements	
• Religion & ethics	
• Many other matters	

Table 2
Advantages of traceability

For a whole chain	Within a company
• Ability to differentiate product and to market high priced materials • Establishes basis for efficient recall procedures • Information used for improved quality & process control • Cost saving by eliminating repeat measurements • Supplier incentive to maintain quality of supply • Marketing of special features • Meets current & future regulatory requirements • Engenders cooperation & trust in the chain	• Improved process control • Improved stock control • Better diagnostics • Direct correlation between raw material & final product • Better planning • Optimal use of incoming materials • Better yield prediction • Avoids uneconomic mixing of high and low quality materials • Aids information retrieval for QA and audits • Improved basis for implementing IT and management systems

to offset any costs against improved returns is greater, the greater the number of parts of the chain that co-operate in the system. Simple systems such as segregation of the catch into size, species and according to date of catch provide opportunities to obtain higher prices for the most desirable products. There are many advantages to be gained directly from traceability (Table 2), to which can be added the business advantages of improved stock control, better financial control, greater efficiencies and faster audits.

It is obvious that the product cannot carry all the possible information about its history on a label and the cost of electronic chips on individual packs are currently prohibitive. This means that systems are required for all the available information. Effectively, all that the product, or traceable unit requires, is a unique identifier on it that can easily and quickly be read. The system can be one in which the information is located centrally in one database, and for an individual company or group, this has several advantages. The first being that it is possible now and it is the cheapest. The disadvantages are its lack of flexibility and that it does not deal with the whole chain. A distributed system provides these aspects and has the capability of dealing with future developments. This distributed system is where information resides in different locations but is capable of being instantly interrogated by approved users. It functions similarly to the Internet but not all information would be available to the enquirer, only as much as the holder of the data allows. This means that information, or access to information, can be sold along with the product.

At the Danish Institute for Fisheries Research a system that can do this has been built up using standard hardware and software and various standard peripheral devices, such as barcode readers for which software was written to send information in wireless mode to a LAN. The system was designed for the fresh fish chain in which recent catches were sorted, graded and weighed on board and the boxes identified by barcodes carrying this information plus details of catch, vessel and position. The information is relayed from ship to shore and the codes are read when the boxes are unloaded to verify details as the fish may be sold while the vessel is at sea. In this chain the information remains with the fish right through to the shop counter in the supermarket where the barcode is scanned and a label printed and placed on the wrapper of the fillet or piece of fish sold. The nature of the information that is placed on the purchased fish is entirely at the discretion of the supermarket shop but the entire history of the cut of fish is all electronically available for display or printout if that is considered desirable. The advantages in marketing, advertising, record keeping and efficiency are obvious. The system has been test proven and validated under actual operating conditions [12].

8. PRODUCT DEVELOPMENT

Systems based on information technology have now been developed to aid in product development. The system archives known formularies, ingredients, compatibilities and incompatibilities, regulatory requirements, sources of raw material, shelflife details, production requirements, special equipment needs, specifications and many other facets [13]. The development specialist thus has at their fingertips all the routine information required, leaving them free to use their real creative talents more efficiently. These systems are particularly useful, since much product development work does not start from a completely new basis but is a further development of existing products (known as line extension). Examples of this can be readily found with basic products such as frozen cod which can be presented in modified forms as fillets, fillets in breadcrumbs, in herbed bread crumbs, in crunchy or curry or Italian or Mexican or Creole crumbs or with a myriad of different sauces and so on. The use of these development databases considerably shortens the lead time from product conception to production since the performance characteristics can be modelled and predicted and, in many cases, pilot batches and storage trials can be obviated depending on the degree of innovation involved and aspects of product safety and shelflife.

9. INFORMATION TECHNOLOGY

Traceability, chain management, TQM, quality systems and product development are all part of the total business systems used by the industry. The links between these different aspects were impractical just a short time ago, but now, with the widespread use of computers and information technology and the developments in peripheral equipment and the Internet browsers, it is only a question of how best to do this and take advantage of the remarkable speed and archiving ability of these tools to improve efficiency.

10. QUALITY

An enormous amount of the research in seafood technology over the last 50 years has gone into the search for indicators of 'quality' and 'freshness' and many promising indicators have been tried and found wanting. There are several reasons for this, the most outstanding being that it has rarely been expressed just what is meant by 'quality' or 'freshness'. In consequence, circular arguments have arisen where the test itself has been used to prove that it works. A different approach where either 'quality' or 'freshness' are defined beforehand has now been presented which avoids the problem [14,15]. The other main problem with quality indicators is that the investigators have not considered what properties an indicator should have to make it effective, what the scope of the indicator should be and in which circumstances it is reliable and in which it is not. Indicators should provide some idea of the current state of a material or product and preferably have predictive value as to what they will be like under sets of different circumstances. Since the properties of seafood products are functions of time and temperature, an indicator should be capable of integrating these effects, otherwise it is only strictly valid at the one temperature: normally 0°C, that of melting ice.

Also it is preferable that an indicator should provide a regular monotonic relationship with period of storage (at a set temperature) but many of the indicators investigated are the products of bacterial metabolism and their production is logarithmic. At best these can only have a limited use. This makes it clear that what is required in an indicator is one which both integrates time-temperature effects and has a linear relationship with storage time. The only system proven capable is that of the Quality Index Method. This method has now been widely applied in Europe and an alliance has been formed to promote its use and provide quality control of its dimension (*www.Qim-eurofish.com*).

The QIM score is readily transferable electronically along with traceability information and when conditions of transport are known (temperature) then the QIM score of received material is predictable from knowledge of the duration of transport. If discrepancies arise then this can be used to detect departures from temperature control or delays in transport.

Knowledge of incoming 'quality' of raw material allows for planning of production in terms of product grades and volumes. QIM is fast, reliable, and has the advantages of predictability and is readily transmittable.

11. SEAFOOD SPOILAGE PREDICTION (SSP) AND SPECIFIC SPOILAGE ORGANISM (SSO)

The concepts of the seafood spoilage predictor software (*www.dfu.min.dk/ssp*) have similar advantages. The relationships are based on the best knowledge of bacterial growth and spoilage currently known. The underlying understanding, arising from years of research, is that under given circumstances there is a specific spoilage organism (SSO) which dominates the microflora, and which causes the deleterious changes in odour, flavour, appearance and texture that we term spoilage. At present the SSO has been fairly well established for a number of product types based on both model studies and validation of these in the products themselves [16]. Knowledge of the SSO is essential if the required shelflife for a product is to be reliably obtained in commercial practice for efficient use of the raw material. When seafoods are first caught or harvested their microflora is representative of the waters from which they come and of their gut flora. The levels of the SSO are generally very low, often undetectable, but the way in which the material is then stored and chilled is selective for the SSO.

A prime example of this occurs if warm water prawns are stored in ice, then the SSO that develops is Pseudomonas fragi, which results in progressive development of fruity, cheesy, yeasty off odours and flavours. But if the same prawns are stored in ice slurry the SSO is Shewanella putrefaciens and development of off odours and flavours appear to be slight for the first few days until the detectable threshhold is reached and unpleasant sulphide-like compounds become very evident [17]. Similar selection occurs if the prawns are packed in oxygen permeable (*P. fragi* becomes the SSO) or oxygen impermeable packs (*S. putrefaciens* becomes the SSO). The processor thus has a choice in the way the prawns are stored and packed as to the type of spoilage that will result and consequently determine the shelflife. This type of knowledge is critical for making decisions about efficiency of utilization and choice of product form for the appropriate market as these techniques such as SSP provide meaningful predictions on products that can be provided as part of the information flow that is transmitted electronically along with traceability.

12. HARVESTING

Some attention has been given to the concepts of harvesting at periods when yields can be optimised. This seems mostly to have occurred with crustacea, which go through a moult cycle. In some instances, such as with soft-shelled crabs, the idea is to harvest them during the moulting period when the carapace is soft and the whole can be cooked and served. But for most prawn fisheries the moult period is to be avoided since both flesh and carapace are soft and cannot withstand harvesting and handling so that yields are low and the product has poorer properties and is less desirable. Less apparent effort has been put into optimising yields from capture fisheries but there has been some research into this, using multivariate methods coupled with neural networks [18].

Many markets pay a premium for product that is live, and there is now a growing body of research directed at the problems of dealing with unfamiliar species or with sustaining them over longer distances to other markets [19].

13. BY-PRODUCTS

These are covered by other authors but it is important to stress that full and proper attention must be given to the care and handling of raw materials used for development of by-products whether these may be flesh pieces such as cheeks or heads for human consumption, or whether they are frames for production of mince, or organs for extraction of biochemicals. The same degree of care is required in handling and keeping them free of contaminants and to ensure minimal growth of spoilage bacteria or opportunities for pathogens to occur or to grow.

14. CONCLUSION

The fisheries and seafood industries face the need to make more efficient utilization of the catch. Much more than just simple stepwise improvements in existing products and processes are necessary, and the industry needs to take steps to catch up with those business concepts and practices that will make it efficient, effective and profitable into the future. Only those companies that recognize that circumstances have changed and that can adopt new techniques and approaches will survive.

15. ACKNOWLEDGMENTS

The author wishes to thank his former employer and colleagues Marco Frederiksen, Steen Silberg, Erling Larsen and Paw Dalgaard at the Danish Institute for Fisheries Research, Dept of Seafood Research, Building 221, Technical University DK-2800 Lyngby, Denmark for stimulating discussions over the years on some of the content in this paper.

REFERENCES

1. Dahlgaard, S.M.P. (1999) The evolution patterns of quality management:some reflections on the quality movement. Total Quality Management, 10: S473-S480.
2. Jarrar, Y.F. and Aspinall, E.M. (1999) Integrating total quality management and business process re-engineering: is it enough? Total Quality Management, 10: S584-S593.
3. Harry, M. and Schroeder, R. (2000) Six Sigma. Doubleday, New York.
4. Bonnell, A.D. (1994) Quality assurance in seafood processing:a practical guide. Chapman & Hall, London.
5. Edvardsen, T., Akse, L., Martinussen, T. and Robertson, R. (1995) Kvalitetsledelse i fiskerærningen. Fiskeri Forskning, Tromsø.
6. Robertson, R., Prytz, K., Løvland, J. and Sørensen, N.K. (1992) Total quality management in the Norwegian fishing industry. In: Quality Assurance in the Fish Industry. Huss, H.H., Jakobsen, M., Liston, J. (eds.), Elsevier, Amsterdam.
7. Jónsdottir, S. M., Moe, T. and Larsen, E. (2000) Quality issues in the Danish seafood industry: results from a survey. J. Aquat. Food Prod. Technol., 9: 17-26.
8. Anonymous. (2001) Processors pay more for 'green' hoki. Fishing News International, June issue: 25.
9. Anonymous. (2001) Pefa.com's fish market trading platform. Eurofish, issue 1: 80.
10. Kyprianou, M-H. (2001) Fisheries and e-commerce. Infofish International, issue 2: 21-24.
11. Frederiksen, M. and Bremner, A. (2001) Fresh fish distribution chains. An analysis of three Danish and three Australian chains. Food Aust., 54: 117-123.
12. Frederiksen, M., Østerberg, C., Silberg, S., Larsen, E. and Bremner, H.A. (2002) Info-fisk: Development and validation of an internet based traceability system in a

Danish domestic fresh fish chain. J. Aquat. Food Prod. Technol., 11: 13-34.
13. Jónsdottir, S., Vesterager, J. and Børresen, T. (2000) Development of a product model for specifying new lines of seafood products. Robotics and Computer Integrated Manufacturing, 16: 465-473.
14. Bremner, H.A. (2000) Towards practical definitions of quality for food science. Crit. Rev. Food Sci. Nutr., 40: 83-90.
15. Bremner, H.A. and Sakaguchi, M. (2000) A critical look at whether 'freshness' can be determined. J. Aquat. Food Product Technol., 9; 5-25.
16. Dalgaard, P. (1997) Predictive microbiological modelling and seafood quality. In: Seafood from Producer to Consumer, Integrated Approach to Quality. Luten, J.B., Børresen, T. and Oehlenschläger, J. (eds.), Elsevier Science BV, Amsterdam., pp. 431-443.
17. Chinivasagam, H.N., Bremner, H.A.,Thrower, S.J. and Nottingham, S.M. (1995) Spoilage pattern of five species of Australian prawns:deterioration is influenced by environment of capture and mode of storage. J. Aquat. Food Prod. Technol., 5: 25-50.
18. Peters, G., Morrissey, M.T., Sylvia, G. and Bolte, J. (1997) Linear regression, neural network and induction analysis to determine harvesting and processing effects on surimi quality. J. Food Sci., 61: 876-880.
19. Paterson, B.D., Goodrick, G.B. and Frost, S. (1997) Controlling the quality of aquacultured food products. Trends in Food Science & Technology, 8: 253-257.

M. Sakaguchi (Editor)
More Efficient Utilization of Fish and Fisheries Products
© 2004 Elsevier Ltd. All rights reserved

OCCURRENCE OF THE SEA URCHIN WITH PULCHERRIMINE, A NOVEL BITTER AMINO ACID

Yuko Murata[a], Masahito Yokoyama[b], Tatsuya Unuma[c], Noriko U. Sata[d], Ryuji Kuwahara[a], Masaki Kaneniwa[a] and Ichiro Oohara[a]

[a]National Research Institute of Fisheries Science, Kanazawa, Yokohama 236-8648, Japan
[b]Japan International Research Center for Agricultural Sciences, Oowashi, Tsukuba, 305-8686, Japan
[c]National Research Institute of Aquaculture, Nansei, Mie 516-0193, Japan
[d]Faculty of Science and Technology, Keio University, Kohoku, Yokohama 223-8522, Japan

Abstract

The sea urchin *Hemicentrotus pulcherrimus* gonads often taste bitter in Iwaki, Fukushima Prefecture, Japan. The bitterness was found to be specific to mature ovaries. From the mature ovaries, a bitter substance was isolated and the structure was determined to be $4S$-(2'-carboxy-2'S-hydroxyethylthio)-$2R$-piperidinecarboxylic acid, being a novel sulfur amino acid, named pulcherrimine (Pul). The relationship among the tri-monthly changes in the gonad index (GI), the frequency of bitter gonads of the sea urchin and their Pul contents in the sea off Iwaki was investigated. The GI values showed a large variation among mature specimens. Many mature individuals of which ovaries taste bitter were found in all months examined. Pul content has been found to be related to a seasonal change, and probably to the maturity of ovaries. From these findings, it was concluded that the presence of mature individuals with a high amount of Pul in all seasons is the major reason why the sea urchins are hardly utilized for food in this area. For the achievement of efficient utilization, the high variability of their reproductive cycle among individuals must be firstly taken into consideration.

1. INTRODUCTION

Sea urchin gonads are one of the most popular sea foods in Japan, due to its peculiar thick taste. The gonads are eaten in a variety of ways: raw, steamed, grilled, salted and so on. The sea urchin *Hemicentrotus pulcherrimus* is widely distributed in the Japanese costal areas. It is one the most important fishery products on the south-west coasts of Japan. It has long been regarded that the sea urchin is the most suitable species as a raw material for salted sea urchin gonads called "Echizen Uni", which is a speciality of Fukui Prefecture [1]. However, the sea urchin of which gonads taste extremely bitter are often found in the catch in the Tohoku area, for example, Fukushima Prefecture. Such sea urchins are not acceptable as food and have no commercial value.

The present study was undertaken for the purpose of efficient utilization of the sea urchin. First of all, the frequency of occurrence of the bitter-tasting sea urchins collected in the sea off Iwaki in Fukushima Prefecture was preliminarily examined [2]. Secondly, a substance responsible for the bitterness was isolated, and subsequently, it was elucidated to be a novel sulfur-containing amino acid by determining of its chemical structure [3]. Thus it was named pulcherrimine (Pul) after the scientific name for the sea urchin *Hemicentrotus pulcherrimus*. Thirdly, tri-monthly examinations were carried out to clarify the seasonal changes in the maturity, the frequency of occurrence of bitter tasting gonads and the pulcherrimine content of the sea urchins in the sea off Iwaki [4]. In this study, Pul was determined by a method which consists of the formation of dimethylaminoazobenzensulphonyl chloride (Dabs-Cl) derivatives of Pul (dabsylation), and the separation of the dabsylate using reversed-phase high-performance liquid chromatography [5].

2. OCCURRENCE OF BITTER-TASTING SEA URCHIN IN THE SEA OFF IWAKI

At first, the frequency of occurrence of the bitter-tasting sea urchins collected in the sea off Iwaki was preliminarily examined [2].

Sea urchins were collected from the sea off Iwaki, Fukushima Prefecture in March 1996 and March 1997, the number of specimens collected were 94 and 99, respectively. Their gonads were dissected out, after the size and the weight were measured. The gonad index (GI) of each sample was calculated by the follows equation:

GI (%) = (gonad weight / body weight) × 100

The bitterness of the sea urchin gonads was tested by a sensory test using a small pieces of the tissue of individual gonads.

Table 1 shows the frequency of occurrence of bitter gonads in the sea urchins collected from the sea off Iwaki, Fukushima Pref. in March 1996 [2]. All specimens were judged to be mature and their sex was readily distingushed by the oozed gametes [6]. Male and female sea urchins were almost identical in the biological data such as test diameter, test height, body weight, gonad weight and gonad index. Ninety-five percent of the female individuals had bitter ovaries, while none of the male individuals had bitter testes. In March 1997, all of the individuals of which the gonads gave a bitter taste were found to be female. These results suggest that the bitterness of the sea urchin is specific to the mature ovaries.

Table 1
The frequency of occurrence of bitter gonads in green sea urchins and other biological data

	Male	Female
Number of specimens (A)	50	44
Bitter gonad (B)	0 (0%)[a]	42 (95%)[a]
Test height (mm)[b]	20.0±2.7	19.7±2.8
Test diameter (mm)[b]	37.9±3.8	37.6±4.0
Body weight (g)[b]	22.1±1.1	21.7±6.4
Gonad weight (g)[b]	2.4±1.1	2.5±1.0
Gonad index (%)[b,c]	10.8±4.5	11.2±4.5

[a]Number of specimens having bitter gonads. Values in parentheses are frequency (B/A×100%).
[b]Mean±standard deviation. [c]Gonad weight / body weight×100

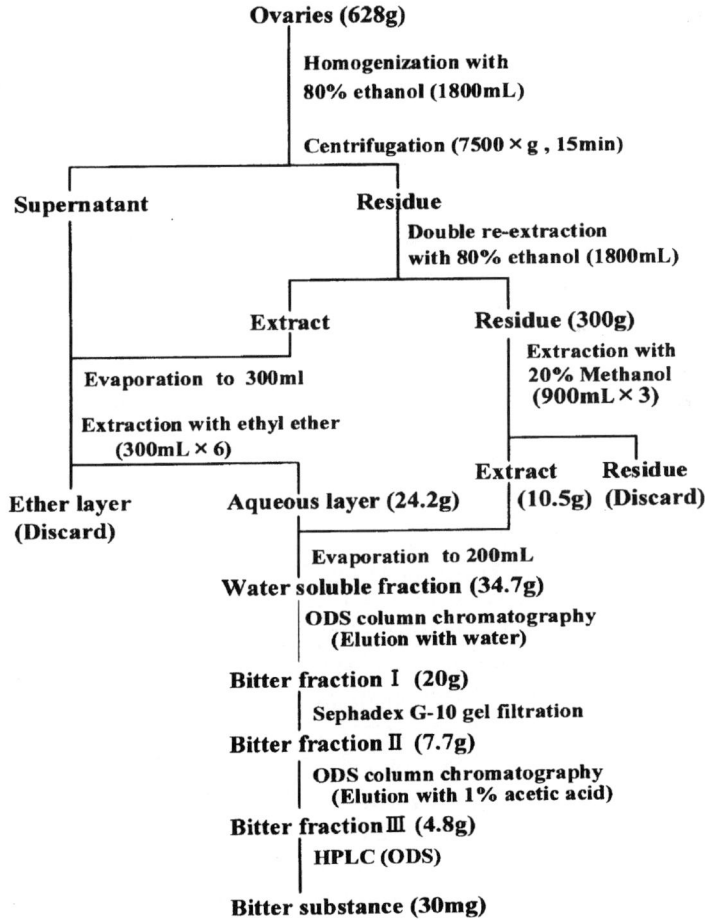

Figure 1. Isolation procedure of pulcherrimine.

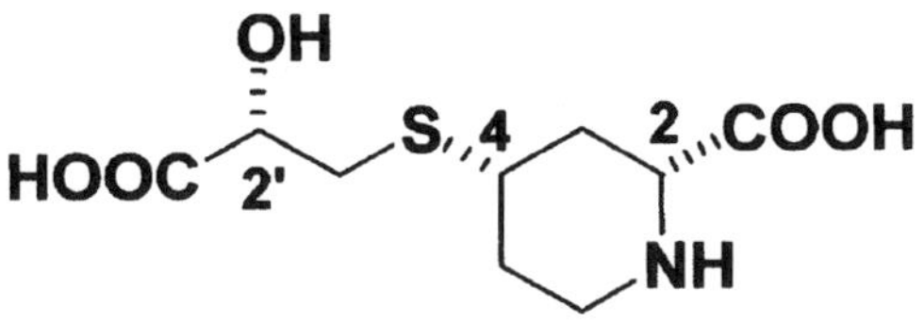

Figure 2. A structure of pulcherrimine (**1**).

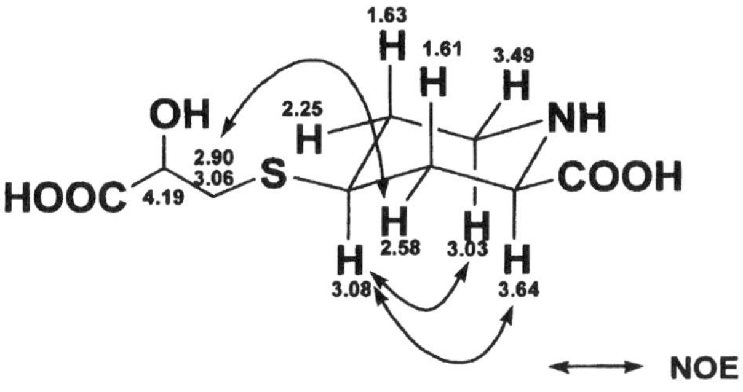

Figure 3. NOE correlations for **1**.

3. ISOLATION AND STRUCTURE ELUCIDATION OF A NOVEL BITTER AMINO ACID, PULCHERRIMINE, FROM THE SEA URCHIN OVARIES

The preliminary investigation showed that the bitterness of the sea urchin gonads was specific to mature ovaries. Secondly, isolation and structure of the bitter substance from the sea urchin ovaries has been elucidated [3]. Figure 1 shows a procedure for isolation of Pul from mature ovaries [3]. A planer structure of Pul (Fig. 2) was determined by HRFABMS data, and 1D and 2D NMR spectral data [3].

Pulcherrimine (**1**): $[\alpha]^{24}_D$-16.5 (c 0.20, H_2O); TLC on silica gel, Rf 0.12 (n-BuOH/AcOH/H_2O, 4:1:2); ^{1}H NMR in D_2O/CD_3OD (40:1) at 600MHz 4.19 (1H, dd, J=6.7, 3.8 Hz, H2'), 3.64 (1H, dd, J=12.7, 3.1, H2), 3.49 (1H, ddd, J=13.1, 4.2, 2.3, H6), 3.08 (1H, m, H4), 3.06 (1H, dd, 13.8, 3.8, H1'b), 3.03 (1H, m, H6), 2.90 (1H, dd, J=13.8, 6.7, H1'a), 2.58 (1H, ddd, J=14.2, 6.2, 3.1, H3), 2.25 (1H, m, H5), 1.63 (1H, m, H5), 1.61 (1H, dd, J=14.2, 12.7, H3); ^{13}C NMR in D_2O/CD_3OD (40:1) at 67.5MHz 180.0s (COOH-2'), 174.2s (COOH-2), 72.3d (C2'), 60.0d (C2), 44.1t (C6), 39.6 (C4), 35.2t (C1'), 34.3t (C3), 29.8t (C5); ESIMS (rel. ext): 248 [24, (M-H)$^-$], 195 (6), 159 (16), 145 (100), 133 (8), 131 (19), 89 (31), 87 (8), 59 (9) ; FABMS (rel. ext): m/z 250 [25, (M+H)+], 204 (15), 162 (22), 128 (48), 82 (98). HRFABMS (matrix: thioglycerol): obsd. (M+H)$^+$ m/z 250.0749 ($C_9H_{16}NO_5S$, +0.5 mmu).

The stereochemistry of Pul was determined by NOE experiments, chiral HPLC analysis and modified Mosher method. From the NOE experiments, the piperidine ring

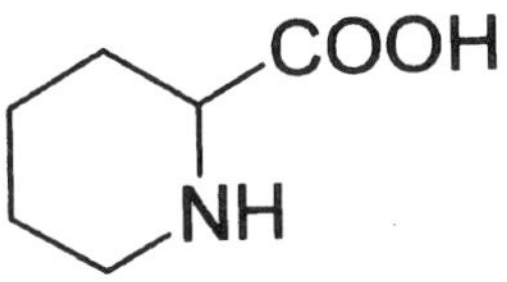

Figure 4. A planer structure of pipecolinic acid.

was in a chair conformation having an ethylthio group on C4 and a carboxyl group on C2 as shown in Fig. 3 [3].

To determine the absolute configuration, **1** was hydrogenolyzed on Raney Ni in H_2O at 60°C to afford pipecolinic (Fig. 4) and lactic acids. Chiral HPLC analysis disclosed that the stereochemistry of pipecolinic acid was in D form [3]. However, the stereochemistry of lactic acid could not be determined due to overlapping of the peak with that of acetic acid in chiral HPLC. Therefore, the modified Mosher method [7-10] was applied to determine the stereochemistry of lactic acid portion. From NMR analysis of MTPA esters of N-Boc-derivative, the stereochemistry at 2' position was determined. 2'S stereochemistry of the lactic acid moiety of pulcherrimine was thus assigned and accordingly, absolute stereochemistry in pulcherrimine was 2'S, 2R and 4S.

4. SEASONAL CHANGES IN THE MATURITY, FREQUENCY OF OCCURRENCE OF BITTER-TASTING GONADS AND THE PUL CONTENT

In the preliminary investigation (March 1996 and 1997), all the individuals collected from the sea off Iwaki were mature. Matsui [11] and Ito [12] found that the sea urchin in Fukui Prefecture and Saga Prefecture has a clearly defined annual reproductive cycle. Therefore, tri-monthly examinations were carried out to clarify the seasonal changes in the maturity, the frequency of occurrence of bitter tasting gonads and the pulcherrimine content of the sea urchins inhabit in the sea off Iwaki in Fukushima Prefecture [4].

At intervals of every three months from November 1998 to November 1999, 100 sea urchins were randomly collected from the sea off Iwaki. This sampling is hereafter called as tri-monthly sampling for convenience. The sampled sea urchins were divided into mature and immature individuals. Mature individuals were defined in this study as those with gametes which ooze from the gonads. Immature individuals were defined as those with gametes which do not ooze from the gonads. The sex of the mature individuals was identified from the oozed ganetes [6]. Four to seven immature individuals collected every three months were histologically observed [13] in order to determine the stage of their gonadal development.

Sample preparation and sensory tests were carried out by the same method as described in Fig. 5. Pul content was measured according to the method described by Murata et al. [5].

Table 2 shows the frequency of mature male, female and immature sea urchins [4]. In November 1998 and February 1999, all of the sea urchins examined were mature, and

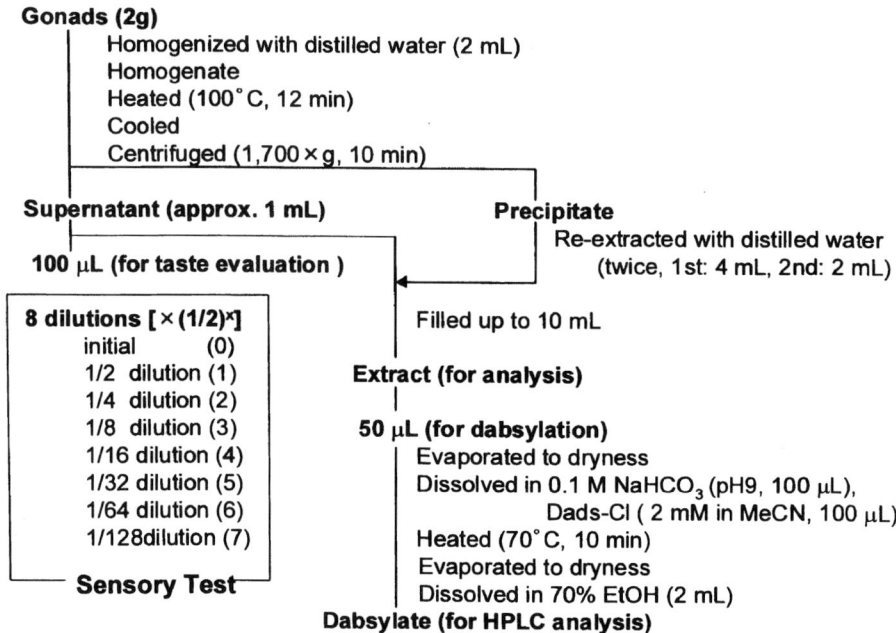

Figure 5. Sample preparation and sensory test. The concentration ranges of extracts for sensory tests included eight dilutions from 1 (initial) to 1/128 of dilution which decreased by a factor of 2 and the initial concentration was the original extract. These eight concentration ranges were converted into eight scales of bitterness. Scales of bitterness were as follows; 0 (initial), 1 (1/2 of dilution), 2 (1/4 of dilution), 3 (1/8 of dilution), 4 (1/16 of dilution), 5 (1/32 of dilution), 6 (1/64 of dilution), 7 (1/128 of dilution). Evaluation was started with the weakest concentration and performed from a weaker to a stronger. The bitterness was defined as the mean value of the scale being converted from the lowest concentration at which five subjects can recognize the bitterness.

Table 2
The freqency of occurrence of mature male and female, and immature sea urchin[a]

	Mature		Immature[b]
	Male	Female	
November 1998	56	44	0
February 1999	57	43	0
May 1999	50	30	20
August 1999	20	21	59
November 1999	56	40	4

[a]As the figures express the number of relevant specimens out of 100 specimens collected at the same time, they mean the percentage of cases. [b]Gametes did not ooze from the gonads and the sex identification was impossible.

Table 3
Test hight, test diameter and body weight of green sea urchin from November 1988 to November 1999

	Month	Test height (mm)[*]	Test diameter (mm)[*]	Body weight (g)[*]
Mature male	Nov. 1998	22.1±2.0	41.6±2.8	21.1±4.9
	Feb. 1999	22.5±2.8	41.3±3.9	24.9±4.5
	May 1999	23.4±2.8	43.3±4.1	26.5±8.7
	Aug. 1999	22.1±2.2	40.6±3.2	23.2±5.3
	Nov. 1999	21.7±1.8	41.6±2.9	23.6±5.6
	mean±SD	22.4±2.4	41.8±3.4	23.9±5.9
Mature female	Nov. 1998	22.1±1.6	41.9±3.3	21.6±4.5
	Feb. 1999	22.3±1.6	40.9±2.8	23.4±4.5
	May 1999	21.8±1.4	41.2±2.1	22.3±3.2
	Aug. 1999	21.0±2.0	39.0±3.4	20.8±4.7
	Nov. 1999	21.9±1.6	40.9±2.6	22.5±4.5
	mean±SD	21.9±1.6	41.0±2.8	22.3±4.3
Immature	May 1999	23.5±2.7	43.4±3.3	27.0±6.7
	Aug. 1999	22.1±2.0	40.9±3.0	23.5±5.7
	Nov. 1999	22.3±0.1	40.5±0.1	24.2±1.6
	mean±SD	22.4±2.1	41.3±3.0	24.2±5.9

[*]Mean±SD

thus the sex of each individual was easily distinguished. The frequency of female sea urchins was 40%; that of the male sea urchins was, naturally, 60%. In May and August 1999, the frequency of immature individuals was relatively high: 20% and 60%, respectively. Seven specimens from each sample out of the immature individuals collected in May and August 1999 were subjected to a histological observations. These observations revealed that these individuals were at the recovering spent stage [6]. Therefore, many immature sea urchins were observed, but, simultaneously, many mature female and male individuals were also observed. This suggests that mature individuals occur in all seasons in the sea off Iwaki. The frequency of mature males was 50% and 20%, in May and August, respectively, and that of mature females was 30% and 20%, respectively. In November 1999, 56% of individuals were identified as male and 40% as female. Four immature individuals were identified as females by microscopic observations: two of them were judged to be at the growing stage, and the other two at the pre-mature stage.

The mean test diameter, test height and body weight of individuals were approximately the same throughout the observations for the five months, i.e. 22 mm, 41 mm and 23 g, respectively (Table 3) [4].

Seasonal changes in the GI of each sex of the sea urchin are shown in Fig. 6 [4]. The GI values showed a large variation among the mature specimens in each season. The fact indicates that the maturation process of the sea urchins might vary among individuals.

The mean GI values of gonads were observed to decrease significantly during the period between February and May 1999, being at the lowest values in May 1999, and to increase thereafter. This change in the GI suggests that the major spawning season is in the period from February to May. The GI of the immature specimens in May, August and

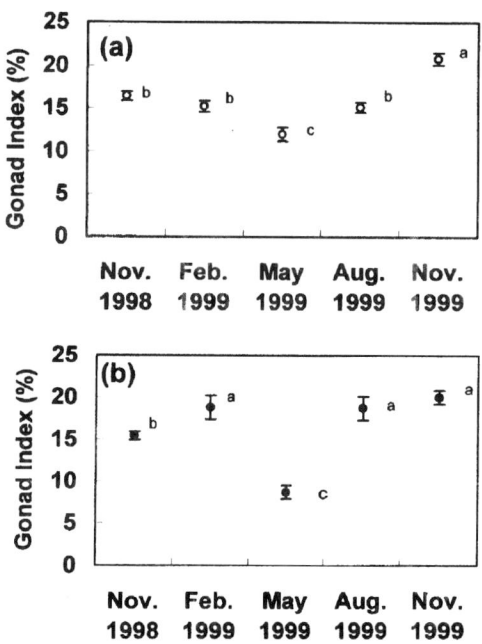

Figure 6. Seasonal changes in the gonad index (mean±SE) of (a) mature male and (b) mature female of the sea urchin inhabits in the sea area off Iwaki. Among supplementary letters, a, b and c, the mean values are significantly differs each other ($p<0.05$).

November 1999 was 8.3 ± 0.8 (%), 12.7 ± 0.5 (%) and 7.5 ± 1.0 (%) (mean ± SE), respectively. These values were lower than that of the mature specimens.

From the variation of the individual GI values and the presence of mature individuals in all seasons, the reproductive cycle of this species of sea urchin is unclear, and the occurrence of the mature stage extends over a long term, at least, in this sea area. The maturation and subsequent spawning behavior are generally considered to be controlled by the sea water temperature [14]. Ito et al. [12] described for the sea urchin that a temperature depression after a high water temperature period accelerates the maturation of gonads. In the sea area off Saga Prefecture, it is observed that spawning starts when the water temperature drops to 15°C. In the sea off Iwaki, the water temperature remains over 20°C from August to October. The water temperature reaches its highest value in September and then declines. From December to June, the water temperature is usually below 15°C [15]. Therefore, the variation of the maturation process and the long term maturation period among individual sea urchins in the sea off Iwaki may be due to the relatively low water temperature.

Figure 7 shows the frequency of occurrence of bitter gonads in each season [4]. More than 95% of the ovaries of sea urchin collected in November 1998, February 1999 and November 1999 tasted bitter. On the other hand, in May and August 1999, 60% of the mature ovaries were bitter. Immature gonads had no bitter taste in May and November,

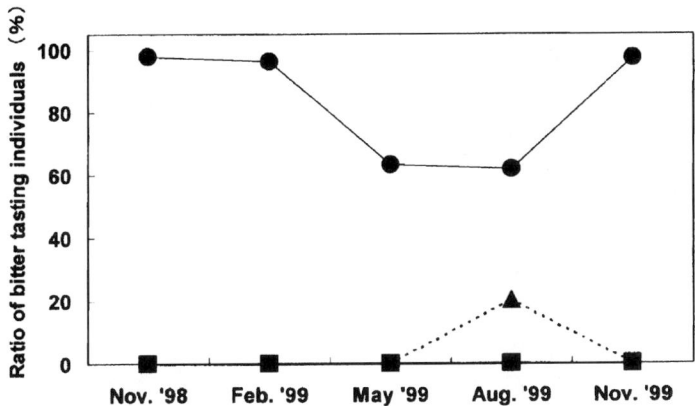

Figure 7. Frequency of bitter tasting gonad observed at intervals of three months; mature males, mature females and immature individuals . ■, mature males; ●, mature females and; ▲, immature individuals.

but 20% of the immature gonads in August 1999 tasted bitter. From these results, the seasonal change in the frequency of occurrence of bitter ovaries agreed well with that of the distribution of mature ovaries.

Twenty specimens were randomly selected from each of mature ovaries and testes and analyzed for Pul content in each month. No Pul was detected in mature testes and they had no bitter taste. Figure 8 shows histograms representing the distribution of Pul content among the mature female individuals in each month [4]. The Pul content distributions had large variances in all months examined and were different among months. Many individuals with Pul levels more than 0.5 mg/100g were found in November 1998, February 1999 and November 1999. Pul levels of mature ovaries were the highest in February. Histograms of February 1999 and November 1999 showed that the mean of Pul content (1.59 and 0.93 mg/100g) was located within the mode column. Also in Nov 1998, the mean Pul content (1.37 mg/100g) located near the mode. Being in contrast, in May and Aug 1999, the mode of each Pul level was at the lowest column, and many mature ovaries with no Pul were found. The distributions were highly skewed for these two months. These findings suggest that the amount of Pul is related to the seasonal change of the sea urchin ovaries. Some of the non bitter ovaries have been found to have Pul, whose content was less than 0.5 mg/100 g. It is probable that the Pul content was lower than its taste threshold level in these gonads.

5. CONCLUSION

The sea urchin is hardly the target for fisheries in the Tohoku area, for example, Fukushima Prefecture because of the bitter-taste of their gonads. This study was undertaken for the purpose of more efficient utilization of the sea urchin.

Since, in a preliminary experiment, the bitterness was found to be specific to mature

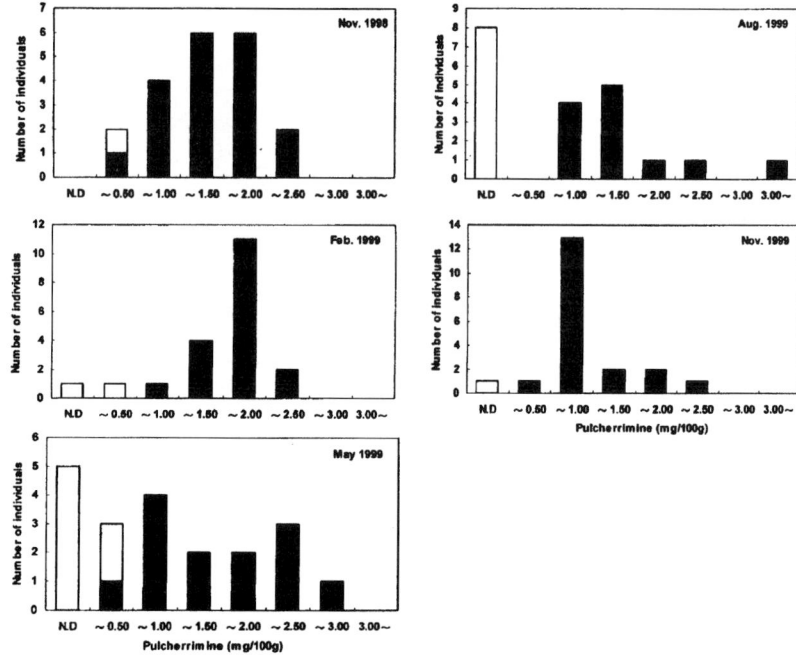

Figure 8. Frequency distribution of pulcherrimine contents among mature female individuals in each season. Open and/or solid areas in a bar, indicate non bitter ovaries and bitter ovaries, respectively. N.D. means that pulcherrimine was not detected.

ovaries, a bitter substance was isolated from the mature ovaries, and the structure was determined to be 4S-(2'-carboxy-2'S-hydroxyethylthio)-2R-piperidinecarboxylic acid, being a novel sulfur amino acid. This substance was named pulcherrimine (Pul) after the scientific name of the sea urchin.

It was found that the GI values showed a considerable variation among the mature specimens in each season, and mature individuals of which ovaries taste bitter were found in all seasons by the tri-monthly examinations. Therefore, the reproductive cycle seems to be unclear and the mature stage extends over a long term in this sea area. These facts may be due to the relatively low water temperature. From these findings, it was concluded that the presence of mature individuals of which ovaries include Pul in all seasons is the major reason for that the sea urchins are hardly utilized for food in the Iwaki area. For the achievement of efficient utilization of this species of sea urchin, their reproductive cycle which varies with individuals must be firstly taken into consideration, although an important question remains still open as to the formation and/or accumulation mechanisms of Pul in the mature ovary of the sea urchin.

6. ACKNOWLEDGMENTS

The author is greatly obliged to Prof. Shiro Konagaya, Department of Home Economics, Kokugakuin Tochigi Junior College, for critical reading of the manuscript. The author is also grateful to Prof. Michio Shigei, Department of Applied Biology, Faculty of Textile Science, Kyoto Institute of Technology, for the species identification. The author is obliged to Mr. Tatsuya Yamamoto, Fukushima Fishery Office for his help in the sample collection.

REFERENCES

1. Kawana, T. (1938) On the propagation of *Strongylocentrotus pulcherrimus*. Suisan Kenkyu Shi, 33: 104-116.
2. Murata, Y., Yamamoto, T., Kaneniwa, M., Kuwahara, R. and Yokoyama, M. (1998) Occurrence of bitter gonad in *Hemicentrotus pulcherrimus*. Nippon Suisan Gakkaishi, 64: 477-478.
3. Murata, Y. and Sata, N.U. (2000) Isolation and structure of pulcherrimine, a novel bitter-tasting amino acid, from the sea urchin (*Hemicentrotus pulcherrimus*) ovaries. J. Agric. Food Chem., 48: 5557-5560.
4. Murata, Y., Yokoyama, M., Unuma, T., Sata, N.U., Kuwahara, R. and Kaneniwa, M. (2002) Seasonal changes of bitterness and pulcherrimine content in gonads of green sea urchin *Hemicentrotus pulcherrimus* at Iwaki in Fukushima Prefecture. Fisheries Sci., 68: 184-189.
5. Murata, Y., Sata, N.U., Yokoyama, M., Kuwahara, R., Kaneniwa, M. and Oohara, I. (2001) Determination of a novel bitter amino acid, pulcherrimine in the gonad of the green sea urchin *Hemicentrotus pulcherrimus*. Fisheries Sci., 67: 341-345.
6. Fuji, A. (1960) Studies on the biology of the sea urchin. I. Superficial and histological gonadal changes in gametogenic process of two sea urchins, *Strongylocentrotus nudus* and *S. intermedius*. Bull. Fac. Fish., Hokkaido Univ., 11: 1-14.
7. Dale, J.A. and Mosher, H.S. (1973) Nuclear magnetic resonance enantiomer reagents. Configurational correlations via nuclear magnetic resonance chemical shifts of diastereomeric mandelate, o-methylmandelate, and -methoxy-trifluoromethylphenylacetate (MTPA) esters. J. Am. Chem. Soc., 95: 512-519.
8. Ohtani, I., Kusumi, T., Kashman, Y. and Kakisawa, H. (1991) High field FT NMR application of Mosher's method. The absolute configurations of marine terpenoids. J. Am. Chem. Soc., 113: 4092-4096.
9. Kamiyama, T., Umino, T., Itezono, Y., Nakamura, Y., Satoh, T. and Yokose, K. (1995) Sulfobacins A and B, novel von Willebrand factor receptor antagonists. II. Structural elucidation. J. Antibiotics, 48: 929-936.
10. Shin-ya, K., Shimizu, S., Kunigami, T., Furihata, K., Furihata, K. and Seto, H. (1995) A new neuronal cell protecting substance, lavanduquinocin, produced by *Streptomyces viridochromogenes*. J. Antibiotics, 48: 574-578.
11. Matsui, I. (1966) The program of the sea urchins. Japan Fisheries Resource Conservation Association, Tokyo.

12. Ito, S., Shibayama, M., Kobayakawa, A. and Tani, Y. (1989) Promotion of maturation and Spawning of sea urchin *Hemicentrotus pulcherrimus* by regulating water temperature. Nippon Suisan Gakkaishi, 55: 757-763.
13. Unuma, T., Konishi, K., Furuita, H., Yamamo, T. and Akiyama, T. (1996) Seasonal changes in gonads of cultured and wild red sea urchin, *Psudocentrotus depressus*. Suisanzoshoku, 44: 169-175.
14. Agatsuma, Y. (1992) Annual reproductive cycle of the sea urchin, *Hemicentrotus pulcherrimus* in southern Hokkaido. Suisanzoshoku, 40: 475-478.
15. Monthly Ocean Report (No. 84) December. (1999) Climate and Marine Department, Japan Meteorological Agency, Japan.

HOW DO HANDLING AND KILLING METHODS AFFECT ETHICAL AND SENSORY QUALITY OF FARMED ATLANTIC SALMON ?

Nils K Sorensen, Torbjorn Tobiassen and Mats Carlehoeg
Norwegian Institute of Fisheries and Aquaculture Ltd., FISKERIFORSKNING, Breivika, 9291 Tromsø, Norway

Abstract
Quality of a fish product is related to several parameters. Lately aspects of ethics and welfare of farmed fish during the slaughter process, have been focused on by fish farmers and consumers. In order to explore means for reducing stress due to restraining, pumping, anaesthetising and killing, four new slaughter methods have been tested for Atlantic salmon. Our main objectives have been to: 1) reduce stress and time spent before death of the fish during the slaughter process and 2) maintain or improve the sensory quality of the fish flesh during iced storage. Experiment one assessed the sensory quality during iced storage. The methods applied were, CO_2 gas (commercial), anaesthesia by Eugenol oil, percussive stun and spiking (concussion and decerebration). All methods were followed by exsanguination by gill cutting. No significant differences in sensory quality between slaughter methods were recorded. The use of Eugenol gave flavour to the flesh. Significantly higher values for texture parameters were registered for day one than for the rest of the storage period. In experiment two, Atlantic salmon were slaughtered by five methods, the four methods mentioned and in addition, cutting the gills only. All fish were gill cut and behavioural indices were recorded to assess when the fish lost consciousness. pH was recorded after 10 minutes as indication of stress level. The commercial method (CO_2) resulted in longer time until unconsciousness and had lowest pH, while percussive stun and Eugenol resulted in shorter periods until unconsciousness and higher pH values. The experiments have demonstrated that slaughter methods for Atlantic salmon can be quick and effective (ethical killing), without any loss of sensory quality and shelf life.

1. INTRODUCTION

Fish farmers and processors have always given priority to fish quality and welfare. The interest in this area has increased much lately, especially related to the killing operation at slaughter [1]. These aspects are important since the volume of farmed Atlantic salmon

(*Salmo salar*) in Norway has increased to more than 450,000 tonnes in 2000, with the consequent demand for a high degree of industrialisation, high capacity and efficiency of slaughter operations. According to regulations, a fish shall not sustain any "avoidable excitement, pain or suffering" during harvesting, and any killing method must result in rapid and irreversible loss of consciousness. That is considered ethical slaughter.

The common method used when killing farmed salmonids in Norway is carbon dioxide narcosis for 5 to 7 minutes, followed by exsanguination (bleeding). This operation implies escape reactions and stress, which may result in reduced fish welfare and fish flesh quality. In order to optimise the fish welfare while maintaining high flesh quality, new killing methods for farmed Atlantic salmon have been tested in two experiments in pilot scale. The idea of acclimating farmed fish before slaughter, either by leaving it for a period near the slaughterhouse, using sedation by chilling and/or anaesthesia, is called rested harvest [2].

2. MATERIALS AND METHODS

In experiment one, Atlantic salmon, of 3-4 kg gutted weight, were taken from a commercial size net pen at Kaarvika research station near Tromso, and kept in a smaller pen for acclimation and starvation for 10-12 days at a water temperature of 5-6 °C. At day of slaughter, 48 fish were taken for further acclimation in tubs for one hour. Fish were killed by four methods, use of anaesthesia: 1) CO_2 or 2) Eugenol-oil and new rapid methods: 3) percussive stun or 4) spiking /Iki-Jime (concussion + decerebration). The fish were gutted and divided into groups of 12, boxed with ice and stored in a chill room until sampling on day 0, 5, 12, 14 and 21. Fillets were cut from two fish and a trained sensory panel of 8 panellists assessed cooked fillet samples in duplicate, at each sampling during the storage period. Fourteen sensory properties were described on an unstructured scale from 0 to 10.

The statistical evaluations were carried out with the statistical analysis system SAS, Windows version 6.12, (SAS Institute INC., USA). For the descriptive sensory test averages over the assessors and replicates were compared for every method and sensory parameter in a two-way ANOVA with interaction and assessors as random effects. A Tukey's test on the 5%-level was performed. The sensory data was also explored by principal component analysis (PCA) using the software The Unscrambler, version 7.5 (Camo A/S Oslo).

In experiment two, salmon was harvested as in experiment one, but kept for 20 hrs in the tubs for acclimation, reducing stress, before killing by five methods, additional was gill cut only, using seven groups of 10 fish each. Eugenol was applied at two concentrations. CO_2 was applied in two ways, as saturated in the tub and added while fish stayed in the tub.

Behavioural indices were assessed according to a detailed description and time recorded until loss of reaction related to self initiated behaviour, stimuli or reflexes [3]. Time of death was estimated for Atlantic salmon by recordings made at 1, 5 and 10 minutes after applying a killing method. The method was verified by measuring brain function after applying the five slaughter methods, assessing the state of unconsciousness

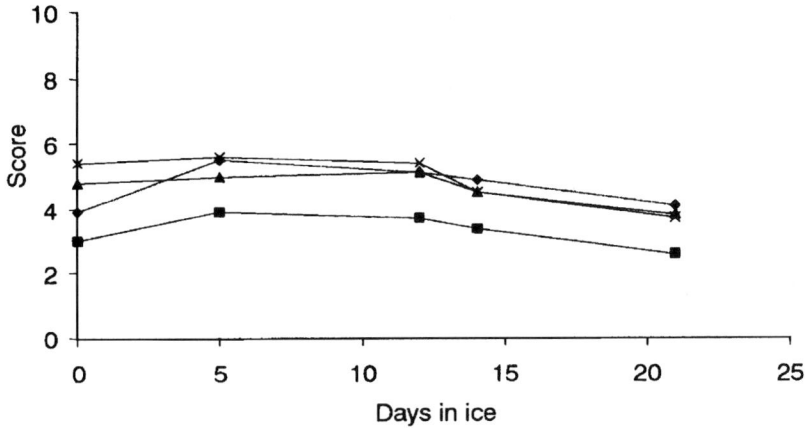

Figure 1. Sensory assessment of Own taste in cooked samples of Atlantic salmon fillets, dependent on killing method and storage time of whole fish in ice. Four fillets from each method at each day of sampling. Assessors, n=8. ♦, CO_2; ■, Eugenol; ▲, Percussive stun; ×, Spiking

and insensitivity of the fish more accurately [4]. pH was measured by a spear electrode in the muscle of individual fish during 40 hrs of storage on ice.

3. RESULTS AND DISCUSSION

The results of the sensory analyses show that there were few and small differences between the slaughter methods during the storage period. Method 2 (Eugenol anaesthesia + gill slit) scored least favourably in general and was different from other methods at all times for the sensory parameters "own smell" and "own taste" (Fig.1). Scores at day zero and day 21 are slightly lower than the other days. This indicates that there is a development from a bland, neutral taste at day zero to a more old, rancid and stale taste and odour at day 21.

This is confirmed by the PCA-plot (Fig. 2), where days 0 and 21 are special, positioned at the upper right side and lower middle part of the plot. The other days, 5, 12 and 14 are positioned in the upper left side of the plot. The PCA-plot explains 74% of the total variation in the material tested, 48% as factor 1 along the x-axis and 26% along the y-axis. The factor 1 describes texture parameters, cohesiveness and fibrousness at the right side, close to day zero, while juiciness is found towards the left side. Factor 2 describes taste and flavour parameters where salmon odour (SALMONO) and − flavour (SALMONF) is found to the upper left, while the more negative descriptors, rancid and stale flavour an odour, are found to the lower right in Fig. 2.

This method of evaluating sensory data has previously proven very helpful, (Sorensen, N.K. and Carlehoeg, M., unpublished data 2000).

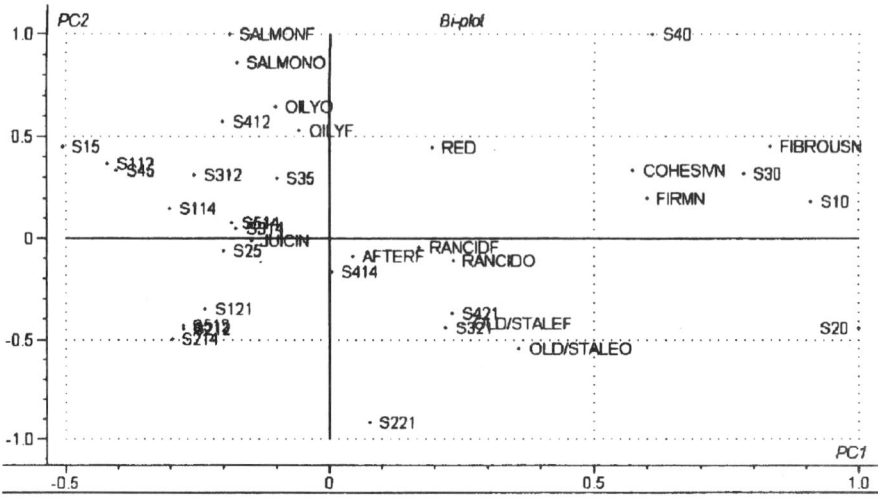

Figure 2. Biplot of principal component analyses (PCA) for the sensory assessments of cooked Atlantic salmon fillets, dependent on four killing methods and storage time in ice. Method 1, CO_2; Method 2, Eugenol; Method 3, Percussive stun; Method 4, Spiking. The method and storage time are indicated by e.g. S114, method 1 and day 14. Salmon odour is labelled SALMONO and salmon flavour is done SALMONF in the biplot. F, flavour; O, odour.

Figure 3 illustrates texture data as the development of fibrousness for the different killing methods during iced storage. Significant changes in other texture parameters, such as firmness and cohesiveness, were also recorded during the storage period. No differences were observed between the killing methods. On day zero high values were recorded, which are related to onset of rigor mortis.

Another important parameter is juiciness, which is related to the feeling in the mouth of juice or water being released over time when eating the fish flesh. Water binding is related to pH, which would indicate that a relation between pH and juiciness should be found. Although there are big differences in pH levels between killing methods, from 7.4 to 6.8 (Fig. 4), there is no change in the level of juiciness according to the sensory analyses. This has been documented before when different killing methods have been related to pH levels and onset and duration of rigor mortis [5], (Sorensen, N.K,. unpublished data, 2000). The initial pH values for both the stunned and the Eugenol treated fish were high, close to physiological levels indicating the effectiveness of the methods.

In the PCA-plot shown in Fig. 2, Eugenol treated fish gave lowest results in factor 2, i.e. the flavour parameters, see S20, S214 and S221, i.e. method 2, day 0, 14 and 21. Differences in smell and taste parameters were not significant between the killing methods during 21 days. The salmon were of acceptable eating quality during the whole period, although the texture became softer by day 21. A practical storage life is estimated

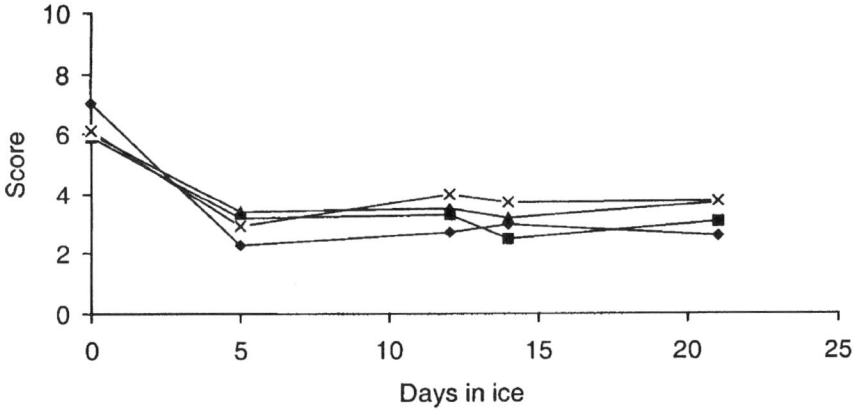

Figure 3. Sensory assessment of fibrousness in cooked samples of Atlantic salmon fillets, dependent on killing method and storage time of whole fish in ice. Four fillets from each method at each day of sampling. Assessors, n=8. ♦, CO_2 ; ■, Eugenol; ▲, Percussive stun; × Spiking.

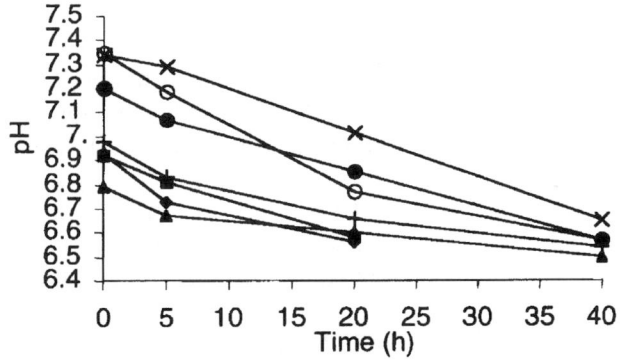

Figure 4. Development of pH in Atlantic salmon flesh during 40 hrs storage in ice, after applying different slaughter methods. Average from 10 fishes in each group is given. pH was measured by spear electrode in each fish. ♦, CO_2; ■, CO_2 added; ▲, Gill cut; ×, Percussive stun; ●, Eugenol 35 ppm; ○, Eugenol 70 ppm; +, Spiking.

to be 16-18 days at 0°C, based on the sensory results and data for bacterial counts. The panel regarded the Eugenol killed fish as "different" and mentioned smell and taste as being spicy. This is understandable because the Eugenol oil is an extract from clove oil. The flavour seemed to be absorbed in the flesh of the salmon making its use questionable as an anaesthesia for commercial practice. It may rather be regarded an additive. The accumulation of taste and flavour from Eugenol have also been experienced in other species e.g. with Atlantic halibut (Akse, L. and Midling, K., unpublished data, 2000).

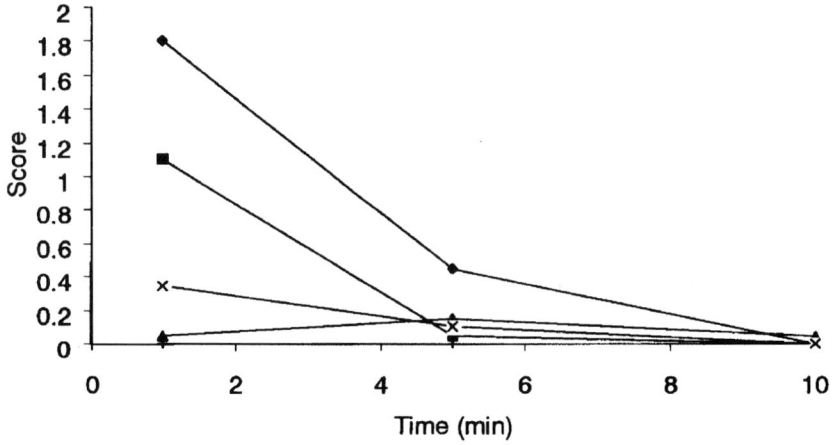

Figure 5. Development of responses to stimuli; pinching the tail, electricity (6V battery) and pricking the nose with a needle, for Atlantic salmon after applying different killing methods. The fish was kept in a tub with seawater and observed for 10 minutes. N=10 in each group. ♦, Gill cut; ■, CO_2 3,5 min; ▲, Eugenol; ×, Spiking; +, Percussive stun.

Results from the second experiment resulted in clear differences between the killing methods in loss of behavioural scores assessed at 1, 5 and 10 minutes after applying the methods. Figure 5 give the results from the behaviour assessment, tested as responses to stimuli by prick by a needle to the nose, pinching the tail of the fish and electrical stimuli from a 6V battery. The best methods are the percussive stun, zero in score and Eugenol because they gave a rapid loss of behavioural indices indicating full anaesthesia or death immediately after they were applied. These methods can therefore be accepted when welfare is considered.

When spiking (Iki-Jime) Atlantic salmon, the results were often difficult to assess because the small brain was not hit correctly at all times. When hit correctly, the method is as good as the percussive stun.

The commercial methods, gill cut and CO_2 + gill cut, do not result in rapid loss of behavioural scores tested either as self-initiated behaviour, stimuli or reflexes. Five minutes after applying the killing method, responses are still recorded which means that the fish are not fully anaesthetised or dead during this period. With regard to fish welfare, these methods should be improved or other methods should be considered. Results from the sensory analysis point out that there are no big differences in eating quality of salmon as a function of the four slaughter methods applied, except that the use of Eugenol oil can be traced in the fish as a special flavour in the cooked flesh.

The project has demonstrated that slaughter methods for farmed Atlantic salmon can be quick and effective (ethical killing), without loss of shelf life or sensory properties.

Differences in pH and onset and development of rigor mortis, depending on killing method, have been recorded, confirming previous results [5,6]. The killing methods chosen in these experiments reflect different degrees of slaughter stress, which also is expected to happen during different pre-slaughter handling methods. We did not find that such differences were critical when the sensory properties of the cooked product were determined. A practical shelf life of gutted Atlantic salmon is around 16 - 18 days in ice, but it is up to the purchaser to decide which degree of freshness is required.

REFERENCES

1. Spedding, C.R.W. (ed.) (1996) Report on welfare of farmed fish. Farm animal welfare council. Surbiton, Surrey, pp. 52.
2. Paterson, B.D., Goodrick, B. and Frost, S.M. (1997) Controlling the quality of aquacultured food products. Trends in Food Science and Technology, 8: 253-257.
3. Kestin, S.C., Van de Vis, J.W. and Robb, D.F.H. (2001) Protocol for assessing brain function in fish and the effectiveness of methods used to stun and kill them. Veterinary Record, 150: 302.
4. Robb, D.F.H., Wotton, S.B., McKinstry, J.L., Sorensen, N.K. and Kestin, S.C. (2000) Commercial slaughter methods used on Atlantic salmon : determination of the onset brain failure by electroencephalography. Veterinary Record., 147: 298-303.
5. Tobiassen, T. (2000) Anaesthesia and killing of salmon and trout. Effects on time before death and quality. MSc Thesis, College of Fisheries, University of Tromsoe, Tromsoe.
6. Robb, D.F.H. (2001) The relationship between killing methods and quality. In: Farmed Fish Quality. Kestin, S.C. and Warriss, P.D. (eds.), Blackwell, Oxford, UK. pp. 220-233.

AUTOXIDATION ACTIVATION ENERGIES OF DOCOSA-HEXAENOIC ACID ETHYL ESTER AND DOCOSAHEXAENOIC TRIGLYCERIDE

Hidefumi Yoshii[a], Takeshi Furuta[a] and Pekka Linko[b]
[a]Department of Biotechnology, Tottori University, Tottori 680-0945, Japan
[b]Department of Chemical Technology, Helsinki University of Technology, Espoo, FIN-02015 HUT, Finland

Abstract

The autoxidation of ethyl docosahexaenoate was investigated at the isothermal condition. The autoxidation of docosahexaenoic triglyceride was investigated with an oxygen sensor. . The activation energy values of DHAEE with the isothermal reaction method were obtained E_a and E_l of 62.5 kJ/mol and 52.1 kJ/mol. The autoxidation activation energy obtained for the DHA oil was 62 kJ/mol. This value is almost same as that of the activation energy for the autoxidation reaction of DHAEE. . This suggests that the activation energies of oxidation do not depend on the chemical structure of DHA.

1. INTODUCTION

The n-3 polyunsaturated fatty acids (PUFAs) have important physiological functions such as antithrombotic, cholesterol depressant, and antiallergenic properties [1]. The application of n-3 PUFAs as food additives is restricted due to their chemically reactive properties. The autoxidation of PUFAs is a free-radical chain reaction, involving initiation, propagation and termination steps. The reaction kinetics of each step has been investigated [2]. For lipid oxidation in food, Labuza [3] has extensively reviewed the kinetic mechanism. Özilgen and Özilgen [4], and Adachi et al.[5] have suggested that the entire oxidation period could be expressed by simple equations of the unoxidized substrate concentration; an autocatalytic reaction and a first order reaction.

In the present study, an oxidative analysis of ethyl docosahexaenoate was carried out with the isothermal oxidation on the basis of the autocatalytic and the first-order kinetic reaction models. The autoxidation kinetics of docosahexaenoic triglyceride (DHA oil) was also investigated using an oxygen sensor and the kinetic parameters of the activation energy of the autoxidation for DHAEE and DHA oil were obtained.

2. MATERIALS AND METHODS

2.1. Materials
DHA oil containing about 47% of DHA and 5% of EPA was obtained from Maruha Corp. Ethyl docosahexaenoate (DHAEE, 92% purity) was from Harima Chemicals, Inc. All other chemicals were of analytical grade.

2.2. Autoxidation of DHAEE at a constant temperature
The isothermal autoxidation of DHAEE was investigated with the gas flow type of equipment. Four µl or DHAEE was carefully measured by a micro-syringe into a flat-bottomed glass bottle (16ϕ × 30mm), from which air had been displaced by nitrogen. These operations were performed in a plastic chamber under a nitrogen atmosphere. The bottles were then held in an air bath at 30-60°C. Oxygen and nitrogen were mixed to an oxygen level of 3-20% and blown into the glass bottles through a Tygon tube (Weyerhaeuser, OH, U.S.A.) at a rate of 40 mL/min. The exit gas was sampled with a gas-tight syringe and analyzed by a TCD gas chromatograph (Shimadzu GC8-A, Kyoto, Japan), using a glass column (2.0 m × 2 mm I.D.) packed with Molecular Sieves 5A (GL Sciences, Tokyo, Japan). The column temperature was 50°C and the filament current 60 mA. Helium was used as the carrier gas at a rate of 40 mL/min. Since it took 2 min to replace the nitrogen in the bottle with the flowing gas mixture, each bottle was connected to the Tygon tube after an interval of 2 min. After a timed period, the glass bottles were taken out, and 4 ml of hexane was added to dissolve the ester. Unoxidized DHAEE was measured by a gas chromatograph (Shimadzu GC14-A) equipped with a flame ionization detector, using a glass column (2.0 m × 2 mm I.D.) packed with 10% Advance-DS on 80/100 mesh Shinchrom A (Shimadzu). The column temperature was 180°C, the detector and injector temperatures were 160°C and 230°C, respectively, and the flow rate of the nitrogen carrier gas was 40 mL/min.

2.3. Analysis of the autoxidation rate constant
A two-step autoxidation mechanism, which can describe the autoxidative process of lipids, has been suggested by Özilgen and Özilgen [4] and Adachi et al.[5] In the first half of the autoxidation period, in which the unoxidized fraction of DHAEE, Y, is greater than 0.5, the autoxidative reaction progresses autocatalytically according to Eq. 1:

$$dY/dt = -k_a Y(1-Y) \tag{1}$$

The integration of Eq.1 gives Eq. 2:

$$\ln\{(1-Y)/Y\} = k_a t + \ln\{1-Y_0)/Y_0\} \tag{2}$$

where k_a is the apparent reaction rate constant at $Y \geq 0.5$, and Y_0 is the initial value of Y. In the latter half of the oxidation process, a first-order reaction with respect to Y was assumed by Adachi et al. [5], as described by Eq. 3:

$$dY/dt = -k_1 Y \tag{3}$$

From Eq. 3, one can obtain Eq. 4:

$$k_a = k_{xa}C_x / (K_a + C_x), \quad k_1 = k_{x1}C_x / (K_1 + C_x) \qquad (4)$$

where k_1 is the apparent reaction rate constant at $Y<0.5$, and t_1 is the time at $Y=0.5$. On the basis of the oxidative reaction scheme, Wong [2], Labuza[3], and Adachi et al. [5] have suggested that reaction rate constants k_a and k_1 were functions of oxygen concentration C_x, which could be expressed by the following Langmuir-type equations (5):

$$\ln 2Y = -k_1(t - t_1) \qquad (5)$$

where K_a and K_1 are the Langmuir parameters for oxygen, and k_{xa} and k_{x1} are intrinsic reaction rate constants, each of which is a function of temperature according to the Arrhenius equation, leading to Eq. 6:

$$k_{xa} = k_{xa0} \exp(-E_a / RT), \quad k_{x1} = k_{x10} \exp(-E_1 / RT) \qquad (6)$$

where E_a and E_1 are the activation energy for k_{xa} and k_{x1}, respectively, and k_{xa0} and k_{x10} are the respective frequency factors. Reaction rate constants k_a in Eq. 2 and k_1 in Eq. 4 can be calculated from the slope of lines by plotting $\ln\{(1-Y)/Y\}$ vs. t and $\ln(2Y)$ vs. t, respectively.

2.4. Measurement of DHA oil with a oxygen sensor

Initially the oxidation rates of Docosahexaenoic Triglyceride (DHA oil) was measured with a oxygen sensor. The reaction vessel (internal diameter, 25mm; vessel height, 135mm) was maintained at a constant temperature with a water bath (Eyela, NTB-211, Tokyo). Three ml of DHA oil was put into the reaction vessel, followed by stirring the sample at 400 rpm with a magnetic stirrer (Vari-Mag, Telemodel 20 P, Osaka). The oxygen concentration would decrease with the progress of the oxidation. The oxygen pressure was measured by a fluorescence oxygen analyzer (FO-900, ASR, Tokyo) connected via an RS-232C communication interface to the computer (Fujitsu, FM-V, Tokyo). The autocatalytic reaction rate constants for DHA oil were obtained at 35, 50, and 70°C.

2.5. Analysis of autoxidation of DHA oil and the emulsified DHA oil with oxygen sensor

The initial change of oxygen pressure in the reaction vessel was considered with the autoxidation kinetic model. The autoxidative reaction was assumed to progress autocatalytically[6] as follows (7):

$$DHA + O_2 \rightarrow Products \qquad (7)$$

In this oxidation reaction, the oxygen pressure in the closed reaction vessel was expressed with s function of the unoxidized fraction of DHA according to Eq. 8:

$$P_{O_2}^0 - P_{O_2} = \frac{RT}{V} DHA_0 (1-Y) \qquad (8)$$

where Y is the unoxidized fraction of DHA, DHA_0 is the initial DHA concentration, V is the closed vessel volume, T is the temperature, and R is the gas constant. On the basis of the oxidative reaction scheme, the Langmuir-type autoxidation kinetic equation can be described as Eq. 9;

$$-dY/dt = \frac{k_a P_{O_2}}{K_a + P_{O_2}} Y(1-Y) \tag{9}$$

where k_a is the apparent reaction rate constant at $Y>0.5$, and K_a is the Langmuir parameter for oxygen (in the reaction system of DHA, K_a used was 1.1 kPa according to Adachi et al. [3]. From Eqs. (8) and (9), the dimensionless oxygen pressure, λ, given by the divided value of the oxygen pressure P_{O_2} by the initial oxygen pressure $P_{O_2}^0$, could be described by the following Eq. (10):

$$-\frac{d\lambda}{dt} = \frac{k_a \lambda}{K_a' + \lambda}(1-\lambda)(1 - \frac{\alpha}{T}(1-\lambda)) \tag{10}$$

where K_a' is $K_a / P_{O_2}^0$ and α is $V/(R \cdot DHA_0)$. The integration of Eq. (10) gives Eq. (11).

$$F(t) = -(1+K_a')\ln(1-\lambda) - \frac{K_a'(T/\alpha)\ln\lambda}{1-(T/\alpha)} + \frac{(1+K_a'-(T/\alpha))\ln(\alpha(-1+T/\alpha+\lambda))}{1-(T/\alpha)} = F_0 - k_a t \tag{11}$$

where F_0 is the initial value of $F(t)$ at time zero. Equation (11) indicates that the reaction rate constant, k_a can be determined by a linear plot of the right-hand side of Eq. 11 against time, t.

3. RESULTS AND DISCUSSION

3.1. Autoxidation of DHAEE under the isothermal condition

Time-course plots of the autoxidation of DHAEE at different temperatures are illustrated in Fig. 1 for DHAEE. The ordinate is the unoxidized fraction of DHAEE with respect to the initial quantity. The autoxidation process had a short induction period, followed by an autocatalytic reaction region in which autoxidation progressed very rapidly. The induction period decreased with increasing temperature. The elevation of temperature markedly affected the autoxidation process, as shown in Fig. 1.

To determine reaction rate constants k_a and k_1, the unoxidized fraction of DHAEE, Y, at time t was plotted according to Eqs. 2 and 4 as shown in Fig. 2. The open symbols represent the data rearranged according to Eq. 2, and the filled symbols to Eq. 4. The experimental values of unoxidized fraction Y correlated well with Eq. 2 for Y larger than 0.5, except for the data during the induction period. The slopes of the solid lines correspond to autoxidation rate constant k_a during the period. The dotted lines in Fig. 2 represent the extended use of Eq. 2 for Y less than 0.5. When Y became less than 0.5 (i.e. the ordinate value is larger than 0 in Fig. 2), the experimental data began to deviate from Eq. 2 as shown by the dotted lines in Fig. 2. For Y less than 0.5, a first-order reaction approximation (Eq. 4) was valid. However, during the final period of the autoxidation

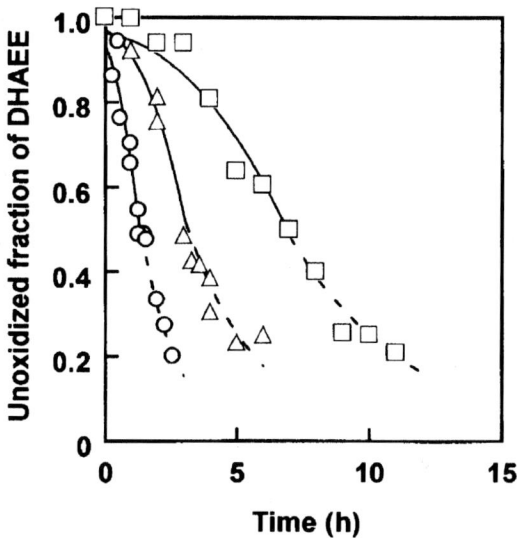

Figure 1. Changes in the unoxidized fraction of DHAEE in autoxidative process for DHAEE at different temperatures. Solid lines and dotted lines are results calculated by Eq. 2 ($Y \geq 0.5$) and Eq. 4 ($Y < 0.5$), respectively. DHAEE: ○ T=313 K; △ 323K; □ 333K; C_x=7.3-7.8 mol/m^3 (20 %).

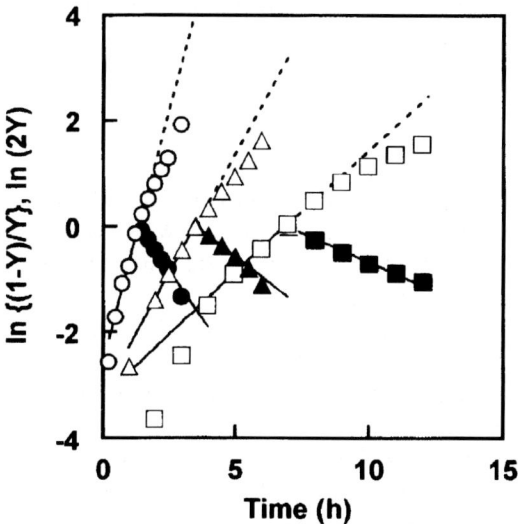

Figure 2. Analysis of the oxidative reaction rate Constants for EPA and DHA at different oxygen levels and temperatures. Rate constants k_a and k_1 correlate well with Eqs. 2 and 4. Open symbols represent $\ln\{(1-Y)/Y\}$ vs. time, and filled symbols $\ln(2Y)$ vs. time. Solid lines are the correlation lines by Eq. 2 ($Y \geq 0.5$) and Eq. 4 ($Y < 0.5$). Dotted lines show the extended use of Eq. 2 at $Y < 0.5$. The oxidation conditions and symbols are the same as those given in Fig. 1.

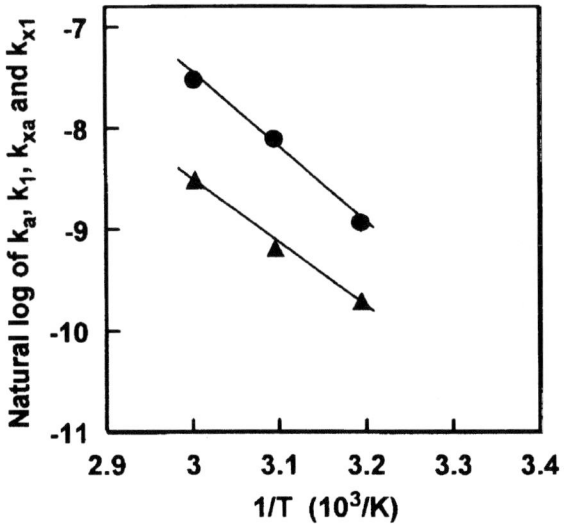

Figure 3. Arrhenius plots of reaction rate constants k_a and k_1 for DHAEE. ●, k_a; ▲, k_1.

process, Y values could not be correlated by Eq. 4, as illustrated in Fig. 2. Reaction rate constant k_1 could be determined from the correlation line in the valid range.

Calculated unoxidized fraction Y against time is illustrated by solid ($Y \geq 0.5$) and by dotted ($Y < 0.5$) lines in Fig. 1, which were constructed by using the reaction rate constants from Fig. 2. The calculated results are in a good agreement with the experimental values.

The activation energy and frequency factor were determined by the slopes of the solid lines in Fig. 3. The plots of k_a and k_1 for DHAEE against $1/T$ are shown by filled symbols at an oxygen level of 20%. In this case, activation energy values E_a and E_1 of 62.5 kJ/mol and 52.1 kJ/mol, respectively, agree well with the results of Adachi et al.[5] Activation energy E_a for DHAEE obtained here was also close to that for tridocosahexanoin (45% purity) reported by Yoshii et al. [7].

3.2. Analysis of autoxidation of DHA oil

The oxidation processes of DHA oil at 308, 323 and 343 K are shown in Fig. 4 and the plot of the right-hand side value, $F(t)$ of Eq. 11 for Fig. 4 is shown in Fig. 5. The induction period at 308 K was observed clearly in the oxidation of DHA oil. The straight lines for the three oxidation temperatures were obtained for the autoxidation of DHA oil. The experimental data in the initial period fitted well with the autoxidation reaction model, and the autoxidation rate constant k_a could be obtained with the Eq. (11). The activation energy obtained for the DHA oil was 62 kJ/mol with Arrhenius plot.

3.3. Comparison of the activation energy of DHAEE and DHA Oil

The activation energy values of DHAEE with the isothermal reaction method were obtained E_a and E_1 of 62.5 kJ/mol and 52.1 kJ/mol. The autoxidation activation energy

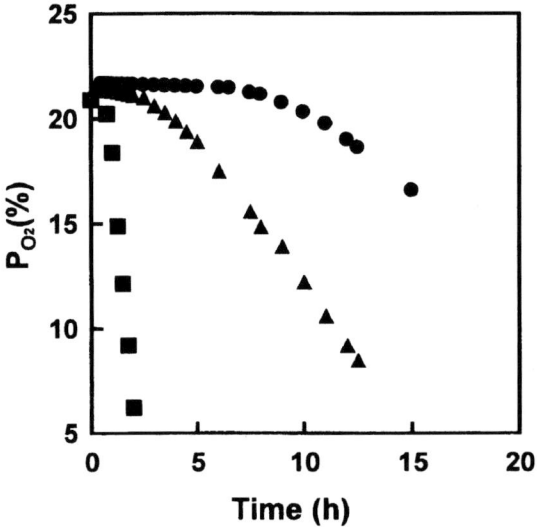

Figure 4. Changes in oxygen pressure during DHA oil oxidation. ●, 308K; ▲, 323K; ■, 343K.

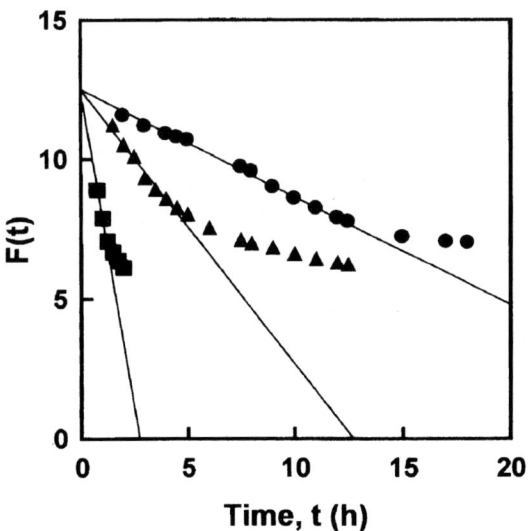

Figure 5. Changes in F(t) during the autoxidation of DHA oil. ●, 308K; ▲, 323K; ■, 343K.

obtained for the DHA oil was 62 kJ/mol with the oxygen sensor. This value is almost same of the activation energy for the autoxidation reaction of DHAEE.

4. CONCLUSION

The autoxidation kinetics of DHAEE and DHA oil were investigated at various temparature. Presented data have respect to autoxidation of DHAEE and DHA oil as polyunsaturated acid compounds. The kinetic data of DHAEE was confirmed with the gas chromatography method. The kinetic data of DHA oil was confirmed with the oxygen sensor method. The activation energies of autoxidation rate constants for DHAEE and DHA oil were about 62 kJ/mol. This suggests that the activation energies of oxidation do not depend on the chemical structure of DHA. The results show that the rate equations used of the autoxidation kinetics for DHA are useful to obtain the autoxidation kinetic constants.

REFERENCES

1. Simopoulos, A. P. (1999) Essential fatty acids in health and chronic disease. Am. J. Clin. Nutr., 70: 560S-569S.
2. Wong, D. W. S. (1989) Mechanism and theory in food chemistry. Van Nostrand Reinhold, New York, pp.1-3.
3. Labuza, T. P. (1971) Kinetics of lipid oxidation in foods. *CRC* Critical Reviews in Food Technology, vol.2: 355-405.
4. Özilgen, S., and Özilgen, M. (1990) Kinetic model of lipid oxidation in foods. J. Food Sci., 55: 498.
5. Adachi, S., Ishiguro, T. and Matsuno, R. (1995) Autoxidation kinetics for fatty acids and their esters. JAOCS, 72: 547-551.
6. Adachi, S., Ishiguro, T. and Matsuno, R. (1995) Thermal analysis of autoxidation of ethyl esters of n-3 and n-6 fatty acids. Food Sci. Technol. Int., 1: 1-4.
7. Yoshii, H., Furuta, T., Kawasaki, K., Hirano, H., Funatsu, Y., Toyomi, A. and Nakayama, S. (1997) Oxidation stability of powdery tridocosahexaenoin included in cyclodextrin and its application to fish meal paste. Biosci. Biotech. Biochem., 61: 1376-1378.

M. Sakaguchi (Editor)
More Efficient Utilization of Fish and Fisheries Products
© 2004 Elsevier Ltd. All rights reserved

HEMOCYANIN-RELATED REACTIONS INDUCE BLACKENING OF FREEZE-THAWED PRAWNS DURING STORAGE

Kohsuke Adachi, Takashi Hirata, Katsunori Nagai, Atsushi Fujio and Morihiko Sakaguchi
Division of Applied Biosciences, Graduate School of Agriculture, Kyoto University, Kitashirakawa, Sakyo, Kyoto 606-8502, Japan

Abstract

Prawns are one of the most important fisheries resources all over the world due to their high market value. They are frozen in many cases in the process of transportation and storage. But their head and tails are sensitive to discoloration after thawing, which severely damages their price. It is generally accepted that this phenomenon is ascribed to melanogenesis resulting from enzymatic oxidation of phenols by phenoloxidase (PO) expressed as its precursor prophenoloxidase from hemocyte. However PO is present in hemocyte at very low concentration (several micrograms per prawn) and inactivated within a week under frozen conditions. On the other hand, hemocyanin, which shows high similarity to prophenoloxidase in primary structure, is present in hemolymph at quite high concentration (several milligrams per prawn). This respiratory pigment is converted into phenoloxidase-like enzyme by hemocyte component. Hemocyanin-derived phenoloxidase (HdPO) is quite stable, maintaining its activity over two months. These results indicate that hemocyanin is a potent inducer of post-harvest blackening and there is a novel approach to develop anti-blackening treatment.

1. INTRODUCTION

Prawns are quite important fishery resources all over the world owing to their high market value. In Japan, however, the supply heavily depends on the imported prawn mainly from Asian countries such as India, Indonesia, Vietnam and Thailand. The total quantity is estimated to be 247,313 t while the output of domestic prawn is 28,436 t in 1999. The value traded for the prawn is 364.4 billion yen, which is the largest among all the imported marine products in Japan (Table 1) [1].

The imported prawns are usually frozen in the process of importation and storage period but they sensitively discolor after thawing, becoming black on their heads and tails. The discoloration severely damages the market value of the prawn while their nutrition and taste do not deteriorate. It has been generally accepted that the phenomenon is

Table 1
Imported amounts of principal marine products

	1989	1994	1995	1996	1997	1998	Growth ratio (%) 96/97	97/98
Quantity (1,000 ton)								
Total	2,292	3,296	3,582	3,450	3,411	3,103	-1.1	-9.0
Sub total:fresh, perishable, cold and frozen products	1,866	2,530	2,540	2,560	2,489	2,355	-2.8	-5.4
Prawns	284	320	312	305	282	251	-7.6	-10.8
Skipjacks and tunas	224	298	306	309	280	316	-9.4	13.0
Salmons and trouts	149	243	203	232	209	224	-10.1	7.1
Crabs	73	124	121	130	124	123	-4.3	-0.5
Cods	148	214	216	185	188	149	1.5	-21.0
Octopuses	112	106	98	95	79	77	-18.1	-2.1
Cuttle fishes	116	116	86	108	96	93	-11.0	-2.4
Flatfishes	89	86	71	89	79	65	-11.5	-17.0
Sub total:salted, dried and smoked products	43	44	42	44	43	34	-2.0	-20.7
Eggs of herrings	10	9	8	10	11	10	14.6	-14.3
Eggs of salmons and trouts	9	11	11	7	7	5	-8.8	-24.4
Sub total:processed products	112	193	224	247	272	254	10.2	-6.6
Eels	21	39	36	46	55	52	21.5	-5.9
Sub total:others	272	529	776	599	607	459	1.4	-24.3
Fish meal	167	379	588	408	432	324	5.9	-25.0
Value (100 million yen)								
Total (A)	14,505	17,091	17,212	19,138	19,456	17,416	1.7	-10.5
Sub total:fresh, perishable, cold and frozen products	11,845	13,710	13,593	14,861	14,967	13,535	0.7	-9.6
Prawns	3,540	3,753	3,686	3,772	3,930	3,644	4.2	-7.3
Skipjacks and tunas	1,379	1,865	1,819	2,204	2,034	2,039	-7.7	0.2
Salmons and trouts	1,300	1,313	995	1,220	1,189	1,240	-2.6	4.3
Crabs	671	1,229	1,278	1,238	1,089	957	-12.0	-12.1
Cods	358	471	487	403	537	352	33.4	-34.4
Octopuses	620	435	499	653	555	455	-15.0	-18.1
Cuttle fishes	624	570	470	572	541	521	-5.3	-3.7
Flatfishes	288	261	239	317	264	209	-16.9	-20.7
Sub total:salted, dried and smoked products	641	516	527	642	514	404	-19.9	-21.4
Eggs of herrings	230	168	173	264	172	144	-34.9	-16.1
Eggs of salmons and trouts	150	157	142	102	86	69	-15.6	-20.5
Sub total:processed products	1,306	2,000	2,109	2,639	2,887	2,493	9.4	-13.6
Eels	539	872	863	1,031	1,152	847	11.7	-26.5
Sub total:others	713	856	893	996	1,089	984	9.3	-9.6
Fish meal	135	202	329	307	356	356	16.0	-8.7
Total value of all imports in Japan (B)	289,786	281,043	315,488	370,934	409,562	366,536	7.8	-10.5
(A)/(B) (%)	5.0	6.1	5.5	5.0	4.8	4.8		

Suisan Nenkan (2000) [1]

ascribed to melanogenesis resulting from enzymatic oxidation by phenoloxidase (PO), which catalyzes the hydroxylation of phenols to o-diphenols and deprotonation of o-diphenols to o-quinones, leading to melanin formation [2].

PO is expressed from their hemocyte as its precursor proPO and functions in various phases of living body such as sclerotization, pigmentation, wound healing on cuticles and defense reactions. In the case of immune response, the stimulation of proPO into PO is regulated by a serine protease called prophenoloxidase activating enzyme (PPAE), which is triggered by the cell wall components (lipopolysaccharides, beta-1,3-glucans or peptidoglycans) of bacteria or fungi [3, 4]. The post-harvest conversion of proPO into PO presumably occurs in the same manner, as observed in the living body and may induce the melanogenesis during storage, although there remains much to be investigated further.

The achievement of PO has been accumulated mainly in insects, which are classified into different class (Insecta) from prawns (Crustacea) although they belong to the same phylum. Indeed they share many physiological characters but several properties clearly differ, for example, insects do not contain oxygen transporter molecule owing to their developed trachea while prawns involve respiratory pigment hemocyanin; insects excrete nitrogen as uric acid in Malpighian tubule while prawn as ammonia in antennal gland; insects hardened their cuticle by cross-linking of protein while prawn by deposition of calcium, etc. Despite these discrepancies, mechanism of insect melanogenesis has been applied to post-mortem blackening of crustacean so far.

There are several reports about melanogenesis of prawn caused by phenol-oxidase [2]. However, the subjects of these reports are mainly focused on crude enzyme from whole heads and cuticle, otherwise purified enzyme from cuticle. No data has been shown about the direct interrelationship of post-harvest blackening and PO from hemolymph. It is well known that PO and its precursor are very unstable protein, which easily aggregate and are inactivated even in the course of experimental procedure. Melanogenesis of prawn, on the other hand, occurs after shipment and storage process, which last for several months. These clearly inconsistent facts stimulate us to investigate the causal relationship between post-mortem melanogenesis and PO.

In this study we investigate the contribution of "phenoloxidase" in prawn to post-harvest melanogenesis in view of enzymatic properties and its primary structure.

2. MATERIALS AND METHODS

2.1. Experimental animals

Kuruma prawns (*Penaeus japonicus*; body weight: 20-30 g, total length: 15-18 cm) were purchased from a wholesale market in Kyoto, Japan. After treatment with 0.1 % sodium sulfamonomethoxine solution for 1 h, they were maintained in a 200-L artificial seawater tank at 20°C.

2.2. Preparation of prophenoloxidase and hemocyanin

Prophenoloxidase was purified from hemocyte using Blue-sepharose, DEAE-cellulose, Hydroxyapatite and Butyl-sepharose. Hemocyanin was isolated from hemolymph, using ultracentrifugational fractionation and sepharose-CL 6B. Details were described previously [5,6].

2.3. Assay for phenoloxidase activity
Phenoloxidase activity was assayed using a modification of Horowitz and Shen [7]. Details were described previously [5,6].

2.4. Molecular cloning of proPO

2.4.1. *Isolation of RNA from the hemocyte*
The collection of hemocyte was performed as described previously. Total RNA was extracted following AGPC method and ultracentrifugation in 5.7 M cesium chloride after disruption of hemocyte by passing through the needle in 4 M guanidine thiocyanate solution. Poly (A) + RNA was selected with Oligotex-dT 30 super (TAKARA).

2.4.2. *Probe for cDNA library screening*
Degenerated oligonucleotide primers pjproPO1 (sense: 5'-CGGGATCCCGTG(CT)(GA)(GA)(ATGC)TG(CT)GG(CGT)TGGCC-3') and pjproPO2 (antisense: 5'-CGGAATTCCG(AG)TC(GA)AA(GC)GG(AG)AT(AT)A(ATG)CCCAT) were designed from highly conserved amino acid sequence of complement-like domain of proPO. *Bam* HI and *Eco* RI cleavage site are added to the 5'-end of pjproPO1 and pjproPO2 respectively, for the insertion of its PCR products to pBluescript. Single-stranded cDNA was synthesized from 1 µg total RNA in 20 µL reaction mixture containing 50 mM Tris-HCl (pH 8.4), 75 mM KCl, 3 mM $MgCl_2$, 1.25 mM each dNTP, 0.5 µg Oligo dT12-18 primer and 200 units of MMLV-RTase. PCR was carried out for 1.5 µL resultant solution as a template in a 10 µL of mixture containing 10mM Tris-HCl (pH 8.3), 50 mM KCl, 1.5 mM $MgCl_2$, 0.1 % TritonX 100, 0.2 mM each dNTP and 0.4 unit Taq DNA polymerase (TOYOBO). The reaction condition was 30 s at 94°C, 30 s at 55°C, and 60 s at 72°C for 35 cycles (ASTEC, Program Temp Control System PC 700). The PCR product was subcloned into pBluescript KS+ and sequenced by ALFred DNA sequencer (Pharmacia). Then we confirmed about 200 bp DNA sequence highly homologous to crayfish proPO. The fragment was labeled with ^{32}P by PCR in a 20 µl of mixture containing template plasmid 1 ng, 120 mM Tris-HCl (pH 8.0), 10 mM KCl, 6 mM $(NH_4)_2SO_4$, 0.1 % Triton X-100, 2 µg BSA, 0.5 µM each primer, 1.25 unit KOD DNA polymerase (TOYOBO), 0.2m M each dNTP except for dCTP and 150 µCi [α-^{32}P] dCTP. The reaction condition was 1 min at 94°C, 1 min at 55°C, and 3min at 72°C for 35 cycles.

2.4.3. *Construction and screening of cDNA library*
cDNA library was constructed from Poly (A) + RNA of hemocyte with using Lamda ZAP II cloning KIT (STRATAGENE) and Gigapack (STRATAGENE) according to manufacturer's instruction and then amplified. After hybridization with the probe mentioned above, the filters were washed once 2 x SSC at 42°C for 15 min and 3 times in 2 x SSC/0.5xSDS for 15 min prior to autoradiography.

2.4.4. Sequence analysis and phylogenetic studies

The deduced amino acid sequence was compared with the database (PIR , SWISS-PROT, DAD, PDB) using FASTA program. Phylogenetic tree construction by neighbor-joining method [8] was done with Clustal W [9], using the amino acid sequences as follows. Accession number to EMBL/GenBank/DDBJ data bank is shown within parenthesis.

Plant polyphenoloxidase lycopo, tomato (*Lycopersicon esculentum*) tyrosinase(z12833); sopo, spinach (*Spinacia oleracea*) tyrosinase (u19270); mdpo, apple (*Malus domestica*) tyrosinase(L29450); papo, apricot (*Prunus armeniaca*) tyrosinase (AF020786).

Vertebrate tyrosinase olty, medaka (*Oryzias latipes*) tyrosinase (D29687); rnty, Japanese pond frog (*Rana nigromaculata*) tyrosinase (D12514); ggty, chicken (*Gallus gallus*) tyrosinase (D88349); mmty, mouse (*Mus musculus*) tyrosinase (DD00440); hsty, human (Homo sapiens) tyrosinase (M27160).

Bacterial and fungal tyrosinase nsty, fungus (*Neurospora crassa*) tyrosinase (M32843); smty, bacterium (*Sinorhizobium meliloti*) tyrosinase (X69526).

Insect hexamerin bdhex, tropical cockroach (*Blaberus discoidalis*) hexamerin (U31328); mdhex, housefly (*Musca domestica*) hexamerin (AF188888); aghex, malaria vector mosquito (*Anopheles gambiae*) hexamerin (U51225).

Hemocyanin pihca, pihcb and pihcc, spiny lobster (*Panulirus interruptus*) hemocyanin subunit a [10], b [11] and c [12]. pvhc1 and pvhc2, white shrimp (*Penaeus vannamei*) hemocyanin subunit 1 (X82502) and 2 (AJ250830); echca-echcg, tarantula (*Eurypelma californicum*) hemocyanin subunit a-g (a; X16893, b; AJ290429, c; AJ227489, d; AJ290430, e; X16894, f; AJ277491, g; AJ277492)

Arthropod prophenoloxidase agppo1-6 malaria vector mosquito (*Anopheles gambiae*) prophenoloxidase (1; L76038, 2; AF04915, 3; AF004916, 4; AJ010193, 5; AJ010194, 6; AJ010195); bmppo1 and bmppo2, silk worm (*Bombyx mori*) prophenoloxidase (1; D49370, 2; 49371); msppo, tobacco horn worm (*Manduca sexta*) prophenoloxidase (L42556); dmppo, fruit fly (*Drosophila melanogaster*) prophenoloxidase (D45835); plppo, crayfish (*Pacifastacus leniusculus*) prophenoloxidase (X83494); pjppo, kuruma prawn (*Penaeus japonicus*) prophenoloxidase (AB065371).

2.5. Storage stability of hemocyanin-derived phenoloxidase

Both Hc and proPO solutions in 10 mM Tris-HCl buffer (pH 7.3) were frozen at -25°C and their residual PO activities were periodically measured for a month. The active forms of both enzymes were also frozen and their activities were measured in the same manner as their inactive forms. ProPO and Hc were activated by 0.05 % and 0.1 % SDS, respectively. When the activity was measured, the solutions were thawed in a water bath at 20°C for 10 min.

3. RESULTS

3.1. Purification, molecular cloning and biochemical characterization of proPO from hemocyte

Prophenoloxidase (proPO) was purified from hemocyte of 100 prawns. Obtained protein was about 0.1 mg [5]. Considering the loss of the experimental step (recovery: 25%), it was supposed that about 4 microgram of proPO was present per prawn. The molecular mass of the proPO is estimated to be 330 kDa by gel filtration [5]. In SDS-PAGE analysis, two closely migrating bands of 78 kDa and 72 kDa were detected under reducing condition, indicating proPO exists as heterotetramer. It has been reported that proPO in insects is activated with the removal of its propeptide by serine protease called prophenoloxidase activating enzyme (PPAE) [13-15]. We have already identified the existence of this activating factor in hemocyte of kuruma prawn, which specifically hydrolyzed synthetic peptide t-butyloxycarbonyl Val-Pro-Arg-MCA and stimulated its activity by beta-1,3 glucan, laminarin [16]. It is supposed that in storage process proPO is also activated by PPAE and catalyzes the oxidation of phenols. ProPO is also effectively activated by SDS treatment, which is supposed to induce the conformational change of the protein without cleavage of propeptide [5]. It is likely that proPO is stimulated by two types of mechanism, one is induced by PPAE and the other is by the change of tertiary structure of the protein. We also tested several physical stresses (heat treatment, freezing and thawing and osmotic pressure) upon the activation of this proenzyme but no influence was observed [5]. PO has its optimum pH and temperature at 9 and 37°C, respectively. However, the activity was suppressed up to about 20 % at the pH of hemolymph (pH 7.5) compared with optimum condition. This enzyme oxidized mono-phenols and o-diphenols indicating PO from hemocyte catalyzes a consecutive reaction, hydroxylation of mono-phenols and deprotonation of o-diphenols [5]. These biochemical properties are quite normal among arthropods' phenoloxidase, however, the experimental steps are severely hampered due to the instability of this enzyme, which was inactivated within few days while black spots proceeds after several months' storage. We focused on this contradicting fact and tested its enzymatic stability under frozen condition. Figure 1

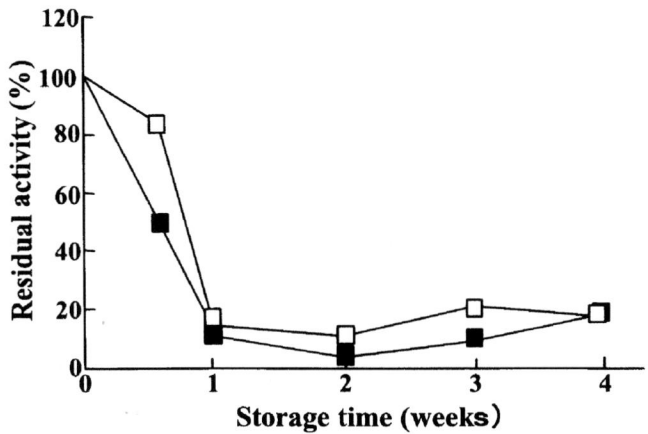

Figure 1. Stability under frozen condition. PO and proPO were frozen and stored at -25 °C. PO was activated by 0.05% SDS. After thawing the sample in water at 20°C for 10 min, residual activity was measured as described in Materials and Methods. □, proPO; ■, PO. Symbols indicate the means of triplet (S.D.< 4).

shows the enzymatic stability of PO and proPO at -25°C, which is the similar temperature to that of transportation and storage. Both samples were thawed in water at 20°C for 10min. and their residual activities were measured. As we expected, both PO and proPO were very unstable and almost completely lost their activity within a week. Taken together with the results that the amount of enzyme is quite little (several microgram per prawn), these results suggest PO from hemocyte is not a main factor of post-harvest blackening and imply the possible contribution of another factor.

Figure 2 shows the nucleotide sequence and deduced amino acid sequence of proPO

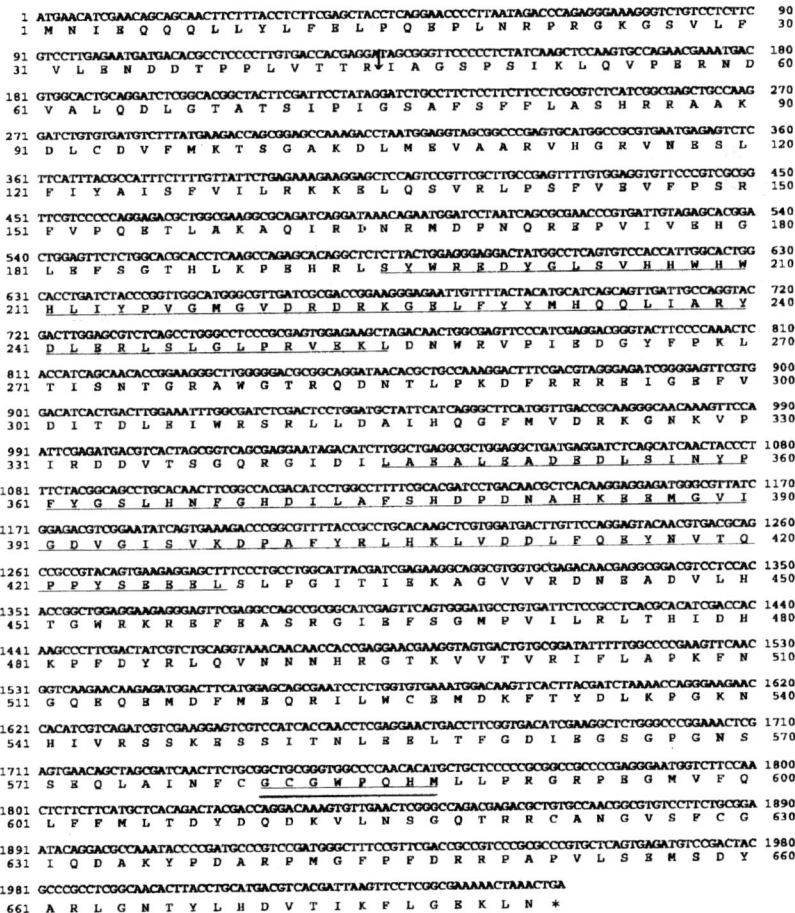

Figure 2. Nucleotide and deduced amino acid sequence of *Penaeus japonicus* proPO. The putative copper-binding domains and activation cleavage site are underlined and marked with arrow, respectively. The predicted N-glycosylation sites are indicated with N. Complement-like domain is double underlined. The accession number to Genbank/EMBL/DDBJ is AB065371.

Table 2
Amino acid sequence identity of *Penaeus japonicus* proPO to other proteins

Proteins	Identity (%)
Pacifastacus leniusculus (crayfush) propheoloxidase	35
Bombyx mori (silk worm) prophenoloxidase	38
Drosophila melanogaster (fruit fly) prophenoloxidase	29
Panulirus interruptus (spiny lobster) hemocyanin subunit A	30
Penaeus vannamei (white shrimp) hemocyanin	38
Eurypelma californicum (tarantula) hemocyanin subunit a	15
Human tyrosinase	13
Medaka tyrosinase	14
Neurospora crassa (fungi) tyrosinase	11
Sinorhizobium meliloti (bacteria) tyrosinase	57

isolated from hemocyte cDNA library of kuruma prawn. This clone contains 2043bp open reading frame which codes for a protein of 681 amino acids with highly similar sequence to marine prawn *Penaeus monodon* [17] including N-terminus region. The computed molecular mass 78091.73 was quite adjacent to the value of one subunit of purified proPO calculated by SDS-PAGE (78k: the other is 72k) [5]. These facts indicate the clone contains the entire coding regions.

Sequence identity to other species' proteins is shown in Table 2. The deduced amino acid sequence shows significant similarity to arthropod prophenoloxidase. The scores also indicate kuruma prawn prophenoloxidase is similar to hemocyanin, an oxygen transporter, rather than every type of tyrosinase from vertebrate, fungi and bacteria which catalyzes the same reaction as phenoloxidase *in vitro*. Figure 3 shows the multiple alignments of two copper binding domains, Cu (A) and Cu (B) which are active center of the enzyme. Three histidine residues in each domain which function as copper binding ligands are perfectly conserved in other species' proPO and hemocyanin. On the other hand, little similarity was observed to vertebrate, bacterial and fungal tyrosinase (data not shown) although histidine residues are conserved. We constructed the phylogenetic tree using prophenoloxidase, hemocyanin, vertebrate tyrosinase, bacterial and fungal tyrosinase, plant polyphenoloxidase and insect hexamerin (Fig. 4). Hexamerin is the storage protein in insect, which belongs to the same family as hemocyanin. Plant polyphenoloxidase is the counterpart of arthropod phenoloxidase and vertebrate tyrosinase. From this tree it is assumed that vertebrate tyrosinase constitute one family while arthropod hemocyanin and prophenoloxidase group with another family. Prophenoloxidase, hemocyanin and hexamerin seem to have evolved from a common ancestor which has no relation with vertebrate tyrosinase. In this family, hemocyanin and prophenoloxidase is closer to each other than to insect storage protein hexamerin. Although crustacean hemocyanins diverse from prophenoloxidases in earlier point than chelicerates hemocyanin, it is assumed that crustacean hemocyanin are closer to prophenoloxidase rather than vertebrate, bacterial and fungal tyrosinase.

Hemocyanin, an oxygen transporter in crustaceans is present in their hemolymph at quite high concentration (about 90% of serum protein) [18] while only a small amount of prophenoloxidase is in hemocyte. ProPO from hemocyte is very unstable even in low

Figure 3. Comparison of putative copper-binding sequences of proPO and hemocyanin. Copper (A) and (B) binding domains are each aligned. Numbers at left of each lane represent the amino acid residues of proteins. Perfectly conserved amino acid residues in sample sequences are enclosed by solid lines. Histidine residues marked with closed circle are predicted ligands for two oxygen-binding copper atoms. Gaps (−) are introduced to optimize alignment. PJPPO, kuruma prawn *Penaeus japonicus* proPO; PLPPO, cray fish *Pacifastacus leniusculus* proPO; DMPPO, fruit fly *Drosophila melanogaster* proPO; ECHCY, tarantula *Eurypelma californicum* hemocyanin subunit a; PVHC, white shrimp Penaeus vannamei hemocyanin; PIHCY, spiny lobster *Panulirus interruptus* hemocyanin subunit A.

temperature storage condition, suggesting this enzyme is probably irrelevant to post-harvest blackening. The author showed the high sequence identity between prophenoloxidase and hemocyanin and their closeness in molecular evolution. If proPO and hemocyanin are originated from the same molecule and retain the same physiological function, both proteins may share the same biochemical function.

As assumed in their primary structural analysis, phenoloxidase activity of hemocyanin has been recently reported in some species [19, 20]. So we assumed that this respiratory pigment, hemocyanin, is the authentic factor responsible for the post-harvest blackening and investigated its biochemical character.

3.2. Isolation and enzymatic characterization of hemocyanin

Hemocyanin (Hc) was purified from hemolymph using ultracentrifugal fractionation and gel filtration chromatography [6]. The author was able to obtain about 0.8 mg protein per prawn with this isolation method. Considering the loss in each step, the total amount of Hc per prawn was estimated to be several mg. This amount was several thousand times

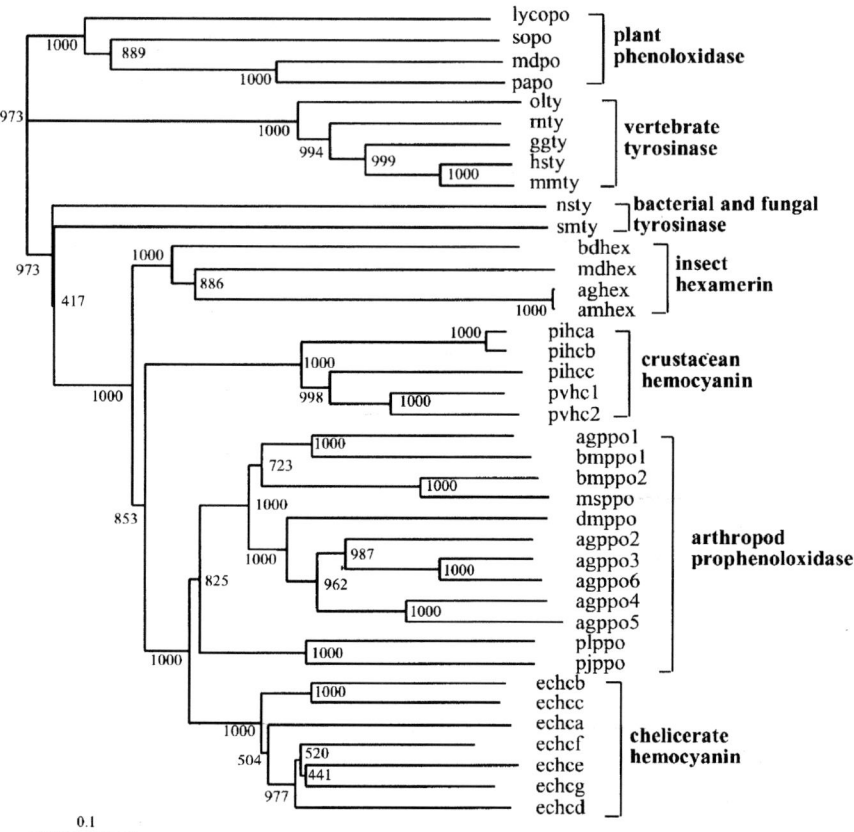

Figure 4. Phylogenetic tree showing the relationship among tyrosinase, hemocyanin, hexamerin and phenoloxidase. The tree was constructed by neighbor-joining method rooted with the plant polyphenoloxidase as a outgroup. The scale bar indicates the length of a branch length of 0.1. The numbers at nodes represent the statistical confidence estimates computed by the bootstrap procedure.

higher than that of proPO from hemocyte. Isolated Hc showed two closed bands in SDS-PAGE analysis, whose molecular weights were 67,000 and 77,000 [6], respectively.

Native Hc itself does not have any enzymatic activity just like proPO from hemocyte. SDS treatment effectively converted Hc into o-diphenoloxidase-like enzyme as well as proPO [6]. No effects on its conversion were observed after freezing-and-thawing treatment (data not shown), indicating this physical process does not directly work on Hc conversion into PO-like enzyme. When 1 unit of enzyme activity was defined as the increase of absorbance 1.0 at 490 nm per mg enzyme per 10 min, the specific activity of HdPO and PO were 18.4U and 80.6U, respectively [6]. The optimum temperature of HdPO was around 40°C. The profile of optimum pH of HdPO had two major peaks, around 4.9 and 8.3, although it was labile to the former condition[15]. HdPO catalyzes a

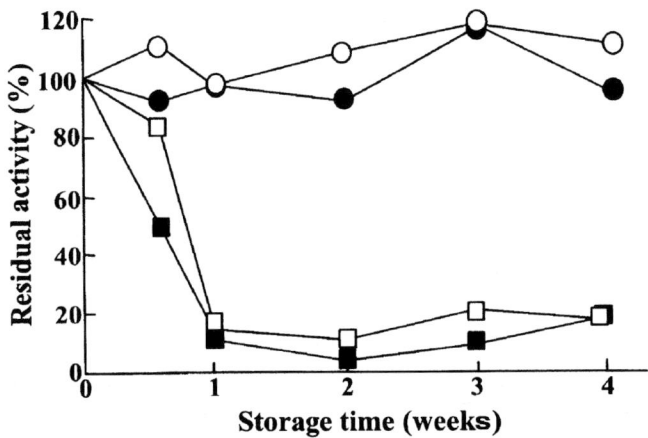

Figure 5. Stability under frozen condition. Both Hc and proPO were frozen and stored at -25°C in their active (HdPO and PO) and inactive forms (Hc and proPO). proPO and Hc were activated by 0.05% and 0.1% SDS, respectively. After thawing the sample (in water at 20°C for 10 min), residual activity was measured as described in Materials and Methods (HdPO) and our previous report (PO). ○, Hc; ●, HdPO; □, proPO; ■, PO. Symbols indicate the means of triplet (S.D.< 4).

consecutive reaction as well as PO: hydroxylation of mono-phenols and deprotonation of o-diphenols [6].

We could observe the several similar biochemical properties to those of PO, however hemocyanin is quite easy to be handled due to its opposite property to proPO, which is quite stable and not inactivated in long experimental step. Figure 5 shows the residual activity of HdPO under frozen condition (-25°C). HdPO completely retained its activity, either in the active form or in the latent form while PO and proPO were almost inactivated within a week. These findings indicate HdPO is a potent inducer of black spot development, which generally proceeds in the prawn when thawed after frozen storage.

4. DISCUSSION

The author here suggested the potential role of PO-like activity of Hc in the black spot development of kuruma prawns. Earlier workers have clarified the high similarity between amino acid sequences of Hc and proPO [3, 4] and the latent function of Hc as a PO-like enzyme [19, 20]. There are many reports about the effects of PO on black spot development in crustaceans [2], but there are no reports focusing on those of HdPO. Most studies have been conducted on PO from cuticle or homogenate of heads, although PO and HdPO were not distinguished under the condition tested. The author succeeded in separating both factors and compared the enzymatic properties of both enzymes from the viewpoint of their contribution to the black spot development for the first time. Several lines of evidence suggested that Hc plays an important role in black spot development in kuruma prawns. The first evidence is that HdPO and PO gave the same profiles of optimum temperature and thermal stability [5, 6]. Also no significant differences were observed on the specificity of substrate and inhibitor, indicating both enzymes share the identical or quite similar scheme in their catalytic reaction [5, 6]. Although both enzymes worked to convert DOPA efficiently into dopachrome at approximately pH 8.0, PO activity was significantly inhibited at pH 7.0, while HdPO activity maintained 40% compared with the activity of the optimum condition [5, 6]. These results suggest that HdPO has its potentiality more effectively in hemolymph, at pH 7.0-7.5, than PO does. Second, Hc occurs in hemolymph at a very high concentration (several mg per prawn), about 1000 times higher than PO in hemocyte (several microgram per prawn) [5, 6]. Although the specific activity of HdPO is about one fifth of PO, the significant difference between the amounts of these proteins suggests the possible importance of HdPO for the black spot development [5, 6]. Third, Hc is very stable under frozen condition in its latent and active forms while proPO and PO are more sensitive (Fig. 5). These findings strongly suggest that after converted into HdPO, Hc induces the melanosis of the freeze-thawed prawn much more rapidly than PO.

HdPO in this study was activated by SDS [6] and its conversion mechanism by endogenous factors remains obscure while the activation of proPO to PO has been definitely clarified [13-15]. The proteolytic enzyme called PPAE (prophenoloxidase activating enzyme) has already been isolated and molecular cloned in some insects. The relation between Hc and PPAE, however, has not investigated yet. Nagai and Kawabata reported that the members of blood coagulation system and antibacterial peptide transform Hc to phenoloxidase in chelicerates, horseshoe crab without cleavage of peptide [21, 22]. We have already identified the nonproteolytic conversion with hemocyte component of kuruma prawn, which is triggered by beta-1,3-glucan, laminarin and is effectively suppressed by several types of protease inhibitors (data not shown). The hemocyte of horseshoe crab, for example, does not contain prophenoloxidase while that of prawn involves it. We cannot conclude here the connection of our results and that of horseshoe crab because prawn and horseshoe crab have different physiological phenol-oxidation system [21, 22].

There are several kinds of anti-blackening treatments for frozen prawn, which intend suppression of phenoloxidase activity from hemocyte but no effective method has been established. In this study we suggest a novel mechanism about post-harvest melanogenesis of prawn, which is induced by hemocyanin-derived phenoloxidase. This enzyme is present at quite high level and very stable under frozen condition, compared to prophenoloxidase from hemocyte. These two types of phenoloxidase shares high identity in primary structure, however, their enzymatic activation schemes are different. Thus there is a novel approach to develop anti-blackening treatment. Further studies, however, are required to confirm whether the conversion of Hc into HdPO proceeds in freeze-thaw process and to investigate how Hc is activated without the removal of peptide.

5. ACKNOWLEDGEMENT

The authors thank Dr. Naoki Yagishita (The Kyoto University Museum) for his kind help in construction of phylogenetic tree.

REFERENCES

1. Suisan Nenkan. (2000) Suisansya, Tokyo.
2. Kim, J., Marshall, M.R. and Wei, C.I. (2000) Polyphenoloxidase. In: Seafood enzymes. Haard, N. F. and Simpson, B.K. (eds), Marcel Decker, New York. pp.271-315.
3. Ashida, M. and Brey, P.T. (1997) Recent advances in research on the insect prophenoloxidase cascade. In: Molecular mechanism of immune responses in insects. Brey, P.T. and Hultmark, D. (eds.), Chapman and Hall, London, pp.135-172.
4. Söderhäll, K., Cerenius, L. and Johansson, M.W. (1996) The prophenoloxidase activating system in invertebrates. In: New directions in invertebrates immunology. Söderhäll, K., Iwanaga, S. and Vasata, G.R. (eds.), SOS Publications, Fair Haven, New Jersey, pp.229-254.
5. Adachi, K., Hirata, T., Nagai, K., Fujisawa, S., Kinoshita, M. and Sakaguchi, M. (1999) Purification and characterization of prophenoloxidase from kuruma prawn *Penaeus japonicus*. Fisheries Science., 65: 919-925.
6. Adachi, K., Hirata, T., Nagai, K., Fujisawa, S. and Sakaguchi, M. (2001) Hemocyanin a most likely inducer of black spots·in kuruma prawn *Penaeus japonicas* during storage. J. Food Sci., 66 :1130-1136.
7. Horowitz, N.H. and Shen, S.C. (1952) *Neurospora tyrosinase*. J. Biol. Chem., 197: 513-520.
8. Saitou, N. and Nei, M. (1987) The neighbor-joining method: A new method for reconstructing phylogenetic trees. Mol. Biol. Evol., 4: 406-425.
9. Thompson, J.D., Higgins, D.G. and Gibson, T.J. (1994) Improved sensitivity of profile searches through the use of sequence weights and gap excision. Comp. Appl. Biosci., 10: 19-29.
10. Bak, H.J. and Beintema, J.J. (1987) *Panulirus interruptus* hemocyanin. The elucidation of the complete amino acid sequence of subunit a. Eur. J. Biochem., 169: 333-338.
11. Jakel, P.A., Bak, H.K., Soeter, N.M., Vereijken, J.M. and Beintema, J.J. (1988) *Panulirus interruptus* hemocyanin. The amino acid sequence of subunit b and anomalous behavior of subunit a and b on polyacrylamide gel electrophoresis in the presence of SDS. Eur. J. Biochem., 178: 403-412.
12. Neuteboon, B., Jakel, P.A. and Beintema, J. J. (1992) Primary structure of hemocyanin subunit c from *Panulirus interruptus*. Eur. J. Biochem., 206: 243-249.
13. Jiang, H., Wang, Y. and Kanost, M.R. (1998) Pro-phenol oxidase activating proteinase from an insect, *Manduca sexta*: A bacteria-inducible protein similar to *Drosophila easter*. PNAS, 95: 12220-2018.
14. Lee, S.Y., Cho, M.Y., Hyun, J.H., Lee, K.M., Homma, K.I., Natori, S., Kawabata, S., Iwanaga, S. and Lee, B.L. (1998) Molecular cloning of cDNA for pro-phenoloxidase-activating factor I, a serine protease is induced by lipopolysaccharide or 1,3-β-glucan in coleopteran insect, *Holotrichia diomphalia* larvae. Eur. J. Biochem., 257: 615-621.
15. Satoh, D., Horii, A., Ochiai, M. and Ashida, M. (1999) Prophenoloxidase-activating enzyme of the silkworm, *Bombyx mori*. Purification, characterization, and cDNA cloning. J. Biol. Chem., 274: 7441-13860.

16. Adachi, K., Hirata, T., Nagai, K., Fujisawa, S. and Sakaguchi, M. (1999) Effects of β-1,3-glucan on the activation of prophenoloxidase cascade in *Penaeus japonicus* hemocyte. Fisheries Science, 65: 926-929.
17. Sritunyalucksana, K., Cerenius, L. and Söderhäll, K. (1999) Molecular cloning and characterization of prophenoloxidase in the black tiger shrimp, *Penaeus monodon*. Dev. and Comp. Immunol., 23: 179-186.
18. van Holde, K.E. and Miller, K.I. (1995) Hemocyanins. Advances in Protein Chemistry, 47: 1-81.
19. Zalateva, T., Muro, P.D., Salvato, M. and Belteramini, M. (1996) The *o*-diphenoloxidase activity of arthropod hemocyanin. FEBS Letter., 84: 251-254.
20. Decker, H. and Rimke, T. (1998) Tarantula hemocyanin shows phenoloxidase activity. J. Biol. Chem., 273: 25889-28892.
21. Nagai, T. and Kawabata, S. (2000) A Link between blood coagulation and prophenol oxidase activation in arthropod host defense. J. Biol. Chem., 275: 29264-29267
22. Nagai, T., Osaki, T. and Kawabata, S. (2001) Functional conversion of hemocyanin to phenoloxidase by horseshoe crab antibacterial peptides. J. Biol. Chem., 276:27166-27170.

SURIMI TECHNOLOGIES

SURIMI GEL PREPARATION AND TEXTURE ANALYSIS FOR BETTER QUALITY CONTROL

Jae W. Park,
OSU Seafood Lab & Dept of Food Science and Technology, Oregon State University,
2001 Marine Drive #253, Astoria, Oregon OR97103, USA

Abstract

Better understanding of the gelation properties of surimi proteins could yield a short cut to improve quality control. Two methods (punch and torsion) have been well described in the Codex Code for Frozen Surimi (FAO/WHO) developed by the US and Japanese governments. The Codex Code recommends buyers and sellers to select either method based on their needs. However, each method contains specific pros and cons. This paper will review the two Codex methods and point out their pros and cons, respectively based on experimental data. In addition, a method for gel preparation and texture analysis for better quality control will be suggested.

Empirical punch test is widely used around the world and provides convenience and simplicity to the user, while its data are specific to the individual machine. Fundamental torsion test is widely used in the US major meat and surimi seafood manufacturers. However, it is time consuming and the maintenance of an exact diameter is extremely difficult. A new gel preparation method using a mold was investigated, which provides significant time saving and convenience while maintaining the accuracy of the torsion method.

In addition, there is a significant discrepancy between the conventional gel cooking method and commercial crabstick manufacturing for their different heating rates. New cooking methods, to address this discrepancy, will be discussed in this presentation as well.

1. INTRODUCTION

Better quality assessment is the key to time savings and better quality control. The quality of surimi is determined based on a number of characteristics, some more important than others. These include gel strength, color, moisture content, impurities, and microbiological counts. Other properties affecting the final quality are pH, protein content, fat content, cryoprotectants, and other food grade additives. Among all properties related to surimi quality, there is no doubt that gel properties, namely gel strength, are of primary interest in surimi production and trade.

The code committee under FAO/WHO asked the governments of Japan and the United States to develop a codex code for frozen surimi [1]. Since 1994, there have been numerous meetings and revisions of the code. This code was particularly developed to assess gel properties with an emphasis on two clearly different methods: the Japanese punch (penetration) test and the US torsion test. We also suggested the buyer and seller to decide, which test would be appropriate depending on their particular interest.

This presentation will cover the pros and cons for these two methods along with a brief description of methodologies. Finally, it will suggest a better method to maintain better quality control. For details of methodologies, please refer to the appendix in the surimi textbook, Surimi and Surimi Seafood [2].

2. PUNCTURE TEST

2.1. Gel preparation

Surimi, partially thawed, is subjected to comminution (chopping) with 3% salt and without moisture adjustment. Upon the completion of chopping, after maintaining 5-8°C, the paste is smashed (to remove air) and stuffed into a plastic casing (3 cm diameter)

The test material is cooked in hot water at $87 \pm 3°C$ for 30 min. For suwari (setting) test, surimi paste is kept in warm water at $30 \pm 2°C$ for 60 min prior to heating in hot water at $87 \pm 3°C$ for 30 min. For modori (gel-softening) test, surimi paste is pre-incubated at 60°C for 30 min prior to heating at $87 \pm 3°C$ for 30 min. Immediately after finishing the heating treatment, the test material is placed in cold water to cool it, and then left at room temperature for 3 h or longer.

2.2 Texture analysis

Texture analysis is performed between 24 and 48 h after cooking. The temperature of gel should be equilibrated to the room temperature. The gel strength (breaking force) and deformability are determined using a punch test unit (rheometer) with a spherical plunger (5 mm diameter) and at a speed of 60 mm/min.

2.3. Advantages and problems associated with punch (penetration) test

This test is most convenient and easy to use. However, it is very specific to the individual unit. Data obtained from various units are not interchangeable [3]. It also possesses numerous fundamental issues in gel preparation and texture analysis.

First, it is the use of 3% salt when gel is prepared. As indicated in Fig. 1, the higher the salt concentration, the higher the gel strength. Based on the processing nature of surimi seafood (i.e., kamaboko or crabstick), the level of salt for commercial practices is between 1.2-2%. The use of 3% in gel testing, as opposed to a maximum level of 2% in the surimi seafood, will apparently show a higher gel value than the value that would be obtained at 1.2-2.0% salt in commercial products.

Second, regardless of the moisture content in surimi, this punch test prepares gels without considering the effect of moisture on the gel value. Moisture contents of

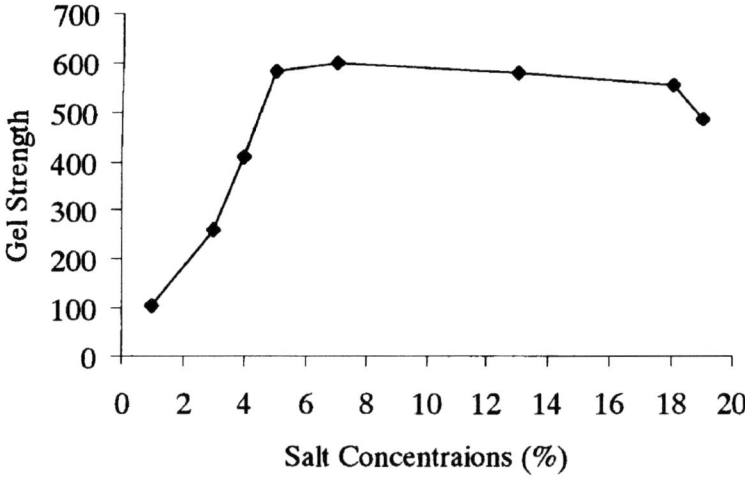

Figure 1. Effect of salt concentration on gel strength. Adopted from Okada, M. (1986) [8].

commercial surimi vary between 72-77% while the majority falls between 73-76%. According to Fig. 2, there is a significant effect of moisture on shear stress (similar to breaking force measured from the punch test). In addition, moisture affects protein content in surimi. As a result, gel values (breaking force) are significantly affected. This approach leaves us a question. Do we want to measure the quality of protein or the quantity of protein (inversely moisture content)? Surimi must be analyzed for its quality as affected by the protein quality while keeping moisture content equal for all surimi.

Third, the problematic nature of gel analysis, simply because it is an empirical method, is quite significant. As indicated in Fig. 3, the shape changes as the plunger pushes down to penetrate into the gel, especially for very elastic gels. This change of the shape often leads to a miscalculation of values. The shaft of the plunger often touches the gel resulting in larger areas of contact. This certainly increases the values. For soft gels, the plunger often passes through the gel without giving a value. The size of the plunger's diameter also affects force and deformation values significantly [4].

2.4. Does the same gel strength mean the same quality of surimi?

The term "gel strength" is always used in surimi trading and is often used in certain scientific communities based on the calculation of breaking force multiplied by deformation.

As indicated in Fig. 4, five different surimi with the same gel strength (960 g*cm) are obviously different. The purchase of proper surimi at the right price must be determined without using the "gel strength" (g*cm), but using breaking force (g) and deformation (cm) independently. The term "gel strength" based on the multiplication of force and deformation is obviously wrong. It does not give any scientifically significant meanings. Gel strength indicates the strength of the gel, or more specifically hardness and breaking force.

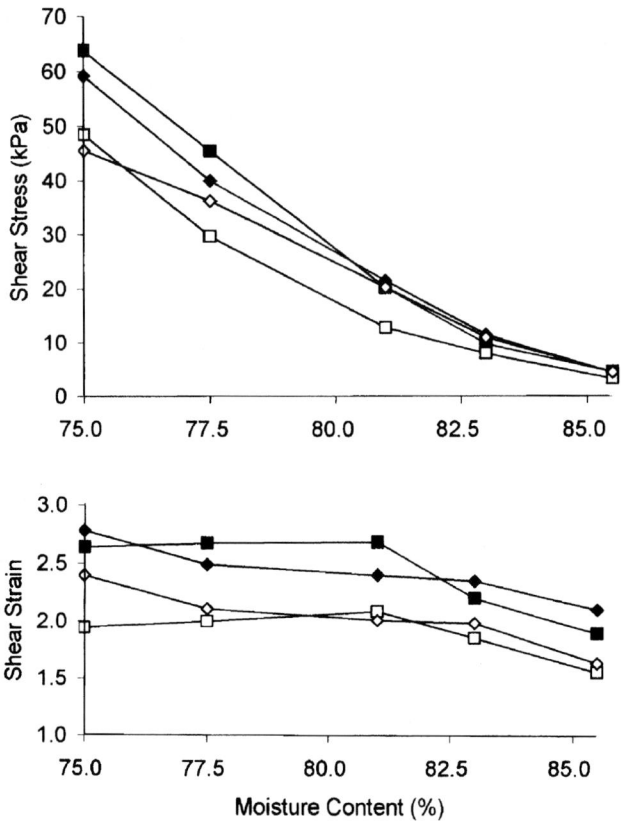

Figure 2. Effects of moisture content on textural properties of surimi.
■, pollock (high); □, pollock (low); ◆, whiting (high); ◇, whiting (low). Adapted from Yoon, W. B., Park, J.W. and Kim, B. Y. (1997)[9].

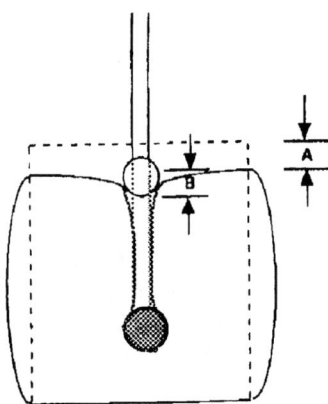

Figure 3. Change in shape of surimi gel cylinder during punch testing. A: squat distance; B: pull-down distance. Adapted from Hamann, D.D. and MacDonald, G. (1992)[4].

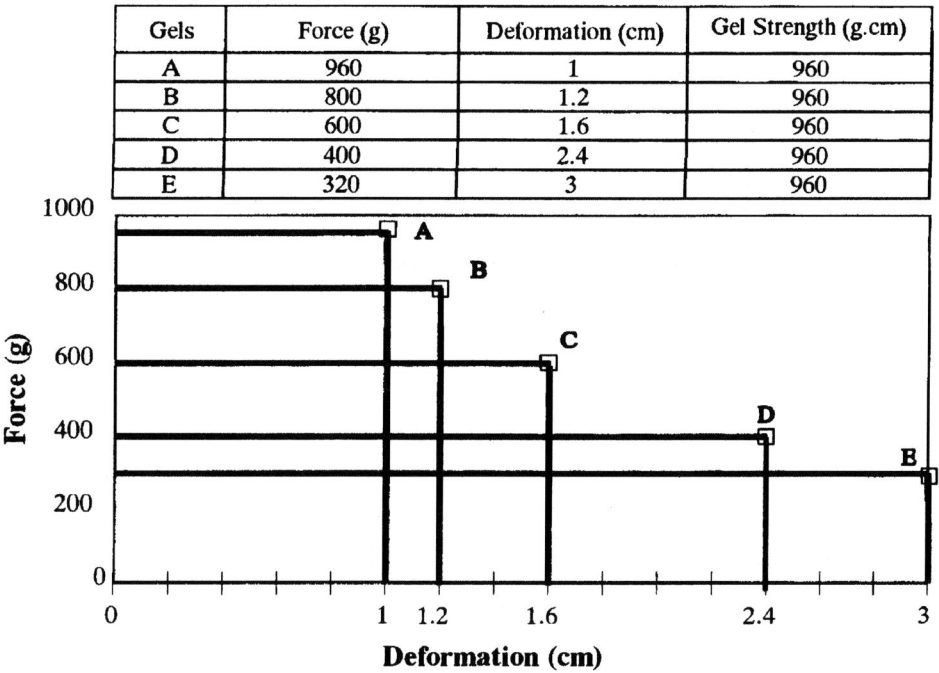

Figure 4. Does the same gel strength mean the same quality of surimi? Adapted from Kim, B.Y. and Park, J.W. (1992)[3].

3. TORSION TEST

3.1. Gel preparation

Surimi paste is prepared using 2% salt and 78% moisture adjustment in a vacuum silent cutter. In addition, the chopping temperature is maintained at 5-8°C. The use of vacuum accelerates comminution and removes air pockets from the sample, which could mislead the value of the surimi. Surimi paste is then stuffed into stainless steel tubes (I.D. 2 cm) with screw caps on both ends. As a result, the uniform density is well maintained. This stainless tube is reusable.

The samples are heated at various temperatures. For low temperature setting, the setting temperature is 0 - 4°C for 12-18 h followed by cooking at 90°C for 15 min. For mid temperature setting, the setting temperature is 25°C for 3 h followed by cooking at 90°C for 15 min. For high temperature setting, the setting temperature is 40°C for 30 min followed by cooking at 90°C for 15 min. For rapid cooking, gels are simply cooked at 90°C for 15 min. For protease analysis, gels are often preheated at 60°C for 30 min followed by cooking at 90°C for 15 min. Upon the completion of cooking, gels are cooled in ice water and kept refrigerated less than 24 h.

3.2. Texture analysis

Gels are equilibrated to room temperature before cutting into pieces 3 cm long. Then

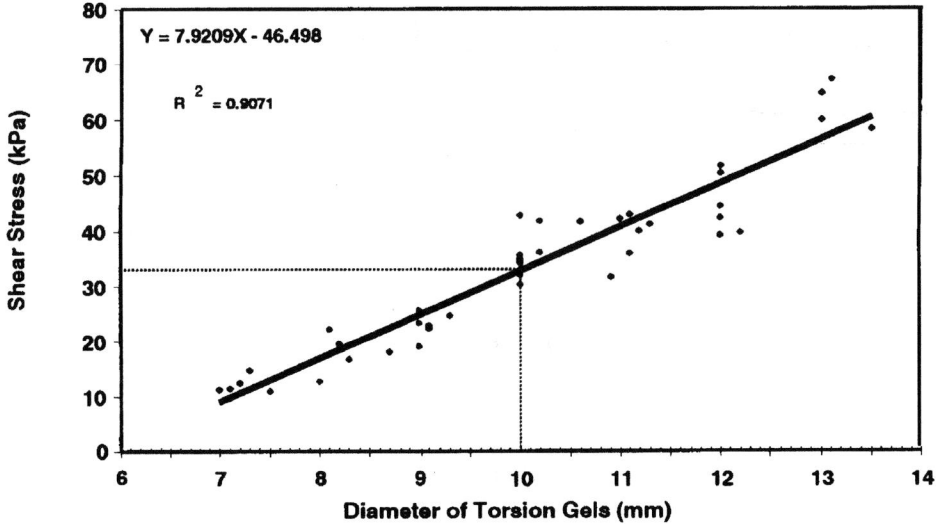

Figure 5. Shear stress values as affected by various diameters. Adapted from Hoffman, J. and Park, J.W. (2001)[5].

they are milled into dumbbell-shape geometry. Using the Hamann gelometer developed using a Brookfield viscometer, shear stress and shear strain values are measured based on the torque value and angular displacement at fracture, respectively. Shear stress indicates the strength of gels, while shear strain denotes the cohesive nature of gels [4].

3.3. Advantages and problems associated with torsion test

This fundamental method delivers accuracy of data because it maintains the shape of the specimen while twisting and it also produces pure shear. Since the principle shear, tension, and compression stresses all have the same magnitude, the values are interchangeable among all the fundamental test units [4]. However, this method is time consuming and needs a high degree of training to maintain the accuracy of the sample geometry, specifically a precise 1.0 cm diameter at the center of the geometry. Effects of various diameters on the gel value are extremely large [5] as shown in Fig. 5. The difference of 1 mm increases or decreases the value of shear stress by 30%. Hoffman and Park [5] also verified the individual variations in measuring diameter. When 12 panelists measured the diameter of one particular sample using the same caliper, they found the individual difference was as high as 1 mm. No matter how the torsion test delivers fundamentally sound data, the control of accuracy, without controlling these variables, appears to be impossible.

4. IMPROVED TORSION TEST

As shown in Fig. 6, molded torsion gels appear to be the answer to overcome the

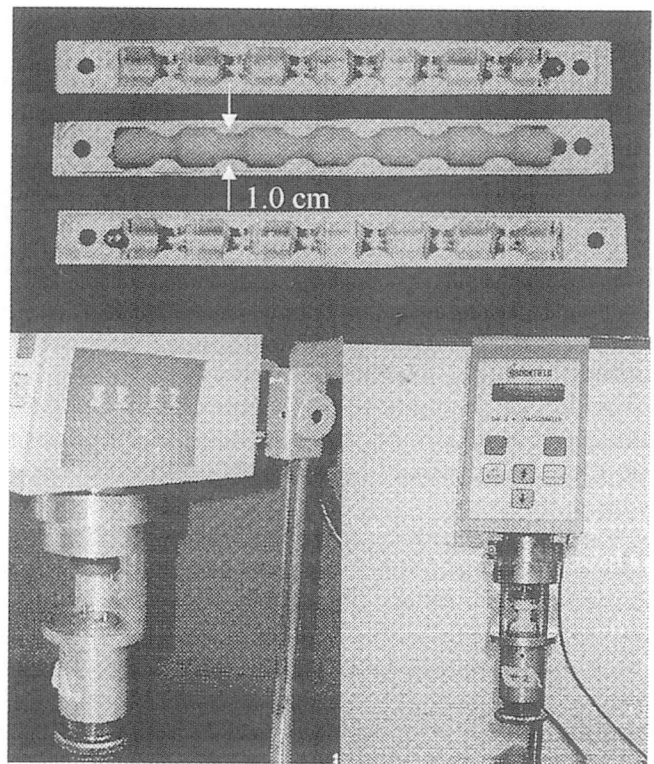

Figure 6. Hamann torsion gelometer and torsion gel mold.

problems resulting from milling. The accuracy of the diameter control is maintained and the individual variations at measurement are eliminated. However, there could be speculation as to whether the use of a mold would induce the formation of a skin that could mislead the data. Hoffman and Park [5] found no increase in stress and strain as cooking time was prolonged. The lack of skin formation was probably due to the use of a lechitin-based spray that was used to coat the internal surface of the mold prior to extruding the surimi paste. Rudan and Barbano [6] also supported these results with their study using mozzarella cheese. They concluded that the use of lechitin-based spray in cooking mozzarella cheese produced a hydrophobic surface coating that prevents moisture loss from the surface and subsequent skin formation.

5. IMPROVED GEL PREPARATION FOR BETTER QUALITY CONTROL

5.1. Chopping temperature

Effect of temperature during chopping on shear stress and shear strain is significant as well [2]. The gelling properties of surimi paste of Alaska pollock chopped at 20°C were extremely low. Maintaining chopping temperatures between 0-5°C, however, provided maximum gelling functionality for Alaska pollock surimi. Our follow-up studies also indicated that the optimum chopping temperature was highly dependent on the fish habitat temperatures.

5.2. Cooking method resembling commercial production

Gel cooking methods currently used, as indicated in the Codex Code, are based on water bath heating. Using stainless steel tubes (2 cm diameter) at 90°C water bath, it takes 12-13 min to obtain the equivalent temperature at the geometric center (Fig. 7). In the case of plastic casings (3.0 cm diameter), it might take 20-25 min to reach 90°C at the center. As indicated in Fig. 7, it is extremely slow cooking that could cause proteolytic proteases to be activated, resulting in significantly lower gel values than what would be obtained during commercial production. However, when temperature profiles were monitored in commercial crabstick lines, the temperature of thin sheet (<2 mm) reaches 90°C easily, within 45-50 s [2], which resembles the heating pattern of ohmic heating (Fig. 7).

The use of rheological values of surimi gels slowly cooked in a water bath may not resemble the formation of gel texture under commercial crabstick production where the heat is conducted very fast. The fast cooking can result in very good gels from low-mid grade surimi and/or enzyme-laden Pacific whiting surimi without enzyme inhibitors [7]. The best method for gel cooking must be determined based on the nature of commercial production. It is highly suggested that the best method for gel cooking resembling crabstick processing is either ohmic heating or microwave. The former is currently developed at the OSU Seafood Lab, while the latter is under investigation at NC State University.

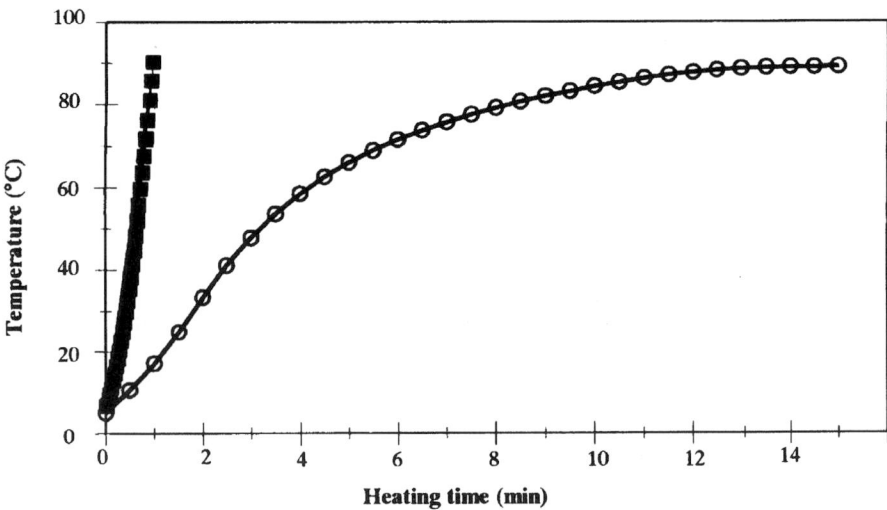

Figure 7. Temperature profiles of two distinctive cooking methods: Ohmic heating (or crabstick cooking) and conventional water cooking. ■, ohmic heating or crabstick cooking; ○, water bath with stainless tube 2.0 cm (I.D.).

5.3. Other factors affecting gel analysis

The use of vacuum during chopping is likely to give a minimum number of air pockets inside the paste, resulting in a reduced standard deviation among the rheological values of samples. It also accelerates the efficiency of chopping, resulting in a shorter chopping time. The use of stainless steel tubes as opposed to conventional plastic casings is also highly recommended. It is difficult to maintain a consistent density of filled casings, which could cause inconsistent results. The stainless steel tubes with screw caps on both ends always give consistent compactness of filled paste. These tubes can be reused continuously. As a result, it is cost saving and no waste is generated.

6. CONCLUSIONS

The short cut to better quality control is to evaluate gel values of surimi accurately and to eliminate variables that affect quality. Molded gels for torsion test are highly suggested to obtain accurate data, which eliminates laborious operation and human errors. Gel cooking methods, using either ohmic heating or microwave, that mimic commercial production will give the most valuable information for better quality control. Other factors to minimize the effects of variables are the control of chopping temperatures, the use of vacuum during chopping, and the use of stainless steel tubes.

REFERENCES

1. FAO/WHO, Codex code for frozen surimi. (2000) In: Surimi and Surimi Seafood. Park, J.W. (ed.), Marcel Dekker, New York, NY., pp. 475-487.
2. Park, J.W. (2000) Surimi seafood: products, market, and manufacturing. In: Surimi and Surimi Seafood. Park, J.W. (ed.), Marcel Dekker, New York, NY, pp. 201-236.
3. Kim, B.Y. and Park, J.W. (2000) Rheology and texture properties of surimi gels. In: Surimi and Surimi Seafood. Park, J.W. (ed.), Marcel Dekker, New York, NY, pp. 267-324.
4. Hamman, D.D. and MacDonald, G. (1992) Rheology and texture properties of surimi-based foods. In: Surimi Technology. Lanier, T.C. and Lee,C.M. (eds.), Marcel Dekker, New York, NY, pp. 429-500.
5. Hoffman, J. and Park, J.W. (2001) Improved torsion testing using molded surimi gels. J. Aquat. Food Prod. Technol., 10:75-84.
6. Rudan, M.A. and Barbano, D.M. (1998) A model of mozzarella cheese melting and browning during pizza baking. J. Dairy Sci., 81:2312-2319.
7. Yongsawatdigul, J. and Park, J.W. (1996) Linear heating rate affects gel formation of Alaska pollock and Pacific whiting. J. Food Sci., 61:149-153.
8. Okada, M. (1986) Ingredients on gel texture. In: Int. Symposium on Engineered Seafoods including Surimi. Martin, R. and Collette, R. (eds), National Fisheries Institute, Washington, DC, pp. 515-530.
9. Yoon, W.B., Park, J.W. and Kim, B.Y. (1997) Linear programming in blending various components in surimi seafood. J. Food Sci., 62:561-564,567.

GELATION OF THREADFIN BREAM SURIMI AS AFFECTED BY THERMAL DENATURATION, TRANSGLUTAMINASE AND PROTEINASE(S) ACTIVITIES

J. Yongsawatdigul[a] and J.W. Park[b]
[a] School of Food Technology, Institute of Agricultural Technology, Suranaree University of Technology, Nakhon Ratchasima, 30000, Thailand
[b] Seafood Laboratory, Oregon State University, Astoria OR 97013, USA

Abstract

Surface hydrophobicity, reactive and total sulfhydryl groups, circular dichroism, and storage modulus (G') revealed that unfolding of actomyosin began at 35°C, while aggregation took place at 45°C. Optimum temperature of crude transglutaminase (TGase) was found at 37 and 55°C and pH 7-7.5. TGase activity was reduced from 99 unit/g dry weight in mince to 47 unit/g dry weight after screw pressing. Gels incubated at 25°C for 4 h and 40°C for 3 h showed an increased force at failure. Degradation of myosin heavy chain (MHC) was evident when incubation time at 40°C was extended. Iodoacetic acid (IAA) was found to be the most effective TGase inhibitor. Autolytic degradation of threadfin bream was highest at 60°C. Salt had no effect on autolytic activity. Force at failure and deformation of surimi gels decreased as incubation time at 60°C was prolonged, corresponding to a decrease of MHC.

1. INTRODUCTION

Thailand is the largest threadfin bream (*Nemipterus* spp.) surimi producer of the world with a total production of over 80,000 metric tons in 2000. Threadfin bream is the second largest resource used for surimi processing after Alaska pollock (*Theragra chalcogramma*). Typically, surimi gel quality is greatly affected by various intrinsic factors, such as thermal denaturation and aggregation of muscle protein, and the presence of proteinase(s) and transglutaminase. Thermal denaturation of muscle proteins typically varies with fish species. Ogawa et al. [1] found that myosin of fish living in the colder temperature is more labile than those living in the warmer temperature.

Transglutaminase (TGase, EC 2.3.2.13) is protein-glutamine: amine γ-glutamyltransferase. The enzyme catalyzes acyl-transfer reaction producing covalent bonds between proteins by the exchange of primary amines for ammonia at the γ-carboxyamide group of peptide-bound glutamine residues [2]. When a peptide-bound lysine residues react with glutamine residues, ε-(γ-glutamyl)lysine bond between proteins is formed. This leads to formation of intra- and inter-molecular covalent bonds. TGase has been reported to increase elasticity and firmness of various food proteins [3]. An increase in gel strength of surimi after *suwari* or setting is also resulted from endogenous TGase activity [4,5]. Therefore, the role of endogenous TGase in gelation of threadfin bream surimi should be elucidated.

Heat stable endogenous proteinase(s) is another important intrinsic factor controlling final surimi gel quality. Hydrolytic reaction of endogenous heat-stable proteinase(s) results in a breakdown of muscle proteins. Therefore, a proper gel network cannot be formed. Cathepsin L is responsible for severe texture degradation of Pacific whiting surimi [6]. Trypsin-like serine protienase is reported to cause poor texture in mehaden surimi [7]. Toyohara and Shimizu [8] demonstrated that serine proteinase was responsible for muscle proteins degradation in threadfin bream (*Nemipterus bathybius*) mince. Kinoshita et al. [9] reported that proteinase(s) inducing textural degradation of threadfin bream presented only in a sarcoplasmic fraction. Based on these results, proteolysis of threadfin bream surimi should be minimal because the responsible proteinase(s) presented in sarcoplasmic fraction would be removed during washing. However, degradation of threadfin bream surimi produced in Thailand has been observed. Thus, presence of proteinase(s) and its role in surimi gelation must be clarified. Objectives of this study were to investigate denaturation pattern of threadfin bream actomyosin. Furthermore, roles of TGase and proteinase(s) in threadfin bream surimi were elucidated.

2. MATERIALS AND METHODS

2.1. Preparation of fish actomyosin (AM)

Threadfin bream (*Nemipterus bleekeri*) was caught off the Gulf of Thailand and immediately transported in ice packed in polystyrene foam boxes to a laboratory at Suranaree University of Technology. Fish bodyweight ranged from 50 to 100 g. AM was prepared from the dorsal muscle according to the method of Ogawa et al. [10] with slight modification as follows. Fish muscle (50 g) was homogenized in 5 volumes of cold 50 mM KCl, 20 mM potassium phosphate buffer (pH 7.0) containing 0.05 mM phenylmethanesulfonyl fluoride (PMSF) in a Polytron (Brinkmann Instruments, Westbury, NY, USA.) for 2 min. The homogenate was centrifuged at $5,000 \times g$ for 10 min at 4°C. The precipitate was homogenized with the same buffer and centrifuged once more. The resultant residue was homogenized in 500 mL 0.6 M KCl, 20 mM potassium phosphate buffer (pH 7.0). The homogenate was centrifuged at $10,000 \times g$ for 5 min at 4°C. The supernatant was collected and diluted with 3 volumes of cold distilled water (4°C). The precipitate was collected by centrifugation at $10,000 \times g$ for 10 min at 4°C and used as actomyosin throughout the study.

2.2. Heat treatment

Actomyosin solution in 0.6 M KCl, 20 mM potassium phosphate buffer (2.5 mg/ml) was aliquoted to 1 mL and heated from 5 to 90°C in a temperature controlled thermal unit (NesLab, Portsmouth, NH, U.S.A.) at a heating rate of 1°C/min. Sample was covered with parafilm and aluminum foil during heating to avoid evaporation. Upon reaching each studied temperature, actomyosin solution was immediately cooled in ice water and centrifuged at 5,000× g for 15 min. The supernatant was collected and analyzed for protein content, surface hydrophobicity, total and reactive sulfhydryls.

2.3. Surface hydrophobicity

Surface hydrophobicity (S_0) of actomyosin was determined using hydrophobic fluorescence probe, 1-anilino-8-napthalenesulfonate (ANS) according to Hayakawa and Nakai [11]. Actomyosin solution (1 mg/mL) was diluted with 0.6 M KCl, 20 mM potassium phosphate buffer (pH 7.0) to obtain a series of protein concentration of 0-1 mg/mL. To 2 mL of each protein solution, 10 μL of 8 mM ANS in 0.1M potassium phosphate buffer (pH 7.0) was added and mixed well. Samples were kept in dark for 10 min. Fluorescence intensity of mixture was measured using a luminescence spectrophotometer (LS 50B, Perkin-Elmer, Beaconsfield, U.K.) at excitation and emission wavelengths of 374 and 485 nm, respectively, and a 5 nm width of both excitation and emission slits. S_0 of each samples was calculated from the slope of the relative fluorescence (R) vs percent (w/v) protein concentration. The relative fluorescence was defined by Monahan et al. [12] as follow:

$$R = (F-F_0)/F_0$$

Where F is the fluorescence of the protein-ANS conjugate and F_0 is the reading of the ANS solution without actomyosin.

2.4. Total and reactive sulfhydryls (SHs)

Total SHs was determined according to Jiang et al. [13]. To 1 mL of actomyosin (4 mg/mL) was added 9 mL of buffer containing 50 mM potassium phosphate buffer, 10 mM ethylenediaminetetraacetic acid (10mM), 0.6 M KCl, 8 M urea (pH 7.0). To 4 mL of the resultant mixture was added 0.4 mL of 0.1% 5, 5'-dinitrobis(2-nitrobenzoic acid). The mixture was incubated at 40°C for 25 min. The absorbance at 412 nm was measured to calculate the total SHs using the extinction coefficient of 13,600 M^{-1} cm^{-1} [14]. Reactive SHs by incubating actomyosin solution at 4°C for 1 h in the absence of urea.

2.5. Circular dichroism (CD)

The purified actomyosin was diluted to a protein concentration of 0.19 mg/mL with 0.6 M KCl, 20 mM phosphate buffer (pH 7.0). CD spectra were taken with a J-720 spectropolarimeter (JASCO, Tokyo, Japan) operating with nitrogen gas purging at 222 nm and a bandpass of 2 nm. The instrument was equipped with JASCO thermal control device (PTC-348W, JASCO, Tokyo, Japan). A 1 mm path length quartz cell was used. Actomyosin solutions were heated from 10 to 80°C at a heating rate of 1°C /min. The instrument was calibrated for wavelength accuracy using helium oxide glass and for intensity accuracy at 290.5 nm and at 192 nm using (1S)-(+)-10-camphorsulfonic acid (Sigma-Aldrich, St. Louis, MO., U.S.A.). Molar ellipticities of actomyosin were determined using a mean residue weight of 115 g/mol as described by (Price, 1996). α-Helicity (%) was estimated from ellipticities at 222 nm, $[\theta]_{222}$, , by the following equation (1)

$$\alpha\text{-Helicity (\%)} = 100 \times \{ [\theta]_{222} / -40,000\}$$

2.6 Dynamic rheological measurement

Development of actomyosin network was measured as a function of temperature using a CS-50 rheometer (Bohlin Instruments, Inc., Cranbury, NJ, U.S.A.). Actomyosin samples (75 mg/mL in 0.6M KCl, 20 mM phosphate buffer, pH 7.0). at 4°C were place between parallel plate (20 mm) with a gap of 1 mm. To avoid sample drying during heating, a plastic cover (trapper) was used. The sample was heated from 10° to 80°C at a heating rate of 1°C/min. Maximum input strain for dynamic analysis was 0.02 at a frequency of 0.1 Hz, a value found to be in the linear viscoelastic region for actomyosin in this study.

2.7. TGase activity assay

TGase activity was measured according to the method of Takagi et al. [15] with slight modifications. The assay mixture contained 1.0 mg/mL N,N'-dimethylated casein, 15 μM MDC, 3 mM DTT, 5 mM $CaCl_2$ and 100 μL crude extract. After incubation at 37°C for 10 min, time found to be in a linear range from preliminary studies, EDTA solution was added to a final concentration of 20 mM to stop reaction. The fluorescence intensity was measured with excitation and emission wavelengths of 350 and 480 nm, respectively, using a Shimadzu spectrofluorometer (RF-1501, Shimadzu Co., Kyoto, Japan). One unit of enzyme was defined as the amount that incorporated 1 nmol of MDC into N,N'-dimethylated casein per minute.

2.8. Temperature and pH profiles of crude TGase

Fish minces were homogenized in a four volumes of extraction buffer (10 mM NaCl, 5mM EDTA, 2 mM DTT, 10 mM Tris-HCl, pH 7.5). The homogenates were centrifuged at 16,000×g for 20 min and supernatants were subsequently centrifuged at 100,000×g for 60 min. The supernatants after ultracentrifugation were used as crude extract. Activitites of crude transglutaminase were determined at temperature ranging from 0-70°C. pH profile of crude extract was examined in 100 mM sodium acetate (pH 4-6.5), 50 mM Tris-HCl (pH 7-7.5), and borate (pH 8-9) buffers.

2.9. Inhibitor study

Frozen threadfin bream surimi was ground with 3% NaCl and moisture content was adjusted to 78%. Phenylmethanesulfonyl fluoride (PMSF), N-ethylmaleimide (NEM), and iodoacetic acid (IAA) was added to surimi paste to contain final concentration of 1 mmol/g surimi. Ethyleneglycol bis(2-aminoethylether) tetraacetic acid (EGTA) was added at concentration 5 mmol/g surimi. Five grams of pastes were incubated at 25 and 40°C for 4 and 1 h, respectively. Enzymatic reaction was terminated by heating samples at 90°C for 30 min. Protein patterns were analyzed using SDS-PAGE.

2.10. Autolytic activity assay

Autolysis at various temperatures was performed as described by Yongsawatdigul et al. [16]. Three grams of threadfin bream surimi were incubated at 0, 25, 40, 60, 55, 60, 65, 70, 80, 90°C for 1 h. Autolysis was stopped by adding 27 mL 5% cold trichloroacetic acid (TCA) solution. The mixture was homogenized and centrifuged. Supernatants were analyzed for TCA-soluble oligopeptides using Lowry assay with tyrosine as a standard. Sample blanks were kept on ice and analyzed to correct for oligopeptides present in fish muscle.

Effect of salt on autolysis was also studied. Surimi paste was prepared to contain 0, 1, 2, 2.5, 3, and 4% NaCl (w/w). Surimi pastes were incubated at optimum temperature for autolytic for 1 h. and oligopeptide contents were determined. Total soluble protein of each paste was determined by solubilizing surimi paste in 5% SDS. Protein content was analyzed using Lowry method with BSA as a standard.

2.11. Surimi gel preparation

Surimi pastes were prepared to contain 3% NaCl and 78% moisture content using a Stephan vacuum silent cutter (UM 5, Germany). Samples were stuffed in a 3-cm. casing and incubated at 4, 25, 40, and 60°C for various times before heated at 90°C for 30 min. The controls were prepared by heating the sample at 90°C without preincubation. Samples were cooled in ice for 30 min and evaluated for breaking force and deformation using a Texture Analyzer (Stable Micro System, Surrey, England) equipped with a 5-mm spherical probe.

2.12. SDS-PAGE

After texture evaluation, samples were used for SDS-PAGE. Degradation of muscle proteins was evaluated on 10% (w/v) acrylamide according to Laemmli [17]. In this case, samples were solubilized according to Yongsawatdigul et al. [16] using 5% SDS solution. Protein cross-links were analyzed using a continuous SDS-PAGE described by Weber and Osborne [18]. The gel contained 3% (w/v) acrylamide, 0.08% (w/v) bis-acrylamide, and 0.25%(w/v) agarose. The running phosphate stock buffer (0.06 M NaH_2PO_4, 0.14 M Na_2HPO_4, 0.2% (w/v) SDS, pH 7.0) was diluted 1:2 with water before use. Solubilization was achieved by homogenizing 0.5 g surimi gel in 10 ml of 2% SDS-8M urea-2% β-mercaptoethanol-20 mM Tris-HCl (pH 8.0) and heated for 100°C for 2 min. The homogenates were stirred continuously for 24 h at room temperature. Supernatant was centrifuged at 10,000× g for 20 min. The protein concentration was measured with the Bradford dye-binding assay [19]. Protein loaded on SDS-PAGE was 15 μg.

3. RESULTS AND DISCUSSION

3.1. Thermal denaturation of threadfin bream actomyosin

Changes in surface hydrophobicity (S_0-ANS) of threadfin bream actomyosin is shown in Fig. 1 [20]. S_0-ANS sharply increase at >30°C, indicating the exposure of buried hydrophobic amino acids to an aqueous environment. This implied that actomyosin of threadfin bream began to unfold at 30°C. Unfolding of actomyosin was also confirmed by an increase of reactive SH (Fig. 2) [20]. In addition, a decrease of total SH indicated formation of disulfide linkages at > 30°C. Circular dichroism (CD) spectra of actomyosin

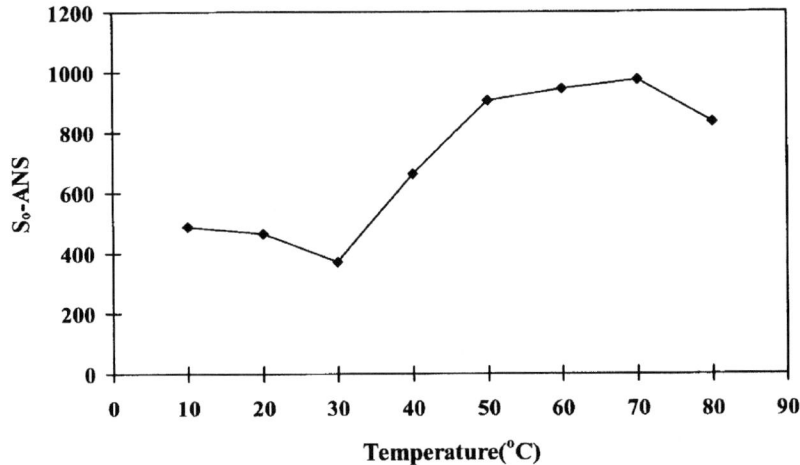

Figure 1. Changes in surface hydrophobicity (S_0-ANS) of threadfin bream actomyosin during heating. Yongsawatdigul, J. and Park, J. W. (2003) [20].

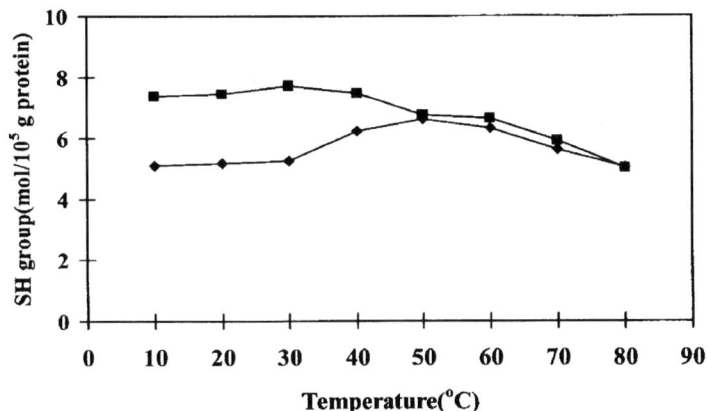

Figure 2. Changes in surface and total sulfhydryl groups of actomyosin during heating. ◆, surface; ■, total. Yongsawatdigul, J. and Park, J. W. (2003) [20].

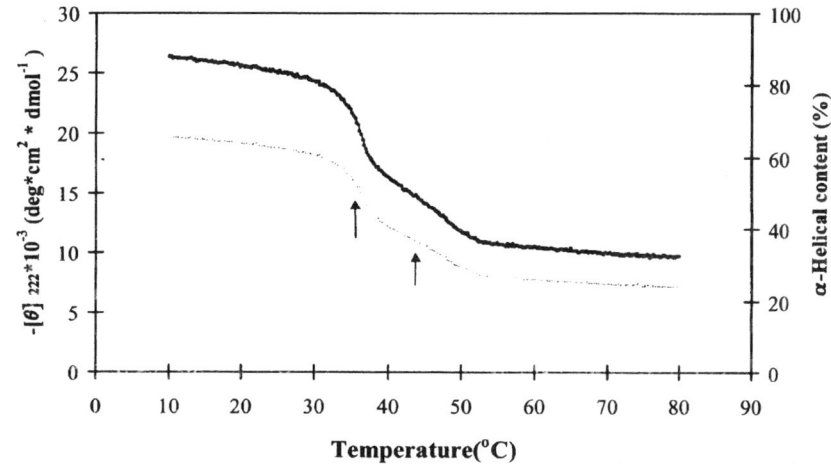

Figure 3. Changes in ellipticity and α-helical content of actomyosin during heating.
●, ellipticity; -----, α-helical content. Yongsawatdigul, J. and Park, J. W. (2003) [20].

is illustrated in Fig. 3 [20]. α-Helical content of actomyosin decreased from 66% at 10°C to about 24% at 80°C. Two distinct transition temperatures were observed at 36.1 and 47.1°C. A decrease of α-helical content was contributed from the unfolding of the rod portion of myosin [21]. It is speculated that transition temperature at 47.1°C resulted from unfolding of F-actin. Changes of dynamic rheological parameters are shown in Fig. 4 [20]. G' of threadfin bream actomyosin started to increase at 34°C and reached the

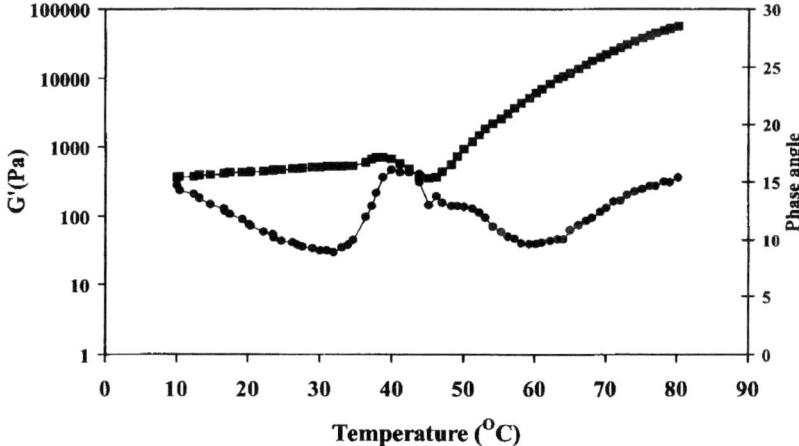

Figure 4. Changes in G' and phase angle of threadfin bream actomyosin during heating.
■, G'; ●, phase angle. Yongsawatdigul, J. and Park, J. W. (2003) [20].

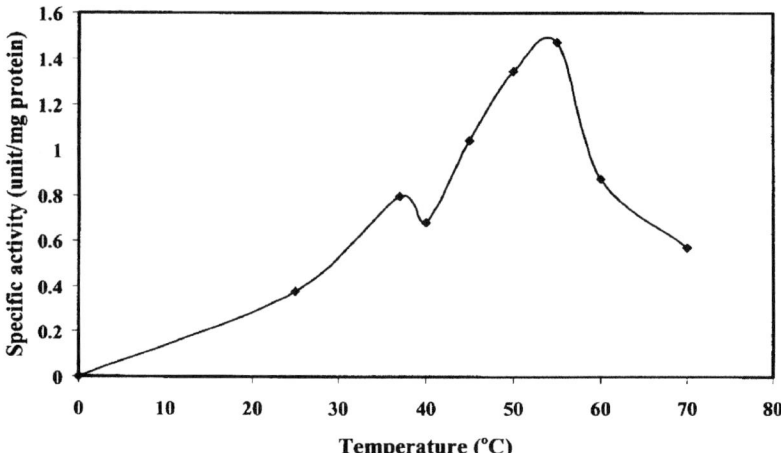

Figure 5. Temperature profile of crude transglutaminase from threadfin bream. Yongsawatdigul, J. and Park, J. W. (2003) [20].

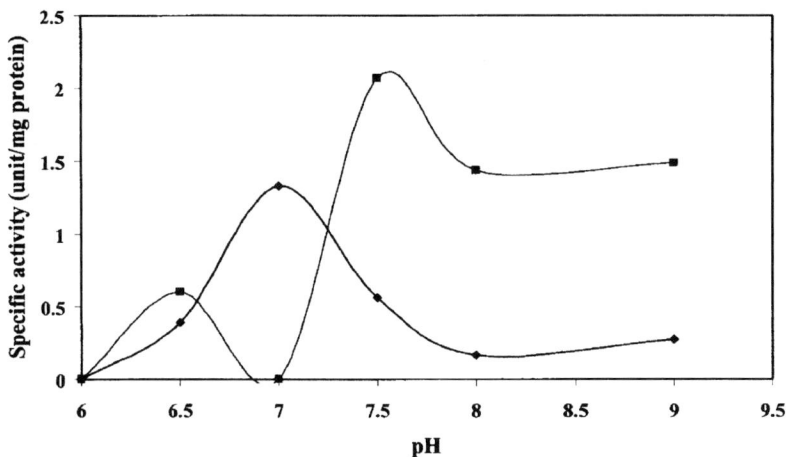

Figure 6. pH profile of crude transglutaminase from threadfin bream measured at 37°C and 55°C. ◆, 37°C; ■, 55°C.

maximum at 38.7°C. An increase in G' indicated the aggregation of actomyosin to gel network structure. A decrease of G' at 38°C was caused by dissociation of actin from myosin in actomyosin complex [22]. All these findings indicated that actomyosin of threadfin bream began to undergo thermal denaturation at about 30°C. Unfolding of actomyosin is a cooperative process including a loss of α-helical structure, the exposure of buried hydrophobic side chain and sulfhydryl groups. Consequently, aggregation of denatured actomyosin by hydrophobic interactions and disulfide linkages led to a formation of gel network.

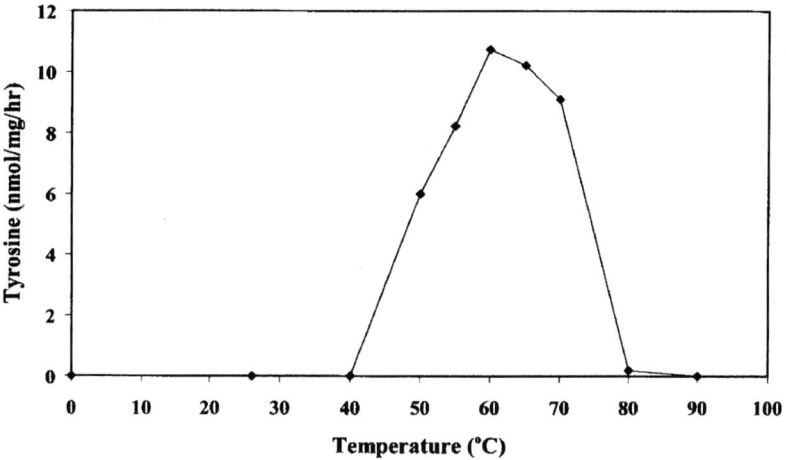

Figure 7. Temperature profile of autolytic activity of threadfin bream surimi.

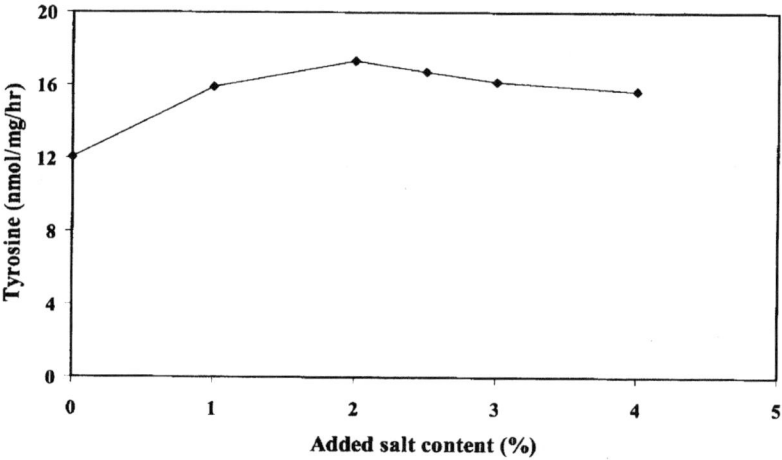

Figure 8. Effect of NaCl on autolytic activity of threadfin bream surimi.

3.2. Effect of pH and temperature on TGase activity

TGase activity in threadfin bream mince was 99.5 Unit/g dry weight, whereas that of surimi (without cryoprotectants) was 32.5 Unit/g dry weight. Washing, dewatering and screw pressing during surimi processing reduced enzyme activity approximately 50%. Temperature and pH profiles of crude TGase are shown in Fig. 5 [20] and 6, respectively. Optimum temperature was found at both 37 and 55°C. Optimum pH at 37 and 55°C was at 7.5 and 7.0, respectively. Kumazawa et al. [23] found that TG −1 and TG-2 from Japanese oyster (*Crassostrea gigas*) exhibited the optimum temperature at 40 and 25°C, respectively. Difference of optimum pH and temperature found in crude TGase might imply the existence of two TGases in threadfin bream muscle. Purification of threadfin bream TGase is under investigation in our laboratory.

3.3. Autolytic activity of threadfin bream surimi

Temperature optimum of autolytic activity was at 60°C as shown in Fig. 7. An et al. [6] reported the optimum temperature for autolytic activity of Pacific whiting at 55°C. Tilapia surimi exhibited autolytic activity at 65°C [16[. Addition of 1-2% salt appeared to increase autolytic activity (Fig. 8). This might be attributed to an increased solubility of myofibrillar proteins by NaCl. A slight decrease in autolytic activity was noticed when salt content increased from 2 to 4%. Toyohara and Shimizu [8] found that proteinase(s) in sarcoplasmic fraction of threadfin bream exhibited high proteolytic activity against myosin heavy chain at pH 7.0 in the presence of 2.5% NaCl.

3.4. Gel quality

Force at failure and deformation of surimi slightly increased after preincubated at 4°C up to 24 hr (Fig. 9A and 9B). The extent of setting in threadfin bream surimi was greatest at 25°C. Four times increase in force at failure resulted in the samples preincubated at 25°C up to 24 h. However, an increased deformation was only observed up to 6 h incubation period. It has been reported that prerequisite of setting at 25°C induced by TGase activity is protein unfolding. However, this study revealed that muscle proteins from threadfin bream, a tropical fish species, could be set at 25°C with minimal conformation changes (Fig. 1-3). High temperature setting (40°C) within 3 h improved gel strength of threadfin bream surimi gels. Prolonged incubation time at 40°C resulted in a decrease in force and deformation. Gel strength of samples incubated at 60°C gradually decreased as incubation period was extended. Soft and mushy texture was observed after incubation at 60°C for 1 h. Kamath et al. [4] reported similar findings on the setting at 25°C of Alaska pollock.

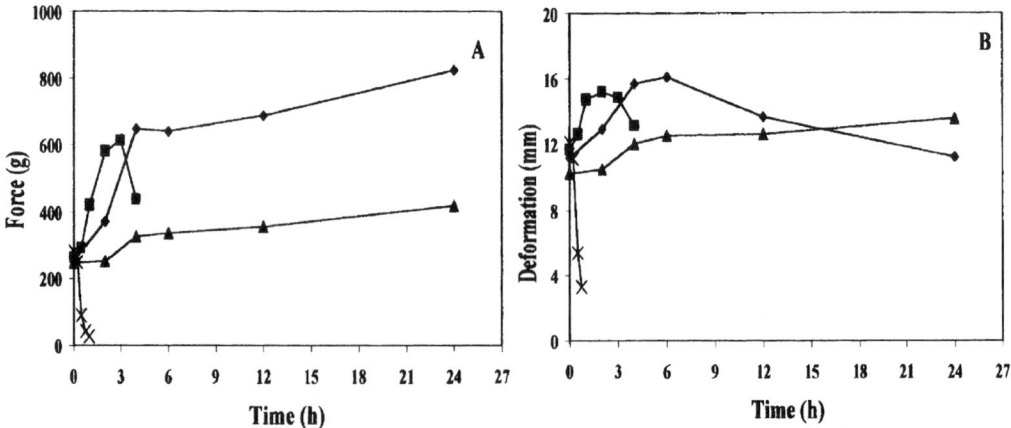

Figure 9. Force (A) and deformation (B) of threadfin bream surimi gels incubated at various time and temperatures: 4°C, 25°C, 40°C and 60°C. ▲, 4°C; ◆, 25°C; ■, 40°C; ×, 60°C.

3.5. SDS-PAGE

Intensity of myosin heavy chain (MHC) of gels incubated at 4°C was comparable at all studied periods (Fig. 10A) [24]. In addition, there was no proteolytic degradation products observed on SDS-PAGE. These results suggested that TGase and proteinase played less important role in gelation of threadfin bream surimi at 4°C. MHC of surimi gels incubated at 25°C decreased with incubation time and completely disappeared when incubated for 12 h (Fig. 10B). MHC cross-links polymer (CP), which were too large to travel through 3% acylamide, were observed on the top of the gel. A decrease of MHC in concomitant with an increased MHC cross-links well corresponded to improvement of textural properties shown in Fig. 9. No evidence of proteolysis was noticed at this temperature. It should be noted that catalysis of TGase was observed at 25°C although unfolding of actomyosin took place to a lesser extent at this temperature. Incubation of

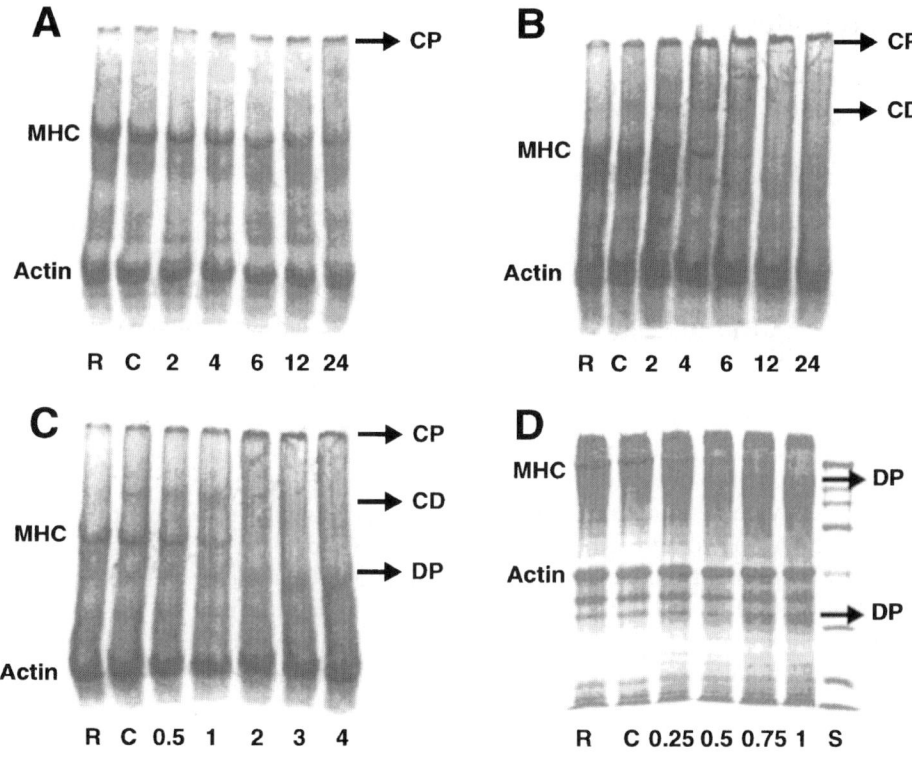

Figure 10. Protein patterns on SDS-PAGE pf threadfin bream surimi gels incubated at 4°C (A), 25°C (B), 40°C (C) and 60°C (D). R, raw sample; C, samples heated at 90°C for 30 min. Numbers indicate incubation time in h. CP, cross-linked polymer; CD, cross-linked dimer; DP, degraded proteins. Reprinted from Yongsawatdigul, J. et al. (2002) [24].

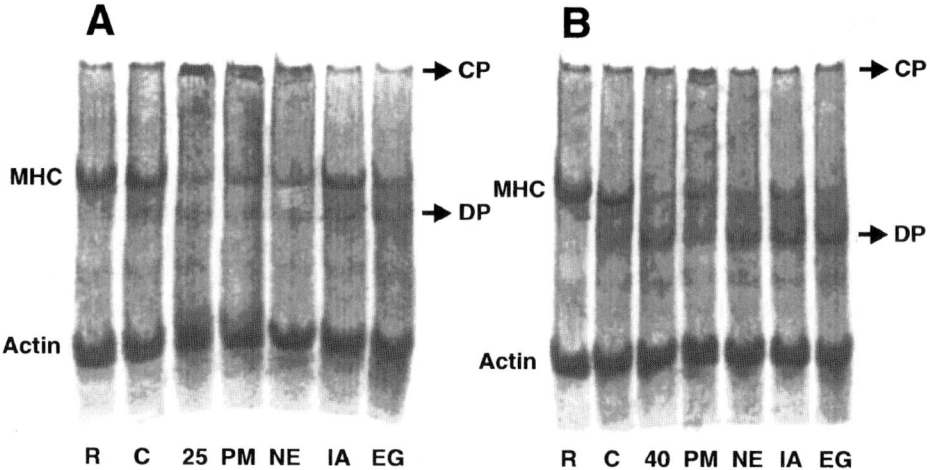

Figure 11. Protein patterns on SDS-PAGE of threadfin bream surimi mixed with various chemicals and incubated at 25°C for 4 h (A) and 40°C for 1 h (B). R, raw sample; C, paste heated at 90°C for 30 min; PM, PMSF; NE, NEM; IA, IAA; EG, EGTA. Reprinted from Yongsawatdigul, J. et al. (2002) [24].

surimi gels at 40°C resulted in a decrease of MHC (Fig. 10C). No MHC was observed in the samples incubated for 2 h or longer. CP was visible at 2 h incubation time. Such changes corresponded with an increased force at failure. It is, therefore, hypothesized that covalent cross-linking strengthened gel network structure, resulting in an increased gel strength. It was likely that TGase played an important role in cross-linking of actomyosin. Fig. 11 showed that addition of TGase inhibitors, particularly IAA, significantly inhibited MHC cross-linking at both 25 and 40°C [24]. IAA is known to react with sulfhydryl group located at active site of TGase. EGTA inhibited TGase activity by chelating Ca^{2+} required for enzyme activation. It should be pointed out that smaller molecular weight proteins appeared to increase when incubated at 40°C for 3 h, suggesting proteolysis of MHC under this condition. Severe degradation of MHC induced by proteinase(s) was noticeable at 60°C (Fig. 10D), the optimum temperature found from autolysis. Proteolysis of troponin and tropomyosin was also observed at 45 min. Proteolysis of myofibrillar proteins led to a poor gel network formation and low gel strength.

4. ACKNOWLEDGEMENT

Authors would like to acknowledge the financial support of Thailand Research fund (TRF), Bangkok, Thailand. Technical assistance of Ms. Wanna Waiprip and Anulak Worratao is greately appreciative.

REFERENCES

1. Ogawa, M., Ehara, T., Tamiya, T. and Tsuchiya, T. (1993) Thermal stability of fish myosin. Comp. Biochem. Physiol., 106B: 517-521.
2. Folk, J.E. (1980) Tranglutaminases. Ann. Rev. Biochem., 49: 517-531.
3. Motoki, M., Nio, N. and Takinami, K. (1984) Functional properties of foodproteins polymerized by transglutaminase. Agric. Biol. Chem., 48: 1257-1261.
4. Kamath, G.G., Lanier, T.C., Foegeding, E.A. and Hamann, D.D. (1992) Nondisulfide covalent cross-linking of myosin heavy chain in "setting" of Alaska pollock and Atlantic croaker surimi. J. Food Biochem., 16: 151-172.
5. Joseph, D., Lanier, T.C. and Hamann, D.D. (1994) Temperature and pH affect transglutaminase-catalyzed 'setting" of crude fish actomyosin. J. Food Sci., 59: 1018-1023, 1036.
6. An, H., Weerasinghe, V., Seymour, T.A. and Morrissey, M.T. (1994) Cathepsin degradation of Pacific whiting surimi proteins. J. Food Sci., 59:1013-1017.
7. Choi, Y.J., Cho, Y.J. and Lanier, T.C. (1999). Purification and characterization of proteinase from Atlantic menhaden muscle. J. Food Sci., 64: 772-775.
8. Toyohara, H. and Shimizu, Y. (1988) Relation between the modori phenomenon and myosin heavy chain breakdown in threadfin-bream gel. Agric. Biol. Chem., 52: 255-257.
9. Kinoshita, M., Toyohara, H. and Shimizu, Y. (1990) Diverse distribution of four distinct types of modori (gel degradation) inducing proteinases among fish species. Nippon Suisan Gakkaishi, 56: 1485-1492.
10. Ogawa, M., Nakamura, S., Horimoto, Y., An, H., Tsuchiya, T. and Nakai, S. (1999) Raman spectroscopic study of changes in fish actomyosin during setting. J. Agric. Food Chem., 47: 3309-3318.
11. Hayakawa, S. and Nakai, S. (1985) Relationships of hydrophobicity and net charge to the solubility of milk and soy proteins. J. Food Sci., 50: 486-491.
12. Monahan, F.J., German, J.B. and Kinsella, J.E. (1995) Effect of pH and temperature on protein unfolding and thiol/disulfide interchange reactions during heat-induced gelation of whey proteins. J. Agric. Food Chem., 43: 46-52.
13. Jiang, S.T., Hwang, D.C. and Chen, C.S. (1988) Denaturation and changes in SH group of actomyosin from milkfish (*Chanos chanos*) during storage at -20°C. J. Agric. Food Chem., 36:433-437.
14. Ellman, G.L. (1959) Tissue sulfhydryl groups. Arch. Biochem. Biophys., 82: 70-77.
15. Takagi, J., Saito, Y., Kikuchi, T. and Inada, Y. (1986) Modification of transglutaminase assay: Use of ammonium sulfate to stop the reaction. Anal. Biochem. 153: 295-298.
16. Yongsawatdigul, J. Park, J.W., Virulhakul, P. and Viratchakul, S. (2000) Proteolytic degradation of tropical tilapia surimi. J. Food Sci., 65: 129-133.
17. Laemmli, Y. (1970) Cleavage of structural proteins during assembly of the head of bacteriophage T4. Nature, 227: 680-685.
18. Weber, K. and Osborn, M. (1969) The reliability of molecular weight determination by dodecyl sulfate polyacrylamide gel electrophoresis. J. Biol. Chem., 224: 4406-4412.

19. Bradford, M.A. (1976) A rapid and sensitive method for quantitation of microgram quantities of protein utilizing the principle of protein-dye binding. Anal. Biochem., 72: 248-254.
20. Yongsawatdigul, J. and Park, J. W. (2003) Thermal denaturation and aggregation of threadfin bream actomyosin. Food Chem., in press.
21. Sano, T., Ohno, T., Otsuka-Fuchino, H., Matsumoto, J.J. and Tsuchiya, T. (1994) Carp natural actomyosin: thermal denaturation mechanism. J. Food Sci., 59: 1002-1008.
22. Xiong, Y.L. and Blanchard, S.P. (1994) Myofibrillar protein gelation: viscoelastic changes related to heating procedures. J. Food Sci., 59: 734-738.
23. Kumazawa, Y., Sano, K., Seguro, K., Yasueda, H., Nio, N. and Motoki, M. (1997) Purification and characterization of transglutaminase from Japanese oyster (*Crassostrea gigas*). J. Agric. Food Chem., 45: 604-609.
24. Yongsawatdigul, J., Worratao, A, and Park, J. W. (2002) Effect of endogenous transglutaminasw on threadfin bream surimi gelation. J. Food Sci., 67: 3258-3263.

OXIDATION DURING WASHING OF FISH MEAT INDUCES A DECREASE IN GEL FORMING ABILITY

Dusadee Tunhun, Yoshiaki Itoh, Katsuji Morioka and Satoshi Kubota
Laboratory of Aquatic Product Utilization, Faculty of Agriculture, Kochi University, Nankoku, Kochi 783-8502, Japan

Abstract

In order to examine the effect of oxidation during washing on the gel forming ability, fish meats were washed with $CuCl_2$ solution and heated in the presence of ethylenediamine tetraacetic acid (EDTA) to prevent further oxidation. By washing with $CuCl_2$ solution, myosin heavy chain (MHC) dimer was formed in the washed meat by disulfide bonds through oxidation of SH groups. The gel strength of gels from oxidized meats was lower than that of control meat. In the gels, polymerization of MHC and actin by disulfide bonds was observed in spite of little oxidation. The oxidized meat paste was mixed with NEM (SH blocking agent) and was heated to prepare gel to examine the effect of inhibiting the polymerization. The resulting gel was still weaker in gel strength than that of the control gel. Moreover, the effect of oxidation during washing was compared with that of oxidation after salt grinding. The gel from oxidized meat after grinding showed almost no decrease in gel strength, though the polymerization of MHC and actin through disulfide bonding occurred accompanying by the oxidation of SH groups. It was suggested that the dimer formation by oxidation during washing influenced to weaken the gel forming ability, being different from the contribution of oxidation in the meat paste after grinding with 3 % salt to gel forming ability. Therefore, meat washing should be careful not to oxidize the meat in order to produce a good quality surimi.

1. INTRODUCTION

We have been studying on the contribution of disulfide bonding to the gel forming ability and on the mechanism of disulfide bond formation upon heating. It was clarified that MHC is polymerized through disulfide bonding [1-3] between myosin S-1 portions upon heating at high temperature above 40°C [4,5]. During ice storage [6] or at low temperature below 30°C [5], the oxidation of SH groups was found to form mainly MHC dimer between myosin rod portions. As mentioned above, the behavior of protein molecules through disulfide bonding has mostly been investigated at high ionic strength such as in 0.6M NaCl, whereas the disulfide bonding of proteins in meat at low ionic strength and its effect on the gel forming ability have not been examined yet.

The structure of myofibril is known to be still kept native at low ionic strength, different from that at high ionic strength [7-10]. Therefore it is supposed that disulfide bond formation through oxidation of SH groups at low ionic strength might occur at the different manner (or at the different portion of proteins) from that at high ionic strength. Furthermore the effect of disulfide bond formation on the gel strength also might be different. Decker et al. [11] examined that the oxidative effect of iron or copper in the presence of ascorbate on the physicochemical properties of turkey myofibrils and reported that the gel strength decreased accompanied by the polymerization of MHC and actin by disulfide bond and the fragmentation of MHC. Some other researchers also reported the similar results on other proteins under the condition of radical regenerating system [12-15].

From the preliminary experiment, we found that washing meat with $CuCl_2$ solution can oxidize at low ionic strength, accompanying MHC dimer formation and no fragmentation of protein. The treated meat with $CuCl_2$ by washing as well as subsequent inhibition of polymerization upon heating was thus used to examine the effect of oxidation at low ionic strength on the gel forming ability. Moreover, it had been pointed out that the oxidation of SH groups into disulfide bonds by adding oxidants at the processing step of salt grinding of meat make strong gels [16,17]. Therefore, the salt-ground meat was treated with $CuCl_2$ and assessed gel properties. In order to make clear whether the effect of $CuCl_2$ washing on the gel forming ability of fish meat is different from that of adding $CuCl_2$ after grinding with salt. Together with this, another oxidant ($KBrO_3$) was employed to ensure the oxidative effect after grinding on gel formation. SDS-PAGE patterns and total SH content were measured to confirm the polymerization through disulfide bonding and the oxidation of SH groups. Additionally, amino acid composition was analyzed to clarify the effect of $CuCl_2$ on muscle protein.

2. MATERIALS AND METHODS

2.1. Meat

Dorsal white meat of carp (*Cyprinus carpio*) and flying fish (*Cypselurus hiraii*), rabbit meat and frozen walleye pollack surimi SS grade (Maruha Co. Ltd, Tokyo Japan) were used.

2.2. Preparation of oxidized meats by washing

The oxidized meats were prepared through washing meat by using two procedures as follows.

2.2.1. *Non-removing $CuCl_2$*

Minced meat was rinsed two times with 20 volumes of 0.3% NaCl and centrifuged at $3,000 \times g$ for 10 min. In order to promote the oxidation, the precipitate was homogenized with 4 volumes of 25 or 50 ppm $CuCl_2 \cdot 2H_2O$ in 0.3% NaCl, pH 7 using non-bubbling homogenizer at 2,000 rpm for 10 min and then centrifuged at $3,000 \times g$ for 10 min.

Control meat was treated with 0.3% NaCl solution without $CuCl_2$. The precipitates were sieved through stainless mesh to remove connective tissue. The resulting meats were referred as $CuCl_2$-treated (oxidized) meat and control (non-oxidized) meat.

2.2.2. Removing $CuCl_2$

After the minced meat was rinsed once with 20 volumes of 0.3% NaCl and was treated by the same method as mentioned in 2.2.1 to promote oxidation, it was rinsed with 0.3 % NaCl containing 25 mM EDTA, pH 7 in order to remove $CuCl_2$ in oxidized meat and to depress the oxidation and then centrifuged at 3,000×g for 10 min.

2.3. Preparation of cooked gel

The meat gels were prepared from non-oxidized and oxidized meats. Each meat was diluted to 3 levels of protein concentration at the interval of 1 % and mixed with 2.7% NaCl to be final concentration 3 % (w/w). For the procedure mentioned in 2.2.1, 25 mM EDTA, pH 7 was added to the meat to terminate the oxidation of meat by chelating copper ion and ground for 20 min. The preparation of meat pastes was performed at 5°C. The various protein concentrations of meat pastes were stuffed into a stainless steel case (30 mm in height and 30 mm in diameter), wrapped with a polyvinylidene sheet, heated at 80°C in a water bath for 20 min, and cooled in ice water. The resulting meat gels were refrigerated overnight at about 5°C until evaluating the gel properties.

2.4. Preparation of meat gel under the inhibition of protein polymerization by SS bond upon heating

The meat gels were prepared from non-oxidized and oxidized meats in the procedure (2.2.1). The various concentrations of 0, 4, 8 and 12 mM N-ethylmaleimide (NEM) were added before grinding to block SH groups and to inhibit polymerization of proteins through disulfide bonding upon heating. The meat pastes were ground and were prepared to gels according to preparation of cooked gel as mentioned above.

2.5. Preparation of oxidized meats gel after grinding

The non-oxidized meat obtained from the procedure (2.2.1) was ground with NaCl at the final concentration of 3% (w/w) for 10 min and then mixed with $CuCl_2 \cdot 2H_2O$ solution (pH 7) to be the final concentration 0, 2, 10, 20 and 30 ppm for 10 min. $KBrO_3$ was also added in salt ground meat as oxidizing reagent instead of $CuCl_2 \cdot 2H_2O$ to confirm the effect of oxidation on gel formation of fish meat. The obtained meat pastes were ground and prepared according to the same method for preparing the cooked gel as mentioned above.

2.6. Assessment of gel properties

The cooled gels were kept at room temperature (20-22°C) for 2 h. Gel properties of each gel were assessed by a Rheometer (Sun Science, Tokyo, Japan) according to the method of Shimizu et al. [18]. Gel strength was expressed as a product of tensile strength S ($g \cdot cm^{-2}$) and breaking elongation e ($\Delta l/l_0$) of gels that were sliced in 5 mm thickness and cut into rings.

2.7. Protein determination

Protein concentration was determined by the Kjeldahl method according to A.O.A.C. [19].

2.8. SDS-polyacrylamide gel electrophoresis (SDS-PAGE)

The meat gel (0.2g) was homogenized with 20 mL of 8M urea-2% SDS-50 mM phosphate buffer (pH 6.8) containing 0.2 mM NEM by using Teflon homogenizer. The homogenate was mixed with one-tenth volumes of 2-mercaptoethanol to prepare reduced sample. Unreduced sample was prepared by adding water instead of 2-mercaptoethanol. SDS-PAGE of these samples was carried out according to the method of Weber and Osborn [20] by using 3% polyacrylamide disc gel.

2.9. Two-step SDS-PAGE

A representative of the first step SDS-PAGE gel of unreduced sample was stained to localize the protein bands. The other gel was used to cut the targeted band. The resulting piece of gel was dipped in 10 % 2-mercaptoethanol solution at 40°C for 60 min to reduce. The reduced cut gel was placed on the top of a new gel column completely without air bubbles, and bromophenol blue marker was added. The second step electrophoresis was carried out according to the same procedure as the first step.

2.10. Total sulfhydryl content

The total SH group content of pastes and gels was determined according to the Ellman method [21] by using 5,5'-dithiobis 2-nitro-benzoic acid (DTNB) with some modification. The meat gels (0.1g) were homogenized with 20 mL of 8M urea-2% SDS-10 mM EDTA in 0.1 M phosphate buffer (pH 6.8) by using A Teflon homogenizer. Four mL of the homogenate was mixed with 0.4 mL of 0.1% DTNB solution and incubated at 40°C for 15 min. The absorbance at 412 nm was measured on a Hitachi U-1000 spectrophotometer. The SH content was calculated from the absorbance using the molar extinction coefficient of 136,000 M^{-1} cm^{-1} for 2-nitro-5-thiobenzoic acid at this wavelength.

2.11. Analysis of amino acid composition

Non-oxidized and oxidized meat samples were hydrolyzed with 6 N HCl at 110°C for 24h. The amino acid analysis was carried out using an amino acid analyzer (Hitachi model L-8500A) with lithium buffer system. The column resin used was a Hitachi custom ion exchange resin No. 2622 SC and the size was 4.6 mm diameter×60 mm length.

3. RESULTS AND DISCUSSION

3.1. Gel forming ability

The gel forming ability of oxidized meat (in which the oxidation was elevated during washing) was evaluated by plotting the logarithm of gel strength against various protein concentrations of meat gels, as shown in Fig. 1. A linear relationship between logarithm of gel strength and protein concentration was observed in both control and oxidized meat

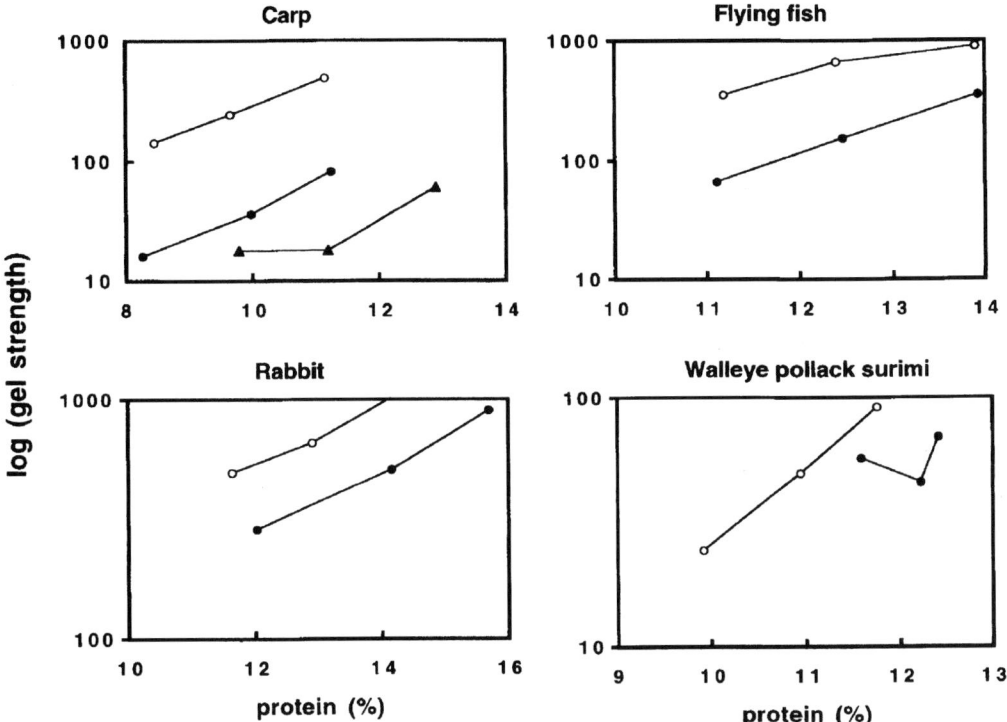

Figure 1. Gel forming ability of meats prepared by washing with various concentrations of $CuCl_2$ solution. ○, 0 ppm $CuCl_2 \cdot 2H_2O$; ●, 25 ppm $CuCl_2 \cdot 2H_2O$; ▲, 50 ppm $CuCl_2 \cdot 2H_2O$.

gel. The treating with 25 ppm $CuCl_2 \cdot 2H_2O$ solution in washing of carp meat affected to reduce the gel forming ability compared with control meat. Furthermore, the treating with higher concentration of 50 ppm $CuCl_2 \cdot 2H_2O$ resulted in the more decreasing in gel forming ability.

In order to confirm the effect of $CuCl_2$-washing on gel forming ability of other meats, flying fish, rabbit, and commercial frozen walleye pollack surimi were treated with the same procedures as carp. Every kind of meat gels showed the linear relationship between logarithm of gel strength and protein concentration in the same manner as carp gels except $CuCl_2$-treated walleye pollack surimi gel, which didn't show straight line. This seems to be due to the shrinkage of surimi gel and the release of drip out of oxidized meat gel, resulting in an increase in protein concentration. A great deal of drip, particularly, in the low concentration of $CuCl_2$-treated meat gels, was assumed to influence highly to the shrinkage of meat gel. Nevertheless, gel forming ability of oxidized walleye pollack surimi at the altered protein concentration after heating was also lower than that of control gel. It is noticeable that the meat washing with $CuCl_2$ solution affects to decrease gel forming ability of the washed meat gel.

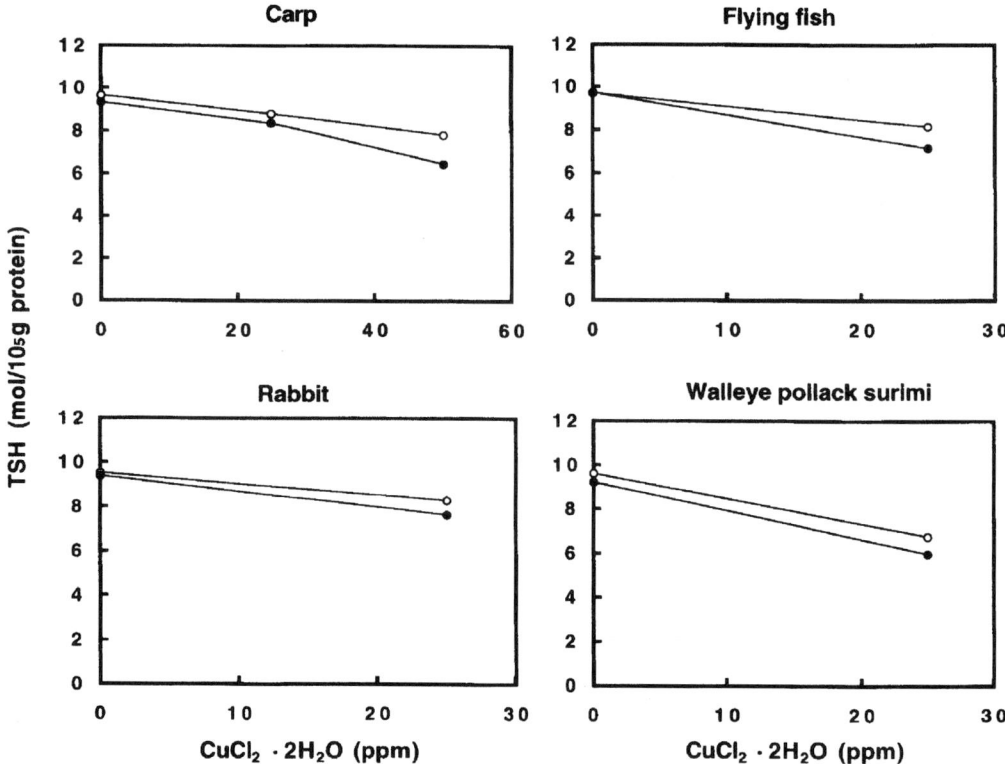

Figure 2. Total SH content of pastes and gels from the meat washed with various concentrations of CuCl$_2$ solution. ○, unheated pastes ; ●, gels heated at 80°C for 20 min.

3.2. Changes in SH group content

Figure 2 illustrates SH group content in the meat paste and the cooked gel. It is obvious that SH group content in carp meat decreased after washing with CuCl$_2$ solution and was not affected by heating. This means that the oxidation of SH groups to disulfide bonds mainly occurred during washing with CuCl$_2$ solution, but not upon heating, suggesting that the adding of EDTA after washing inhibited the oxidation owing to its chelating role for copper ion. In addition, the other three kinds of meat indicated similar decreasing trend in SH group content as carp.

3.3. SDS-PAGE patterns

To examine the polymerization behavior of proteins in meats by washing with CuCl$_2$ solution and by the following heating, SDS-PAGE was carried out. Unreduced samples of oxidized meats showed a drastic decrease in band intensity of MHC (Fig. 3) and concurrently, new thick band double molecular weight size of MHC appeared. In the reduced samples, this new band disappeared. This suggests that MHC was oxidized to MHC dimer by disulfide bonding. Along with consideration from two-step SDS-PAGE

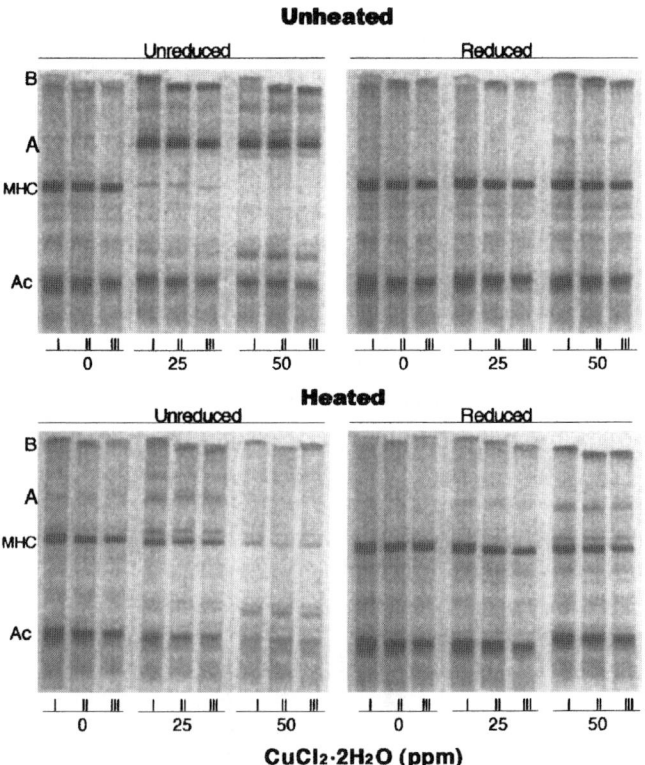

Figure 3. SDS-PAGE patterns of pastes and gels prepared from carp meat oxidized by washing with 0, 25 and 50 ppm $CuCl_2 \cdot 2H_2O$ solutions. I, II, and III indicate high, middle and low level of protein concentration samples, respectively. MHC, Myosin heavy chain; Ac, Actin; Unreduced and Reduced, samples of pastes or gels solubilized without and with 10% 2-mercaptoethanol, respectively. Unheated, pastes ; Heated, gels heated at 80°C for 20 min; A and B, unknown bands.

analysis of this new band, it was clearly showed that the new band was composed of MHC monomer, which is formed through disulfide bonds. (Fig. 4a) In addition, the top polymer on SDS gel was found along with decreasing in actin band in unreduced samples. Since this polymer was dissociated into myosin and actin by 2-mercaptoethanol, it was assumed that it was formed from intact myosin molecules and actin through disulfide bonding. On the contrary, control meat showed a different SDS-PAGE pattern. No change of MHC band without formation of any new band was observed.

After heating, unreduced samples of control meat gel indicated a slight decrease in MHC and actin together with the formation of new three bands, a band just above MHC, dimer band and top polymer. While unreduced samples of the $CuCl_2$-treated meat gel showed a markedly decrease in MHC dimer along with the disappearance of actin. Concomitantly, the new band above MHC and thicker band of top polymer were

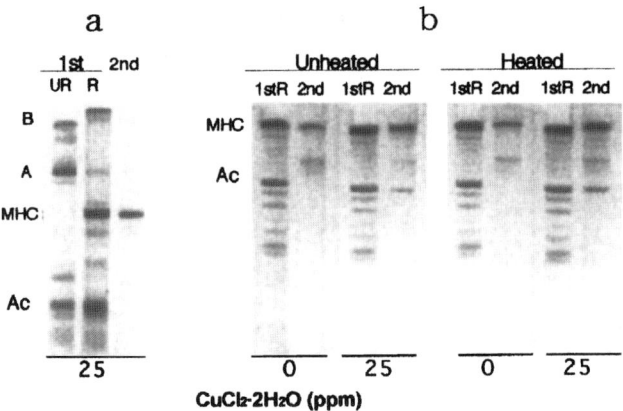

Figure 4. Two step SDS-PAGE (2nd) patterns of A and B bands on the first step SDS-PAGE (1st) patterns of carp meat gels oxidized by washing with 0 and 25 ppm $CuCl_2 \cdot 2H_2O$ solutions. a, two step SDS-PAGE of A band using 3 % polyacrylamide gel; b, two step SDS-PAGE of B band using 10 % polyacrylamide gel; UR and R, samples of pastes or gels solubilized without and with 10% 2-mercaptoethanol, respectively.

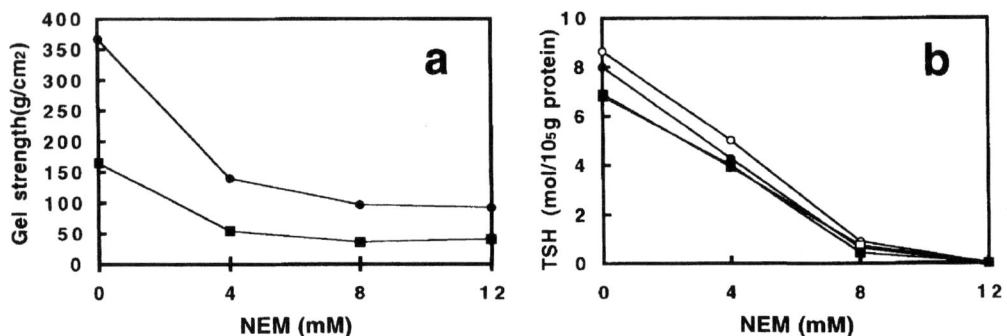

Figure 5. Gel forming ability (a) and total SH content (b) of pastes and gels prepared from carp meat that was washed with 0 and 25 ppm $CuCl_2 \cdot 2H_2O$ solutions and subsequent blocking SH groups by addition of various concentrations of NEM. ○,● and □, ■: 0 and 25 ppm $CuCl_2 \cdot 2H_2O$, respectively; Open and closed symbols, unheated pastes and gels heated at 80°C for 20 min, respectively.

observed. The polymerization of MHC was more induced in the higher concentration of oxidizing agent (50 ppm $CuCl_2 \cdot 2H_2O$). The top polymer of SDS-polyacrylamide gel (Fig. 4b) and a band just above MHC (data not shown) in unreduced samples were analyzed by the two-step SDS-PAGE, indicating that it mainly comprises MHC monomer and actin. This polymer was formed through disulfide bonding upon heating. The band above MHC was observed on the heated-meat gel, while it was invisible in the meat paste before heating. This seems to be a combined substance of myosin and actin according to the mobility.

The other kinds of employed meats demonstrated the oxidative effect on SDS-PAGE patterns in the same manner as carp. (Fig. 5) In every reduced sample of various oxidized meat gels, fragmentation of protein was not observed. This is different from the results in the case of free radical regenerating system [12-15]. Therefore, the decrease in gel forming ability is not due to the fragmentation of protein.

3.4. Effect of $CuCl_2$ -removing from oxidized meat on gel forming ability

As shown above, the SH content of oxidized meats decreased slightly after heating at 80°C (Fig. 2). However, it was observed that more compacted top polymer (polymer at the top of SDS-PAG) from MHC and actin by disulfide bonding was formed, whereas such behavior was not observed in control meat (Fig. 3). The query is that enough amount of EDTA added could not perfectly terminate the oxidation and the polymerization of MHC and actin. Because the molar concentration of $CuCl_2 \cdot 2H_2O$ (25ppm) and EDTA concentration is 0.15 mM and 25mM, respectively.

In order to confirm the effect of remaining copper ion in oxidized meat, removing copper ion by rinsing with 25 mM EDTA solution was considered as a further step after washing with $CuCl_2$ solution in the preparation of oxidized meat (data not shown). The gel strength of meat gels prepared from EDTA-washed meat was also weaker in gel forming ability than the control meat. SH groups decreased a little after heating in the EDTA-washed meat. SDS-PAGE patterns of EDTA-washed meat gels also showed MHC dimer formation through disulfide bonding in the unheated meat paste and showed MHC polymer formation through disulfide bonding accompanied by the polymerization of actin in the heated gels, as was similar as shown in Fig. 3. These results indicate that the copper ion remaining in the $CuCl_2$-treated meat does not affect the polymerization of MHC and actin upon heating at 80°C. Therefore, the MHC polymerization which occurred considerably upon heating and more strongly in $CuCl_2$-treated meat gel than in the control meat gel is not due to the disulfide bonding through oxidation of SH groups, but might be due to disulfide interchange reaction because of little decrease in SH content upon heating. That is, the intramolecular disulfide bonds might be initially formed in MHC and later rearranged by interaction with free SH groups on the other protein molecules during heating. [22-23]

From the above results, another question is that although the washing with $CuCl_2$ solution promoted MHC dimer formation by disulfide bonding through the oxidation of SH groups and the heated-gel of the oxidized meat showed much more amount of polymer formation of MHC and actin compared with non-oxidized meat, the gel forming ability of the oxidized meat was lower than the control meat. Hence, the MHC polymer might prevent the gel formation upon heating.

3.5. Effect of inhibition of polymerization in meat upon heating

In order to confirm the effect of MHC dimer, which was abundantly formed in $CuCl_2$-washed meat on gel forming ability, gels were prepared by adding NEM to modify SH groups in meat paste prior to heating and to inhibit the polymerization of MHC and actin through disulfide bond that occurs even in the presence of EDTA in the $CuCl_2$-washed meat. The gel forming ability, as shown in Fig. 5a, became markedly weak by adding of 4 mM NEM in both control meat and $CuCl_2$-washed meats, and subsequently slightly decreased by adding of 8 and 12 mM NEM. The interesting result was that the $CuCl_2$-

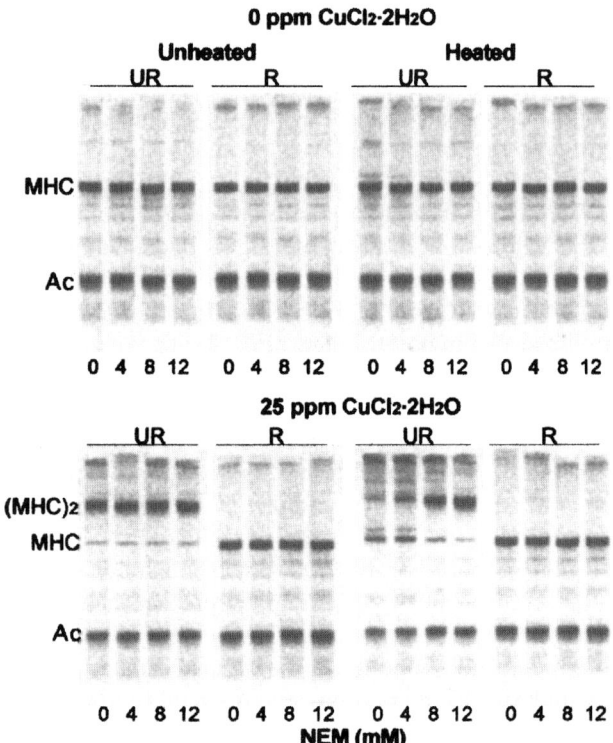

Figure 6. SDS-PAGE patterns of pastes and gels prepared from carp meat that is oxidized by washing with 0 and 25 ppm CuCl$_2$·2H$_2$O solutions and subsequent blocking SH groups by addition of various concentrations of NEM. MHC, Myosin heavy chain; Ac, Actin; UR and R, the same as described in Fig. 4.

washed meat is lower in gel forming ability than the control meat at every NEM concentration.

SH content of gels was determined to confirm the blocking of SH groups by NEM. (Fig. 5b). SH content decreased about 4 and 8 mol/10^{5}g protein by the addition of 4 and 8 mM NEM in both control and CuCl$_2$-washed meats, besides, the addition of 12 mM NEM showed complete blocking of SH groups prior to heating.

SDS-PAGE pattern of gels was showed in Fig. 6. In CuCl$_2$-washed meat paste, MHC dimer through disulfide bonds was observed in all samples, although MHC dimer was not observed in the control samples. By heating, MHC dimer in the NEM non-added gel from oxidized meat was polymerized by disulfide bonding. However, MHC dimer through disulfide bonding remained more along with the increase in NEM amount, indicating that the polymerization of MHC dimer through disulfide bonding was inhibited by adding NEM. In the NEM-added control samples, polymerization of proteins was not observed.

Consequently, it was assured that the lowering in gel forming ability of CuCl$_2$-washed meat is due to the formation of MHC dimer by disulfide bonding through oxidation of SH groups.

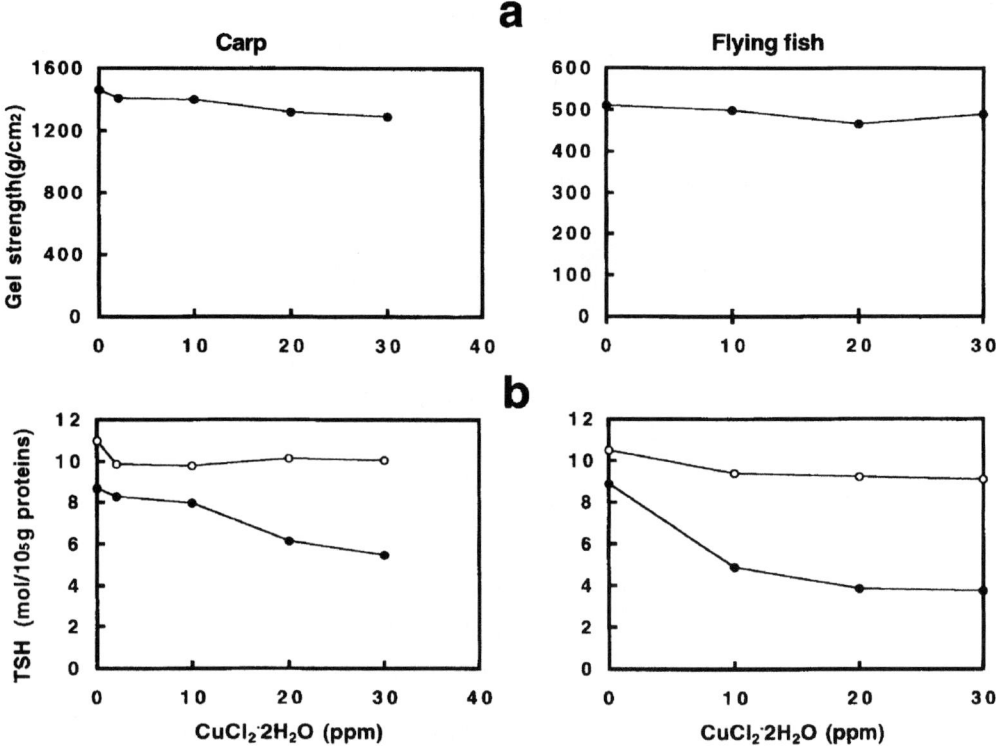

Figure 7. Gel forming ability (a) and total SH content (b) of pastes and gels prepared from carp, flying fish meat oxidized by adding various concentration of $CuCl_2 \cdot 2H_2O$ solutions after salt grinding. ○, unheated pastes ; ●, gels heated at 80°C for 20 min.

It has been reported that the formation of disulfide bonds in the presence of high salt concentration such as 0.6 M NaCl is beneficial to strengthening the gel network during thermal gelation [7,24]. However, since the structure of muscle protein at low salt concentration is different from that at high salt concentration, it is supposed that the portions of disulfide bonding in muscle protein are also different between at high salt and low salt solutions. The disulfide bonding in washed meat before heating might hinder interactions of proteins upon heating and, hence, inhibit the gel formation [13].

Besides, as the addition of oxidant into salt-ground meat before heating increase the gel strength [16], the addition of $CuCl_2 \cdot 2H_2O$ into non-oxidized meat after grinding with salt is one more interesting to make clear the difference in the oxidative effect on gel forming ability and polymerization behavior of protein molecules, from the washing by $CuCl_2$ solution

3.6. Effect of adding $CuCl_2 \cdot 2H_2O$ after grinding

The effect of oxidation by adding various concentration of $CuCl_2 \cdot 2H_2O$ (pH 7) after grinding on the gel formation of carp and flying fish meats were examined and the results

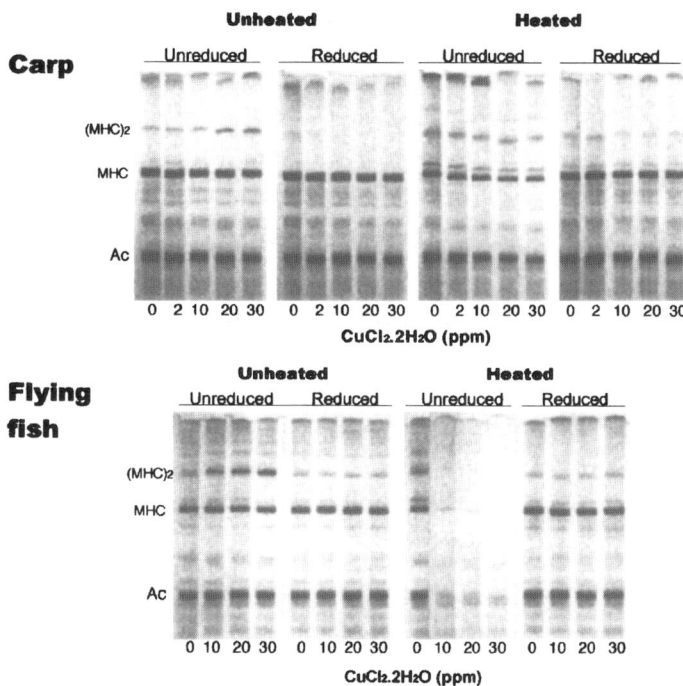

Figure 8. SDS-PAGE patterns of pastes and gels prepared from carp and flying fish meat oxidized by adding various concentration of $CuCl_2 \cdot 2H_2O$ solutions after salt grinding. MHC, Myosin heavy chain; Ac, Actin; UR and R, the same as described in Fig. 4.

were shown in Fig. 7a. The gel strength of carp meat gel did not change greatly by adding $CuCl_2 \cdot 2H_2O$, slightly decreased and that of flying fish was almost constant. These results indicate that the effect of $CuCl_2 \cdot 2H_2O$ on gel forming ability is identical to both carp and flying fish meats, and that the effect of addition of $CuCl_2 \cdot 2H_2O$ is not the same one as the effect of washing with $CuCl_2$ solution.

SH groups of meat prior to heating showed almost unchanged in spite of adding $CuCl_2 \cdot 2H_2O$. In the case of carp meat, adding of 2 ppm $CuCl_2 \cdot 2H_2O$ decreased SH content a little and thereafter up to 30 ppm SH content was constant, and in the case of flying fish, SH content decreased gradually a little along with the increase in $CuCl_2 \cdot 2H_2O$ concentration (Fig. 7b). In contrast, after heating the meat pastes of carp and flying fish, SH groups decreased clearly along with rising of $CuCl_2 \cdot 2H_2O$ concentration, although the decrease in SH content was slightly bigger in flying fish than in carp. These results suggest that SH groups are oxidized during heating in the presence of oxidant, although SH groups in the meat pastes before heating are hard to be oxidized even in the presence of oxidant. In addition it was suggested that the oxidation by $CuCl_2 \cdot 2H_2O$ after grinding and during heating does not affect the gelation of carp and flying fish meats.

Figure 8 shows SDS-PAGE patterns before and after heating of the meat pastes to which $CuCl_2 \cdot 2H_2O$ were added. In the unreduced samples of unheated meat pastes, MHC dimer increased only a little along with the increase in $CuCl_2 \cdot 2H_2O$ concentration and this band was not observed in the reduced samples. This slight formation of MHC dimer is due to disulfide bonding. MHC dimer formation was slightly stronger in flying fish than carp. A little formation of MHC dimer coincides well with a little decrease in SH content in both meat pastes. This behavior was different from that in the meat washed by $CuCl_2$ solution where MHC dimer was abundantly formed.

The polymer observed at the top of SDS-PAG in the unreduced samples increased after heating, agreeing with the decrease in MHC and actin band intensity. In the reduced samples, MHC and actin bands appeared again like before heating. This means that MHC and actin were polymerized through disulfide bonding by heating and promoted in the presence of $CuCl_2 \cdot 2H_2O$. Polymerization of MHC and actin was also stronger in flying fish than carp, suggesting that SH groups of flying fish meat paste are easier to be oxidized than that of carp meat paste according to Fig. 7b.

Anyhow it was noted that the oxidation of SH groups are accelerated by heating rather than in the salt-ground meat pastes in the presence of $CuCl_2 \cdot 2H_2O$. Furthermore it was made clear that the addition of $CuCl_2 \cdot 2H_2O$ into the salt-ground meat paste promotes mainly the formation of MHC and actin polymer during heating rather than that of MHC dimer through disulfide bonding before heating. However, gel strength did not increase by adding $CuCl_2 \cdot 2H_2O$, although polymerization of protein through disulfide bonding occurred, accompanied by the oxidation of SH groups.

3.7. Effect of adding $KBrO_3$ after grinding

$KBrO_3$ was selected as an oxidant instead of $CuCl_2 \cdot 2H_2O$ to confirm the effect of oxidation after grinding on gel formation, since this reagent had been used as an improver for elasticity of kamaboko [16] and bread in the previous days [25], later it has been banned to apply in food processing because of carcinogenic property [26-27].

The meat was mixed with various amount of $KBrO_3$ after grinding with 3% NaCl, and was heated to prepare gel (data not shown). Less than 0.005% of $KBrO_3$ showed no effect on gel forming ability and adding more than 0.01% affected to decline slightly. Breaking strength and elongation showed almost similar patterns as gel strength. However gel strength did not decrease so much as in the case of oxidized meat by washing, indicating that the effect of adding $CuCl_2 \cdot 2H_2O$ into meat paste on the gel strength is quite similar to that of adding $KBrO_3$.

SH content in the meat paste before heating did not decrease, and after heating decreased proportionally with the amount of $KBrO_3$, as seen in the case of adding $CuCl_2$.

From SDS-PAGE patterns was observed that intensity of MHC and actin decreases with an increase in $KBrO_3$ concentration in the unreduced samples of heated gels and appeared again in the reduced samples, although MHC dimer formation is a little before heating. This indicates that the polymerization of MHC and actin through disulfide bonding was promoted by heating in the presence of $KBrO_3$. The polymerization extent of MHC and actin seems to correspond well to the decrease in total SH content.

It was confirmed that the effect of mixing of $CuCl_2 \cdot 2H_2O$ with the salted meat paste is almost similar to that of mixing of $KBrO_3$. This suggests that the function of $CuCl_2 \cdot 2H_2O$ added was the same as that of $KBrO_3$. In this research, $KBrO_3$ as well as $CuCl_2 \cdot 2H_2O$ did not increase gel strength, which conflicted with the already known results that $KBrO_3$

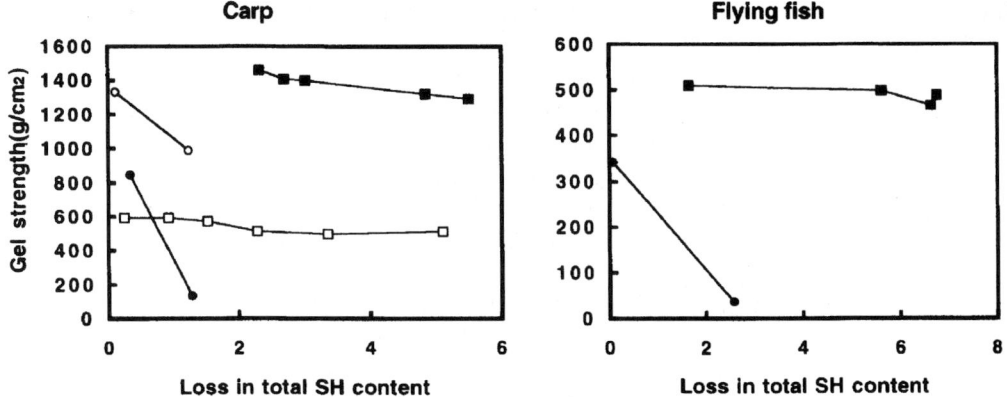

Figure 9. Relationship between gel strength and loss in total SH content of various gels from carp and flying fish meats. ○, gels prepared without removing $CuCl_2$ after washing with $CuCl_2$ solution; ●, gels prepared with removing $CuCl_2$ after washing with $CuCl_2$ solution; □, gels prepared by adding $CuCl_2$ to the meat washed with 0.3% NaCl; ■, gels prepared by adding $KBrO_3$ to the meat washed with 0.3% NaCl.

strengthen the heated gel of denatured frozen surimi.[16] The exact reason of different characteristic is not clear now, but might possibly be due to species difference.

3.8. Relationship between gel forming ability and oxidation of SH groups

Figure 9 is that the relationship between gel strength and loss of TSH content of oxidized meat gels during washing and after grinding and heating were plotted in order to compare the oxidative effects between washing with $CuCl_2 \cdot 2H_2O$ and adding of $CuCl_2 \cdot 2H_2O$ or $KBrO_3$. Since the gel strength of $CuCl_2$-washed meat gels in the previous data is not one at the same protein concentrations, it cannot be compared directly. Therefore, gel forming ability of oxidized meats by washing in the previous data was recalculated at the certain protein concentration by using the linear regression equation of gel strength and protein concentration. The correlation coefficients (R^2) of all equations were higher than 0.9. Protein concentration used in calculation for carp and flying fish meat gels, which was oxidized during washing, were 14 % and 11 %, respectively in order to make the same concentration as that of the oxidized meats after grinding and heating in each fish species.

In the case of adding $KBrO_3$ and $CuCl_2 \cdot 2H_2O$, gel strength did not decrease so much, and changed little despite the proceeding of oxidation of SH groups. On the contrary, in the case of washing with $CuCl_2 \cdot 2H_2O$, gel strength decreased largely along with less oxidation of SH groups. Therefore, it was showed that the gel forming ability of the oxidized meat by washing is lower than that of the meat paste oxidized after grinding even at the same oxidation level. This means that the oxidation of SH groups in the washed meat before grinding with salt inhibits the gel formation to make weaker gel, different from that in the salt-ground meat paste.

3.9. Effect of $CuCl_2$ on the amino acid composition in meat pastes and gels heated at 80°C

It has been reported that the metal-catalyzed oxidation like a free radical regenerating system ($Fe-H_2O_2$, Fe-AsA) modifies amino acid side chains and break peptide bonds to fragments [28-29]. From SDS-PAGE patterns as shown in Figs. 3 and 8, fragmentation of proteins was not observed in the gels that were prepared in the presence of $CuCl_2$. Accordingly, in order to confirm whether the modification of amino acid side chains occurred or not during the preparation of gels in the presence of $CuCl_2$, the amino acid compositions of gels prepared from the control and the $CuCl_2$-added meat pastes were examined. Each meat paste was heated at 80°C for 20 min in the absence and presence of 25 mM EDTA (data not shown).

The content of every amino acid except cysteine and half-cystine did not change by heating in the presence of $CuCl_2$. The total amount of cysteine and half-cystine was almost constant and furthermore cysteic acid content did not increase by heating in the presence of $CuCl_2$. Therefore, it was clarified that the oxidative effect of $CuCl_2$ is mild. That is, oxidation is only limited to the oxidation of sulfhydryl groups to disulfide bonds, and amino acid side chains other than sulfhydryl groups are not modified by $CuCl_2$. It was suggested that the gel weakening by the oxidation during washing is not due to the modification of amino acid residues.

REFERENCES

1. Itoh, Y., Yoshinaka, R. and Ikeda, S. (1979) Behavior of the sulfhydryl groups of carp actomyosin by heating. Nippon Suisan Gakkaishi, 45: 1019-1022.
2. Itoh, Y., Yoshinaka, R. and Ikeda, S. (1980) Formation of polymeric molecules of protein resulting from intermolucular SS bonds formed during the gel formation of carp actomyosin by heating. Nippon Suisan Gakkaishi, 46: 621-624.
3. Itoh, Y., Yoshinaka, R. and Ikeda, S. (1979) Effect of the sulfhydryl reagents on the gel formation of carp actomyosin by heating. Nippon Suisan Gakkaishi. 45: 1023-1025.
4. Kishi, A., Itoh, Y. and Obatake, A. (1997) The subfragment responsible for the polymerization of myosin heavy chain through SS bonding during the heating of carp myosin. Nippon Suisan Gakkaishi, 63: 237-241.
5. Runglerdkriangkrai, J., Itoh, Y., Kishi, A. and Obatake, A. (1999) Responsibility of myosin S-1 and rod for the polymerization of myosin heavy chain through SS bonding upon heating of actomyosin. Fisheries Sci., 65: 310-314.
6. Sompongse, W., Itoh, Y. and Obatake, A. (1996) Responsibility of myosin rod for the dimer formation of myosin heavy chain through SS bonding during ice storage of carp actomyosin. Fisheries Sci. , 62: 473-477.
7. Ishioroshi, M., Samejima, K. and Yasui, T. (1979) Heat-induced gelation of myosin: factors of pH and salt concentrations. J. Food Sci., 44: 1280-1283.
8. Boyer, C., Joandel, S., Ouali, A. and Culioli, J. (1996) Ionic strength effects on heat-induced gelation of myofibrils and myosin from fast-and slow-twitch rabbit muscles. J. Food Sci., 61:1143-1148.

9. Lin, T.M. and Park, J.W. (1998) Solubility of salmon myosin as affected by conformational changes at various ionic strengths and pH. J. Food Sci., 63: 215-218.
10. Sano, T., Noguchi, S.F., Matsumoto, J.J. and Tsuchiya, T. (1990) Effect of ionic strength on dynamic viscoelastic behavior of myosin during thermal gelation. J. Food Sci. 55: 51-54 & 70.
11. Decker, E.A., Xiong, Y.L., Calvart, J.T., Crum, A.D. and Blanchard, S.P. (1993) Chemical, physical, and functional properties of oxidized turkey white muscle myofibrillar proteins. J. Agric. Food Chem., 41:186-189.
12. Xiong, Y.L. and Decker, E.D. (1995) Alterations of muscle protein functionality by oxidative and anti-oxidative process. J. Muscle Foods, 6: 139-160.
13. Martinaud, A., Mercier, Y., Marinova, P., Tassy, C., Gatellier, P. and Renerre, M. (1997) Comparison of oxidative processes on myofibrillar protein from beef during maturation and by different model oxidation systems. J. Agric. Food Chem., 45: 2481-2487.
14. Srinivasan, S. and Hultin, H.O. (1997) Chemical, physical and functional properties of cod proteins modified by a non enzymic free-radical generating system. J. Agric. Food Chem., 45: 310-320.
15. Liu, G., Xiong, Y.L. and Butterfield, D.A. (2000) Chemical, physical, and gel-forming properties of oxidized myofibrils and whey-and soy-protein isolates. J. Food Sci., 65: 811-818.
16. Okada, M. and Nakayama, M. (1961) The effect of oxidants on jelly strength of kamaboko. Nippon Suisan Gakkaishi, 27: 203.
17. Yoshinaka, R., Shiraishi, M. and Ikeda, S. (1972) Effect of ascorbic acid on the gel formation of fish meat. Nippon Suisan Gakkaishi, 38: 511-515.
18. Shimizu, Y., Machida, R. and Takenami, S. (1981) Species variation in the gel-forming characteristics of fish meat paste. Nippon Suisan Gakkaishi, 47: 95-104.
19. A.O.A.C. (1995) Official Methods of Analysis of the Association of Official Analytical Chemists, 16th ed., A.O.A.C., Washington, DC.
20. Weber, K. and Osborn, M. (1969) The reliability of molecular weight determination by dodecyl sulfate-polyacrylamide gel electrophoresis. J. Biol. Chem., 244: 4406-4412.
21. Ellman, G.L. (1959) Tissue sulfhydryl groups. Arch. Biochem. Biophys., 82: 70-77.
22. Ryle, A.P. and Sanger, F. (1955) Disulphide interchange reactions. Biochem. J., 60: 535-540.
23. Huggins, C., Tapley, D.F. and Jensen, E.V. (1951) Sulphydryl-disulphide relationships in the induction of gels in proteins by urea. Nature, 167: 592-593.
24. Smyth, A.B., Smith, D.M. and O'Neill, E. (1998) Disulfide bonds influence the heat-induced gel properties of chicken breast muscle myosin. J. Food Sci. 63:584-588.
25. Munz, E. and Brabender, C.W. (1940) Extensograms as a basis of predicting baking quality and reaction to oxidizing agents. Cereal Chemistry., 17: 313-332.
26. Fisher, N., Hutchinson, J. B., Berry, R., Hardy, J., Ginocchio, A.V. and Waite, V. (1979) Long-term toxicity and carcinogenicity studies of the bread improver potassium bromate 1. Studies in rat. Food Cosmet. Toxical., 17: 33-39.
27. Kurokawa, Y., Maekawa, A., Takahashi, M. and Hayashi, Y. (1990) Toxicity and carcinogenicity of potassium bromate-a new renal carcinogen. Environ. Health

Perspect., 87: 309-335.
28. Amici, A., Levine, R.L., Tsai, L. and Stadtman, E.R. (1989) Conversion of amino acid residues in proteins and amino acid homopolymers to carbonyl derivative by metal-catalyzed oxidation reactions. J. Biol. Chem., 264:3341-3346.
29. Uchida, K., Kato, Y. and Kawakishi, S. (1992) Metal-catalyzed oxidative degradation of collagen. J. Agric. Food Chem., 40: 9-12.

GEL-FORMING CHARACTERISTICS OF VARIOUS FISH CAUGHT IN TOSA BAY

Akira Nomura[a], Atsushi Obatake[b] and Yoshiaki Itoh[c]
[a]Kochi Prefectural Industrial Technology Center, Nunosida 3992-3, Kochi 781-5101, Japan
[b]The former, Kochi University
[c]Laboratory of Aquatic Product Utilization, Faculty of Agriculture, Kochi University, Nankoku, Kochi 783-8502, Japan

Abstract

Gel forming ability of miscellaneous fish caught in Tosa Bay off Kochi was examined to find good materials or utilize these fish. Nineteen species of fish used were classified into 4 groups according to the effect of meat washing on the modori appearance. Type I: both unwashed meat and washed meat show no modori and an excellent gel forming ability (3 species), Type II: unwashed meat shows modori around 60°C and washed meat do not show modori (3 species), Type III: both unwashed meat and washed meat show modori around 60°C (3 species), and Type IV: unwashed meat shows modori at 60°C and washed meat shows modori around 40°C (10 species), which was a type newly found. Preheating of washed meat of Type IV should not be carried out principally around 40°C. Mixing with the meat of Type I or II improves the gel-forming ability. A 40°C modori-inhibiting factor should exist in sarcoplasmic fraction. In addition modori appearance in washed meat of Type IV was due to the activation of myofibril type protease by removing the modori-inhibiting factor that might inhibit protease activity.

1. INTRODUCTION

Generally Alaska pollack frozen surimi has been used for fish paste products such as kamaboko and chikuwa in Japan. As a results, taste or texture of the products has been similar everywhere in Japan, and local characteristics of the products are now going to disappear. However in Kochi Prefecture, Shikoku District in Japan, some kamaboko makers use raw fish as materials for good quality of products. Especially lizardfish *Saurida undosquamis* is a popular and good material. Now this species of fish has been decreasing in catching amount off Kochi. The other kinds of fish are thus necessary for kamaboko making.

We have been studying the gel forming characteristics of not only lizardfish [1-4] but the miscellaneous fish [5-10] caught in Tosa Bay off Kochi to understand the characteristics of lizardfish and to find good materials or utilize the miscellaneous fish for kamaboko making. We report here the gel forming characteristics of the miscellaneous fish. Gel forming characteristics were examined on the unwashed and washed meats, since the meat washing is currently a usual procedure in kamaboko processing.

2. MATERIALS AND METHODS

2.1. Fish

Following 19 fish species caught in Tosa Bay, Japan were used. Princes small porgy (himekodai) *Chelidoperca hirundinacea*, saddled weever (kurakagetoragisu) *Parapercis sexfasciata*, blacktipped greeneye (tsumaguroaomeeso) *Chlorophthalmus nigromarginatus*, thorny flathead (matsubagochi) *Rogadius asper*, silverbelly seaperch (wakiyahata) *Malakichthys wakiyae*, bigeyed greeneye (aomeeso) *Chlorophtalmus albatrossis*, broad alfonsino (nanyoukinme) *Beryx decadactylus*, mirror dory (kagamidai) *Zenopsis nebulosa*, blackmouth cardinalfish (sumikuiuo) *Synagrops janicus*, snubnose brotula (yoroiitachiuo) *Hoplobrotula armata*, Japanese dragonet (yomegochi) *Calliurchthys japonicus*, white edged jawfish (sokoamadaimodoki) *Owstonia grammodon*, sea conger (gotenanago) *Anago anago*, hilgendorf saucord (yumekasago) *Helicolenus hirgendorfi*, seawedged perch (ara) *Niphon spinosus*, redbanded searobin (kanado) *Lepidotrigla guentheri*, deep sea smelt (nigisu) *Glossanodon semifasciatus*, *Chaunax* sp. (midorifusaanko) *Chaunax abei* and blackthroat seaperch (akamutsu) *Doederleinia berycoides*. All of them were in early stage of rigor-mortis. Table 1 shows the details of fish used. The moisture of meats was determined after drying at 105°C.

2.2. Measurement of meat pH

Minced meat (5 g) was homogenized with 95 mL of 3% NaCl solution in a non-bubbling homogenizer (Nihon Seiki, Tokyo, Japan). The pH of resulting homogenate was measured using a glass electrode pH meter (Horiba, Kyoto, Japan).

2.3. Estimation of gonad somatic index of fish

Gonad somatic index (GSI) as a maturity index was estimated by dividing gonad weight with fish body weight.

2.4. Preparation of meat paste

Unwashed meat paste was prepared by grinding the minced meat (moisture 84%, pH 6.9-7.0) with 3% NaCl. Washed meat paste was prepared by washing the minced meat with water at pH 6.9-7.0 and with 0.3% NaCl solution and by grinding with 2.7% NaCl. All procedures were carried out in a cold room at 5°C.

2.5. Preparation of kamaboko gel

Kamaboko gel was prepared by stuffing meat pastes into stainless steel tube (ø3.1×3.0 cm), wrapped with polyvinylidene chloride at 30-90°C for 20 min and 2 h to determine the temperature-gelation curves.

Table 1
Body length, body weight, moisture content, pH and GSI of fish used for preparing meat paste

Species	Body length(cm)[a]	Body weight(g)[a]	Moisture of muscle(%)	pH of muscle	GSI[b]	Prepared date
Princes small porgy	14.0±0.4	32.3±3.2	73.6	6.60	0.44	Jan.26
Saddled weever	17.7±0.3	66.8±2.8	78.1	6.60	0.15	Nov.6
Blacktipped greeneye	23.7±0.5	134.5±9.7	76.9	6.36	0.68	Apr.19
Thorny flathead	16.6±0.9	38.9±4.3	79.3	6.08	2.47	Aug.11
Silverbelly seaperch	17.3±0.3	72.1±4.4	77.4	6.56	3.33	Sep.18
Bigeyed greeneye	12.1±0.2	13.4±0.6	81.1	6.57	–	Oct.28
Broad alfonsino	21.2±1.6	111.1±4.4	79.9	6.27	–	Sep.2
Mirror dory	32.0±0.6	505.0±24.7	79.7	6.04	0.41	Dec.21
Blackmouth cardinalfish	20.3±0.5	92.7±4.9	77.9	6.03	0.05	Oct.26
Snubnose broutula	31.5±0.5	206.6±8.1	81.5	6.31	0.22	Sep.9
Japanese dragonet	38.8±1.0	38.6±1.5	80.4	6.02	–	Feb.8
White edged jawfish	23.9±0.6	72.4±1.4	82.6	6.67	0.64	Mar.1
Sea conger	53.8±3.1	404.8±69.6	79.9	5.88	11.61	Jun.7
Hilgendorf saucord	21.4±0.4	184.0±10.4	83.4	6.25	0.72	Dec.8
Redbanded searobin	18.3±0.3	80.2±5.3	74.3	5.93	0.25	Oct.19
Deep sea snelt	19.1±0.2	43.6±1.1	78.0	6.37	–	Nov.17
Sawedged perch	28.8±0.7	287.8±21.0	77.7	6.28	0.09	Feb.2
Chaunax sp.	22.5±0.7	180.5±20.1	83.0	6.07	0.52	Nov.28
Blackthroat seaperch	23.1±0.3	177.5±3.0	79.7	6.23	0.60	Jan.21

[a]Mean±S.D (n=20-100) ; [b]Gonado-somatic index (mean value of 10 fish)

2.6. Preparation of sarcoplasmic protein

Meat was mixed with 6 volumes of water, and then the meat was filtrated with a nylon cloth. After the floating lipid was removed, the resulting filtrate was centrifuged at 10,000×g for 20 min, followed by packing into a dialysis cellulose tube to concentrate with polyethyleneglycol 20,000.

2.7. Measurement of gel strength

Gel strength was evaluated by the strength that was expressed as the products of breaking strength and elongation obtained by stretching test according to the method of Shimizu et al.[11].

2.8. SDS-polyacrylamide gel electrophoresis

The polymerization and degradation behavior of protein molecule by heating was examined by SDS-PAGE analysis. Meat gel (0.5g) was homogenized with 10 mL of 8 M urea - 2% SDS - 50 mM phosphate buffer (pH6.8) and mixed with 1 mL of 2-mercaptoethanol. The resulting solution was applied to SDS-PAGE using 3 % polyacrylamide gel according to the method of Laemmli [12].

3. RESULTS AND DISCUSSIONS

3.1. Temperature-gelation curves

The temperature-gelation curves of the unwashed and washed meats from 19 species were examined [5]. These species were classified into 4 groups according to the effect of meat washing on the modori appearance, as the following Type I to IV and the typical temperature-gelation curves of each type were shown in Fig. 1.

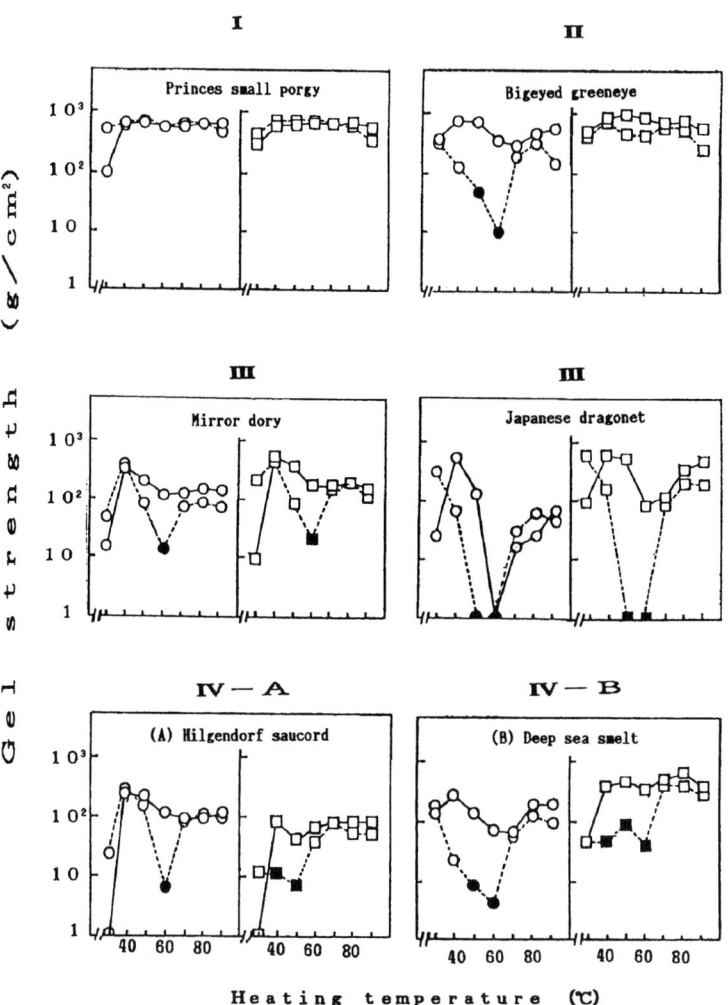

Figure 1. Temperature-gelation curves of unwashed and washed meat pastes prepared from fish caught off Kochi. ○, unwashed meat paste ; □, washed meat pastes; ——, 20min heating; ······, 2h heating ; ●, ■, modori-gel. Nomura, A. et al. (1993) [5].

Type I : Both unwashed meat and washed meat show no modori and an excellent gel forming ability.(princes small porgy, saddled weever and blacktipped greeneye).

Type II : Unwashed meat shows modori around 60°C and washed meat do not show modori (thorny flathead, silverbelly seaperch and bigeyed greeneye)

Type III : Both unwashed meat and washed meat show modori around 60°C (broad alfonsino, mirror dory, blackmouth cardinalfish, snubnose brotula, Japanese dragonet, and white edged jawfish)

Type IV : Unwashed meat shows modori at 60°C and washed meat shows modori around 40°C (Type IV of fish is a newly found type. Furthermore this group can be divided into 2 types. One is named Type IV-A, that is, the modori at 60°C disappears and the modori at 40°C appears after washing (sea conger and hilgendorf saucord). The other is named Type IV-B, the modori at 60°C does not disappear and the modori at 40°C appears after washing (seawedged perch, redbanded searobin, deep sea smelt, Chaunax sp. and blackthroat seaperch).

3.2. Distribution of gel strength at 80°C of four Types

To compare the gel forming ability at 80°C among 4 groups 19 species, breaking strength was plotted against breaking extension as shown in Fig.2. This plotting indicates that Type I is the highest in gel forming ability in both unwashed and washed meats; the next is Type II and III, and Type IV was the lowest. After washing, gel forming ability of Type II and III was improved to some extents but Type IV was not done [5].

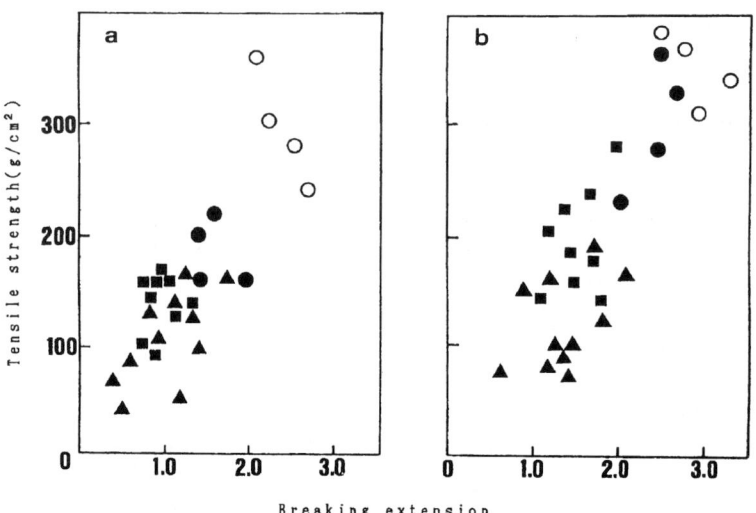

Figure 2. Plotting of tensile strength against breaking extension ($\Delta l/l_0$) of gels prepared by 80°C 20min heating from unwashed meat pastes (a) and washed meat pastes (b) of Type I – IV. ○ , Type I ; ●, Type II ; ■, Type III ; ▲, Type IV. Nomura, A. et al. (1993) [5].

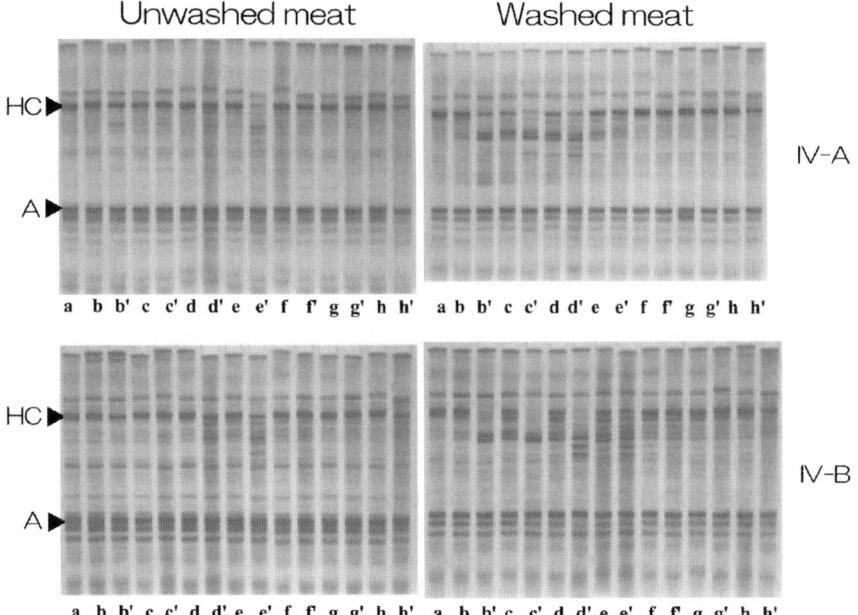

Figure 3. SDS-polyacrylamide gel electrophoresis patterns of gels prepared from unwashed meat paste and washed meat paste from Type IV. Type IV−A, Hilgendorf saucord ; Type IV-B, Redbanded searobin. HC, myosin heavy chain ; A, actin. a, unheated ; b, 30°C; c, 40°C; d, 50°C; e, 60°C; f, 70°C; g, 80°C; h, 90°C for 20min : b', 30°C; c', 40°C; d', 50°C; e', 60°C; f', 70°C; g', 80°C; h', 90°C for 2h. Nomura, A. et al. (1993) [5].

3.3. SDS-PAGE patterns of Type IV gel

Protein degradation in Type IV gel was examined by SDS-PAGE analysis to investigate the effect of washing on the modori occurrence of Type IV meat [5]. Figure 3 shows SDS-PAGE patterns of unwashed and washed meat gels from Type IV. The unwashed meat did not show the degradation of myosin heavy chain below 40°C, while the washed meat showed the degradation of myosin heavy chain below 40°C. This evidence suggests that modori phenomenon of washed meat at 40°C is due to degrading myosin heavy chain by myofibril-binding type protease.

Modori in the unwashed meat of Type II seems to be due to a sarcoplasmic type of protease, since the modori and the degradation of myosin heavy chain are depressed after washing. Modori of Type III also seems to be due to a myofibril binding type protease, because the modori and the myosin degradation are not depressed in spite of washing (data not shown) according to the classification of modori-inducing protease by Kinoshita et al. [13].

Type I has very weak or almost no activity in myosin heavy chain degradation. However when the unwashed meat pH of this type is lowered to pH 6 by adding lactic acid, the modori at 40°C was induced by washing, suggesting that the pH lowering of Type IV meat is responsible to induce the modori at the temperature [8].

3.4. Effect of two-step heating

Since two-step heating is usually used to increase the gel strength of cooked gel, at the next, we examined the effect of two-step heating on gel strength of the 4 types [6]. Gel was prepared by heating meat paste at 40°C for various time, prior to heating at 80°C for 20min. Gel strength of unwashed and washed meats from Type I *(Chelidoperca hirundinacea)*, II *(Chlorophtalmus albatrossis)* and III *(Zenopsis nebulosa)* was increased by two-step heating (data not shown). As shown in Fig. 4 and Fig. 5, washed meat from Type IV-A *(Helicolenus hirgendorfi)* and IV-B *(Doederleinia berycoides)* showed the negative effect of two-step heating along with the preheating time at 40°C, although unwashed meat showed the positive effect of two-step heating. This finding suggests that the two-step heating of washed meat of Type IV should not be carried out principally; very short time heating at 40°C could be done.

3.5. Effect of mixing washed meat of Type IV with meats of Type I and II

In order to utilize washed meat of Type IV, unwashed and washed meats of Type I and II, both of which have high gel forming abilities and polymerization abilities of myosin

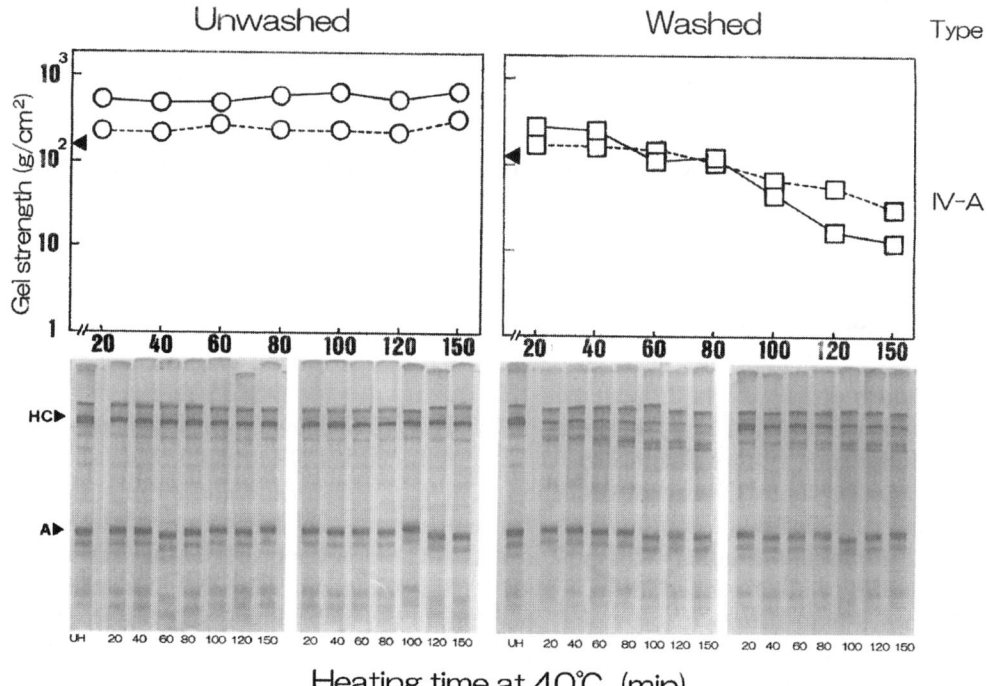

Figure 4. Effect of preheating at 40°C on the gel strength of Type IV-A gel cooked at 80°C and the SDS-polyacrylamide gel electrophoresis patterns. HC, myosin heavy chain; A, actin. ○, unwashed meat paste ; □, washed meat paste ; ——, heated at 40°C; ········, heated at 40°C prior to 80°C 20min. ▲, heated only at 80°C 20min. IV-A, Hilgendorf saucord. Nomura, A. et al. (1994) [6].

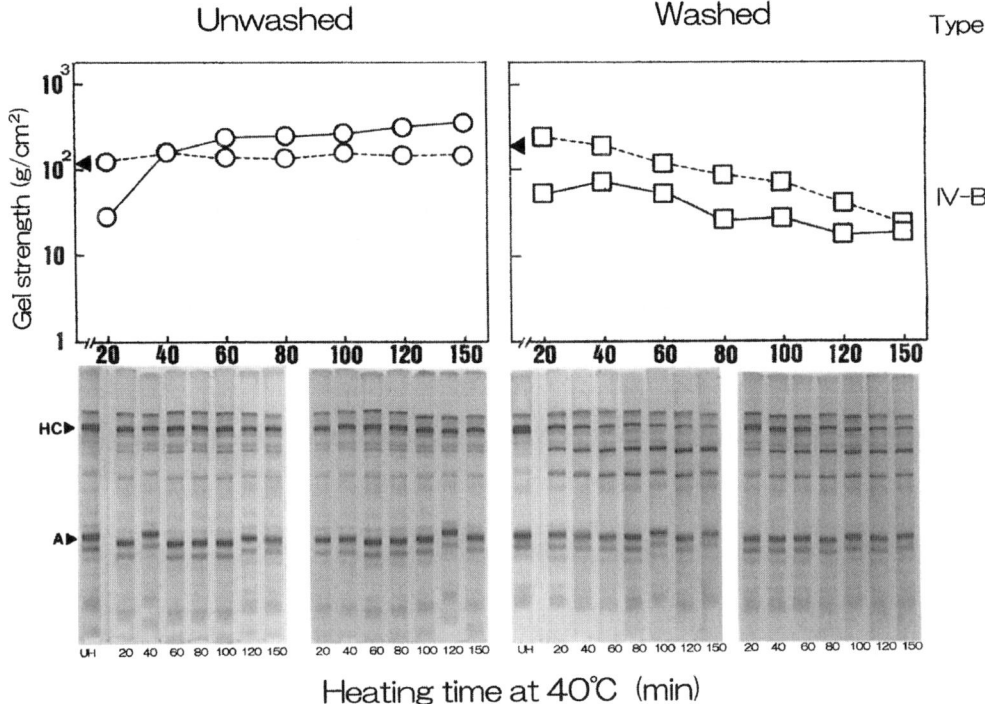

Figure 5. Effect of preheating at 40°C on the gel strength of Type IV-B gel cooked at 80°C and the SDS-polyacrylamide gel electrophoresis patterns. HC, myosin heavy chain; A, actin. ○, unwashed meat paste ; □, washed meat paste ; ——, heated at 40°C; ·······, heated at 40°C prior to 80°C 20min. ▲, heated only at 80°C 20min. IV-B, Blackthroat seaperch. Nomura, A. et al. (1994) [6].

heavy chain, were mixed with the meat of Type IV (*Helicolenus hirgendorfi*) washed meat, and then temperature-gelation curve was examined [7]. The results were shown in Fig. 6. By adding unwashed meat of Type I (*Chelidoperca hirundinacea*) into Type IV washed meat, modori was not observed at any temperatures. By adding unwashed meat of Type II (Chlorophtalmus albatrossis), modori at 40°C did not appear, but that at 60°C did as shown in Fig. 7. The washed meat of Type I and II showed the similar effects that both meats improve the gel formation at 30°C, but the modori at 40°C was still observed. The improvement of gel forming ability at 30°C seems to be due to transglutaminase activity in the meat, since myosin heavy chain was probably polymerized by covalent bonding. These results suggest that a modori-inhibiting factor might exist in the sarcoplasmic fraction other than transglutaminase. We confirmed that sarcoplasmic fraction of Type IV meat inhibits the modori of washed meat of Type IV at 40°C (data not shown). It was concluded that the modori-inhibiting factor exists in sarcoplasmic fraction. Furthermore the modori appearance in washed meat of Type IV is possibly due to the activation of myofibril type protease through removing the modori-inhibiting factor that might be

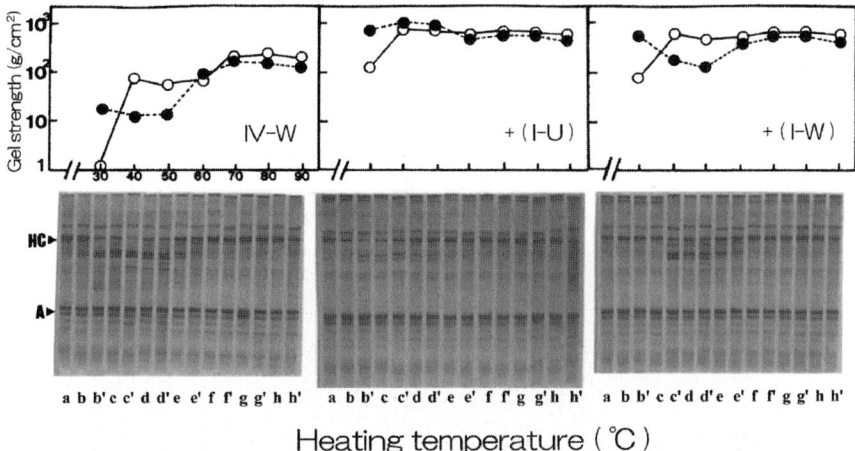

Figure 6. Temperature-gelation curves and SDS-polyacrylamide gel electrophoresis patterns of gels that were prepared from the mixture of Type IV washed meat mixed with unwashed and washed meats of Type I. ○, 20min heating ; ●, 2h heating ; HC, myosin heavy chain ; A, actin. IV-W, washed meat paste of Type IV (Hilgendorf saucord); I-U and I-W, unwashed and washed meat of Type I (Princes small porgy), respectively; a, unheated ; b, 30°C; c, 40°C; d, 50°C; e, 60°C; f, 70°C; g, 80°C; h, 90°C for 20min : b', 30°C; c', 40°C; d', 50°C; e', 60°C; f', 70°C; g', 80°C; h', 90°C for 2h. Nomura, A. et al. (1995) [7].

protease inhibitor(s). We investigated the characteristics of modori-inhibiting factor, purified it in the further experiments and found that it is an inhibitor for serine type protease and a monomeric protein that is 80,000 in molecular weight [9, 10].

4. CONCLUSION

It was found that there exist 4 types of gel-forming characteristics among 19 fish species tested according to the effect of meat washing on the modori appearance.

Type I: Both unwashed and washed meats show no modori and the most excellent gel forming ability (3 species).

Type II: Unwashed meat shows modori around 60°C and washing is effective to remove modori (3 species).

Type III: Unwashed meat shows modori around 60°C as seen in Type II but washing is not effective to remove modori (6 species).

Type IV: Unwashed meat shows modori at 60°C and washed meat shows it around 40°C (7 species). Both meats are the lowest in gel-forming ability. This type of fish is a newly found type. In Type IV, there are 2 types: one is Type IV-A in which modori disappears and modori appears after washing. Another is Type IV-B where 60°C modori does not disappear and 40°C modori appears after washing.

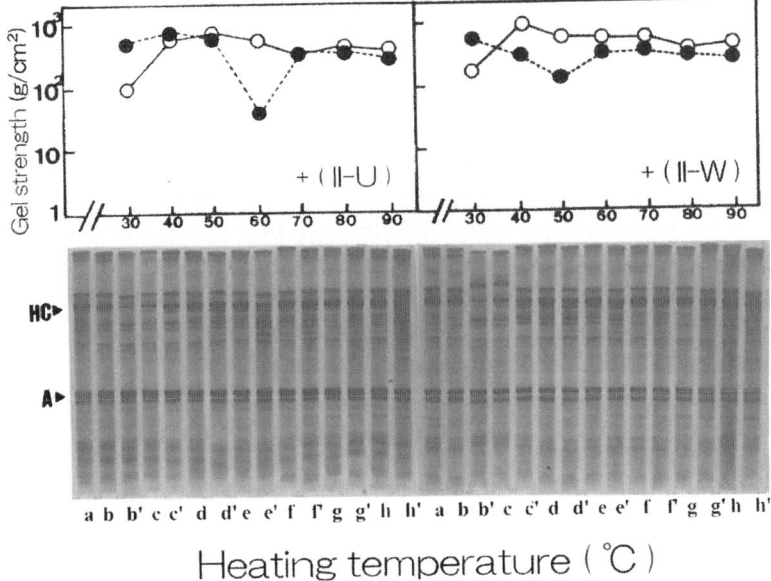

Figure 7. Temperature-gelation curves and SDS-polyacrylamide gel electrophoresis patterns of gels that were prepared from the mixture of Type IV washed meat mixed with unwashed and washed meats of Type II. ○, 20min heating ; ●, 2h heating ; HC, myosin heavy chain ; A, actin. II-U and II-W, unwashed and washed meat of Type II (Bigeyed greeneye), respectively; a, unheated ; b, 30°C; c, 40°C; d, 50°C; e, 60°C; f, 70°C; g, 80°C; h, 90°C for 20min : b', 30°C; c', 40°C; d', 50°C; e', 60°C; f', 70°C; g', 80°C; h', 90°C for 2h. Nomura, A. et al. (1995) [7].

These results suggested that it is important to examine the gel forming characteristics of not only unwashed meat but washed meat.

Two-step heating of washed meat of Type IV should not be carried out principally, although very short time heating at 40°C is positive. Mixing with Type I or II improves gel-forming ability. Sarcoplasmic fraction could have 40°C modori-inhibiting factor. Furthermore, modori appearance in washed meat of Type IV seems to be due to the activation of myofibril type protease through removing the modori-inhibiting factor that might probably be a protease inhibitor.

REFERENCES

1. Suwansakornkul, P., Itoh, Y., Hara, S. and Obatake, A. (1993) The gel forming characteristics of lizardfish. Nippon Suisan Gakkaishi, 59: 1029-1037.
2. Suwansakornkul, P., Itoh, Y., Hara, S. and Obatake, A. (1993) Identification of proteolytic activies of gel-degrading factors in three lizardfish species. Nippon Suisan Gakkaishi, 59: 1039-1045.

3. Itoh, Y., Maekawa, T., Suwansakornkul, P. and Obatake, A. (1995) Seasonal variation of gel-forming characteristics of three lizardfish species. Nippon Suisan Gakkaishi, 61: 942-947.
4. Itoh, Y., Maekawa, T., Suwansakornkul, P. and Obatake, A. (1997) Effect of meat-washing on the gel formation and on the polymerizing and degrading behavior of lizardfish meat proteins at 40 and 60°C. Nippon Suisan Gakkaishi, 63: 286-290.
5. Nomura, A., Itoh, Y., Soen, T. and Obatake, A. (1993) Effect of washing the meat of fish species caught in Tosa Bay on the appearance of modori-phenomena (disintegration of gel). Nippon Suisan Gakkaishi, 59: 857-864.
6. Nomura, A., Itoh, Y., Nishikawa, S. and Obatake, A. (1994) The gel strengthening effect of two-step heating in the meat pastes from different fish in modori-inducing conditions. Nippon Suisan Gakkaishi, 60: 667-673.
7. Nomura, A., Itoh, Y., Toyoda, H. and Obatake, A. (1995) Inhibitory effect of sarcoplasmic protein fraction of fish meat upon modori-phenomenon induced at about 40°C by washing meat. Nippon Suisan Gakkaishi, 61: 744-749.
8. Nomura, A., Itoh, Y., Yamamoto, T. and Obatake, A. (1997) The pH of fish meat and the appearance of modori-phenomena around 40°C in the washed meat paste. Nippon Suisan Gakkaishi, 63: 103-104.
9. Nomura, A., Itoh, Y., Osaka, Y., Kitamura, Y., Miyazaki, H. and Obatake, A. (1998) A factor inhibiting myosin heavy chain degradation of washed meat paste around 40°C, in the sarcoplasmic protein of fish meat. Nippon Suisan Gakkaishi, 64:744-749.
10. Nomura, A., Itoh, Y., Yahata, K., Taniwaki, N. and Obatake, A. (2000) The properties and esolation of a myosin heavy chain degradation inhibitor in the sarcoplasmic protein of fish meat. Nippon Suisan Gakkaishi, 66:731-736.
11. Shimizu, Y., Machida, R. and Takenami, S. (1981) Species variations in the gel-forming characteristics of fish meat paste. Nippon Suisan Gakkaishi, 47: 95-104.
12. Laemmli, U. K. (1970) Cleavage of structural proteins during the assembly of the head of bacteriophage T4. Nature, 227:680-685.
13. Kinoshita, M., Toyohara, H. and Shimizu, Y. (1990) Diverse distribution of four distinct types of modori (gel degradation)-inducing proteolysis among fish species. Nippon Suisan Gakkaishi, 56:1485-1492.

ACCEPTABILITY AND ITS IMPROVEMENT OF KAMABOKO GEL DERIVED FROM SILVER CARP SURIMI

Xichang Wang[a], Takashi Hirata[b], Yutaka Fukuda[c], Masato Kinoshita[b] and Morihiko Sakaguchi[b]

[a]College of Food Science, Shanghai Fisheries University, Yangpu, Shanghai 200090, China
[b]Division of Applied Biosciences, Graduate School of Agriculture, Kyoto University, Kyoto 606-8502, Japan
[c]JIRCAS, 1-1 Ohwashi, Tsuba, Ibaraki 305-8686, Japan

Abstract

Sensory evaluations were conducted on kamaboko gels prepared from silver carp (SC) surimi. Different kamaboko gel specimens, produced with or without the extract from walleye pollack (WP) muscle (Ewp) or trimethylamine N-oxide (TMAO), were tested using a panel of university students for their odor, flavor, texture, whiteness, and overall acceptability. According to the sensory evaluation results done in Kyoto, there were significant differences ($p < 0.005$) between SC kamaboko gel and WP kamaboko gel for all sensory evaluation indices. When compared with WP kamaboko gel, kamaboko gel derived from SC surimi was unacceptable to the Japanese students especially because of its odor. A slight increase in their sensory scorings of kamaboko gels occurred when Ewp was added into SC surimi, especially when scoring the odor of SC kamaboko gels. In addition, the acceptability of kamaboko gel with the addition of TMAO has a higher scoring in its odor as compared with the control (no addition of TMAO), which is presumed as a result of a little of trimethylamine derived from TMAO additive.

1. INTRODUCTION

Kamaboko, one of the most popular and traditional fish products in Japan, has made a great progress along with frozen surimi successfully developed for utilizing a large amount of walleye pollack since the 1960s. Surimi is originally a Japanese term, and frozen surimi is an intermediate foodstuff with a high potential for long shelf-life, for distribution over a wide area, and for producing various texturized products. Along with a tremendous growth in production and consumption worldwide, surimi and surimi-based products have been established as internationalized goods, and there are various kinds of surimi-based products available in current world food markets. Frozen surimi technology has created an extremely effective method for the exploitation of so-called low-value and high-yield fish products nowadays. A vast quantity of cultured freshwater fish in China is becoming an important resource that is to be utilized more efficiently as material for processing. Therefore, it is clear and important that cultured freshwater fish should be developed and utilized as one of the new materials in surimi and surimi-based products in the near future.

Freshwater fish is an ordinary table fish in China, and has been established as a main dish after the ideas of healthy food, high yield, cheap price and various cooking and seasoning methods were introduced recently, especially in the inland area of China. Marine fish has always been a primary fish consumed by the Japanese because of such factors as its pleasant flavor, good texture, abundant varieties, and freshness, in contrast, freshwater fish was consumed in much smaller extent in Japan. In addition, kamaboko, which is made from marine fish meat, is a traditional food as well as one of the most popular fish processed products in Japan.

Although there are some differences between China and Japan in the processing, utilization and consumption of fisheries resources, little information about these is available [1]. It is of interest to compare the acceptability of kamaboko gels derived from freshwater fish and/or marine fish meat. The purpose of this study was to obtain some comparative data on sensory evaluation, and to guide the product development of freshwater fish [2].

2. MATERIALS AND METHODS

2.1. Materials

Silver carp *Hypophthalmichthys molitrix* surimi, which is made from live cultured fish according to the conventional process technique of Japanese frozen surimi, was manufactured and supplied by the laboratory of the Sino-Japan collaborated research project Development of Technology for Utilization of Freshwater Fisheries Resources in Shanghai, China.

Walleye pollack *Theragra chalcogramma* surimi (SA grade), made by Nichiro Co. Ltd (Tokyo, Japan), was supplied by the Industrial Research Center of Ehime prefecture in Japan.

Extract of walleye pollack meat was prepared using the hot water extraction method. Fresh white dorsal meat was homogenized with three volumes of distilled water, and extracted for 30 min in a boiling water bath. After being cooled, the supernatant was centrifuged (10, 000 ×g for 15 min) and filtered. The final extract was made up to equal quantities of raw meat with evaporation.

Three concentration 665mM, 443mM and 222mM of trimethylamine *N*-oxide (TMAO) solutions were separately prepared by TMAO standard (Aldrich chemical company, Inc. USA) in advance.

2.2. Preparation of kamaboko gel

Half-thawed surimi was chopped in a mini kitchen-cutter (Yamada FP-1S; Mitsubishi Co. Ltd, Tokyo, Japan), and to compare the gel-forming properties, the final moisture content of all samples was adjusted to 79% by adding some water except for the addition of 2.5% NaCl and 10% (surimi weight basis) extract solution or TMAO solution in present study. After becoming sticky, the resulting meat sol was stuffed into folded polyvinylidene chloride tube casings, which measured 35 mm in diameter and approximately 150 mm in length. Both ends of each tube casing were tied with cotton thread, and all of the proceeding operations were accomplished at 10°C or below. Next, the filled casing samples were preincubated (suwari in Japanese) at 30°C for 60 min and subsequently heated (cooked) at 90°C for 30 min in water bath incubators. Finally, the heated casing gels were taken out and immediately cooled in an ice-water bath to room temperature and then stored at 4°C in cold-storage overnight until required for the sensory evaluation.

2.3. Sensory evaluation

Sensory evaluation of kamaboko gels was conducted by more than 60 university students specializing in food science and studying either at the Faculty of Agriculture, Kyoto University, Japan, Shanghai Fisheries University, Shanghai (coastal area in China), and Huazhong Agricultural University, Wuhan (inland area in China), respectively. For testing, casing gels were cut into bite-sized (3 mm) samples and, after being warmed slightly at about 25°C), all samples were served to the students randomly and at times other than during meal hours. All panels were asked to score four characteristics (odor, taste, texture, and whiteness), as well as overall desirability using a 5-point hedonic scale (1, dislike extremely; 2, dislike moderately; 3, neither like nor dislike; 4, like moderately; and 5, like extremely) [3-5]. A score of 3 was the division between acceptable and unacceptable.

2.4. Instrumental measurement of gel properties

The breaking force and breaking strain of a heat-induced gel were measured using a rheometer (RE-3305; Yamaden Co., Ltd, Tokyo, Japan) with a spherical plunger measuring 5 mm in diameter. For the measurement, the kamaboko gel was cut into 22 mm×30 mm cylindrical test specimens after removing the casing film, and the penetration speed of the plunger was established as 60 mm/min.

Prior to conducting the penetration test, whiteness of the cylindrical cross-sected samples were measured immediately with a spectrophotometer (CM-2002; Minolta Co. Ltd; Osaka, Japan) for the values of L^*, a^* and b^* (CIE Laboratory system) to the first decimal place. Whiteness [6], as an index for the general appearance of the test specimen, was calculated as:

Whiteness = $100 - [(100 - L^*)^2 + a^{*2} + b^{*2}]^{0.5}$

2.5. Determination of TMA and TMAO by gas chromatography

According to the peak area on GC recorded chart, and was calculated the amount of TMA contained in the sample by the standard TMA (Nacalai Tesque, Inc. Japan) solution curve directly; in addition, the amount of TMAO contained in the sample was then calculated by subtracting the amount of the amount of TMA from the amount of both TMAO and TMA [7].

2.6. Statistical analysis

The StatView for Windows (version 4.58; Abacus Concepts Inc., Cary, NC, USA) was used for data analysis. Analysis of variance was performed to compare the mean hedonic scores of all sensory indices among the different surimi gels. In all cases, the criterion for statistical significance was set at $p < 0.05$.

3. RESULTS AND DISCUSSION

3.1. Acceptability comparison of kamaboko gels prepared from silver carp surimi and walleye pollack surimi between Chinese and Japanese consumers

The sensory evaluation results indicated that there were significant differences ($p < 0.005$) for all sensory characteristics (odor, taste, texture and whiteness), as well as overall desirability between silver carp kamaboko gel and walleye pollack kamaboko gel (Fig.1a), when evaluated by the Japanese. The mean scores of silver carp kamaboko gel were all less than 3, whereas those of walleye pollack kamaboko gel were all more than 3, whereby silver carp kamaboko gel was rated much lower in terms of odor and overall acceptability than walleye pollack kamaboko gel. Kamaboko gel derived from freshwater fish surimi was unacceptable to the Japanese, most probably because of its odor.

The same sensory tests accomplished in Shanghai (coastal area in China) and Wuhan (inland area in China) showed that the kamaboko gel from walleye pollack surimi also obtained higher sensory scores than that from silver carp surimi (Fig. 1b,c) [8]. However, silver carp kamaboko gel obtained higher sensory scores in China than in Japan especially in its odor and overall desirability (Fig. 1a,b,c), that is, kamaboko gels made from both walleye pollack and silver carp surimi were acceptable for Chinese in Shanghai and Wuhan. The Chinese who live in the inland area especially have been accustomed to and/or like to eat the freshwater fish, and the above findings indicate that there were some differences in the acceptability of fish food products, which is apparently influenced by peoples' eating habits and where they live.

3.2. Effects of fish muscle extracts on the acceptability of kamaboko gels

In view of the aforementioned results, odor and taste were considered to be the principal factors influencing the acceptability of kamaboko gels. It could be predicted that different results would be obtained if kamaboko gel specimens were formed with or without the additive of muscle extract from walleye pollack (Ewp). As listed in Table 1 [8], there was a slight increase in the sensory scores given in terms of odor, taste, texture, whiteness, and overall desirability of kamaboko gels when extract was added into silver carp kamaboko gels. The notable results were the significantly higher odor scorings ($p < 0.05$) given to silver carp surimi gels in which the extract of walleye pollack had been added as compared with the control (no additive of extract). In general, to improve the acceptability of silver carp kamaboko gels, the addition of its original and/or marine fish muscle extract is necessary.

According to the results of the mechanical measurements done on samples' gel properties (Table 2) [8], there were also significant differences ($p < 0.05$) between silver

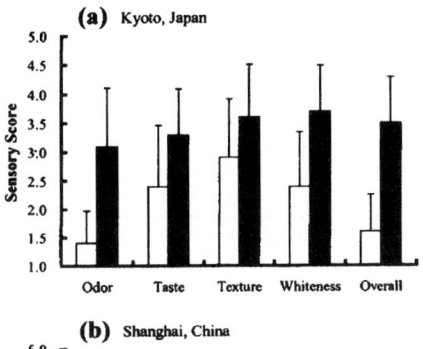
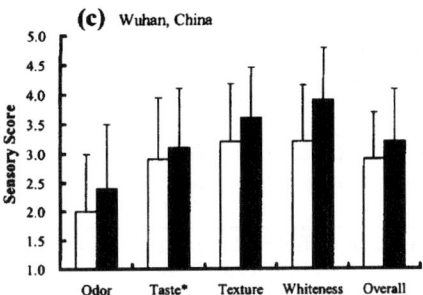
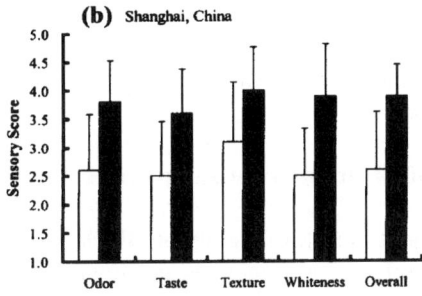

Figure 1. Acceptability comparison of kamaboko gels derived from silver carp surimi and from walleye pollack surimi between Chinese and Japanese. □, silver carp; ■, walleye pollack. Values are shown as the mean ±SD. *No significant difference ($p > 0.05$) between silver carp and walleye pollack surimi gels was observed. Wang, X. et al. (2002) [8].

Table 1
Effect of walleye pollack extracts on the acceptability of kamaboko gels derived from silver carp surimi and from walleye pollack surimi in Kyoto, Japan

Sample name	Odor	Taste	Texture	Whiteness	Overall desirability
SC-0	1.4±0.6[a]	2.4±1.1[a]	2.9±1.0[a]	2.4±0.9[a]	1.6±0.6[a]
SC-Ewp	1.9±1.0[b]	2.5±0.8[a]	3.0±0.9[a]	2.9±0.9[b]	2.4±0.8[b]
WP-0	3.1±1.0[c]	3.3±0.8[b,c]	3.6±0.9[b,c]	3.7±0.8[c]	3.5±0.8[c]
WP-Ewp	3.1±0.9[c]	3.2±1.0[c]	3.6±1.0[c]	3.8±0.7[c]	3.4±0.8[c]

[a–c] Means with different letters are significantly different ($p < 0.05$), and those with the same letter represent no significant difference ($p > 0.05$) within a subgroup of a column. Data are shown as mean ±SD. SC, silver carp surimi; WP, walleye pollack surimi; 0, no additive of extract; Ewp, walleye pollack extracts added. Wang, X. et al. (2002) [8].

carp kamaboko gels and walleye pollack kamaboko gels in terms of the breaking force and whiteness, whereby WP sample gels had higher values for both. Furthermore, a higher breaking force was observed in silver carp surimi gels to which the extract of

Table 2
Effect of walleye pollack extracts on the texture and whiteness of kamaboko gels derived from silver carp surimi and from walleye pollack surimi

Item	SC-0	SC-Ewp	WP-0	WP-Ewp
Breaking force (g)	478±48[a]	482±41[a]	587±62[b]	568±57[b]
Breaking strain (mm)	13.1±0.9[a]	12.8±0.9[a]	12.9±0.7[a,b]	12.2±0.6[b]
Whiteness	77.2±0.8[a]	77.6±1.1[a]	81.1±0.6[b]	80.2±0.5[b]

[a,b]Means with different letters are significantly different ($p < 0.05$), and those with the same letter represent no significant difference ($p > 0.05$) within a subgroup of a row. Data are shown as mean ± SD. Abbreviations are the same as those given in Table 1. Wang, X. et al. (2002) [8].

walleye pollack had been added compared with the control. These physical results are in accordance with the aforementioned organoleptic data.

3.3. Gel-forming property and acceptability of silver carp surimi gels as affected by TMAO addition

TMAO is well known as a particular component is rich in walleye pollack muscle and other marine products, and TMA as a representative component is dominant in marine fishy odor [9]; in contrast, there is no TMAO detected in silver carp muscle, and other components different from TMA were related to freshwater fishy odor. It is of interest to survey the effect of TMAO addition on the quality of freshwater fish surimi gels.

As listed in Table 3, it is obvious that TMAO is rich in extracts of walleye pollack muscle; besides, frozen surimi produced by fully washing technique still contained some TMAO, moreover heated gels also remained a little TMAO which has been converted into TMA partially. On the other hand, as to freshwater fish like silver carp, TMAO and TMA were no detectable in its muscle, extracts, and frozen surimi. Gels prepared by the addition of some quantities of TMAO were affirmed the occurrence of TMA even though its amount is not so high, which is available to overcome the deficiencies of freshwater fish in odor. That is, the TMA content of silver carp kamaboko gels added into TMAO was detectable by GC analysis.

The results (Table 4) show that the acceptability of kamaboko gel with the addition of TMAO has a higher scoring in its odor as compared with the control T-0 (no addition of TMAO), which is presumed as a result of a little of TMA derived from TMAO additive. Meanwhile, breaking force of the former gels occurred a higher values than the control, especially T-2 addition is more effective than other addition conditions, that is an available for the improvement of acceptability of TMAO to be added.

Baskakov et al. [10] reported that the extraordinary capability of organic osmolytes, TMAO, forces two thermodynamically unfolded proteins to fold to native-like species having significant functional activity, maintains the structure and function cellular proteins in organisms exposed to denaturing environmental stresses. In addition, Anthoni et al. [11] also illustrated that both TMAO and taurine have been shown to stabilize enzymes against thermal denaturation and the former acts as a cryoprotectant during frozen storage. All of these maybe give some information to explain the effect of TMAO on heat-induced gelation.

Table 3
Contents of TMAO and TMA in walleye pollack (muscle extracts, frozen surimi and heated gels) and silver carp surimi gels

Item	TMAO	TMA
Walleye pollack		
Muscle extracts	1311	18.10
Frozen surimi	79.01	0.14
Heated gels	69.11	0.43
Silver carp surimi gels		
T-0	—	—
T-1	126.40	0.38
T-2	215.71	1.47
T-3	443.87	6.20

Abbreviative notations T-0, T-1, T-2 and T-3 mean amounts of TMAO additive in 0, 167, 333 and 500 mg/100g (surimi weight basis), respectively. Heating conditions of gels were at 30°C setting for 1 hour and subsequently at 85°C heating for 30min. TMAO and TMA were determined by GC.

Table 4
Effects of TMAO addition on the quality of silver carp surimi gel

Addition of TMAO	T-0	T-1	T-2	T-3
Gel strength				
Breaking force (g)	431±96.2[a]	473±78.3[a]	490±89.3[b]	473±119.0[a]
Breaking strain (mm)	11.70±1.25	11.67±1.35	11.87±1.07	11.06±1.65
Sensory evaluation				
Odor	2.1±0.93[a]	2.4±0.53[a]	3.0±0.50[b]	2.4±0.73[a]
Taste	2.7±0.71[a]	2.7±1.12[ac]	2.9±1.05[bc]	3.3±1.22[ac]
Texture	3.4±1.24[ab]	3.7±1.32[ab]	2.7±0.87[a]	3.2±0.83[b]
Whiteness	3.1±0.93[a]	3.8±0.97[a]	2.7±0.71[b]	3.6±0.73[a]
Overall desirability	2.3±1.00	2.6±0.88	2.9±0.93	2.9±0.93

Abbreviative notations are the same as given in Table 3. Data are shown as mean ±SD. [a–c]Means with different letters are significantly different ($p<0.05$), and the same letter represents no significant difference ($p>0.05$) within a subgroup of a row.

4. CONCLUSION

The present study demonstrated that the acceptability of kamaboko gels differs between Chinese and Japanese consumers. In particular, odor was a principal factor; kamaboko gel derived from the freshwater fish surimi was unacceptable to the Japanese mainly because of its odor, whereas it was acceptable to the Chinese, notably those from inland China. It is likely that the addition of marine fish extracts (e.g. walleye pollack, rich in TMAO) and/or the partial mixing with marine fish surimi were a considerable treatment to improve the acceptability of kamaboko gel derived from freshwater fish surimi, meanwhile to enhance the gel strength to some extent.

5. ACKNOWLEDGMENTS

We are grateful to Professors S. Lin and X. Luo for their support during this study and to all the panel students (Huazhong Agricultural University of China) for their active and earnest participation in the sensory evaluation study. Special thanks are also due to Drs. H. Oka and Y. Hiraoka (Industrial Research Center of Ehime prefecture, Japan) for providing the surimi samples.

REFERENCES

1. Chambers, E. and Robel, A. (1993) Sensory characteristics of selected species of freshwater fish in retail distribution. J. Food Sci., 3: 508-512, 516.
2. Chia, S., Baker, R. and Hotchkiss, J. (1983) Quality comparison of thermoprocessed fishery products in cans and retortable pouches. J. Food Sci., 5: 1521-1525.
3. Yoshikawa, S. (1969) Sensory evaluation of flavor. New Food Industry, 4: 51-54.
4. Furukawa, H. (1994) Scoring method. In: Determination Related to Deliciousness: Practice on Food Sensory Test. Furukawa, H. (ed.), Saiwai Bookshop Press, Tokyo, pp. 29-39.
5. Sheng, M., Brochetti, D., Duncan, S. and Lawrence, R. (1996) Hedonic ratings of reduced-fat food products: Factors affecting ratings from female and male college-age students. J. Nutr. Recipe Menu Dev., 2: 31-40.
6. Park, J. W. (2000) Evaluation of functional ingredients. In: Surimi and Surimi Seafood. Park, J. W. (ed.), Marcel Dekker, Inc., New York. pp. 381-382.
7. Bystedt, J., Swenn, I. and Aas, H.W. (1959) Determination of trimethylamime oxide in fish muscle. J. Sci. Food Agric., 10:301-304.
8. Wang, X., Hirata, T., Fukada, Y., Kinoshita, M. and Sakaguchi, M. (2002) Acceptability comparison of kamaboko gels derived from silver carp surimi and from walleye pollack surimi between the Chinese and Japanese. Fish. Sci., 68: 165-169.
9. Tokunaga, T. (1982) Chemistry of fishy flavour. In: Gyoshu Chikunikushu. Ohta, S. (ed.), Koseisha-koseikaku, Tokyo, pp. 29-88.
10. Baskakov, I. and Bolen, D.W. (1998) Forcing thermo-dynamically unfolded proteins to fold. J. Biol. Chem., 273: 4831-4834.
11. Anthoni, U., Børresen, T., Christophersen, C., Gram, L. and Nielsen, P.H. (1990) Is trimethylamine oxide a reliable indicator for the marine origin of fish? Comp. Biochem. Physiol., 97B: 569-571.

SEAFOOD PROCESSING AND SAFETY

INNOVATIVE TECHNIQUES FOR TRADITIONAL DRIED FISH PRODUCTS

Tiong Ngei Kok, Soon Eong Yeap, Woan Peng Lee, How Kwang Lee and Guat Theng Aug
Marine Fisheries Research Department, 2 Perahu Road, Off Lim Chu Kang Road, Singapore 718915, Republic of Singapore

Abstract
Three innovative techniques were developed for producing value-added traditional fish products, namely hot-air oven and vacuum oven drying to process dried shrimp instead of sun-drying and puff machine to cook the fish cracker instead of deep-frying. Shrimp was dried for 5 hours at 85°C in the hot-air oven and 2.5 hours at 130°C in the vacuum oven. The moisture content for hot-air dried shrimp was 7.9% and water absorption ability was 48%. The vacuum dried shrimp had a moisture content of 3.4% and a water absorption ability of 118%. Using the vacuum oven drying, drying time was greatly reduced and dried shrimp with a low moisture content, high water absorption ability and a new crispy texture was obtained. The vacuum dried shrimp is suitable to be eaten directly as a snack food and can also be used as an ingredient for instant noodles. Storage studies were also conducted on the dried shrimp products produced by both techniques under two packaging conditions, i.e. air-pack and vacuum-pack. Apart from the vacuum-dried air-pack shrimp which gave off an ammoniacal smell, all other shrimp samples were found to remain acceptable after 7 months of storage at 28 ± 2°C. Fish cracker cooked by the puff machine had a greater amount of linear expansion than those cooked by deep-frying with the added advantage that no oil was used in the puffing process resulting in a healthier product.

1. INTRODUCTION

Dried shrimp and fish cracker (keropok) are fish products that are widely processed in the region for domestic consumption and export. Traditional techniques of processing involved sun drying of shrimp and deep-frying of keropok. Although theses techniques have the advantages of being straightforward and low cost, there are several inherent problems. Sun-drying of shrimp is time consuming and unpredictable (dependent on weather conditions). It is difficult to obtain a low moisture content and the products are prone to contamination by insects and pests. It also results in products with a generally short storage-life, ranging from approximately 2 or 3 to 6 months [1]. Deep-frying of fish cracker results in a product with high oil content that is not healthy. This paper describes two studies carried out by Marine Fisheries Research Department (MFRD) as part of its on-going regional project on improvement of traditional fish products, namely, the use hot-air oven or vacuum oven in place of sun-drying for processing dried shrimp and puff machine in place of deep fryer for cooking fish cracker.

2. MATERIALS AND METHODS

2.1. Materials

Fresh Indian white shrimp, *Penaeus indicus* (count about 95 pieces/ kg) was purchased from Jurong Fish Market. The fresh shrimp was washed with tap water and boiled in 5% brine solution for 15 min (shrimp:brine−1:1.5). The boiled shrimp was drained, cooled and frozen at -18°C. The frozen shrimp was divided into 2 lots − one lot was dried in the hot air oven and the other dried in the vacuum oven (Fig. 1).

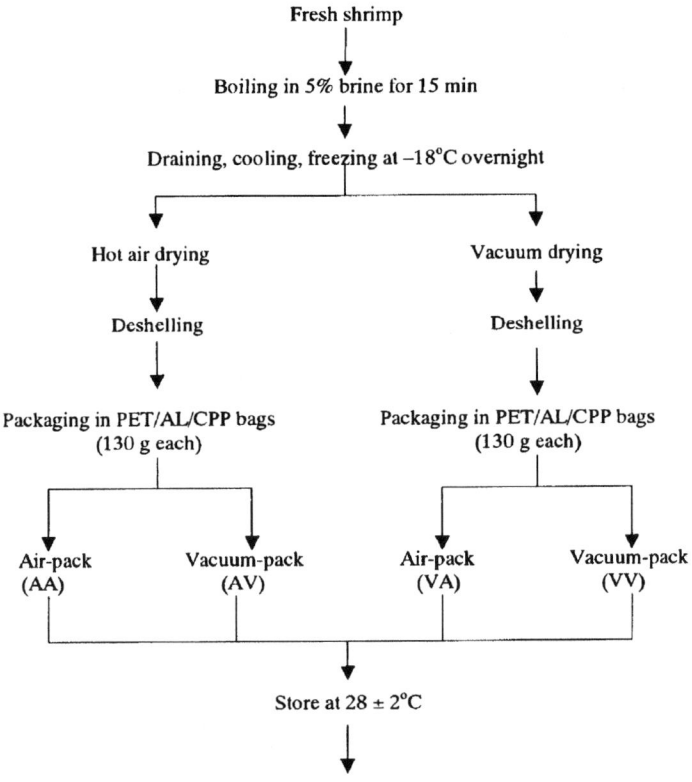

Figure 1. Dried shrimp processing.

Table 1
Formulation of fish cracker (Keropok)

Ingredients	Composition (%)	Weight (g)
Fish (Torpedo, hard-tail)	39.0	1800
Tapioca flour	39.0	1800
Salt	2.1	95
Sugar	1.0	48
MSG	1.0	48
Pepper	0.5	24
Ice water	17.3	800
Total	100	4615

Torpedo, hard-tail (*Megalaspsis cordyla*) was purchased from wet market. Tapioca flour, MSG, pepper, sugar and salt were used as ingredients of the fish cracker (Keropok), as shown in Table 1.

2.2. Drying Procedures

Hot air drying and vacuum drying were performed as follows: the shrimp was dried for 5h at 85°C in the SMA-112 Smoke Master by Hanaki Manufacturing Co., Ltd. and for 2.5h at 130°C in the NRD-150E vacuum drier by Ebara Engineering Pte Ltd., respectively. After drying, the shells of the shrimp were removed using a Krups Rotary 500 food processor and a sieve.

2.3. Packaging and storage procedures

The deshelled shrimp was packed in lots of 130g into Polyethylene terephthlate (PET)/ Aluminium (AL)/ Non-oriented polypropylene (CPP) bags. Dried shrimp was vacuum packed with an ENA-1-AGUDS vacuum sealer (Nippon Polycello Industry Co., Ltd) which had been evacuated to 650 mmHg prior to sealing. The sealing of bag for air packing was done with a F1-K300 Fuji Impulse Sealer. The sealed samples were stored in the Memmert electronic incubator BE 40 at 28±2°C. Physical, chemical and sensory analysis were carried out on the dried shrimp at 0, 1, 3, 5 and 7th month of storage.

2.4. Physical analysis

For the measurement of water absorption, dried shrimp (5g) was soaked in 70mL distilled water at 25°C for 15min. Adhering water was removed with blotting paper. Water absorption is expressed as the percentage increase in weight of the dried shrimp.

Measurement of linear expansion of fish cracker was made by the method of Yu [6].

$$\text{Linear expansion (\%)} = \frac{(\text{Length after puffing} - \text{Length before puffing}) \times 100}{\text{Length before puffing}}$$

2.5. Chemical analysis

Astaxanthin was determined as follows: two grams of dried shrimp (pulverized) was extracted with 50mL acetone overnight. The mixture was filtered with Whatman filter paper No. 41 and the optical density of the filtrate read at 437.5nm using the Perkin Elmer model lambda 14P spectrophotometer. Astaxanthin content is calculated from the equation as follows: Astaxanthin content (mg/100g dried shrimp) = O.D. 437.5 nm×50 mL×0.4×50.

Volatile basic nitrogen content, peroxide value and moisture content were measured by the methods of Yamagata [2], Low [3] and Ng [4], respectively. Water activity of pulverized dried shrimp was measured using a thermoconstanter HUMIDAT-TH2 with a measuring chamber set at 25°C.

2.6. Sensory analysis

The dried shrimp was rated for attribute liking by 5 experienced panelists using a hedonic scale of 9 for "like extremely" and 1 for "dislike extremely" [5]. The attributes were as follows: appearance, color, flavor, odor, texture and overall acceptability. Sensory analysis was conducted every month for the seven months of storage.

2.7. Preparation of fish cracker

Deboned fish was thoroughly mixed with the ingredients, using a Stephan mechanical blade mixer at ambient temperature (25°C). The mixture was stuffed into 50mm diameter casing and tied at both ends. The stuffed rolls were steamed for 90min at atmospheric pressure and cooled in ice water. They were frozen at -18°C overnight, then sliced with an A250 Sirman mechanical slicer to thickness of 2mm. The slices were dried in the SMA − 112 Smoke Master by Hanaki Manufacturing Co., Ltd for 2.5h at 40 − 55°C. Moisture content of the products was determined [4] and the products were then separated into 2 lots. One lot was fried in oil at 200, 210, 220 and 230°C, until expansion is complete; and the other lot was puffed in the puff machine by Fujico Pte Ltd, which was set at 200, 210, 220 and 230°C for 1, 2, 3 and 4 seconds (Fig. 2).

3. RESULTS

3.1. Quality characteristics of dried shrimp by hot-air oven and vacuum oven drying

Table 2 shows the comparison of the characteristics of dried shrimp by hot-air oven drying and vacuum oven drying. The texture of shrimp produced by the hot-air drying technique was hard and brittle while the vacuum drying technique was soft and crispy. The texture of commercial dried shrimp was usually soft and chewy. The moisture content of vacuum dried shrimp is also about 2.5 times lower than hot-air dried shrimp. Correspondingly, water absorption ability of vacuum oven dried shrimp was approximately 2.5 times higher, indicating good reconstitutability. It also had a slightly higher astaxanthin content (better color) and a lower VBN content.

3.2. Physical and chemical characteristics of dried shrimp during storage

After an initial increase of 1 to 2% during the first month of storage, moisture content of both the hot-air and vacuum dried shrimp remained relatively constant at 8% and 5% respectively (Fig. 3). Correspondingly, water absorption ability remained relatively constant at 109% and 40% respectively after a drop of about 137% and 121% in the first

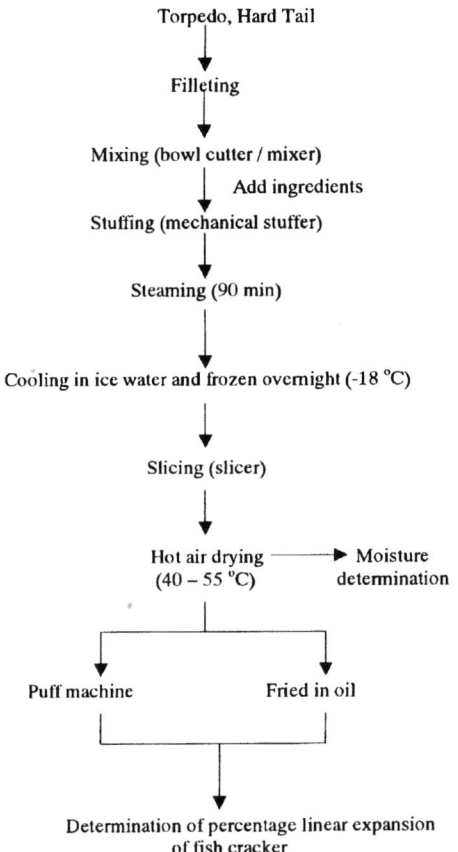

Figure 2. Fish cracker (Keropok) processing.

Table 2
Comparison of the quality characteristics of dried shrimp from hot-air and vacuum drying

Qualities characteristic	Hot air drying	Vacuum drying
Colour	Brick red	Orange
Texture	Hard and brittle	Soft and crispy
Moisture content	7.91%	3.36%
Water absorption ability	47.91%	118.14%
Astaxanthin	176.8 mg/100g	234.1 mg/100g
VBN	11.11 mg/100g	2.59 mg/100g

month of storage (Fig. 4). The water activity of the both hot-air and vacuum dried shrimp was in the range 0.26-0.34.

The major color pigment found in dried shrimp is astaxanthin [7]. After the first month of storage, astaxanthin content of all samples decreased significantly, but slowed down during the subsequent months (Fig. 5). As reported by Biede et al. [8], the vacuum

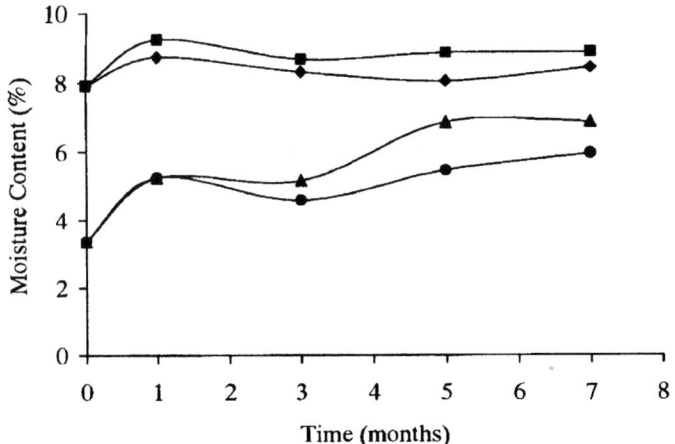

Figure 3. Changes in moisture content during storage of dried shrimp. ◆, Air-dried, Vacuum packed; ■, Air-dried, Air packed; ▲, Vacuum-dried, Vaccum packed; ●, Vaccum dried, Air-packed.

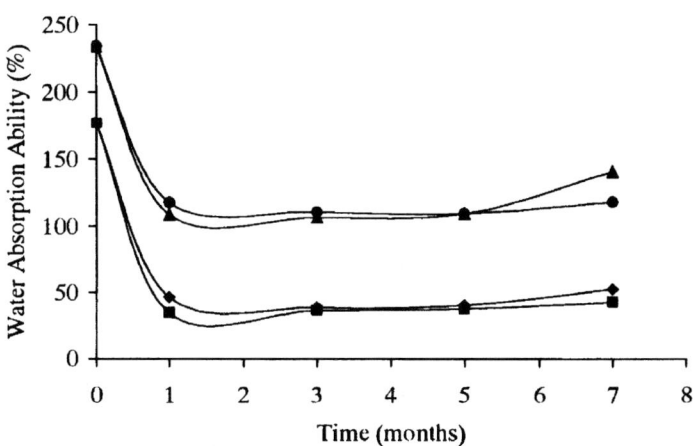

Figure 4. Changes in water absorption ability during storage of dried shrimp. ◆, Air-dried, Vacuum packed; ■, Air-dried, Air packed; ▲, Vacuum-dried, Vaccum packed; ●, Vaccum dried, Air-packed.

packaging proved to be more effective in slowing down the rate of astaxanthin oxidation (in both hot-air and vacuum dried shrimp) compared to air packaging.

VBN content of all samples remained relatively constant for the first 5 months, but the hot-air oven dried vacuum packed samples increased sharply by the 7th month of storage (Fig. 6). Peroxide was not detected in all shrimp samples throughout the 7 months of storage, indicating that the shrimp did not turn rancid.

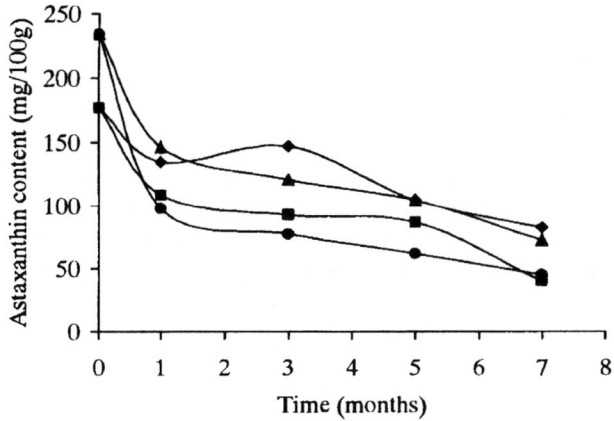

Figure 5. Changes in astaxanthin content during storage of dried shrimp. ◆, Air-dried, Vacuum packed; ■, Air-dried, Air packed; ▲, Vacuum-dried, Vaccum packed; ●, Vaccum dried, Air-packed.

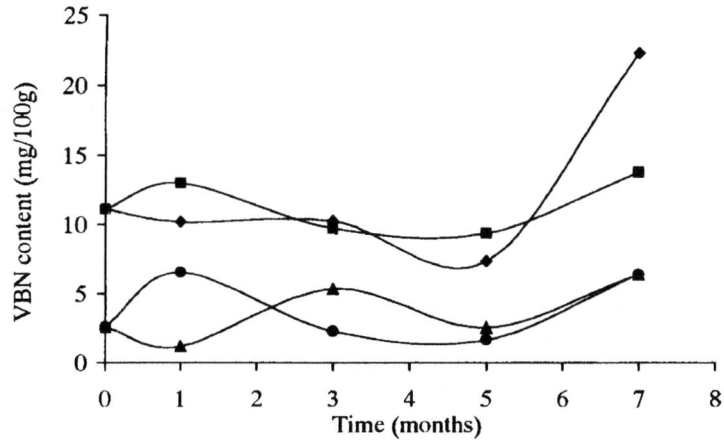

Figure 6. Changes in VBN during storage of dried shrimp. ◆, Air-dried, Vacuum packed; ■, Air-dried, Air packed; ▲, Vacuum-dried, Vaccum packed; ●, Vaccum dried, Air-packed.

3.3. Sensory characteristics of dried shrimp during storage

At the end of 7 months of storage at $28 \pm 2°C$, all the samples remained acceptable as judged by its appearance, color, flavor, odor, texture and overall acceptability, except for the vacuum-dried, air-pack shrimp sample which had an ammoniacal smell.

3.4. Physical characteristic of fish cracker (keropok)

Moisture content of the dried fish cracker was found to be about 11%. When the fish cracker was cooked using the puff machine, the amount of expansion was found to increase with temperature and time of puffing (Table 3). A maximum percentage linear expansion of 166.9% was obtainable at 230°C and 4s. With the deep-fryer, the time required for maximum expansion was reduced with increase in temperature whilst the amount of expansion obtained increased with increase in temperature (Table 4). A maximum percentage linear expansion of 120.9% was obtained at 230°C and 4s. This was lower than that obtained by the puffing technique, indicating that it can be used as an alternative to deep frying in oil to produce fish cracker with a greater degree of linear expansion. Fish cracker cooked by the puff machine was also crispier than that prepared by deep-frying.

4. DISCUSSION

The use of hot-air oven or vacuum oven in place of sun-drying significantly reduces the time required for drying shrimp from days to hours. Using the two drying techniques, the conditions for drying could also be better controlled to give a more consistent quality, and reducing the risk of contamination by dust, flies and other insects during drying. Both hot-air and vacuum drying techniques produce moisture content lower than the 20-30% moisture content found in commercial dried shrimp [8]. Vacuum drying produces dried shrimp with a soft and crispy texture, which makes it possible to eat the vacuum dried shrimp directly as a snack food. In vacuum drying, the moisture within the shrimp evaporates rapidly in the first 30min. This produces a porous texture within the shrimp,

Table 3
Linear expansion of fish cracker cooked by puff machine at various temperatures and times (%)

Temperature (°C)	1s	2s	3s	4s
200	82.3	77.1	74.0	89.8
210	85.3	84.8	82.2	103.7
220	87.0	94.7	139.2	146.1
230	143.5	148.7	155.2	166.9

Table 4
Linear expansion of fish cracker cooked by deep-frying at various temperatures and times

Temperature (°C)	Time for maximum expansion (s)	Linear expansion (%)
200	9.0	81.9
210	8.5	102.2
220	5.5	106.4
230	4.0	120.9

and improved water absorption ability. As a result, vacuum oven dried shrimp had a new, light and crispy texture and a high reconstitutability. The vacuum dried shrimp can be used as an ingredient for instant noodles in place of freeze dried shrimp. The vacuum drying technique can therefore be used to add value to traditional dried shrimp in that the vacuum dried shrimp can be eaten directly as a snack or as an ingredient for instant noodles.

The puff machine uses high heat and pressure to force the moisture within the fish cracker to escape out rapidly, thereby expanding and cooking it. The puff machine can be used to replace deep-frying to produce fish cracker with a greater amount of linear expansion, and a healthier product as no oil is used in the puff machine. Fish cracker is used in this study to illustrate the use of the puffing technique to add value to dried fish product. MFRD is also studying the use of the puff machine to produce value-added products from other dried fish products such as dried anchovies, shrimp, cuttlefish/squids and fish maw.

5. AKNOWLEGEMENTS

The authors would like to thank the CEO of Agri-Food and Veterinary Authority, Director, Agrotechnology Division, and Chief, MFRD, for their permission to publish this paper.

REFERENCES

1. Ng, M.C., Chia, G.H. and Lee, H.K. (1996) Southeast Asian Fish Products (3rd Ed). Marine Fisheries Research Department, Southeast Asian Fisheries Development Center, Singapore.
2. Yamagata, M. and Low, L.K. (1992) Determination of volatile basic nitrogen, trimethylamin oxide nitrogen and trimethylamine nitrogen by Conway's microdiffusion method (barium hydroxide method). In: Laboratory Manual on Analytical Methods and Procedures for Fish and Fish Products (2nd Ed.)., Miwa, K. and Low, S.J. (eds.), Marine Fisheries Research Department, Southeast Asian Fisheries Development Center, Singaporem, B-8.1-B-8.6.
3. Low, L.K. (1992) Determination of peroxide value. In: Laboratory Manual on Analytical Methods and Procedures for Fish and Fish Products. (2nd Ed.). Miwa, K. and Low, S.J. (eds.), Marine Fisheries Research Department, Southeast Asian Fisheries Development Center, Singapore, C-7.1-C-7.4.
4. Ng, M.C. (1992) Determination of moisture. In: Laboratory Manual on Analytical Methods and Procedures for Fish and Fish Products. (2nd Ed.)., Miwa, K. and Low, S.J. (eds.), Marine Fisheries Research Department, Southeast Asian Fisheries Development Center, Singapore, A-1.1-A-1.2.
5. Larmond, E. (1970) Methods for sensory evaluation of food (m.p.) Department of Agriculture, Canada. 57.
6. Yu, S.Y. (1991) Technology for fish cracker (Keropok) Production. In: Proceedings

of the Seminar on Advances in Fishery Post-Harvest Technology in Southeast Asia. Singapore, 6 - 11 May, 1991, Hooi, K.K., Miwa, K. and Mohamed, B.S. (eds.), Marine Fisheries Research Department, Southeast Asian Fisheries Development Center, Singapore, pp. 43-150.
7. Goodwin, T.W. (1960) Biochemistry of pigments. In: The Physiology of Crustacea. 1. Metabolism and Growth. Academic Press, New York.
8. Biede, S.L., Himelbloom, B.H. and Rutledge, J.E. (1982) A Research Note: Influence of storage atmosphere on several chemical parameters of sun-dried shrimp. J. Food Sci., 47: 1030-1031.

FUNCTIONAL FISH PROTEIN ISOLATES PREPARED USING LOW IONIC STRENGTH, ACID SOLUBILIZATION /PRECIPITATION

S.D. Kelleher[a], Y. Feng[a], H. Kristinsson[b], H.O. Hultin[a] and D.J. McClements[a]
[a]University of Massachusetts/Amherst, Marine Station, Gloucester, MA 01930 USA
[b]University of Florida, Department of Food Science and Nutrition, Gainesville, FL 32611, USA

Abstract

A technique was developed, whereby proteins from animal muscle tissue were solubilized under conditions of low ionic strength and low pH, that retained high yields of proteins, which when precipitated and neutralized were capable of producing gels with good functional properties. Gels made from acid-solubilized proteins commenced gelation at a substantially lower temperature when compared to washed muscle. Fluorescence data suggests extensive protein unfolding at pH values between 5.0 and 3.0, whereupon further acid addition in the range 3.0-2.0 resulted in a re-folding of the proteins. Extensive unfolding of the protein appears to be crucial to the development of a gel upon precipitation and neutralization.

1. INTRODUCTION

The interactions that proteins have either with their surrounding solvent or with adjacent proteins determine their ability to function. Solubility of proteins has historically been divided into three main categories, water-soluble, non-soluble and high salt-soluble [1]. The high salt soluble fractions of proteins are those that become solvated at an ionic strength of approximately greater than 0.3M and less than the amount needed to cause salting-out. In this fraction are the myofibrillar proteins, which represent about 50% of the total proteins [1,2] and are responsible in vivo for the contractile forces and locomotion of animals.

The terms soluble and insoluble are all related to solubility characteristics under conditions near those of neutral pH and at physiological ionic strengths. As the pH of the solvent surrounding the protein molecules is lowered into the acid range, the electrical charges of the proteins undergo a change. Hamm [3] found that bovine muscle proteins, between pH 3-4.5, lost negative charges on glutamic and aspartic acid side-chains creating an overall net positive charge, along with strong inter- and intra-repulsive forces. These forces incur a high degree of solubility. Capitalizing on the facts that myofibrillar proteins from most animal species appear to be quite soluble in low pH solutions of low ionic strength, and that these solutions are low in viscosity, a technique was investigated that separates the proteins from undesirable components prior to their recovery [4].

2. MATERIALS AND METHODS

2.1. Materials

Fresh chilled Atlantic cod (Gadus morhua) and Atlantic mackerel (Scomber scombrus) were obtained from local processors.

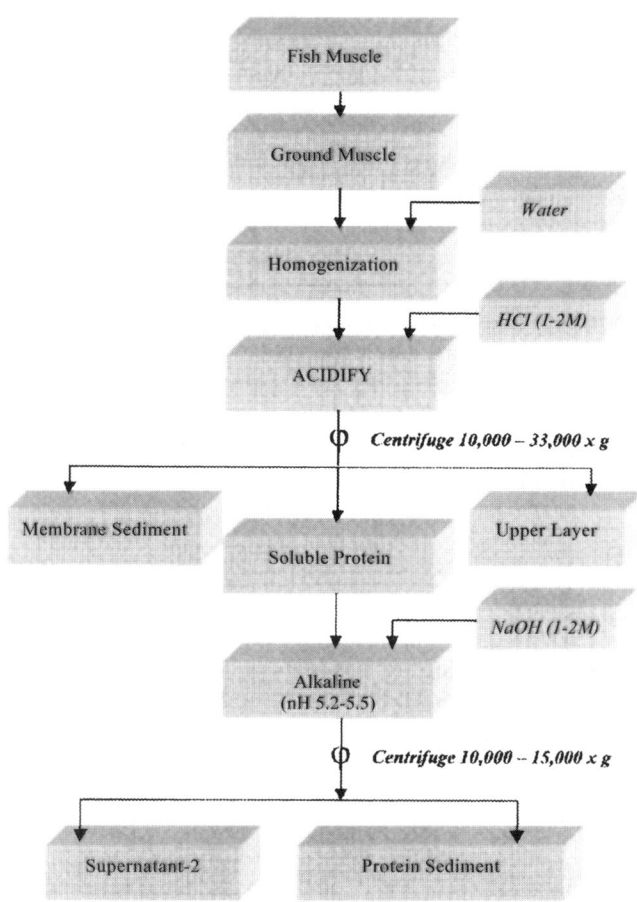

Figure 1. Steps in the acid solubilization/precipitation technique.

2.2. Methods

2.2.1. Acid solubilization/precipitation technique

Muscle tissue was combined with water at a ratio of 1:9. The mixture was homogenized using a Polytron at speed 76 for 1 minute. The resultant homogenate was adjusted to pH 2.0-3.5 using 1-2 M hydrochloric acid added drop-wise. The homogenate was centrifuged for 20-30 minutes at a g-force between 10,000-33,000. Three distinct phases were produced by the centrifugation step: a top layer, a bottom sediment layer and a middle aqueous layer, which contained the soluble proteins. The aqueous layer was separated from the other layers and adjusted to its isoelectric point (approximately pH 5.2-5.5) using 1-2 M NaOH added drop-wise. The aggregates, which were approximately 95% moisture, were de-watered using centrifugation at 10,000-15,000 g-force for 20-30 min. The resultant de-watered precipitate was mixed with cryoprotectants (4% sucrose, 4% sorbitol, and 0.3% sodium tripolyphosphate) and frozen at either -40° C or -80° C in polyethylene, Whirl-Pak bags. A generalized flow diagram of the technique is shown in Fig. 1.

2.2.2. Manufacture of surimi and gel values

Precipitated protein from acid solubilized (iso-electric) and water-washed material was dewatered to approximately 75% moisture using an Ultracentrifuge Model L-65B

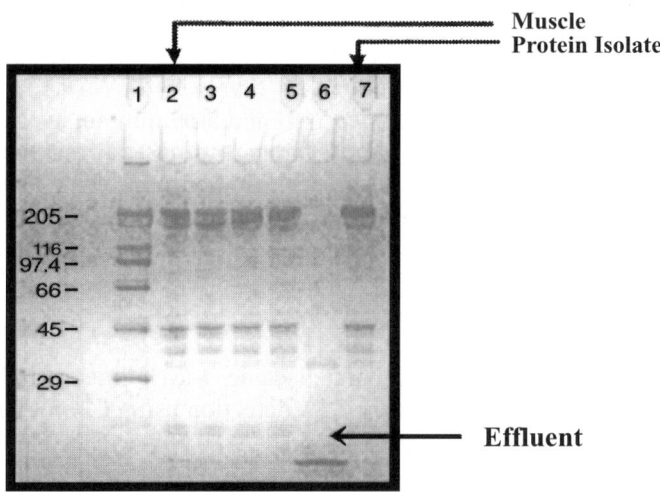

Figure 2. SDS-PAGE (4-20%, linear gradient) of Atlantic mackerel light muscle, at selected stages of the acid solubilization/precipitation technique. Lane 1, Molecular weight standards; Lane 2, Atlantic mackerel light muscle proteins; Lane 3, Proteins from homogenate, pH 6.5; Lane 4, pH 3.0 adjusted homogenate proteins; Lane 5, Proteins from the supernatant layer after centrifugation (10,000 × g for 30 min); Lane 6, Proteins from the supernatant layer after precipitation at pH 5.5; Lane 7, Proteins from final sediment precipitated at pH 5.5. Protein was applied to lanes at 15 μg/lane.

(Beckman, Palo Alto, CA) and a No. 19 rotor at 10,000 RPM for 30 min (15,000 × g). Cryoprotectants (4% sucrose, 4% sorbitol, and 0.3% sodium tripolyphosphate) were blended into the dewatered mince using a Model R301 chilled food processor (Robot Coupe, Inc., Ridgeland, MS) for 30 seconds. The pH was adjusted to between 6.8-7.0 using 1 M NaOH or 250 mM NaHCO$_3$ prior to freezing at either -40°C or -80°C. Gels were produced according to Lanier et al [5]. Stress and strain of gelled material at structural failure was evaluated using the torsion techniques of Wu et al [6].

2.2.3. Storage modulus and gel point determination

The dynamic gelling properties of a protein suspension or paste were characterized with a Rheometer (Bohlin CS-10, Bohlin Instruments, Cranbury, NJ) using a pair of coaxial cylinders. The diameter of the inner cylinder (bob) was 25 mm and the internal diameter of the outer cylinder (cup) was 27.5 mm. Determination of gel point was accomplished using a suspension at 3% protein concentration. The protein suspension or its paste (13.7 g) was loaded in the gap between the two cylinders. A thin layer of mineral oil was placed on top of the sample to prevent evaporation during the trials. The sample was heated by circulating temperature-controlled water around the cup supplied from a programmable water bath, which was controlled by Bohlin software. The actual temperature was measured by a thermocouple. For gel point determination the rheometer was set to the oscillation mode with a fixed frequency of 0.05 Hz and a target strain of 0.01. Temperature increased at a rate of 1.5°C/min from 15-65°C.

2.2.4. Protein, lipid, phospholipid, thiobarbituric acid (TBARS) and hydrophobicity

Lipid was determined using solvent extraction according to Lee et al [7]. Protein was estimated using the biuret procedure of Torten and Whitaker [8]. Phospholipid was estimated using a dry ashing technique [9] and assuming an average molecular weight of 750 daltons for phosphatidylcholine. Thiobarbituric acid reactive substances (TBARS) were determined using the whole muscle method of Lemon [10]. Hydrophobicity was estimated using 30 μM ANS with 10,000 fold diluted, washed cod muscle Richards et al, [11] incubated at varied pH for 1 hr, using excitation (350nm) and emission (470nm) wavelengths.

2.2.5. Electrophoresis

Proteins and peptides were separated using 4-20% linear gradient SDS-PAGE using the Laemmli [12] method and stained for detection using Coomassie blue.

3. RESULTS AND DISCUSSION

A component mass balance through the acid solubilization/precipitation technique using Atlantic mackerel light muscle, found that 15.1 g of protein was recovered in the final protein isolate from an initial tissue content of 16.8 g. Slight protein losses were found at all steps in the technique, with the largest loss occurring in the first centrifugation upper layer. Emulsification of proteins to the lipid in this very high neutral lipid layer may explain the protein loss. Lipid content found in the initial tissue was 9.4

g/100 g and in the final protein isolate, 0.1 g/100 g, which was substantially phospholipid. The upper layer of the first centrifugation was found to contain a majority of the lipid (74.5%) using the technique. The elimination of reactive membrane lipid (phospholipid) was found to occur mostly in the first centrifugation sediment (31.7%) and in the second centrifugation supernatant (17.9%). The final protein isolate retained 11.1% of the phospholipid from the initial muscle.

Viscosity of the homogenate was highest at pH 7, approaching 100 mPa.s. and rapidly decreased (< 20 mPa.s) as the pH transited from neutral pH to the isoelectric point (pH 5.5). Between pH 5.5 to around 4.0 the viscosity increased possibly due to increased hydration of the unfolding protein. Lowering the pH further to pH 2.5-3 resulted in a continual decrease in viscosity reaching a minimum (5-15 mPa.s).

Atlantic mackerel polypeptides, separated by SDS-PAGE, found at each of the selected steps in the technique are shown in Fig. 2. When a comparison is made between lane 2, mackerel light muscle, and lanes 3-5 and 7, no significant differences are found. This suggests that all the proteins found in intact muscle proceed through the technique to the final sediment. The final protein sediment contains a large percentage of myosin heavy chain (205 kDa), which plays a major role in protein functionality [13]. The supernatant layer after adjustment to pH 5.5 (lane 6) is totally devoid of any peptide greater than approximately 35kDa. Significant amounts of very low molecular weight peptides account for a majority of the protein found at this step in the technique.

Stress and strain at structural failure (torsion) was performed on gels manufactured from the protein isolate adjusted to neutral pH. Mackerel light muscle formed gels that were capable of being folded twice in the traditional fold test and had a strain value of 2.6

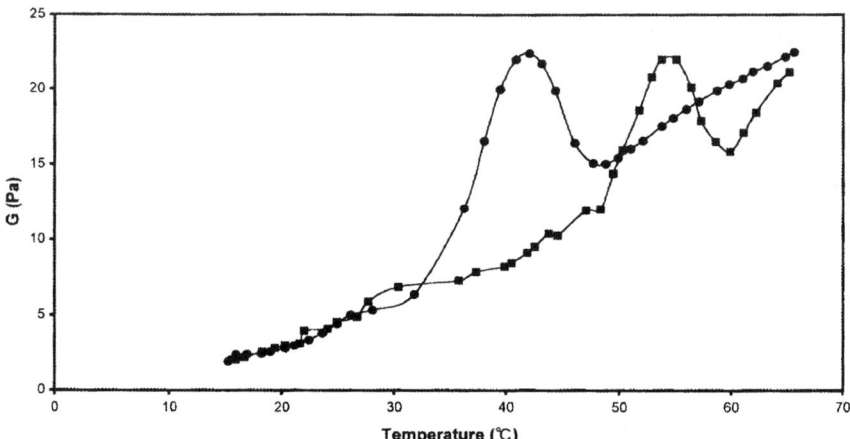

Figure 3. Storage modulus of Atlantic cod (acid solubilized/precipitated and water-washed) protein during a temperature scan. Bohlin Rheometer CS-10, fixed frequency 0.05 Hz, target strain 0.01. Temperature was scanned from 15 - 65 degrees C at a rate of 1.5 degrees C/min. Protein (3%) was suspended in 25 mM phosphate buffer with 450 mM NaCl, pH 7.0. ●, acid solubilized; ■, washed muscle.

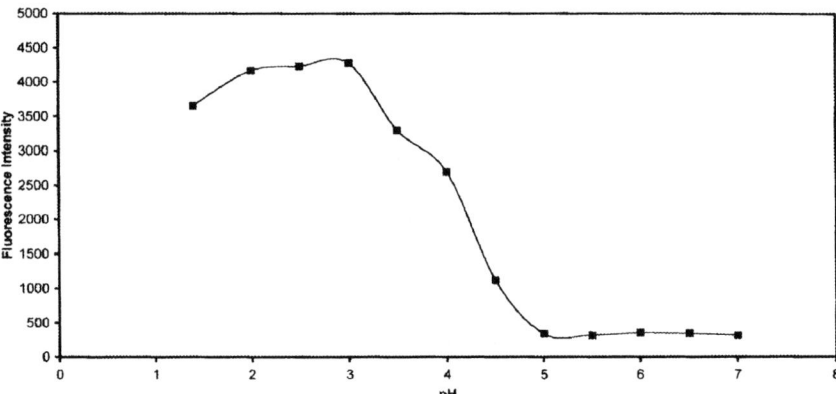

Figure 4. ANS-fluorescence binding of Atlantic cod myofibrillar proteins. Proteins were diluted 10,000 fold and mixed with 33 μM ANS for 15 min prior to excitation at 350 nm and emission at 470 nm.

± 0.1 and a stress value of 31.1 ± 3.8 kPa. Atlantic cod gels had a strain value of 2.8 ± 0.9 and a stress value of 22.0 ± 2.0 kPa. Gels could not be produced from isolates where homogenate pHs' were adjusted between 4.0-5.5. The texture of these non-gelling isolates was found to be somewhat coarse and sand-like.

Oxidation values, as assayed by thiobarbituric acid reactive substances method, was found to be approximately 1/3 lower in the acid solubilized samples stored frozen for six (6) months in polyethylene, Whirl-Pak bags, compared to controls made using standard surimi manufacturing techniques (29.8 versus 10.3 mmol/kg tissue). Richards et al. [11] found the value range between 12-20 mmol/kg tissue to be the lower threshold of rancidity.

Small deformations under dynamic thermal conditions were examined on protein isolates adjusted to neutrality using a Bohlin rheometer. Atlantic cod was chosen because all its lipid is in the form of membrane lipid, thus creating a simple system. The transition from sol to gel occurred between 30-40°C in the acid solubilized sample, as indicated by a large increasing rate of change in G' (Fig. 3), and at approximately 50-55°C in the washed muscle control. Feng and Hultin [14] found setting of Atlantic mackerel light muscle began around 53°C. Isolated beef myofibrils were found to commence gelling between 43-56°C [15] and setting of beef heart surimi was first detected between 50-55°C [16].

The classical mechanism for the conversion of native protein to a heat induced irreversible gel was described by Ferry [17] as a two-stage process.

$$XP_N \rightarrow xP_D \rightarrow (P_D)_x$$

The first stage involves denaturation of the protein, where the protein undergoes Conformational changes at the secondary, tertiary, and quaternary levels. The second stage involves the aggregation of the denatured proteins, which align and interact to form the three-dimensional protein-water network.

The sol-to-gel transition that occurred in the acid solubilized proteins, at a lower temperature compared to washed muscle at neutral pH suggests that less energy is required for denaturation-aggregation reactions to occur. Evidence from the literature and viscosity data, points to protein unfolding at low pH, thereby lowering the energy required for the two-stage reactions as described by Ferry [17].

Fluorescence data using the hydrophobic surface active agent 1-anilo naphthalene δ sulfonate (ANS) showed increased ANS binding from around the isoelectric point (pH 5.0) to approximately pH 3.0, followed by a decrease as the pH was further lowered (Fig. 4). The data from ultra-violet (232 nm) spectroscopy also followed a somewhat similar trend. Homogenate samples adjusted in the pH 5.5-4.0 range produced neutralized isolates that were not capable of gelling. ANS binding upon further decreasing pH from 3 to 2 suggests that the proteins are refolding and gaining structure. Neutralized protein isolates made from homogenates adjusted in this pH range form gels with good functional properties. Possible refolded structures could be an a-helix, which has been shown to be very stable in acid environments or an intermediate molten-globule structure [18].

REFERENCES

1. Morrissey, P.A., Mulvihill, D.M. and O'Neill, E.M. (1987) Functional properties of muscle proteins. In: Developments in Food Proteins. Hudson, B.J.F. (eds.), Elsevier Appl. Sci., London,UK, pp.195-255.
2. Suzuki, T. (1981) Fish and krill protein. Applied Science Pub., London, UK.
3. Hamm, R. (1994) The influence of pH on the protein net charge in the myofibrillar system. Reciprocal Meat Conference Proceedings, 47:5-9.
4. Hultin, H.O. and Kelleher, S.D. (1999) Process for isolating a protein composition from a muscle source and protein composition. U.S. Patent Number 6,005,073. Issued Dec. 21.
5. Lanier, T.C., Hart, K. and Martin, R.E. (eds.) (1991) A Manual of Standard Methods for Measuring and Specifying the Properties of Surimi. Univ. of North Carolina Sea Grant Program, Raleigh, NC.
6. Wu, M.C., Hamann, D.D. and Lanier, T.C. (1985) Rheological and calorimetric investigations of starch-fish protein systems during thermal processing. J. Tex. Studies, 16:53-74.
7. Lee, C.M., Trevino, B. and Chaiyawat, M. (1996) A simple and rapid solvent extraction method for determining total lipids in fish tissue. J. AOAC Intl., 79:487-492.
8. Torten, J. and Whitaker, J. R. (1969) Evaluation of the biuret and dye-binding methods for protein determination in meats. J. Food Sci., 29:168-174.
9. Kovacs, M.I.P. (1986) Determination of phosphorus in cereal lipids. Anal. Biochem., 154:420-423.
10. Lemon, D.W. (1975) An improved TBA test for rancidity. New Series Circular No. 51, Halifax Laboratory, Halifax, Nova Scotia, Canada.

11. Richards, M.P., Kelleher, S.D. and Hultin, H.O. (1998) Effect of washing with and without antioxidants on quality retention of mackerel fillets during refrigerated and frozen storage. J. Agric. Food Chem., 46:4363-4371.
12. Laemmli, U.K. (1970) Cleavage of structural proteins during the assembly of the head of bacteriophage T4. Nature, 227:680-685.
13. Park, J.W., Lin, T.M. and Yongsawatdigul, J. (1997) New developments in manufacturing of surimi and surimi seafood. Food Rev. Int., 13:577-610.
14. Feng, Y. and Hultin, H.O. (1997) Solubility of the proteins of mackerel light muscle at low ionic strength. J. Food Biochem., 21:479-496.
15. Samejima, K., Egelandsdal, B. and Fretheim, K. (1985) Heat gelation properties and protein extractability of beef myofibrils. J. Food Sci., 50:1540-1546.
16. Srinivasan, S. and Xiong, Y.L. (1996) Gelation of beef heart surimi as affected by antioxidants. J. Food Sci., 61:707-711.
17. Ferry, J.D. (1948) Protein gels. Adv. Protein Chem., 4:1-78.
18. Ptitsyn, O.B. (1995) Molten globule and protein folding. Adv. Protein Chem., 47:83-229.

CONTROL OF SUGAR CONTENT IN FISH FILLETS BY SOAKING AND COLD PRESERVATION

Tooru Ooizumi, Katsuhiko Tsuruhashi, Yohei Miono and Yoshiaki Akahane
Department of Marine Bioscience, Faculty of Biotechnology, Fukui Prefectural University, Obama, Fukui 917-0003, Japan

Abstract

To control the sugar content in fish fillets, the permeability of various sugar compounds into fillets by soaking together with their dispersing ability inside the soaked fillets were investigated. Irrespective of the kind of sugars, an increase in the osmotic pressure given by sugars in the soaking solution caused the increase in the content of sugars and the decrease in the moisture of the soaked fillets. Comparing the permeability and the dewatering effect of sugars in connection with their molecular weights, lowering the molecular weights of sugars caused an increase in the permeability and a decrease in the dewatering effect. Furthermore, the content of sugars in the external parts were several times higher than those in the internal parts, even when the fillets were soaked for more than 100 h. In contrast, the moisture remained higher in the internal parts than in the external parts of the soaked fillets. Following preservation of the soaked fillets at 4°C for over 50 h was necessary to disperse sugars and moisture in the fillets. Moreover, lowering the molecular weights of sugars also caused an increase in their dispersing ability.

1. INTRODUCTION

Sugar compounds are widely applied to processing of various seafoods. To cite one example, sugars are used in processing of seasoned and dried products not only to confer a taste but also to reduce water activity. Moreover, it is already known that frozen surimi is processed on the basis of the protective action of sugar compounds against the protein denaturation during frozen storage [1]. The development of the process of the frozen surimi production enabled us to preserve fish mince without loss of the functional properties through suppressing protein denaturation for a long period. However, long-term storage of fish fillets having the functional properties of myofibrillar protein has not been successful though it will be of great advantage to more efficient utilization of fish flesh with unstable myofibrillar protein as a material of various processed foods.

Apart from minced meat products such as frozen surimi, the soaking of fish fillets in a sugar solution is a conventional method to make sugars permeate into it. To control the rate of the protein denaturation in the fish fillets during frozen storage as well as to confer a taste by using sugar compounds, it is necessary to adjust their concentrations in the fillet. However, very few papers addressed the permeability of sugar compounds. Although we have recently revealed the permeability of only three kinds of sugar alcohols with different molecular weights [2], there is no information about the permeability of other sugars. Thus, the relation between the permeability of sugars and their chemical formulae including molecular weights is not fully elucidated. Moreover, no attempt has been carried out to clarify the dispersing ability of sugars in the soaked fillets. Lack of the understanding of the dispersing ability of sugars prevents us from predicting their distribution in the soaked fillets.

In the present study, we aimed to compare the permeability of eleven kinds of sugar compounds into fish fillets and their dewatering effects in connection with the osmotic pressure level of the soaking solution. Furthermore, to follow the dispersion of sugar compounds and the migration of moisture inside the soaked fillets, changes in the level of sugars and the moisture content in the different parts of the soaked fillets were investigated.

1. MATERIALS AND METHODS

2.1. Soaking of fish fillet

Fish fillets with a uniform size (1×1×6 cm) were prepared from dorsal muscle of fresh yellowtail caught in Wakasa bay or Toyama bay and were soaked in double the weight of sugar solutions with various concentrations containing 0.25 M NaCl at 4°C for up to 120 h. Eleven kinds of sugars used in this study were as follows; glycerol, xylitol, sorbitol, mannitol, lactitol, maltitol, xylose, glucose, fructose, trehalose, and sucrose. Among them, xylitol, sorbitol, lactitol, and maltitol (Towa Chemical Industry Co., Japan) as well as trehalose (Hayashibara Industry Co., Japan) were food additive grade. While, the others were reagent grade (Nacalai Tesque Inc., Japan). At appropriate intervals, the soaked fillets were taken out of the solutions, wiped with a filter paper to remove the surface fluid, and minced to homogenize them. Subsequently, the levels of sugars and the moisture contents in the soaked fillets were measured.

On the other hand, to follow the dispersion of sugars and the migration of moisture inside the fillets during soaking, changes in the content of sugars and moisture in the different parts of the fillets were examined. For this purpose, the fillets with a uniform size (2×3×6 cm) were soaked in 2.0 M glycerol, sorbitol, or maltitol in the same manner as described above. The soaked fillets were taken out of the solutions and were immediately separated into four layers with a certain thickness from the surface toward the center. The contents of these components in each part were then measured. Furthermore, the dispersion of sugars and the migration of moisture inside the soaked fillets during subsequent cold preservation at 4°C for up to 110 h were also investigated in the same manner as in the case of soaking.

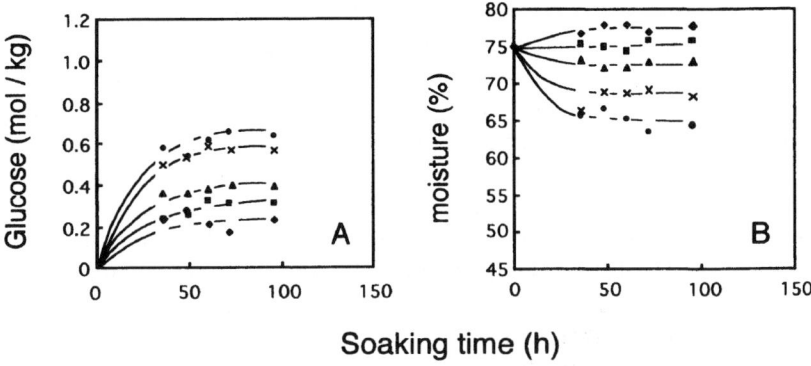

Figure 1. Changes in the contents of glucose and moisture in fish fillets during soaking in various concentrations of glucose solutions. Yellowtail fillets (1×1×6 cm) were soaked in 0.5, 0.75, 1.0, 1.5 and 2.0 M of glucose solution containing 0.25 M NaCl. At appropriate intervals, the fish fillets were taken out from the solution for the measurement of the glucose level and the moisture content. Glucose contents measured by the method of Yasui et al. [3] (A) and moisture content estimated by measuring the weight loss after drying at 110°C for 20 h (B) were plotted as a function of the soaking time. ◆, 0.5; □, 0.75 ; ▲, 1.0 ; ×, 1.5 ; ●, 2.0 M.

2.2. Determination of sugars and moisture

Sugars were determined with HPLC using a TSK Gel Amide 80 packed column (TOSOH Co. Ltd., Japan) according to the method of Yasui et al. [3] with a slight modification. Moisture contents in the soaked fillets were estimated by measuring the weight loss after drying at 110°C for 20 h.

2.3. Calculation of osmotic pressure

Osmotic pressure of sugars in the soaking solutions was calculated according to the equation presented by Ogawa [4].

3. RESULTS AND DISCUSSION

3.1. Permeation of various sugar compounds and changes in moisture content by soaking

Fish fillets with a uniform size were soaked in the glucose solution with a fixed concentration from 0.5 M to 2.0 M. At adequate intervals, the soaked fillets were taken out and the changes in the contents of glucose and moisture in the fillets were examined. Fig.1A demonstrated that the rapid permeation in the early period of the soaking was followed by the slower one in the latter period irrespective of the glucose concentrations of the soaking solutions. Finally, the level of glucose reached plateaus after soaking for more than 50 h. The same trends were observed in the permeation of other sugar compounds (data not shown). Therefore, the levels of sugar compounds in the fillets after soaking for more than 50 h could be regarded as maximum permeation.

The changes in moisture contents of the soaked fillets as a function of the soaking time were shown in Fig.1B. The soaking of the fillets in 0.5 M glucose solution slightly increased the moisture content and no change in the moisture content was observed by soaking in 0.75 M glucose solution. However, further increase in the glucose concentrations of the soaking solution decreased the moisture contents of the fillets. Similarly to the permeation of sugars, the soaking for more than 50 h hardly changed the moisture contents in the fillets any more, regardless of the glucose concentrations of the soaking solutions. Though the data were not shown, analogous trends were observed by soaking in the solutions of the other monosaccharides and their sugar alcohols. However, the moisture changes by soaking in disaccharides solutions were somewhat different. The soaking of the fillets in disaccharides solution with lower concentrations rather decreased the moisture content. In addition, though the marked moisture decrease was caused by soaking in higher concentrations of disaccharides solution in the early period of the soaking, the moisture contents continued to decrease even in the latter period of the soaking.

Since Nanbu et al. [5] reported that the permeation of sorbitol was promoted in the presence of a small quantity of NaCl, we added 0.25 M NaCl at a final concentration to all of the soaking solution. Konno et al. [6] mentioned that sugar compounds accelerated the solubilization of myofibrillar protein induced by NaCl. Therefore, the increase in the moisture content caused by soaking in the lower concentrations of monosaccharides solution was considered to be due to the increase in the water holding capacity of myofibrillar protein accompanied with their solubilization.

As shown in Fig.1, both the permeation of glucose and the change in the moisture content were dependent on the concentration of the soaking solutions. Accordingly, the

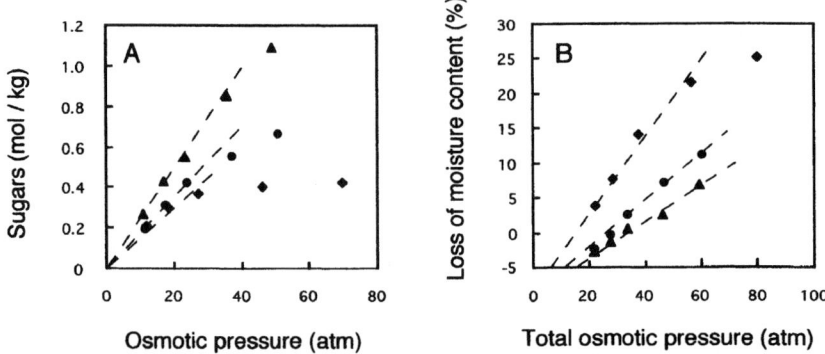

Figure 2. Effects of osmotic pressure of sugars in soaking solution on their permeation and loss of moisture content of the fillets. Fish fillets were soaked in the various concentrations of glycerol, glucose and sucrose in the same manner as shown in Fig.1. The levels of sugars (A) and the loss of moisture content (B) in the fillets after soaking for more than 50 h were plotted against the osmotic pressure of the soaking solutions. ▲, glycerol; ●, glucose; □, sucrose.

maximum permeation of sugars was analyzed in connection with the osmotic pressure of the soaking solution (Fig.2A). Figure 2A demonstrated the osmotic pressure dependency of the permeation of glucose together with glycerol and sucrose. The increase in the osmotic pressure of the soaking solution accelerated the permeation of these sugars into fillets. At the same osmotic pressure of the soaking solution, the level of glycerol was the highest among three and was followed in descending order by glucose and sucrose. These data supported our previous results [2], which suggested that the permeation of sugar alcohol depended on not only the osmotic pressure level but also on the kind of sugar alcohol.

At the same time, the moisture decrease caused by soaking of the fillets in the solutions of these three kinds of sugars were examined in connection with the osmotic pressure. The loss of the moisture content in the fillets after soaking for more than 50 h was plotted as a function of the total osmotic pressure level of the soaking solution (Fig.2B). The negative values of the loss of moisture content imply the increase of the moisture in the fillets. The total osmotic pressure is sum of the osmotic pressure of sugars and NaCl in the solutions. The increase in the total osmotic pressure accelerated the moisture decrease. At the same osmotic pressure of the soaking solution, sucrose caused the most significant moisture decrease among three followed in descending order by glucose and glycerol. These data suggested that the osmotic pressure of the soaking solution as well as the permeability of sugar compounds involved moisture decrease.

Irrespective of the kind of sugar compounds, there were linear relationships between the osmotic pressure of the soaking solution and the maximum permeation or the moisture decrease within the lower osmotic pressure range. From these relations, the amount of the permeation of sugar compounds and the loss of the moisture content given by 1 atm of the osmotic pressure were estimated as the permeation rate and the dewatering rate, respectively. The permeation rates and the dewatering rates of all sugar compounds examined in this study were summarized in Fig.3. The molecular weights of the sugar compounds were shown in the parentheses of Fig.3. Regardless of the difference in the chemical formulae between sugars and sugar alcohols, structural isomers, or stereoisomers, the permeation rates decreased whereas the dewatering rates increased with an increase in the molecular weight. The permeation rates and the dewatering rates of sugar compounds with similar molecular weights were almost constant. These results indicated that the permeation rates and the dewatering rates of sugar compounds strongly depended on their molecular weights. Thus, it was suggested that the permeation rates as well as the protective effects [7,8] of sugars against the protein denaturation were necessarily to be taken into consideration for practical use of sugars to prevent protein denaturation in the fillets.

3.2. Distribution of sugar compounds and moisture in soaked fillets

The permeation rates and the dewatering rates mentioned above were analyzed by measuring the contents of sugars and moisture in the whole of the soaked fillets. Therefore, distribution of sugars and moisture inside the soaked fillet is still unclear. Thus, the fillets soaked in 2.0 M sorbitol were immediately separated into four layers from the surface toward the center and these components in each part were determined. The results were demonstrated in Fig. 4. The sorbitol content in the external 3 mm from

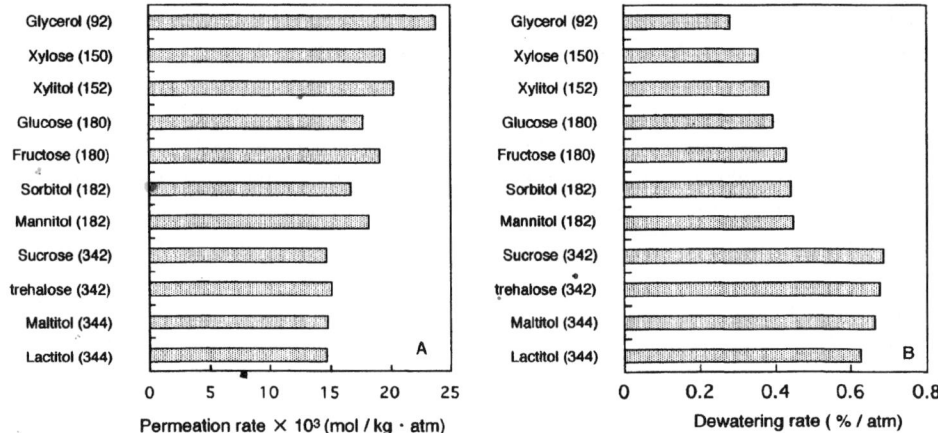

Figure 3. Permeation rates and dewatering rates of various sugar compounds with different molecular weights. The permeation rates (A) were defined as the amount of permeation of sugars (mol / kg) caused by 1 atm of osmotic pressure of the soaking solution and were estimated from the relations between the permeation of sugars and the osmotic pressure as in Fig.2A. The dewatering rates (B) were defined as the loss of the moisture content (%) induced by 1 atm of osmotic pressure and were estimated from the relations between the loss of moisture and the osmotic pressure as in Fig.2B. The molecular weights of sugars are shown in the parentheses.

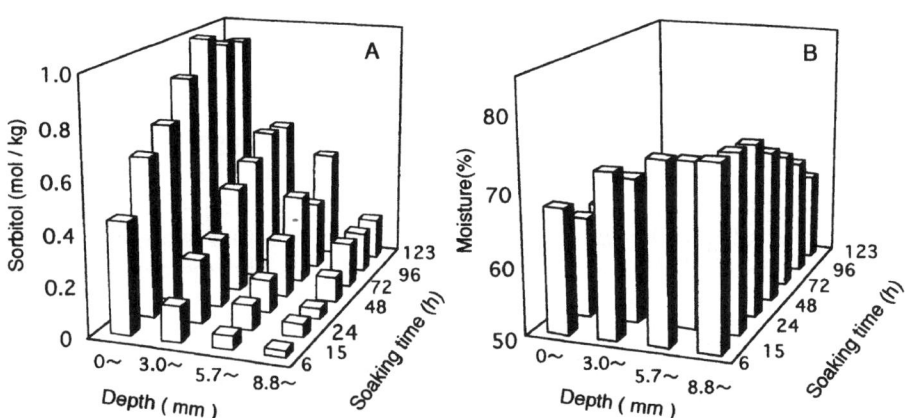

Figure 4. Changes in the content of sorbitol and moisture in the different parts of fish fillets during soaking. Fish fillets (2×3×6 cm) were soaked in 2.0 M sorbitol in the same manner as in Fig.1. At appropriate intervals, the fillets were taken out from the solution and were immediately separated into four layers from the surface toward the center. The mean depth of the layers from the surface of the fillets were as follows; 0-3.0 mm, 3.0-5.7 mm, 5.7-8.8mm, and deeper than 8.8 mm. Sorbitol levels (A) and the moisture (B) of these parts were determined.

the surface rapidly rose whereas slight increase in the sorbitol content was observed in the most internal part (Fig. 4A). The sorbitol content in the most internal part was around 0.2 mol / kg even in the fillets soaked for 123 h. On the contrary, the moisture content in the external part dropped in the early period of the soaking whereas that in the most internal part hardly changed after soaking for 48 h and gradually decreased thereafter (Fig. 4B). Consequently, the level gradients of both components were generated inside the soaked fillet from the surface toward the center during soaking.

3.3. Dispersion of sugars and moisture inside the soaked fillets by subsequent cold preservation

Subsequently, we investigated the dispersion of sorbitol and the migration of moisture during the cold preservation of the fillets after soaking in 2.0 M sorbitol for 25 h (Fig.5). The sorbitol content in the center of the soaked fillet before the cold preservation was considerably lower than that in the external parts as shown in Fig. 4. The sorbitol contents in the external parts of the soaked fillets were gradually decreased while those in the internal parts increased with the cold preservation, suggesting that the dispersion of sorbitol proceeded inside the fillets. In contrast, moisture migrated from the center to the external parts of the fillets during the cold preservation. Thus, the cold preservation promoted the dispersion of sorbitol as well as the migration of moisture.

Accordingly, it is natural to assume that the osmotic pressure of the soaking solution rather prevented the dispersion of sorbitol and the migration of moisture during the soaking period. However, the dispersion of sorbitol and the migration of moisture proceeded so slowly during the following cold preservation that the levels of these components in the external part were still different from those in the internal one even after the preservation for 110 h.

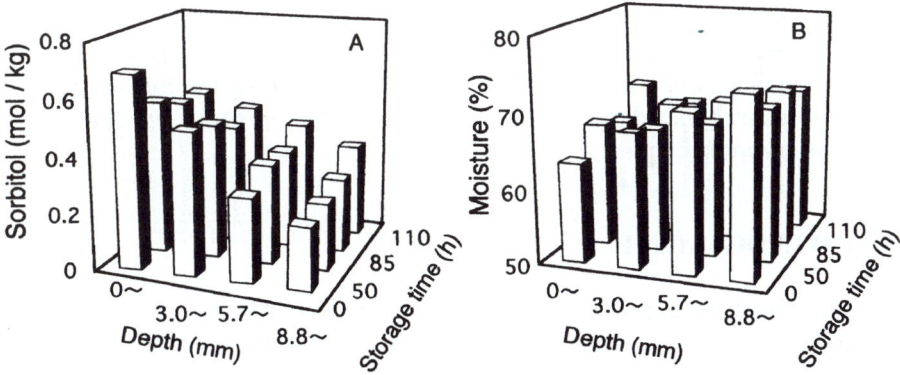

Figure 5. Changes in the contents of sorbitol and moisture of the different parts of soaked fillets during following cold preservation. Fish fillets (2×3×6 cm) soaked in 2.0 M sorbitol for 25 h were preserved at 4°C. After appropriate intervals, the fillets were separated into four layers from the surface toward the center in the same manner as in Fig.4. Sorbitol content (A) and the moisture (B) of these parts were then determined.

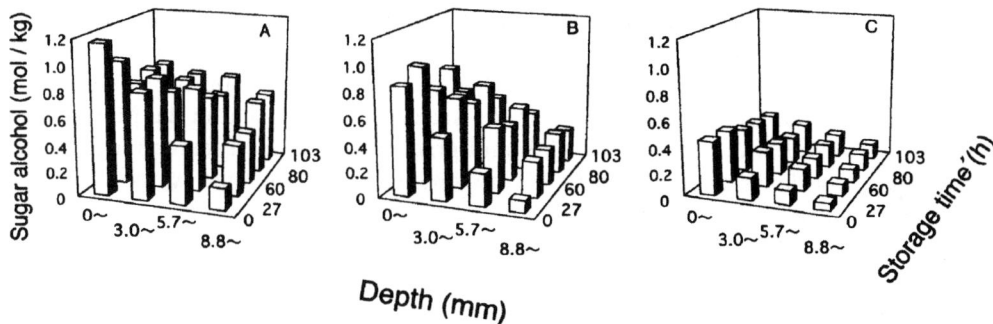

Figure 6. Comparison of dispersing ability of three kinds of sugar alcohols with different molecular weights. Fish fillets (2×3×6 cm) soaked in 2.0 M of glycerol (A), sorbitol (B), or maltitol (C) for 25 h were subsequently preserved at 4°C. Dispersion of the sugar alcohols inside the soaked fillets during the cold storage was followed in the same manner as in Fig.5.

Finally, we compared the dispersing ability of three kinds of sugar alcohols with different molecular weights. Figure 6 demonstrated the changes in the level of sugar alcohols in the different parts of the fillets after soaking in 2.0 M glycerol, sorbitol, or maltitol during the subsequent cold preservation. The cold preservation caused rapid dispersion of glycerol in the soaked fillets. However, the dispersion of sorbitol and maltitol inside the soaked fillets proceeded more slowly. In particular, the increase in the level of maltitol in the internal part was very slight. These results suggested that the molecular weight of sugars also affected their dispersing ability.

As a conclusion, it was obvious that the cold preservation of the soaked fillets was effective to disperse sugars and moisture equally to every part of the fillets. The results obtained in this study are considered useful for not only conferring a taste to the fillet but also the development of the frozen fish fillet production having the functional properties through suppressing the denaturation of myofibrillar protein by applying sugar compounds as cryoprotectants.

4. ACKNOWLEDGEMENT

The authors wish to express their sincere gratitude to Dr. Ken-ichi Arai, National Surimi Manufacturers Association, for his valuable suggestions and critical reading of the manuscript. The authors also thank to Towa Chemical Industry Co. Ltd. for providing a part of sugar compounds used in this study.

REFERENCES

1. Kawashima, K., Arai, K. and Saito, T. (1973) Studies on muscular protein of fish-X.

The amount of actomyosin in frozen surimi from Alaska pollack. Nippon Suisan Gakkaishi, 39: 525-532.
2. Ooizumi, T., Tamagawa, H., Akahane, Y. and Kaneko, Y. (2000) Permeability of sugar alcohols with different molecular weights into fish meat. Fish. Sci., 66:974-979.
3. Yasui, K., Furukawa, T. and Hase, S. (1980) High performance liquid chromatographic determination of saccharides in dairy products. Nippon Shokuhin Kogyo Gakkaishi, 27: 358-362.
4. Ogawa, T. (1989) Tsukemono seizougaku, Kohrin, Tokyo pp. 39-46.
5. Nanbu, S., Kiuchi, H., Yamamoto, Y., Kawamori, Y., Funatsu, Y. and Arai, K. (1995) Effect of sorbitol on change in myofibrillar protein of meat block from walleye pollack during soaking in NaCl solution. Nippon Suisan Gakkaishi, 61: 225-231.
6. Konno, K., Yamanodera, K. and Kiuch, H. (1997) Solubilization of fish muscle myosin by sorbitol. J. Food Sci., 62: 980-984.
7. Ooizumi, T., Hashimoto, K., Ogura, J. and Arai, K. (1981) Quantitative aspect for protective effect of sugar and sugar alcohol against heat denaturation of fish myofibrils. Nippon Suisan Gakkaishi, 47: 901-908.
8. Matsumoto, I. and Arai, K. (1986) Comparison of protective effects of several sugars on thermal and freeze denaturations of fish myofibrillar protein. Nippon Suisan Gakkaishi, 52: 901-908.

PROPERTIES OF PROTEASES RESPONSIBLE FOR DEGRADATION OF MUSCLE PROTEINS DURING ANCHOVY SAUCE FERMENTATION

Yeung Joon Choi[a], Min-Soo Heu[a], Heung-Rak Kim[b] and Jae-Hyeung Pyeun[b]
[a]Division of Marine Bioscience/Institute of Marine Industry, Gyeongsang National University, Tongyeong 650-160, Korea
[b]Faculty of Food Science and Biotechnology, Pukyong National University, Pusan 608-737, Korea

Abstract

The protease playing an important role in anchovy sauce fermentation was attempted to isolate, and properties of the three proteases such as trypsin, chymotrypsin and cathepsin L-like enzyme purified from anchovy, *Engraulis japonica*, were investigated. The molecular weights of trypsin, chymotrypsin and cathepsin L-like enzyme purified from anchovy viscera and muscle were estimated by SDS-PAGE to be 25.6 kDa, 26.1 kDa and 25.8 kDa, respectively. The trypsin and chymotrypsin had their maximal activity at pH 8.0 and 45°C for synthetic substrates, while cathepsin L-like enzyme showed maximal activity at pH 6.0 and 50°C for N-benzoyl-D,L-arginine-β-naphthylamide. Trypsin, chymotrypsin and cathepsin L-like enzyme had higher Km values for myofibrillar proteins than those for casein. The kcat of chymotrypsin and cathepsin L-like enzyme for myofibrillar proteins were higher than that of trypsin, and also chymotrypsin and cathepsin L-like enzyme caused higher hydrolysis in myofibrillar proteins of anchovy. Proteolytic activities were decreased with increase in sodium chloride concentration. Cathepsin L-like enzyme and chymotrypsin were more responsible for the autolysis of muscle proteins from fish and fermentation during fish sauce processing than trypsin.

1. INTRODUCTION

Anchovy, *Engraulis japonica*, is a main source of fermented fish sauce in Korea. The anchovy sauce contains amino acids, peptides and organic compounds produced from complete hydrolysis of tissue by enzymes from fish intestine and muscle. High proteolytic activity in anchovy intestine and muscle tissue accelerates autolytic degradation of tissue proteins and may shorten the fermentation time during production of anchovy sauce. Trypsin and chymotrypsin are quantitatively important in the digestive system due to their high proteolytic activities. Hence, understanding properties of trypsin and chymotrypsin is important in the degradation of muscle protein and may support an information for the production of anchovy sauce.

A survey of proteolytic digestive enzymes in various species of fish has revealed that a serine protease is widely distributed in fish intestine. Trypsin has been characterized thoroughly as to their physicochemical and enzymatic properties from the intestine of crayfish [1,2]. anchovy [3], dogfish [4,5], mackerel [6] and capelin [7]. Chymotrypsin from the pancreas of carp has been found to be similar to mammalian enzymes in both physical properties [8] and kinetic properties [9]. However, anionic chymotrypsin isolated from dogfish [4,5] and from mackerel [6] have greater relative activities against casein than bovine chymotrypsin. We reported the isolation and characterization of several alkaline proteases responsible for the post-mortem autolytic degradation of abdominal tissues by the proteases from the intestine of anchovy [10], menhaden [11], and skipjack [12].

Protein hydrolysis occurs in fish sauce fermentation via autolytic activity [13,14]. Trypsin and chymotrypsin and other digestive enzyme are principally responsible for autolysis [15,16]. Trypsin-like activity in paties fermentation increased and reached a maximum in the first month and then dramatically declined [17]. Cathepsin A and D were found to be responsible for the protein hydrolysis in paties formation as trypsin and chymotrypsin. However, cathepsin B and D minimally affected protein degradation in paties [18]. When Pacific whiting and its surimi by-products, after being mixed with high salt concentrations up to the level of 25%, were subjected to autolysis at 50°C, cathepsin L-like enzyme and metalloproteases played a significant role in hydrolyzing proteins [19]. In traditional fish sauce fermentation, the rate of production depends only on the activity of enzymes in the fish [17].

The objective of this study was to purify and compare the properties of proteases responsible for autolysis and fermentation of fish sauce of anchovy.

2. MATERIALS AND METHODS

2.1. Extraction of proteases

Fresh anchovy, *Engraulis japonica*, was harvested from the Southern East Sea of Korea and transported on ice to the laboratory. 500 g of separated viscera were chopped and homogenized in 1500 mL of 20 mM Tris-HCl, pH 7.0, containing 5 mM $CaCl_2$ and 0.02% sodium azide with an Ultra-Turrax type tissue grinder (H-AM type Kokusan, Japan). After incubation at 40°C for 3 hr, the crude enzyme solution was obtained by taking the supernatant after centrifugation (8,500×g, 20 min). In cathepsin, 1% NaCl containing 1 mM EDTA-2Na and 0.02% sodium azide solution was used as extractive buffer.

2.2. Trypsin purification

Tryptic activity was monitored with hydrolysis of N-α-benzoyl-DL-arginine-p-nitroanilide(BAPNA) during purification. The crude enzyme solution was subjected to 30-70% ammonium sulfate (AS) fractionation. The AS fraction was dissolved in an appropriate amount of 20 mM Tris-HCl, pH 7.0, and dialyzed against the same buffer. The dialyzate was loaded onto a benzamidine-Sepharose 6B column (1.1×12 cm) equilibrated with the above buffer. Unabsorbed protein was washed with the equilibration buffer. The trypsin-like enzyme was eluted from the column using 20 mM Tris-HCl, pH 7.0, containing 125 mM benzamidine and 1% NaCl. The trypsin fractions were dialyzed against 10 mM sodium phosphate, pH 7.0, containing 100 mM NaCl and concentrated with ultrafiltration. The concentrate was applied on a Sephadex G-75 column (2.6×75 cm) equilibrated with the same buffer. The trypsin fraction was dialyzed against 20 mM Tris-HCl, pH 7.0. The dialyzate was applied on a DEAE-Sepharose A-50 column (1.6×30 cm) and the column was eluted with a 700 mL linear gradient ranging from 0 to 0.4 M sodium chloride. The trypsin fraction was concentrated and dialyzed against 10 mM sodium phosphate buffer, pH 7.0, containing 100 mM NaCl, and loaded on a Sephacryl S-100 column (0.9×55 cm). Trypsin was concentrated by ultrafiltration and stored at −40°C until used in subsequent characterization studies.

2.3. Chymotrypsin purification

The unabsorbed fraction of the benzamidine-Sepharose 6B column in trypsin purification was used as the sample for chymotrypsin purification. The effluent from the column was concentrated with ultrafiltration and dialyzed against 10 mM sodium phosphate, pH 7.0, containing 100 mM NaCl. The concentrate was applied on a Sephadex G-75 column (2.6×75 cm) equilibrated with the same buffer. Chymotrypsin fraction was dialyzed against 20 mM Tris-HCl, pH 7.0. The dialyzate was applied on a DEAE-Sephacel column (3.0×30 cm) and the enzyme was eluted with a 700 mL linear gradient ranging from 0 to 0.6 M NaCl. After the chymotrypsin fraction was dialyzed, the dialyzate was rechromatographed with DEAE-Sephacel column (1.6×30 cm), concentrated and redialyzed against 10 mM sodium phosphate buffer, pH 7.0, containing 100 mM NaCl. The dialyzate was loaded on a Sephacryl S-100 column (0.9×55 cm) equilibrated with the same buffer, pH 7.0. The chymotrypsin solution was concentrated by ultrafiltration and stored at −40°C until used in subsequent characterization studies.

2.4. Purification of cathepsin L-like enzymes

The crude enzyme solution was subjected to 30-80% AS fractionation. The AS fraction was dissolved in 20 mM sodium acetate, pH 6.0, containing 0.1 M NaCl and 1 mM cysteine and dialyzed against the same buffer. After centrifugation (8,500×g, 20 min), the supernatant was loaded onto a Sephadex G-75 column (2.6×75 cm) equilibrated with the above buffer. The cathepsin fractions were pooled and concentrated with ultrafiltration, and then reloaded on the Sephadex G-75 column as above. Fractions with high catheptic activity were dialyzed against 10 mM sodium acetate, pH 6.0, containing 1 mM cysteine. The dialyzate was applied on a CM-Sephadex C-50 column (2.6×40 cm) and the column was eluted with a 1000 mL linear gradient ranging from 0 to 0.3 M NaCl. Cathepsin fractions were pooled and rechromatographed with CM-Sephadex C-50 column (2.6×25 cm) after dialysis. Fractions with high catheptic activity were concentrated and stored at −40°C until needed for use in subsequent characterization studies.

2.5. Proteolytic activity assay

Enzymatic activity for myofibrillar protein was determined according to the method of Pyeun and Kim [20]. One unit of activity was defined as that releasing 1 μmole of Tyr per min. Hydrolyse of N-α-benzoyl-DL-arginine-p-nitroanilide (BAPNA) and N-Succinyl-(Ala)$_2$-Pro-Phe-p-nitroanilide (SAAPPNA) was measured using the assay method of Erlanger et al. [21,22]. Activity toward benzoyl-L-tyrosine ethyl ester (BTEE) and N-toluenesulfonyl-L-arginine methyl ester (TAME) were monitored by the method of Hummel [23]. Activities toward acetyl-L-tyrosine ethyl ester (ATEE) was measured according to the method of Schwert and Takenaka [24]. Activity for N-benzoyl-DL-arginine-β-naphthylamide (BA-Nap), N-carbobenzoxy-Phe-Arg-4-methoxy-β-naphthylamide (ZPA-MNap), N-carbobenzoxy-(Arg)$_2$-4-methoxy-β-naphthylamide (ZAA-MNap), and N-carbobenzoxy-(Gly)$_2$-Arg-4-methoxy-β-naphthylamide (ZGGA-MNap) were measured using the assay method of Barrett [25,26].

2.6. Preparation of myofibrillar protein

A myofibrillar protein was extracted by the modified method of Jiang et al.[27]. 25 g of minced anchovy muscle was homogenized in 250 mL of 40 mM Tris-HCl, pH 7.0 containing 25 mM KCl and 4.5 mM EDTA for 2 min and centrifuged (600×g, 15 min). After repeating 3 times to remove sarcoplasmic protein, precipitates were suspended in 40 mM Tris-HCl, pH 7.0 containing 0.1 M KCl and 4.5 mM EDTA and filtered through four layer of cheese clothes to remove connective tissues. Filtrates were used as myofibrillar protein.

2.7. Protein concentration and electrophoresis

Protein was determined according to the method of Lowry et al. [28] with bovine serum albumin as a standard. Absorbance at 280 nm was used to follow protein concentration during the purification. Analytical polyacrylamide gel electrophoresis (PAGE) was performed by the method of Davis [29] at pH 8.3 using a 7.5% gel.

2.8. Determination of molecular weight

The molecular weights of the proteases were determined using sodium dodecyl sulfate polyacrylamide gel electrophoresis (SDS-PAGE) by the method of Laemmli [30] using 5% stacking gel and 10% separating gel and also using Sephacryl S-100 column (0.9×55 cm) gel filtration according to the method of Whitaker [31]. The proteins used as molecular weight standards for SDS-PAGE were bovine albumin (Mr 66,000), egg albumin (Mr 45,000), glyceraldehydes-3-phosphate dehydrogenase (Mr 36,000), carbonic anhydrase (Mr 29,000), trypsinogen (Mr 24,000) and α-lactoalbumin (Mr 14,200). Bovine albumin (Mr 66,000), carbonic anhydrase (Mr 29,000), cytochrome c (Mr 12,400) and aprotinine (Mr 6,500) were used as the marker proteins for gel chromatography.

2.9. Determination of amino acid composition

The amino acid compositions of proteases were determined with a amino acid analyzer (LKB-Biotech, 4150, Sweden) after hydrolysis in 6 N HCl at 110°C for 24, 48 and 72 hr, respectively. Tryptophan levels were determined by using the method of Hugli and Moore [32]. Cysteine was assayed according to the method of Spencer and Wald [33].

2.10. pH and temperature dependence

The protease activities were determined using casein and BAPNA as the substrates for trypsin, BTEE for chymotrypsin, and BA-Nap for cathepsin at pH values ranging from 3.0 to 12.0 at a temperature of 30°C. Activities were determined in different buffers with overlapping pH points to exclude the possibility of an influence exerted by the nature of the buffers. The optimum temperatures of the proteases against synthetic substrates were measured at pH 8.0 for trypsin and chymotrypsin, and pH 6.0 for cathepsin over a temperature range of 20-70°C.

2.11. Determination of kinetic parameters

Apparent Michaelis-Menten constant (Km) and maximum velocity (Vmax) were determined by least-squares regression analysis. Kinetic data with myofibrillar protein were obtained with concentration range of 0.01-0.40% at pH 6.0 and 40°C.

2.12. Degradation of myofibrillar protein

The reaction mixture consisted of myofibrillar protein and the enzyme at a protein ratio of 200:1 (w:w) in 20 mM sodium acetate, pH 6.0, containing 100 mM NaCl and 0.02% sodium azide. The reaction was carried out at 20°C and then terminated after 10, 60, and 180 min by addition of SDS-PAGE sample buffer solution. The proteolytic activities of trypsin, chymotrypsin, cathepsin, and their mixture for myofibrillar protein as substrate were measured by the amount of tyrosine released from reaction mixture after incubating for 60 min at 30°C.

2.13. Effect of NaCl concentration on proteolytic activities

The reaction mixture consisted of myofibrillar protein containing a range of 0-25% NaCl and proteases at a protein ratio of 200:1 (w:w) in 20 mM sodium acetate, pH 6.0, containing 100mM NaCl. The reaction was carried out at 30°C for 3 hr and their activities were measured as the method of Pyeun and Kim [20].

3. RESULTS AND DISCUSSION

3.1. Purification of trypsin, chymotrypsin and cathepsin L-like enzyme

The purification of trypsin and chymotrypsin from anchovy viscera based on hydrolysis of BAPNA and BTEE, respectively, are summarized in Table 1. As shown in Fig. 1, high molecular weight proteins were effectively removed with purification steps. After Sephacryl S-100 gel filtration, purity of trypsin was established from analytical PAGE at pH 8.3, and a single band was obtained as evidence for the homogeneity (Fig. 1). For chymotrypsin, the enzyme was efficiently separated by DEAE-Sephacryl chromatographies. After Sephacryl S-100 gel filtration as a final step, chymotrypsin was homogeneous based on the pattern of analytical PAGE (Fig. 1). Purities of trypsin and chymotrypsin were increased 78 and 119 fold, respectively, compared with crude extract.

The purification of cathepsin L-like enzyme from anchovy muscle based on the hydrolysis of BA-Nap was summarized in Table 2. After second CM-Sepahdex C-50 ion-exchanger column, cathepsin L-like enzyme was homogeneous in analytical PAGE (Fig.1) and purity increased by 83 fold compared to crude enzyme.

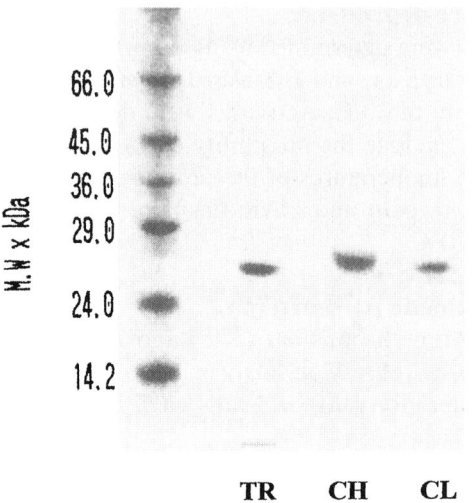

Figure 1. Analytical polyacrylamide gel electrophoresis of purified trypsin (TR), chymotrypsin (CH) and cathepsin L-like enzyme (CL) from anchovy.

Table 1
Purification of trypsin and chymotrypsin from anchovy viscera

Fraction	Trypsin			Chymotrypsin		
	Specific Act. (U/mg)	Yield (%)	Purity (Fold)	Specific Act. (U/mg)	Yield (%)	Purity (Fold)
Crude extract	0.05	100	1.0	0.82	100	1.0
Ammonirum sulfate fraction	0.34	5.94	3.5	6.62	22.2	8.0
Benzamidine-Sepharose 6B						
Absorbed	1.36	7.1	3.0			
Unabsorbed				6.14	13.2	7.5
Sephadex	1.51	4.3	33	21.3	12.4	25.9
1'st DEAE-Sephacel	2.38	3.2	52	41.7	5.8	50.7
2'nd DEAE-Sephacel				46.41	3.8	56.4
Sephacryl S-100	3.56	3.2	78	98.27	3.4	119

Yields and purities of trypsin and chymotrypsin were calculated from activity using BAPNA or BTEE, respectively.

3.2. Molecular weight

Molecular weights were estimated to be 25.6 kDa for trypsin, 26.1 kDa for chymotrypsin, and 25.8 kDa for cathepsin L-like enzyme by SDS-PAGE (Fig. 2). The molecular weight determined by SDS-PAGE and gel filtration had no difference (Fig. 3). The molecular weights of three proteases were similar to those of fish and mammalian. The molecular weight of cathepsin L was reported as 28.8 kDa for Pacific whiting [34], 30- and 37 kDa for chum salmon [35]. Cathepsin L would be associated with protease inhibitor, which regulates for protein turnover and/or for stabilization of active enzyme in muscle [36]. However, in this purification process, low-molecular-weight protease inhibitor was not identified.

Table 2
Purification of cathepsin L-like enzyme form anchovy muscle

Fraction	Specific Activity (U/µg)	Yield (%)	Purity (fold)
Crude extract	1.5	100	1.0
Ammonium sulfate fraction	4.8	31.9	3.2
First Sephadex G-75 fraction	6.0	4.0	4.0
Second Sephadex G-75 fraction	21.7	3.4	14.5
First CM-Sephadex C-50	51.2	3.1	34.1
Second CM-Sephadex C-50	125	2.9	83.3

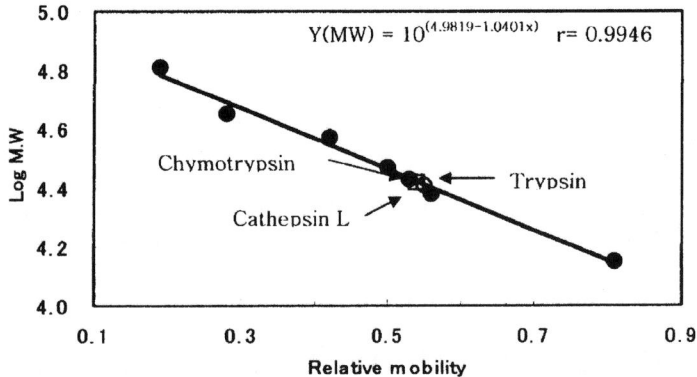

Figure 2. Estimation of molecular weights of purified trypsin, chymotrypsin, and cathepsin L-like enzyme from anchovy by SDS-PAGE. ○, Trypsin; ●, chymotrypsin; □, cathepsin L-like enzyme.

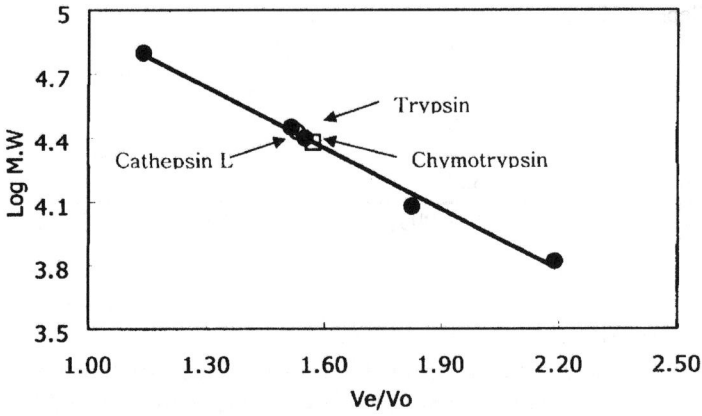

Figure 3. Estimation of molecular weights of purified trypsin, chymotrypsin, and cathepsin L-like enzyme from anchovy by gel chromatograpohy. ○, Trypsin; ●, chymotrypsin; □, cathepsin L-like enzyme.

Table 3
Amino acid composition of trypsin, chymotrypsin, and cathepsin L-like enzyme from anchovy (residue/mol-protein)[a]

Amon Acid		Trypsin	Chymotrypsin	Cathepsin L
Non-polar				
	Ala	11	12	22
	Ile	10	10	9
	Leu	15	14	13
	Met	3	2	1
	Phe	5	4	12
	Pro	11	15	9
	Trp	2	6	5
	Val	16	19	10
Polar				
Negative charge				
	Asp	32	25	29
	Glu	24	21	22
Positive charge				
	Arg	6	9	19
	Lys	6	6	10
Uncharged at pH 7				
	Cys	5	19	26
	Gly	28	17	14
	His	9	10	6
	Ser	30	21	16
	Thr	14	21	12
	Tyr	8	3	1
Total residues		234	234	235
Total amount		30605	30024	29629
Corrected M.W		26411	25830	25399
Av. hydrophobicity[b] (kcal/residue)		0.97	0.97	0.86

[a]Molecular weights of the trypsin, chymotrypsin and cathepsin L adapted in the analysis of amino acid composition were based on the values determined by SDS-PAGE.
[b]Average hydrophobicity was calculated by the method of Bigelow [44].

3.3. Amino acid composition

The amino acid composition of purified trypsin, chymotrypsin, and cathepsin L-like enzyme were presented in Table 3. The amino acid composition of trypsin was similar to that of anchovy trypsin from *Engraulis encrasicholus* [3]. The amino acid profile of anchovy chymotrypsin was similar to that of bovine chymotrypsin. But histidine and lysine content in anchovy chymotrypsin were significantly different from those in bovine chymotrypsin. Trypsins from marine animals have high ratio of acidic-to basic amino acids [37]. This is quite different from mammalian trypsins which have basic isoelectric points. Cathepsin L-like enzyme contains high level of basic amino acid residues compared with anchovy trypsin and chymotrypsin. The cationic nature of the cathepsin L-like enzyme reflects its amino acid composition, which shows a higher proportion of basic amino acid residues.

3.4. pH and temperature optima

The optimum pH for hydrolysis of BAPNA by trypsin was pH 8.0~9.0 and that for BTEE by chymotyrpsin at 30°C was pH 7.5~8.5, while that for BA-Nap by cathepsin L-like enzyme at 30°C was pH 6.0 (Fig. 4). At physiological pH, relative activities of trypsin and chymotrypsin for synthetic substrates were approximately 60% of its maximum activities, while that of cathepsin L-like enzyme for BA-Nap showed approximately 90% of its maximum activity. The pH optimum for synthetic substrate was similar to that of two cathepsin L from Pacific whiting, which had an optimal pH at 5.5 and 5.6[34], chum salmon, at pH 5.6 [35], and to that of cathepsin L and cathepsin L-like enzyme from mackerel, at pH 5.0 and 5.5, respectively [38]. These results suggest that cathepsin L-like enzyme is more important than trypsin and chymotrypsin in the initial step of fermentation during fish sauce processing.

Anchovy trypsin, chymotrypsin, and cathepsin L-like enzyme had maximum activity for synthetic substrates at 45°C (Fig. 5). The relative activity of trypsin at 25°C was about 25% of its maximum activity, however that of chymotrypsin was comprised at about 50%, which indicates that considerable activity is maintained even at ambient temperature. The relative activity of cathepsin L-like enzyme showed 10% of its maximum activity.

3.5. Substrate specificity

Trypsin from anchovy had activities toward ester and amide substrates containing arginine residue, but displayed no activity against chymotrypsin substrate, which indicates that trypsin is not contaminated with chymotrypsin-like enzyme (Table 4). Trypsin showed higher activity against ester substrates than against amide substrates except for the ZGGA-MNap. Chymotrypsin had high activities against BTEE and SAAPPNA, but no activity toward trypsin substrate. The cathepsin-like enzyme had activity on ZPA-MNap, ZAA-MNap and L-arginine-naphtylamide (Arg-Nap), which are

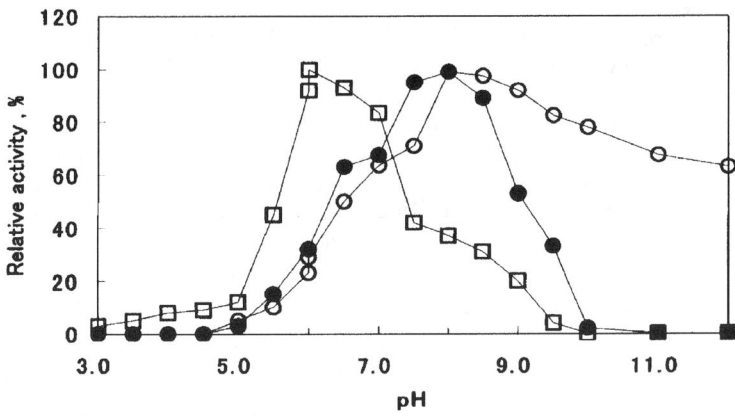

Figure 4. pH optima of purified trypsin, chymotrypsin, and cathepsin L-like enzyme from anchovy for synthetic substrates. ○, Trypsin; ●, chymotrypsin; □, cathepsin L-like enzyme.

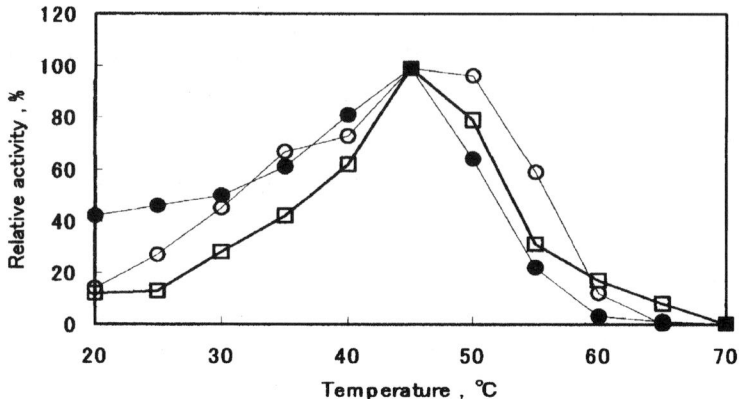

Figure 5. Temperature optima of purified trypsin, chymotrypsin, and cathepsin L-like enzyme from anchovy for synthetic substrates. ○, Trypsin; ●, chymotrypsin; □, cathepsin L-like enzyme.

considered to be the best synthetic substrate for cathepsin L, B and H, respectively [39]. With results, the enzyme could be classified into cathepsin L.

3.6. Kinetic properties on natural substrate

Km values of trypsin and chymotrypsin for myofibrillar protein were lower than that of cathepsin L-like enzyme (Table 5). The Km of chymotrypsin and cathepsin had higher than that of trypsin, while the Vmax showed highest values in chymotrypsin. Trypsin showed a similar affinity and degradation activity between casein and myofibrillar protein. Chymotrypsin showed a similar degradation activity for casein and myofibrillar protein, but did not show a difference between affinity for both substrates. The Km of cathepsin L-like enzyme for casein and myofibrillar protein showed 0.23 U/mg and 0.56 U/mg, respectively, while the kcat/Km for both substrate showed 49.45min^{-1}/% and 7.77 min^{-1}/%. The results suggested that cathepsin L-like enzyme showed a higher degradation for myofibrillar protein compared to casein. The proteolysis for natural substrate showed a quite difference in the reaction compared with reaction for synthetic substrate. The real values of Km and Vmax showed a difference by comparison with the apparent values because product from proteolytic reaction contributed to values of kinetic parameters [40,41]. The results suggested that chymotrypsin and cathepsin L were more responsible to autolysis of muscle protein after death and in fermentation.

3.7. Degradation of myofirillar protein

After myofibrillar protein from anchovy was incubated with trypsin for 10 min, The myosin heavy chain was degraded slowly, and produced fragment of 150 kDa on SDS-PAGE, while the actin and troponin-T was not degraded. Fragments of 80 kDa and 60 kDa was gradually increased with incubating time, Fragment of 150 kDa from myosin heavy chain was remained after 180 min (Fig. 6). In chymotrypsin, The degradation of

Table 4
Specificity of the trypsin, chymotrypsin and cathepsin L-like enzyme for synthetic substrates (specific activity, U/mg)

Substrate(0.5 mM)	Trypsin	Chymotrypsin	Cathepsin L
ATEE	0	3.56	0
BTEE	0	93.36	1.94
Suc-(Ala)$_2$-Pro-Phe-NA	0	27.27	0
BAEE	22.14	0	3.72
TAME	30.56	0	2.59
BA-p-NA	3.00	0	0.36
BA-Nap	5.54	0	0.28
ZPA-MNap	1.57	0	1.33
ZAA-MNap	3.28	0	0.61
ZGGA-MNap	23.28	0	0.65
Arg-Nap	0	0	0.35
Hi-Phe	0	0	0
Hi-Arg	0	0	0

Table 5
Kinetic parameters of the trypsin, chymotrypsin and cathepsin L-like against myofibrillar protein

Parameter	Myofibrillar protein		
	Trypsin	Chymotrypsin	Cathepsin L
Km (μM)	49.28	396.77	73.42
Vmax (U/mg)	3.55	217.54	1.42
Kcat (min^{-1})	90.88	5,656.04	36.64
Kcat/Km (min^{-1}/μM)	1.84	14.26	0.50

myosin heavy chain was occurred, and fragment of 150 kDa was increased rapidly. After 60 min, the fragments of 160 kDa, actin, tropomyosin, and troponin-T band were gradually degraded, and all proteins composed of myofirillar protein were degraded after incubating for 180 min.

After cathepsin L-like enzymes was incubated with myofibrillar protein from anchovy, myosin heavy chain and actin were greatly degraded, and produced 150 kDa, 130 kDa, 100 kDa, 37 kDa, 29 kDa and 15 kDa. Most of actin was degraded, and fragments from myosin heavy chain and actin also were disappeared gradually after 60 min. Actin and tropomyosin were degraded completely after 180 min. The results suggested that trypsin, chymotrypsin and cathepsin L-like enzymes mainly hydrolyzed myosin heavy chain. Chymotrypsin had the highest contribution to degradation of anchovy muscle. The autolysis of muscle protein was responsible for proteases from viscera and muscle, and promoted with interaction of each protease.

3.8. Effect of NaCl concentration on proteolytic activities

The activities of trypsin and chymotrypsin were not affected up to 5% NaCl, but that

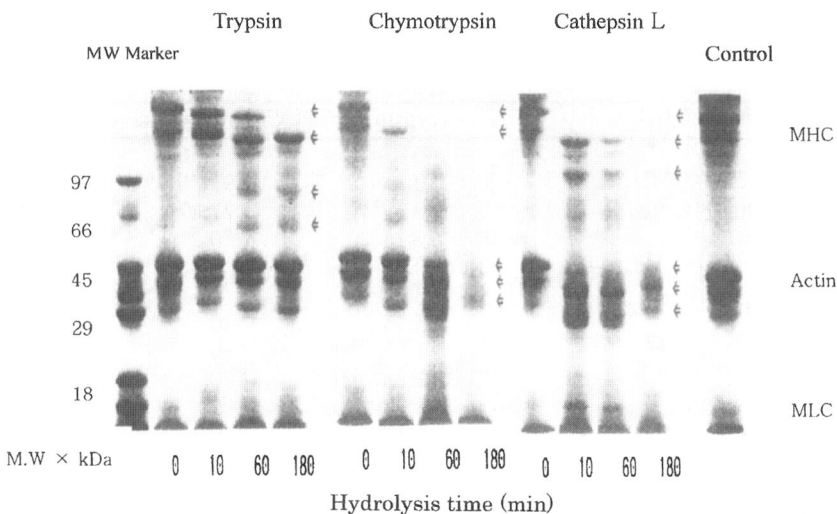

Figure 6. Changes of myofibrillar protein from anchovy during hydrolyzing with trypsin, chymotrypsin, and cathepsin L-like enzyme on SDS-PAGE.

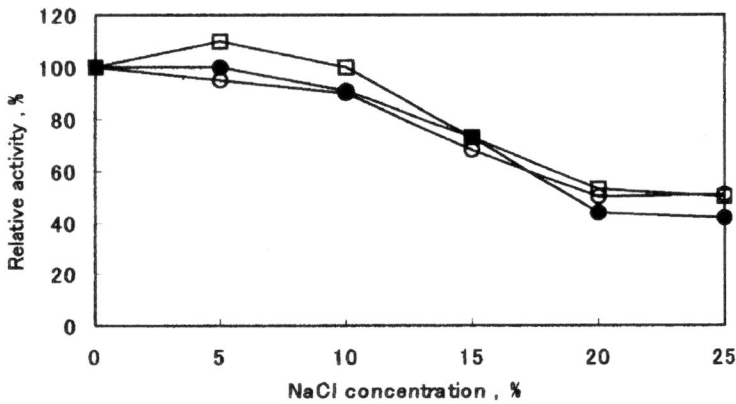

Figure 7. Effect of NaCl concentration on the hydrolysis of myofibrillar protein by trypsin, chymotrypsin, and cathepsin L-like enzyme from anchovy. ○, Trypsin; ●, chymotrypsin; □, cathepsin L-like enzyme.

of cathepsin L-like enzyme was increased to about 10% (Fig. 7). The activities of proteases were decreased with increase in NaCl concentration, the activities were remained 50% in 25% of NaCl. The decrease of protease activities with NaCl was due to increasing of ionic strength in reaction mixture, and conformation changes of proteases [42]. Under high salt concentration such as preparation of fish sauce, the remaining activities contributed to hydrolysis of fish muscle. The growth of microorganisms was completely inhibited above 20% of NaCl [43]. The results suggested that activities of proteases from viscera and muscle were more important than those of proteases from microorganisms in fermentation during fish sauce processing.

REFERENCES

1. Kim, H.R., Meyers, S.P. and Godber, J.S. (1992) Purification and characterization of anionic trypsins from the hepatopancreas of crayfish, *Procambarus clarkii*, Comp. Biochem. Physiol.,103B: 391-398.
2. Kim, H. R., Meyers, S. P., Pyeun, J. H. and Godber, J. S. (1994) Enzymatic properties of anionic trypsins from the hepatopancreases of crayfish, *Procambarus clarkii*. Comp. Biochem. Physiol., 107B: 197-203.
3. Martinez, A., Olsen, R.L. and Serra, J.L. (1988) Purification and characterization of two trypsin-like enzymes from the digestive tract of anchovy *Engraulis encrasicholus*. Comp. Biochem. Physiol., 91B: 677-684.
4. Ramakrishna, M., Hultin, H.O. and Atallah, M.T. (1987) A comparison of dogfish and bovine chymotrypsins in relation to protein hydrolysis. J. Food Sci., 52: 1198-1202.
5. Ramakrishna, M., Hultin, H. O. and Racicot, W. (1987) Some kinetic properties of dogfish chymotrypsin. Comp. Biochem. Physiol., 87B: 25-30.
6. Kim, H.R. and Pyeun, J.H. (1986) The protease distributed in the intestinal organs of fish 2. Characterization of the three alkaline proteinases from the pyloric caeca of mackerel *Scomber japonicus*. Bull. Korean Fish. Soc., 19: 547-557.
7. Hjelmeland, K. and Raa, J. (1982) Characteristics of two trypsin type isozymes isolated from the arctic fish capelin (*Mallotus villosus*). Comp. Biochem. Physiol., 71B: 557-562.
8. Cohen, T., Gertler, A. and Birk, Y. (1981) Pancreatic proteolytic enzymes from carp (*Cyprinus carpio*)-I. Purification and physical properties of trypsin, chymotrypsin, elastase and carboxypeptidase B. Comp. Biochem. Physiol., 69B: 639-646.
9. Cohen, T., Gertler, A. and Birk, Y. (1981) Pancreatic proteolytic enzymes from carp (*Cyprinus carpio*)-II. Kinetic properties and inhibition studies of tyrpsin, chymotrypsin and elastase. Comp. Biochem. Physiol., 69B: 647-653.
10. Heu, M.S., Pyeun, J.H., Kim, H.R. and Godber, J.S. (1991) Purification and characterization of alkaline proteinases from the viscera of anchovy *Engraulis japonica*. J. Food Biochem., 15: 51-66.
11. Pyeun, J.H., Kim, H.R. and Godber, J.S. (1990) Comparative studies on the enzymatic properties of two trypsin-like enzymes from menhaden, *Brevoortia tyranus*. Bull. Korean Fish Soc., 23: 12-20.
12. Pyeun, J.H., Kim, H.R. and Heu, M.S. (1988) The proteinase distributed in the intestinal organs of fish. 3. Purification and some enzymatic properties of the alkaline proteinases from the pyloric caeca of skipjack, *Katsuwonus vagans*. Bull. Korean Fish Soc., 21: 85-96.
13. Beddows, C. G., Ismail, M. and Steinkraus, K. H. (1976) The use of bromelain in the hydrolysis of mackerel and the investigation of fermented fish aroma. J. Food Technol., 11: 379-388.
14. Beddows, C.G., Ardeshir, A.G. and Daud, W.J. (1979) Biochemical changes occurring during the manufacture of Budu. J. Sci. Food Agric., 30: 1097-1103.
15. Alm, F. (1965) Scandinavian anchovies and tidbits. In: Fish as Food. Borgstrom, G. (ed.), Academic Press, New York, pp.195-198.

16. Voskresensky N.A. (1965) Salting of herring. In: Fish as Food. Borgstrom, G. (ed.), Academic Press. New York., pp. 107-?.
17. Orejana, F.M. and Liston, J. (1982) Agents of proteolysis and its inhibition in patis (fish sauce) fermentation. J. Food Sci., 47: 198-203,209.
18. Del Rosario, R.R. and Maldo, S.M. (1984) Biochemistry of Patis formation I. Activity of cathepsins in Patis hydrolysates. Phil. Agr., 67: 167-175.
19. Lopetcharat, K. and Park, J.W. (1999) Autolytic activity of enzymes in Pacific whiting and surimi by-product (SBP) during fish sauce production. Abstract book of 1999 IFT annual meeting, p. 149
20. Pyeun, J.H. and Kim, H.R. (1986) The proteinase distributed in the intestinal organs of fish 1. Purification of the three alkaline proteinases from the pyloric caeca of mackerel, *Scomber japonicus*. Bull. Korean Fish. Soc., 19: 537-546.
21. Erlanger, B.F., Kokowsky, N. and Cohen, W. (1961) The preparation and properties of two new chromogenic substrates of trypsin. Arch. Biochem Biophys., 95: 271-278.
22. Erlanger, B.F., Edel, F. and Cooper, A.G. (1966) The action of cymotrypsin on two new chromogenic substrates. Arch. Biochem. Biophys., 155: 206-210.
23. Hummel, B.C. (1959) A modified spectrophotometric determination of chymotrypsin, trypsin and thrombin. Can. J. Physiol., 37: 1393-1399.
24. Schwert, G.W. and Takenaka, Y. (1955) A spectrophotometric determination of trypsin and chymotrypsin. Biochim. Biophys. Acta., 16: 570-575.
25. Barrett, A. J. (1972) A new assay for cathepsin B_1 and other thiol proteinases. Anal. Biochem., 47: 280-293.
26. Barrett, A.J. (1976) An improved color reagent for use in Barrett's assay of cathepsin B. Anal. Biochem., 76: 374-376.
27. Jiang, S.T., Tsao, C.Y., Wang, T. and Chen, C.S. (1990) Purification and characterization of protease from milkfish muscle (*Chanos chanos*). J. Agric. Food Chem., 38: 1458-1463.
28. Lowry, O.H., Rosebrough, N.J., Farr, A.L. and Randall, R.J. (1951) Protein measurement with Folin phenol reagent. J. Biol. Chem., 193: 265-275.
29. Davis, B.J. (1964) Disc-electrophoresis II, method and application to human serum protein. Ann. N.Y.Acad. Sci., 121: 404-427.
30. Laemmli, U.K. (1970) Cleavage of structural proteins during the assembly of the head of Bacteriophage T4. Nature, 227: 680-686.
31. Whitaker, J.R. (1963) Determination of molecular weights of proteins by gel filtration on Sephadex. Anal. Chem., 35: 1950-1953.
32. Hugli, T.E. and Moore, S.J. (1972) Determination of tryptophan content of protein by ion exchange chromatograph of alkaline hydrolysates. J. Biol. Chem., 247: 2828-2834.
33. Spencer, R.L. and Wald, F. (1969) A new convenient method for estimation of total cystine-cysteine in proteins. Anal. Biochem., 32: 185-190.
34. Seymour, T. A., Morrissey, M. T., Peters, M. Y. and An, H. (1994) Purification and characterization of Pacific whiting proteases. J. Agric. Food Chem., 42: 2421-2427.
35. Yamashita, M. and Konagaya, S. (1990) Participation of cathepsin L in extensive softening of the muscle of chum salmon caught during spawning migration. Nippon Suisan Gakkaishi, 56: 1271-1277.

36. Mason, R.W., Taylor, M.A. and Etherington, D.J. (1984) The purification and properties of cathepsin L from rabbit liver. Biochem. J., 217: 209-217.
37. Simpson, K. (2000) Digestive proteinases from marine animals. In: Seafood Enzymes. Haard, N.F. and Simpson, B.K. (eds.), Marcel Dekker, New York, pp. 411-450.
38. Lee, J.J., Chen, H.C. and Jiang, S-T. (1993) Purification and characterization of proteinases identified as cathepsin L and L-like (58Kda) proteinase from mackerel (*Scomber australasicus*). Biosci. Biotechnol. Biochem., 57: 683-690.
39. Barrett, A.J. (1980) Enzyme Regulation and Mechanism of Action. Oxford, Pergamon.
40. Doke, S., Ninjoor, V. and Nadkarni, G. B. (1980) Characterization of cathepsin D from the skeletal muscle of fresh water fish, Tilapia mossambicca. Agric. Biol. Chem. 1980; 44: 1521-1528.
41. Boneto, M. J., Manjon, A., Liorca, F. and Iborra, J. L. (1984) Acid proteinase activity in fish-II. Purification and characterization of cathepsin B and D from *Mujil auratus*. Comp. Biochem. Physiol., 78B: 207-213.
42. Dixon, M. and Webb, E.C. (1979)   In:Enzymes. Longman, London.
43. Pelczar, M.J. and Chan, E.C.S. (1981) Elements of Microbiology. McGraw-Hill Books Co, New York.
44. Bigelow, C.C. (1967) On the average hydrophobicity of protein and the relation between it and protein structure. J. Theoret. Biol., 16: 187-211.

M. Sakaguchi (Editor)
More Efficient Utilization of Fish and Fisheries Products
© 2004 Elsevier Ltd. All rights reserved

MICROBIAL RISK ASSESSMENT OF PERSIAN CAVIAR DURING PROCESSING AND COLD STORAGE

Vadood Razavilar[a] and Sohrab Rezvani[b]
[a]University of Tehran, Faculty of Veterinary Medicine, Department of Food Hygiene, Tehran, Iran.
[b]Fishery Research Institute of Iran, Tehran, Iran

Abstract
Study was conducted to find any potential microbial hazard in the processing and storage of Persian caviar in 2 processing plants in Mazandaran region of Iran. Totally 80 samples were assessed for 12 microbial examinations of caviar (in 3 stages of unprocessed, processed and cold storage) and processing environment (in 11 items). The 12 microbial examinations were including of 2 standard plate count (SPC) at 25 and 37°C, Fungi, Coliforms, Staphylococci, Enterococci (for microbial count) and *Salmonella, Plesiomonas, Vibrio, Listeria, Clostridium botulinum* and *C. perfringens* (for detection only). In unprocessed caviar the microbial count (per gram) was zero for all cases of bacterial examinations, except for Fungi and SPC (0 to 4×10^1), whereas in processed caviar (granulated and salted), 1-2 logs increase was observed (9.2×10^1 to 9.3×10^2) in SPC and fungal count. In the cold stored processed caviar (at -3°C after 6 months) there was another 1-2 logs increase (5×10^2 to 1.2×10^4) in SPC and fungl count. Coliforms, Enterococci and Staphylococci were isolated in very low numbers ($<10/g$). Except for *C. perfringens*, the results of other microbial examinations were negative. From the 11 items of environmental samples analysed for microbial indices, fingers of caviar handlers, cutting knife and the caviar screen sweeper, were determined the most risky items for microbial contamination of caviar. Microbial growth may occur due to the time/temperature abuse during processing and storage. Overall the results indicated a good microbial condition in most cases and for the risky environmental points during processing, an appropriate corrective action needed to improve the microbial quality of the final product.

1. INTRODUCTION

Iran is one of the major supplier and exporter of sturgeon caviar in the world. Of the 26 sturgeon species found in the world, 4 Caspian species are the main source for world production of sturgeon caviar: Beluga (Huso huso), Osetra (*Acipenser gueldenstaedti*), Sevruga (*A. stellatus*) and Ship (*A. nudiventris*). No matter which species is used, caviar always consists of detached, lightly salted, uncooked fish eggs. Therefore from the earliest times the main concern has been the preservation of caviar from spoilage and health hazard.

Risk assessment is one method that is playing a major role in helping us to further improve food safety and in planning further initiatives. It has the potential to improve our ability to reduce food spoilage and food-borne illness in a number of ways. Risk assessment also will help industry to develop more effective HACCP Plans [1-6].

This study was conducted to find the potential microbial hazards in the processing and storage of Persian caviar in 2 plants of Mazandaran region of Iran. The results obtained from this study will help caviar plants to scientifically develop HACCP plans for the future.

2. MATERIALS AND METHODS

Two caviar Processing plants in Mazandaran region was selected for 12 microbial examinations of caviar and caviar processing environment.

2.1. Caviar product

Three stages of caviar 1) unprocessed (just after extraction from fish), 2) processed (after granulation, preserving and canning) and 3) processed and cold stored caviar (after 6 months of storage at $-3°C$) were used for sampling in 2 caviar processing plants (plant A and B). Totally 8 samples with 2 replicates were analysed for 3 different stages of caviar. For the other microbial examinations, 5g samples were used except for Salmonella (with 25g sample).

2.2. Caviar processing environment

Following 11 environmental items were selected: fishing wharf, fish cutting table, fish cutting knife, caviar skimmer, caviar screen, caviar screen sweeper, caviar processing table, caviar can, water, caviar handlers fingers and processing room air samples. Swab and rinse tests were used for microbial examinations of environmental samples [7].

2.3. Microbial examinations of caviar and caviar processing environment

Twelve microbial examinations were used for both caviar and environmental samples. These were included: 2 standard plate counts (SPC at 25 and 37°C), Fungal count, Coliform count, Staphylococcal count and Enterococcal count, along with 6 microbial detections of *Salmonella, Plesiomones, Vibrio, Listeria, Clostridium botulinum* and *C. perfringens*. Except for Coliform count (with MPN technique), for the other microbial counts and detections, surface plating were used [7,8].

Overall combination of 1632 ($8\times3\times2\times12=576$ for caviar and $8\times11\times12=1056$ for processing environment) microbial examinations were done for this study.

3. RESULTS AND DISCUSSION

Due to the limitation and exclusivity of caviar production in the world (mainly limited to the Caspian sea area), there is very limited studies about microbial qualifcation of this product. In some cases these limited studies are not even published. At the present time due to EC requirement for exportation of caviar to European countries we are implementing HACCP system of product control, in our caviar processing plants and this study is part of a research work for microbial evaluation of caviar using HACCP system.

Table 1 shows the average results of 6 microbial count examinations of Persian caviar for 3 different stages of processing in 2 caviar processing plants (A and B). In all of 3 stages [Unprocesed (1), Processed (2) and stored at $-3°C$ for 3 months (3)], the average number of microbial count examinations (SPC_{37}, $SPC_{25°C}$, Fungi, Coliforms, Staphylococci and Enterococci) were ranged from 2/g (for fungi in stage 1) to 1.2×10^4/g (for SPC_{37} in stage 3).

The average count of Coliforms, Staphylococci and Enterococci were zero for all 3 stages of caviar except for Coliforms (5 and 7/g in stage 3 for plant A and B, respectively), and Staphylococci (8/g in stage 3 of plant A). The results of other 6 microbial detection (*Salmonella, Plesiomonas, C. botulinum* and *Listeria, Vibrio* and *C. perfringens*) were negative for all except for *C. perfringens* in caviar stage 3 of plant B, which was positive (not shown in Table 1). As shown in Table 1, there are 1–2 logs increase in the counts of SPC_{37}, SPC_{25} and fungal counts in each stages of caviar processing. This microbial increase can be due to time/ temperature abuse during processing and storage of caviar, which must be considered for corrective action.

Table 2 represents the average results of 6 microbial count exaiminations for 11 environmental surfaces from 2 different caviar processing plants (A and B). Of the 11 different items of processing environment for both plant A and B, the caviar screen sweeper was the most risky point of caviar contamination. This point needs an urgent

Table 1
The average results of 6 microbial count[a] examinations of Persian caviar (in 3 stages of processing) in 2 different processing plants (A and B)

Microbial count	Unprocessed caviar (stage 1)		Processed caviar (stage 2)		Processed and stored at $-3°C$ for 6 months (stage3)	
	A	B	A	B	A	B
SPC (at 37°C)	3×10^1	4×10^1	9.3×10^2	9.2×10^1	1.2×10^4	3.6×10^3
SPC (at 25°C)	1.5×10^1	3.8×10^1	3.1×10^2	2.7×10^2	8.6×10^3	4.6×10^3
Fungi	2	5	3×10^1	2.3×10^1	5×10^2	6.3×10^2
Coliforms	—[b]	—	—	—	5	7
Staphylococci	—	—	—	—	8	—
Enterococci	—	—	—	—	—	—

[a]The other 6 microbial examinations (detection only) were negative, except for *C. perfringens* which was positive in the processed and stored caviar of plant B.
[b]Negative for all samples at the highest concentration (1/10 of 5g sample).

Table 2
The average results of 6 microbial count[a] examinations of the 11 environmental surfaces from 2 different caviar processing plants (A and B)

Plant	Environmental samples	SPC (37°C)	SPC (25°C)	Fungi	Coliforms	Staphylococci	Enterococci
A	Fishing wharf	10×10^4	1.2×10^4	3	<1	—	—
	Fish cutting table	4.5	1.2×10^1	<1	<2	—	—
	Fish catting knife	2.1×10^2	3.5×10^2	<1	1.6×10^1	—	—
	Caviar skimmer	1.3×10^2	1.1×10^3	—[b]	—	—	—
	Caviar screen	2.3×10^3	1.8×10^2	<1	—	—	—
	Caviar screen sweeper	8×10^3	1.3×10^4	1.8×10^2	4×10^2	8	<1
	Caviar processing table	<1	3.7	<1	<1	—	—
	Caviar can	<1	1.7×10^1	<1	—	—	—
	Water	5	4	<1	—	—	—
	Caviar hadlers fingers	4×10^2	5×10^2	<3	1.6×10^1	8	—
	Air	<1	<1	—	—	—	—
B	Fishing wharf	1.5×10^4	1.7×10^4	3.3×10^2	6.5×10^3	—	—
	Fish cutting table	4.2	5×10^1	—	<1	—	<1
	Fish cutting knife	1.3×10^2	1×10^2	—	—	—	—
	Caviar skimmer	<1	<1	<1	—	—	—
	Caviar screen	5.4×10^2	6.5×10^2	6.5	6.7×10^2	—	<1
	Caviar screen sweeper	1.1×10^4	1.3×10^4	<1	<1	—	<1
	Caviar processing table	7	1×10^1	—	—	—	—
	Caviar can	<1	7	—	—	—	<1
	Water	1.5×10^1	7.5×10^1	5	—	—	—
	Caviar handlers fingers	2.5×10^2	1×10^1	—	—	—	—
	Air	<1	<1	—	—	—	—

[a] The other 6 microbial examinations (detection only) were negative, except for Listeria (*L. denitrificans*) in caviar screen sweeper (plant B), *C. perfringens* on caviar handlers fingers (plant A) and *Vibrio fluvialis* on fish cutting table (plant B), which was positive.
[b] Negative for all samples at the highst concentration (1/10 of 5 g sample)

corrertive action to improve the microbial quality of the final caviar product. The other 2 risky points were caviar handler fingers (plant A) and fishing wharf with relatively more and higher microbial contaminations.

The results of other 6 microbial examinations (*Salmonella, Plesiomonas, Vibrio, C. perfringens, C. botulinum* and Listeria) were negative for all except for Listeria (*L. denitrificans*) in caviar screen sweeper of plant B, *C. perfringens* in hands of caviar handlers of plant A, and *Vibrio* (*V. fluvialis*) on fish cutting table of plants B (not shown in Table 2).

There are very limited number of studies about microbial evaluation of caviar (fish roe) and caviar processing environment [9–14], and even an internationally approved microbial standard is lacking for this product. Based on the report of Sternin and Dore [1], the aerobic SPC count of 10^4 to 10^6/g considered as low microbial quality for this product.

As Table 1 shows, for all 3 stages of Persian caviar (even after 6 months of storage), the SPC count (at 25 and 37°C) ranged from 1.5×10^1 to 8.6×10^3/g, except for one case of stage 3 caviar in plant A, which the SPC count reached to 1.2×10^4/g. This means that the microbial quality of Persian caviar even afther 6 months of storage is in a good condition.

Based on Basby et al. [9] on the microbial spoilage of salted lumpfish caviar at the level of 3.5 to 5% salt (without preservatives), The bacterial groups of Enterobacteriacea, Lactobacilacea and *Vibrio* were determined as the agents of microbial spoilage. In another study Marsilla et al. [12] reported that *C. perfringens* was one of the majoer microbial contamination of salted fish roe.

In our study bacterial groups of Micrococcacea, Pseudomonadacea, Bacillacea and Enterobateriacea were isolated from the SPC cultures.

Although *C. botulinum* was not isolated from caviar and caviar processing environment, this does not mean that it is always safe. Since caviar is a hermetically processed food, contamination and growth of *C. botulinum* in caviar is possible under anaerobic conditions and preventive measures must be considered [13,16–19].

Overall since caviar is an uncooked and lightly salted fish eggs and due to the various microbial contamination reported in this product (1,10,11), and also based of the results of our study (Table 1 and 2), 3 major action needed to be considered for the safety and quality improvement of the product (1) proper corrective action in the risky points of processing environment (fishing wharf, caviar handlers and caviar screen sweeper points in our study), to prevent contamination of caviar from these risky points (2), implementation of correct time/ temperature conditions to prevent growth of microorganisms in the product and its processing environment and (3) use of a proper and permited preservative to keep the product under a relyable safe condition. These 3 items are part of our aims for the future studies.

4. ACKNOWLEDGEMENT

This study is a result of research , sponsored by the University of Tehran (Faculty of Veterinary Medicine) and Fishery Research Institute of Iran. The senior author expresses his gratitude to Dr. H. Rostami, Mr. A. Salmani for their most valuable cooperation.

Gratitude is also expressed to Mr. A.H. Shodjai, Mr. R. Safari and Mrs. M. Nayyerani for their excellent technical assisstance.

REFERENCES

1. Sternin, V. and Dore, I. (1993) Caviar. Cultra, Moscow, Russia.
2. Flick, C.J. and Hackney, C. R. (1995) HACCP training for seafood processors . J. Food Prot. , 58:61-68.
3. Haby, M.G., Rippen, T.E., Coale, C.W., Miget, R. J. and Sutton, H.C. (1996) Status of seafood quality and safety management systems in retail seafood departments. IFT annual meeting: Book of Abstracts, p. 148.
4. Huss, H.H. (1992) Development the HACCP concept in fish processing. Int. J. Food Microbiol., 15: 33-44.
5. Melanson, J. (1995) The Canadian experience with seafood inspection-fish inspection in Canada. Journal of the Association of Food and Drug Officials, 59:49-57.
6. WHO (1996) Safe and sanitary processing of seafood-United State of America.WHO, Geneva, Switzerland, 47:203.
7. Harrigan, W.F. and Mc Cance, M.E. (1990) Laboratory Methods in food and Dairy Microbiology., Academic Press, London.
8. Vanderzant, C. and Splittatoesser, D.F. (1992) Compendium of Methods for the Microbiological Examination of Foods. American Public Health Association, Washington, D.C.
9. Basby, M., Jeppesen, V.F. and Huss, H. H. (1998) Spoilage of lightly salted lumpfish (*Cyclopterus lumpus*) roe at 5 degree C. J. Aq. Food Prod. Technol., 7(4):23-34 .
10. Brunner, B., Marx, H. and Stolle, A. (1995) Compositional and hygienic aspects of commercial caviar. Archiv. fuer Lebensmittelhygiene, 46:80-85.
11. Jokinen, M., Martensson, B., Sahlman, M. and Malmheden, I. (1992) Many bacteria in fish roe. Var-Foeda, 44:209-212.
12. Marsilla-del-pascual, B.A., Munoz-Sanchez, S., Guillin-Marco, A. and Martinez-del-olmo, A. (1987) Study of hygiene in the salt fish and dried saltfish product industries in the Murcia region of Spain. Alimentaria, 41:43-46.
13. USSR Gosudarstvennyi Komitet SSSR Po Standartam. (1979) Canned sturgeon caviar. Technical requirements. Soviet-Standard, GOST, 7442-79.
14. Ushakova, R.F. (1974) Microbiological study of manufacture of pasturized grainy. Caviar Trudy-Vsesoyuznogo-Nauchno-Issledovatel,101: 196-203.
15. Dolman, C.E. and Iida, H. (1963) Type E botulism: its epidemiology, prevention and specific treatment. Can. J. Pub. Health, 54: 293-308.
16. Fukuda, T., Kitao, T., Tanikawa, H. and Sakaguchi, G. (1970) An outbreak of type B botulism occuring in Miyazaki Prefecture. Japan. J. Med. Sci. Biol., 23: 243-248.
17. Health and Walfare Canada. (1997) Botulism in Canada-Summary for 1976. Can. Dis. Weekly Rep., 3:46-47.
18. Pourtaghva, M., Machoun, A., Fatollah-Zadeh, A., Khodadoust, H. and Farzam, Z. (1975) Le botulisme en Iran. Med.Malad. Infect, 5:536-539.
19. Sebald, M. (1970) Sur le botulisme en France de 1956 a 1970. Bull. Acad. Nat. Med., 154:707.

SOME BACTERIAL PATHOGENS IN THE INTESTINE OF CULTIVATED SILVER CARP AND COMMON CARP

Afshin Akhondzadeh Basti, Taghi Zahrae Salehi and Saied Bokaie
Department of Food Hygiene, Faculty of Veterinary Medicine, University of Tehran, P.O.Box: 14155-6453, Tehran, Iran

Abstract

There is an increasing demand for cultivated fish in all over the world. Unfortunately all of these fish are sold in whole states with viscera in the fish markets in Iran. The flora of water influences intestinal flora of the live fish. Fish captured in lakes, rivers and fish farms often carry a wide range of pathogenic microorganisms in their intestine due to contamination of the water with sewage, land run-off, etc. Therefore a total of 120 cultivated fish (Common carp and Silver carp) were bacteriologically investigated for pathogenic organism in their intestine, using FDA bacterial isolation methods. Food-borne pathogens were obtained in 65% of the examined specimens including 65% with *Escherichia coli*, 4.2% with *Salmonella* spp., 15% with *Shigella* spp., 16.7% with *Vibrio parahaemolyticus* and 3.3% with *Listeria monocytogenes* serotype 4b. Serological typing of *Salmonella* and *Shigella* revealed the detection of *Salmonella typhimurium* (1 strain), *S. newport* (4 strains) and *Shig. flexneri* (13 strains), *Shig. dysentriae* (7 strains) and *Shig. boydii* (7 strains). Biochemically identified *E. coli* were serologically investigated. The sera: O2K1, O119B14, O127B8 and O114K9 were detected by the serologic method.

1. INTRODUCTION

In all over the world a dramatic increase of food-borne diseases has been observed [1-4]. Food such as fish and shellfish are capable of transmitting many of the established food-borne infections and intoxications [5]. Fish can be carriers of pathogenic microorganisms. The following microorganisms have been found in the alimentary tract of fish: *Salmonella paratyphi* A and B, *S. suipestifer*, *S. entertidis*, *Escherichia coli* [6].

The intestinal flora in fish intestine may reflect the condition of water [7]. The risk of *Salmonella* and *Shigella* infections was lower in marine fish than in freshwater fish which might carry these groups of organisms if caught in polluted water, river or lakes due to contamination of water with sewage, land run-off, etc [7-9]. Also It has been reported the colonies of presumptive Aeromonas were detected in gill, gut and at the end of shelf life in the surface of abdominal cavity of the cultivated rainbow trout [10].

There is an increasing demand for cultivated fish in Iran. Unfortunately all of these fish are soled in whole states with viscera in the fish markets in Iran. Therefore a total of 120 cultivated fish (silver carp and common carp) were bacteriologically investigated for pathogenic organism in their intestine.

2. MATERIAL AND METHODS

One hundred twenty of whole ungutted cultivated fresh silver carp (*Hypophtalmichthys molitrix*) and common carp (*Cyprinus carpio*) were caught over 6 months in 2 fish farms near the city and polluted river which were fertilized by large animal manure, in Tehran and Gilan province of Iran. Each fish was transported in a sterile plastic bag containing ice to the laboratory and bacteriologically investigated for the presence of *Salmonella* spp., *Shigella* spp., *Listeria monocytogenes, Vibrio parahaemolyticus* and *E. coli* in their gut, using FDA bacterial isolation methods [11]. In brief, the entire gut of each fish was removed and fecal samples obtained aseptically, inoculated into 225 mL of brain heart infusion broth (BHI, Merck) and incubated at 35°C for 24 h. After the incubation, 10 mL aliquots of the BHI were inoculated into 10 mL of selenite cystine enrichment broth (Merck), gram negative broth (GN, Merck), MacCONKEY broth, alkaline pepton water (Vibrio enrichment broth, Merck) and Listeria enrichment broth (LEB, Merck) with potassium thiocyanate (37.5 g/L, Merck) and nalidixic acid (50 µg/mL, Sigma). All of these selective broths except LEB were incubated at 35°C for 24 h, while LEB was incubated at 25°C for 48 h. After the incubation, the selenite cystine broth subcultured onto *Salmonella Shigella* agar (SS, Merck), while GN broth subcultured onto xylose lysine desoxycholate agar (XLD, Merck) and SS agar and the MacCONKEY broth subcultured onto violet red bile agar (VRBA, Merck), eosin methylene-blue lactose sucrose agar (EMB, Merck) and brilliant-green phenol-red lactose sucrose agar (BPLS, Merck). Also the alkaline peptone water subcultured onto thiosulfate citrate bile salt sucrose agar (TCBS, Merck) and LEB subcultured onto *Listeria* selective agar (Merck) with nalidicxic acid (50 µg/mL). Suspected bacterial colonies from the selective agars were screened morphologically and biochemically, and when suggestive of *Salmonella* spp., *Shigella* spp, *L. monocytogenes, V. parahaemolyticus* and *E. coli*, were confirmed serologically with commercial antisera (Difco and Mast).

3. RESULTS AND DISCUSSION

Table 1 shows the frequency of bacterial pathogens isolated from intestinal contents of the cultivated silver carp and common carp. *Salmonella* spp., *Shigella* spp., *L. monocytogenes* serotype 4b, *V. parahaemolyticus* and *E.coli* were isolated from 4.2%, 15%, 3.3%, 16.7% and 65% of samples respectively. Similar observations have been registered by many researchers [6,8,12-28].

Serologically typing of *Salmonella* and *Shigella* revealed the detection of *S. typhimurium* (1 strain), *S. newport* (4 strains), *Shig. flexneri* (13 strains), *Shig. dysentriae* (7 strains) and *Shig. boydii* (7 strains). These results are summarized in Table 2. Also biochemically identified *E. coli* were serologically investigated. The sera: O157H7 was not detected and the sera : O2K1, O119B14, O127B8 and O114K9 were detected by the serological investigation. Nearly similar observations were obtained by many workers [7,13,17,21]. Occasional isolation of *V. parahaemoliticus* has been made from the samples of one the fish farms in this study. In these cases, it is assumed that salinity has risen locally [29].

Salmonella spp., *Shigella* spp., *E. coli*, *V. parahaemoliticus* and *L. monocytogenes* have been implicated in food poisoning outbreaks [1,2,4,7]. Findings show that the intestinal flora of fish is correlated with the degree of contamination of food and water intake by fish, agreeing with the results obtained by many researchers [8,14,27,30,31].

In order to obtain high quality fish, following points can be recommended: a) prevent pollution of fresh water with excreta, b) hygienic procedures should be adopted in fish meal processing, c) sterilization methods of infected fish meal, d) practical methods for measuring Salmonella cells number in fish meal, and e) a summary of good house keeping practices [32].

Table 1
Frequency of bacterial pathogens in 120 specimens from the intestinal content of cultivated fish

Organisms	Number of isolate	Percentage
Escherichia coli	78	65.0
Shigella spp.	18	15.0
Salmonella spp.	5	4.2
Listeria monocytogenes	4	3.3
Vibrio parahaemolyticus	20	16.7

Table 2
Serotypes of isolated *Salmonella* and *Shigella* species

Serotypes	Number of isolates	Percentage
Salmonella typhimurium	1	0.8
Salmonella newport	4	3.3
Shigella flexneri	13	10.8
Shigella dysentriae	7	5.8
Shigella boydii	7	5.8

4. ACKNOWLEDGMENT

The authors express their sincere appreciation and thanks for the financial support received from University of Tehran, Faculty of Veterinary Medicine for this research.

REFERENCES

1. Altekruse, M. L. and Swerldlow, D. L. (1997) Emerging Foodborne disease.

Emerging Infectious Disease, 3: 285-293. Center for disease control and prevention, Atlanta, Georgia, USA.
2. Foster, E. M. (1969) Historical overview of key issues in food safety. Emerg. Infect. Dis., 3 : 481-482. Madison Wisconsin, USA.
3. Gerigk, K. (1992) WHO surveillance program for control of foodborne infection and intoxication, In: Europe. Proc. 3rd World Congress of Foodborne Infections' and Intoxications, Berlin.
4. Tauxe, R. V. (1997) Emerging foodborne diseases: An Evolving Public Health Challenge. Emerging Infectious Disease, 3, Center for disease Control and Prevention, Atlanta, Georgia, USA.
5. Huss, H. H. (1992) Fresh fish quality and quality changes. FAO publications, 103-104.
6. Prost, M. (1977) Fish as a source of infections and parasitic infestations in man. Med. Weter., 33 : 641-646.
7. Youseff, H., EL-Tiammy, A. K. and Ahmed, S. (1992) Role of aerobic intestinal pathogens of fresh water fish in transmission of human diseases. J. Food Prot., 55: 739-740.
8. Sugita, H., Tamura, M. and Deguchi, Y. (1985) Enterobacteria present in the gastro intestinal tract of freshwater in the Tama river. Bull. Coll. Agric. Vet. Med. Nihon Univ. Nichidai Nojuho, 42: 211-215.
9. Zutte, D. (1979) The role of surface waters in the spread of Salmonella. Vlaams Diergeneeskd. Tijdschr., 48: 79-88.
10. Gonzales, C. J., Cardenal, D. P., Prieto, M., Otero, A. and Lopez, M. L. G. (1992) Microbiological quality of fresh rainbow trout (Oncorhynchus mykiss) and microbial evolution during its chill storage. Actas Del IV Congreso Nacional De Acuicultura, 563-567.
11. Speck, M. L. (1997) Compendium of methods for microbiological examination of foods, 3rd edition American public health association.
12. Aoki, T., Jo, Y. and Egusa, S. (1980) Frequent occurrence of drug resistant bacteria in ayu (*Plecoglussus altivelis*) culture. Fish Pathol., Tokyo., 15: 1-6.
13. Baker, D. A., Smitherman, R. O. and Mc Caskey, T. A. (1983) Longevity of Salmonella typhimurium in Tilapia and water pools fertilized with swine waste. Appl. Environm. Microbiol., 45: 1548-1554.
14. Berg, R. W. and Anderson, A. W. (1972) Salmonella and Edwardsiella tara in gull faces a source of contamination. Appl. Microbiol., 24 : 503.
15. Fernandes, C. F., Flick, G. J. and Thomas, T. B. (1998) Growth of inoculated psychrotrophic pathogens on refrigerated fillets of aqua cultured rainbow trout and channel catfish. J. Food Prot., 61: 313-317.
16. Foltz, V. D. (1969) Salmonella ecology. J. American oil Chemist Society, 40: 222-224.
17. Goda, FFM., Shouman, MT., Wasef, NA. and Farid, AF. (1992) Bacterial examination of Nasser lake fish Tilapia nolitica sold at Giza and Cairo governmental fish centers. Veterinary Medical J., 28: 387-395.
18. Lartseva, L. V., Rogatkina, I. Y. U. and Bormotova, S. V. (1996) Enteric bacteria isolated from marketable fishes of the Volga-Caspian basin. Gigiena I. Sanitariya, 3:

22-23.
19. Leitao, M. F. F., Teixeira, A. R. F. and Baldini, V. L. S. (1996) Bacterial flora of river and lake fish in Sao Paulo state. Coletanea do Instituto de Technologia de Alimentos, 15: 91-111.
20. Morse, E. V. and Duncan, M. A. (1996) Salmonella as monitors of fecal pollution in the aquatic environment. J. Environ. Sci. Health, Part A. 11: 591-601.
21. Nedoluha, P. C. and Westhoff, D. (1996) Microbiological flora of aquacultured hybrid striped bass. J. Food Prot., 56: 1054-1060.
22. Niemi, M., I. (1982) Fecal indicator bacteria at fresh water rainbow trout (Salmo gairdneri) farms. Vesientutkimuslat. Julk. Helsinki Publ Water Res. Inst. Helsinki. 1985; pp. 64, 49.
23. Niewolak, S. and Tucholski, S. (1996) Sanitary and bacteriological study of common carp reared in ponds supplied with biologically pretreated sewage. Arch. Ryb. Pol. Arch. Pol. Fish., 3: 203-215.
24. Ogbondeminu, F. S. (1996) The Occurrence and distribution of enteric bacteria in fish and water of tropical aquaculture ponds in Nigeria. J. Aquacultt. Trop. 8: 61-60.
25. Pal, D. and Gupta, C. (1992) Microbial pollution in water and its effect on fish. J. Aout. Anim. Health, 4: 32-39.
26. Pivorarov, Y. V. P. and Podyasener, S. V. (1996) Aerobic microflora of freshly caught fish and frozen fish. Gigiena I. Sanitaria, 6: 80-81.
27. Twiddy, D. R. and Reilly, P. J. A. (1995) Occurrence of antibiotic-resistant human pathogens in integrated fish farms. Research of contributions Presented at the Ninth Session of the Indo Pacific Fishery Commission Working party on fish technology and marketing. Cochin-India. 7-9 March, No. 514 Suppl.: 23-37.
28. Wyatt, L. E. (1973) Microbiological profiles of fresh water catfish. Dissertation Abstracts International B, 38: 5206.
29. Varnam, A. H. and Evans, M. G., (1991) Foodborne Pathogens. Wolf Science Book, The Netherlands.
30. Beckers, H. J. (1986) Incidence of foodborne disease in the Netherlands: annual summary. J. Food Prot., 49: 924-931.
31. Wyatt, L. E., Nikelson, R. I. I. and Vanderzant, C. (1979) Occurrence and control of Salmonella in fresh water catfish. J. Food Prot., 44: 1067-1069.
32. The fish Inspector (1991) Control of Salmonella in fish meal. No.14, January 1991, Published and printed by INFOFISH, PO Box 10899, 50728 Kuala Lumpur, Malaysia.

PARTICIPANTS

T. Abe	Hokkaido Food Research Center	JAPAN
K. Adachi	Kyoto University	JAPAN
Y. Akahane	Fukui Prefectural University	JAPAN
A. Akhondzadeh Basti	University of Tehran	IRAN
N. M. Akhtar	Central Institute of Fisheries Education	INDIA
M. Ando	Kinki University	JAPAN
S. Benjakul	Prince of Songkla University	THAILAND
T. Borresen	Danish Institute for Fisheries Research	DENMARK
H. A. Bremner	Allan Bremner and Associates	AUSTRALIA
Y. Choi	Gyeongsang National University	KOREA
Z. Dawo	Australian Marine College	AUSTRALIA
H. Einarsson	Icelandic Fisheries Laboratories	ICELAND
M. Espe	Institute of Nutrition	NORWAY
A. Fujimoto	Hokkaido University	JAPAN
M. Fujitsuka	Ajinomoto Co., Inc.	JAPAN
Y. Fukuda	National Research Institute of Fisheries Science	JAPAN
Y. Funatsu	Toyama Prefectural Food Research Institute	JAPAN
A. Gelman	Kimron Veterinary Institute	ISRAEL
A. Haga	National Institute of Agrobiological Sciences	JAPAN
Y. Hasegawa	Muroran Institute of Technology	JAPAN
S. Hirano	Maruhachi Muramatsu, Inc.	JAPAN
T. Hirata	Kyoto University	JAPAN
M. Hogstad	University of Tromsoe	NORWAY
Y. Huang	University of Georgia	USA
Y. Ishikawa	Tottori University	JAPAN
Y. Itoh	Kochi University	JAPAN
K. Itou	Fukui Prefectural University	JAPAN
H. Kakimoto	National Coastal Fisheries Development Association	JAPAN
H. Kanehiro	Tokyo University of Fishery	JAPAN
M. Kaneniwa	National Research Institute of Fisheries Science	JAPAN
S. Kanoh	Mie University	JAPAN
K. Katayama	Kaiyoh Kensetsu Co., Ltd.	JAPAN
H. Kawaguchi	Ajinomoto Co., Inc.	JAPAN
T. Kawai	Shiono Koryo Kaisha Ltd.	JAPAN

M. Kawasaki	Riken Vitamin Co., Ltd.	JAPAN
S. Kelleher	University of Massachusetts	USA
K. Kikuchi	Yaizu Suisankagaku Industry Co., Ltd.	JAPAN
J. S. Kilpatrick	Nutreco Aquaculture Division	CANADA
Y. Kinoshita	Hakodate Regional Industry Promotion Organization	JAPAN
T. N. Kok	Marine Fishries Research Department	SINGAPORE
F. Kow	Australian Marine College	AUSTRALIA
T. Kurebayashi	Maruhachi Muramatsu, Inc.	JAPAN
E. Langmyhr	SSF-Norwegian Herring Oil & Meal Industry Research Institute	NORWAY
C. Lee	University of Rhode Island	USA
B. Liaset	Institute of Nutrition	NORWAY
K. Maeyama	Mikimoto Pharmaceutical Co., Ltd.	JAPAN
Y. Makino	Kagawa Prefectural Industrial Technology Center	JAPAN
Y. Makinodan	Kinki University	JAPAN
H. Matsue	Daimatsu Co., Ltd.	JAPAN
N. Matsue	Daimatsu Co., Ltd.	JAPAN
Y. Matsue	Maruhachi Muramatsu, Inc.	JAPAN
S. Matsunaga	The University of Tokyo	JAPAN
T. Miura	Hokkaido University	JAPAN
S. Mochizuki	Oita University	JAPAN
K. Morioka	Kochi University	JAPAN
N. Morita	National Insutitute of Advanced Industrial Science and Technology	JAPAN
M. T. Morrissey	Oregon State University	USA
M. Murata	Education University of Nara	JAPAN
Y. Murata	National Research Institute of Fisheries Science	JAPAN
H. Nakagawa	Hiroshima University	JAPAN
H. Namba	Shida Kanzume Co., Ltd.	JAPAN
P. Nichols	CSIRO Marine Research	AUSTRALIA
F. S. Noguchi	MARUHA Corporation	JAPAN
Y. Nomura	Tokyo University of Agriculture and Technology	JAPAN
K. Ohishi	Numazu Industrial Research Institute of Shizuoka Prefecture	JAPAN
H. Oka	Industrial Research Center of Ehime Prefecture	JAPAN
H. Okada	Cosmo Foods Co., Ltd.	JAPAN
M. Okada		JAPAN
N. Okada	Kashimaya Co., Ltd.	JAPAN
T. Okada	Cosmo Foods Co., Ltd.	JAPAN
T. Okano	Kaiyoh Kensetsu Co., Ltd.	JAPAN
T. Okayama	Fukui Prefectural University	JAPAN
S. Ono	Hokkaido University	JAPAN

T. Ooizumi	Fukui Prefectural University	JAPAN
J. Park	Oregon State University	USA
J. Pyeun	Gyeongsang National University	KOREA
V. Razavilar	University of Tehran	IRAN
H. W. Rehbein	Federal Research Centre for Fisheries	GERMANY
M. Sakaguchi	Kyoto University	JAPAN
K. Sakai	Yaizu Suisankagaku Industry Co., Ltd.	JAPAN
J. Santoso	Tokyo University of Fisheries	JAPAN
K. Sato	Kyoto Prefectural University	JAPAN
A. Seki	Mitsubishi Chemical Safety Institute Ltd.	JAPAN
I. Shih	Da-yeh University	TAIWAN
S. Sigurdardottir	Icelandic Fisheries Laboratories	ICELAND
X. Song	Kyoto University	JAPAN
I. Sorensen	University of Tromsoe	NORWAY
N. K. Sorensen	Norwegian Institute of Fisheries and Aquaculture Ltd.	NORWAY
T. Suzuki	Tokyo University of Fisheries	JAPAN
S. Svenning	University of Tromsoe	NORWAY
M. Tahara	Kaiyoh Kensetsu Co., Ltd.	JAPAN
K. Takahashi	Hokkaido University	JAPAN
A. Takiguchi	Chiba Prefectural Research Center of Marine Industry	JAPAN
M. Tanaka	Tokyo University of Fisheries	JAPAN
M. Tejada	Instituto del Frio	SPAIN
D. Thomas		NEW ZEALAND
Y. Tsukamasa	Kinki University	JAPAN
K. Tsumura	Kaiyoh Kensetsu Co., Ltd.	JAPAN
K. Tsuruhashi	Fukui Prefectural University	JAPAN
D. Tunhun	Kochi University	JAPAN
S. Ueno	Sun Food Laboratory Inc.	JAPAN
S. Urabe	Urabe corporation	JAPAN
W. Visessanguan	National Center for Genetic Engineering & Biotechnology	TAHILAND
S. Wada	Tokyo University of Fisheries	JAPAN
S. Wang	Da-Yeh University	TAIWAN
X. Wang	Shanghai Fisheries University	CHINA
Y. Watanabe	National Research Institute of Fisheries Science	JAPAN
J. Weerasinghe	Food Science Australia	AUSTRALIA
U. Wijkstrom	FAO	ITALY
A. Worratao	Suranaree University of Technology	THAILAND
C. Xue	Ocean University of Qingdao	CHINA
M. Yamagishi	Taikichou Local Products Research Center	JAPAN
M. Yamazawa	National Research Institute of Fisheries Science	JAPAN
Y. Yen	Da-Yeh University	TAIWAN

M. Yokoyama	Maruhachi Muramatsu, Inc.	JAPAN
J. Yongsawatdigul	Suranaree University of Technology	THAILAND
H. Yoshii	Tottori University	JAPAN
K. Yoshioka	Nakamura Gakuen University	JAPAN
T. Yoshioka	Hakodate Regional Industry Promotion Organization	JAPAN
B. Yoshitomi	Nippon Suisan Kaisha, Ltd.	JAPAN
D. Zhou	Shanghai Fisheries University	CHINA
P. Zhou	Shanghai Fisheries University	CHINA

SPONSORS AND DONORS

Sponsors

Ministry of Education, Culture, Sport, Science and Technology, Japan (A Grant-in-Aid for Symposium Holding)

Kyoto University, Japan (University Foundation)

The Fisheries Agency, Ministry of Agriculture, Forestry and Fisheries, Japan

Department of Agriculture, Foresty and Fisheries, Kyoto Prefecture, Japan

Japan Fisheries Association

National Federation of Fisheries Cooperative Association, Japan

Donors

Ajinomoto Co., Inc., Food Research & Development Laboratories

All-Japan Kamaboko Makers' Association

Beckman Corlter K.K.

Daiichi-chinmi Co., Ltd.

Daikin Corporation

Daimatsu Co., Ltd.

N.T.S. Inc.

Hakodate Shipbuilding Company Cooperative Association

Hamayoshiya Inc.

Japan Food Research Laboratories Osaka Branch

Kaiyo Kensetsu Co., Ltd.

Kanmonkai Co., Ltd.

Kashimaya Co., Ltd.

Kenkama Co., Ltd.

Khono Corporation

Kochi Kamaboko Young-man Association

KOGAKU Co., Ltd.

Kojima Instruments Inc.

Koyo Chemical Co., Ltd.

Kyowa Hakko Kogyo Co., Ltd.
Food Development Department

Kyowa Tecnos Co., Ltd.

Maekawa Taste Co., Ltd.

Marino-Forum 21

Maruha Corporation, Central Research Institute

Maruhachi Muramatsu Inc.
Dept. Development Research

Marukyu Co., Ltd.

Matsuoka Kamaboko Co., Ltd.

Matsuoka Suisan Co., Ltd.

Mikimoto Pharmaceutical Co., Ltd

Mizoguchi Jigyoh Co., Ltd.

Mutoh Co., Ltd. Hakodate Branch

National Cooperative Association of Squid Processors

Nestle Science Promotion Committee

Nichimo Co., Ltd.

Nichirei Corporation, Research & Development Center

Nippon Chemical Feed Co., Ltd.

Nippon Formula Feed MFG. Co., Ltd.
Central Laboratory

Nippon Lever K.K.

Nippon Marine Oil Association

Nippon Suisan Kaisha, Ltd.
Central Research Laboratory

Nissin Food Products Co., Ltd.
Central Research Institute

Oguraya Yamamoto Co., Ltd.

Riken Vitamin Co., Ltd.

Sakaguchi Unlimited Partnership

Sakami Co., Ltd.

Senmi Ekisu Co., Ltd.

Shida Kanzume Co., Ltd.

Shiono Koryo Kaisha Ltd.

Sun Food Laboratory Inc.

Takeda Food Products Co., Ltd.

Tokyo Fisheries Promotion Foundation

Tosakatsuo Suisan Co., Ltd.

Tosasyoku Co., Ltd.

Towa Chemical Industry Co., Ltd., Center for Technical Development and Application

Tsukugon Co., Ltd.

Uchida Accounting Office

Wallac Berthold Japan Co., Ltd.

Yaizu Suisankagaku Industry Co., Ltd.

Yamachu Co., Ltd.

Yamaki Co., Ltd.

AUTHOR INDEX

Abe, T. 203
Adachi, K. 317
Akahane, Y. 415
Akhondzadeh Basti, A. 447
Ang, G. T. 397
Aoyama, M. 179

Bokaie, S. 447
Børresen, T. 3
Bremner, H. A. 275

Cai, Y. 139
Carlehoeg, M. 301
Chechic, K. 75
Chen, L. 139
Choi, Y. J. 425

Drabkin, V. 75

Elliott, N. G. 115

Fang, Y. 139
Feng, Y. 407
Fujio, A. 317
Fujita, M. 107
Fukuda, Y. 387
Funatsu, Y. 193
Furuta, T. 309
Fusetani, N. 131

Gabay, I. 75
Gelman, A. 75
Glatman, L. 75

Hasegawa, Y. 233
Heu, M. 425
Hirata, T. 209, 317, 387

Hultin, H. O. 407

Iida, K. 179
Itoh, Y. 357, 375

Kakimoto, H. 263
Kanehiro, H. 253
Kaneniwa, M. 289
Kanoh, S. 179
Katayama, K. 263
Kawai, T. 209
Kawasaki, K. 193
Kelleher, S. D. 407
Kikuch, K. 97
Kim, H. 425
Kinoshita, M. 387
Kitahashi, T. 159
Kok, T. N. 397
Konagaya, S. 193
Kristinsson, H. 407
Kubota, S. 357
Kuwahara, R. 289

Lee, H. K. 397
Lee, W. P. 397
Li, Z. 139
Lin, H. 139
Linko, P. 309
Liu, Y. C. 233

Mae, M. 179
Maeyama, K. 179
Matahira, Y. 97
Matsunaga, S. 131
McClements, D. J. 407
Miono, Y. 415
Miura, T. 55

Mooney, B. D. 115
Morioka, K. 357
Morrissey, M. T. 37
Murata, N. 159
Murata, Y. 289

Nagafusa, K. 107
Nagai, K. 317
Nakagawa, H. 243
Nakamura, Y. 159
Natsui, N. 233
Nichols, P. D. 115
Niizeki, N. 209
Nishimura, K. 159
Niwa, E. 179
Noguchi, S. F. 63
Nomura, A. 375
Nomura, Y. 147

Obatake, A. 375
Ohtsuki, K. 159
Oohara, I. 289
Ooizumi, T. 415

Park, J. W. 333, 343
Pyeun, J. 425

Qi, X. 223

Razavilar, V. 441
Rezvani, S. 441

Sachs, O. 75
Sakaguchi, M. 209, 317, 387
Sakaguchi, Y. 107
Sakai, K. 97
Santoso, J. 169
Sata, N. U. 289
Sato, K. 159
Seki, A. 107
Shichinohe, K. 159
Shimizu-Suganuma, M. 159
Shimomura, K. 179
Song, X. 209
Sorensen, N. K. 301
Suzuki, T. 169

Sylvia, G. 3

Tachibana, N. 203
Tahara, M. 263
Takagi, K. 179
Takahashi, K. 87
Takahashi, T. 179
Tanaka, M. 203
Tanaka, R. 179
Tobiassen, T. 301
Tsumura, K. 263
Tsuruhashi, K. 415
Tsutsumi, M. 159
Tunhun, D. 357

Uchiyama, K. 233
Ueda, S. 179
Unuma, T. 289

Wada, S. 107
Wang, X. 387
Watanabe, M. 179
Watanabe, Y. 25
Wijkstrom, U. 9

Xue, C. 139

Yamada, A. 107
Yeap, S. E. 397
Yokoyama, M. 289
Yongsawatdigul, J. 343
Yoshie, Y. 169
Yoshii, H. 309
Yoshioka, K. 107
Yoshitomi, B. 45

Zahrae Salehi, T. 447
Zheng, X. 223
Zhou, P. 223

KEYWORD INDEX

Acceptability 390
Acid solubilization /precipitation 410
Actin depolymerizing agents 131
Activation energy 309
Actomyosin 344
Alaskan pollock 41
Amino acid 290
Analysis 142
Anchovy 426
Angiogenesis 160
Anserine 97
Anti-fatigue effect 103
Antioxidant 172
Aquaculture 10, 75
Aquaculture feed 48
Aquatic resources 4
Artificial bait 59
Artificial reef 263
Australia 116
Autolysis 435
Autoxidation 310
Auxis rochei 194

Bacterial pathogens 448
Bacteriological change 80
Bioactive metabolites 131
Bioactive substance 234
Bioinformatics 5
Bioprocessing 4
Bitterness 290
Black spot 327
Bone metabolism 111
Bone mineral density 109, 154
Bone modeling 108
Bonito extracts 98
Brown seaweed 140
By-catch 123

Bycatch species 31
By-products 121

Calcium 108
Calyculins 133
Cancer 159
Capture 10
Carnosine 97
Carp 210
Cartilage 161
Cathepsin L-like enzyme 427
Caviar 442
Cell culture matrix 151
Chain management 278
Chelate 172
Chemical composition 79
Chemical recycling 255
Chemiluminescence 140
Chitin 91
Chondroichin sulfate 162
Chopping temperature 339
Chymotrypsin 427
Cold preservation 421
Cold storage 441
Collagen 92, 149, 160
Column chromatography 224
Common carp (cultivated) 448
Consumption 10
Cooking method 340
Cosmetic materials 180
Cosmetics 72, 239
Cuttlefish 108, 224
Cyprinus carpio 210, 448
Cytotoxin 131

DHA 88, 108, 116, 310
DHAEE 310

DNA 92
Deep-sea fish 63
Demand 11
Dewatering effect 416
Discard amount (Japan) 27
Discard amount (world) 26
Discard species 33
Discoloration 317
Dispersing ability 421
Disposal 255
Disulfide bond 358
Dried shrimp 400
Dried skipjack 204

Economic benefits 37
Electronic trading 280
Energy resource recovery 256
EPA 116
Epidermal keratinocytes 187
Ethical slaughter 302
Euphasia superba Dana 45
Euthynnus pelamis 210
Exercise performance 98
Expanded polystyrene 259
Extrinsic factors 38
Extractive components 193
Extractive nitrogen 204
Extracts 390

Feed 70
Feed additive 243
Fibroblast 181
Fish 9
Fish cracker 400
Fish fillet 416
Fish meat 358
Fish sauce 194, 426
Fisheries 10
Fisheries management 38
Fisheries products 63
Fishing (lantern fish) 64
Fishing net 254
Flounder 210
Food 11, 51
Food safety 442
Food supplement 149

Free radicals 139, 234
Frigate mackerel 194
Fucoidan 139
Functional fish protein 407
Future 15

GC/MS 210
Gel forming ability 360, 379
Gel strength 379
Gelation 344, 407
General composition 47
Glycan 91
Glycogen 180
Gonad 290
Gonad index 290
Growth 244

Habitat 268
HACCP 442
Hemicentrotus pulcherrimus 289
Hemocyanin 319
Hepatic tissues 210
Hot-air drying 398
Human blood 103
Hypophytalmichthys molitrix 448

Imidazole dipeptide 97
Incineration 255
Information technology 283
Inhibitor 225
Ink (cuttlefish) 223
Ink (squid) 60, 223
Inorganic phosphate 109
Intestine 448
Intrinsic factors 37

Kamaboko 375
Kamaboko gel 388
Katsuobushi 204
Killing methods 302
Kinetics 309
Km value 230
Krill 45
Kuruma prawn 319

Land fill 255

Lantern fish 63
Lipid 117
Lipid metabolism 109, 250
Lipid peroxidation 143
Liver 210
Liver function 246

MHC 343, 362
MMP 151, 160
MMP-inhibitor 160
Marine invertebrate 131
Marine oils 116
Marine resources 108
Material recycling 255
Meat washing 361, 376
Melanogenesis 317
Microarray 4
Microbial contamination 442
Microbial examination 442
Mirobial hazard 442
Microwave heating 340
Modori 378
Modori-inhibiting factor 382
Moisture content 417
Muscular tissues 210
Mycaloides 132

Natural apatite 72
Natural seasonings 69
Nitrogen balance 56
Nursery 263
Nutrigenomics 5
Nylon 254

Odor 211
Ohmic heating 340
Omega-3 PUFA 116
Organoleptic evaluation 79
Osmotic pressure 416
Ovaries 290
Over-matured chum salmon 203

Paralichthys olivaceus 210
POV 172
PUFA 309
Pacific salmon 41

Pacific whiting 38
Packing 399
Pacu 75
Pearl oyster 180
Penaeus japonicus 319
Peptide 52
Permeability 419
Phenoloxidase 319
Phospholipid 88, 180
Piaractus mesopotamicus 75
Pinktada fucata 180
Plant 442
Polyethylene 254
Polyphenol oxidase 223
Polyphenolic compound 172
Porphyra meal 244
Post-genomic techniques 4
Prawn 317
Processing 442
Product development 283
Profitability 281
Pro-oxidant 172
Prophenoloxidase 319
Protease 48, 151, 236, 426
Protease (myofibril type) 382
Protein deposition 244
Protein phosphatases 137
Proteoglycan 162
Proteolysis 204, 344
Puffing 399
Pulcherrimine 290
Punch test 334
Purification 226
Purple laver 243

Quality 38, 283
Quality management 276

Radical scavenging 100
Radical scavenging effects 142
Rate constant 310
Recycling method 256
Red sea bream 244
Risk assessment 442

SH group 362

SPME 210
Sakebushi 205
Salmon 91, 203, 302
Scallop shells 233
Scavenging 172
Scomberomorus niphonius 210
Sea urchin 290
Seafood 116
Seasonal variation 47
Seaweed 170
Sensory evaluation 389
Sensory quality 301
Sepia esculeuta 224
Shark 147, 161
Shells 263
Silver carp 388, 448
Skin fibroblast cells 235
Skipjack 210
Slurying 66
Smoking 76
Soaking 416
Spanish mackerel 210
Spoilage prediction 284
Sponges 132
Squid 55, 91
Steaming 205
Storage modulus 411
Stress 302
Stress response 244
Sturgeon 442
Sugar compounds 417
Superoxide anion 234
Supply 11
Surfactant 181
Surimi 194, 347, 388
Surimi gel 334
Sustainable development 3

Texture 334
Threadfin bream 343
Torsion test 337
Traceability 281
Transglutaminase 343
Trimethylamine 389
Trimethylamine oxide 389
Trypsin 426

Tuna 123

Utilization 233

Vacuum drying 398
Value-adding 115
Vitamin D 108
Volatiles 211

Waste 58, 194, 243, 255, 263, 353
Wax 65
Welfare 302

Zero emission 3, 55
Zymography 152